GROUND WATER
CONTAMINATION

GROUND WATER CONTAMINATION

TRANSPORT AND REMEDIATION

Philip B. Bedient
Rice University

Hanadi S. Rifai
Rice University

Charles J. Newell
Groundwater Services, Inc.

PTR Prentice Hall
Englewood Cliffs, New Jersey 07632

Library of Congress Cataloging-in-Publication Data

BEDIENT, PHILIP B., (date)
    Ground water contamination : transport and remediation / Philip B.
  Bedient, Hanadi S. Rifai, Charles J. Newell.
        p.   cm.
    Includes bibliographical references and index.
    ISBN 0-13-362592-3
    1. Groundwater—Pollution. 2. Groundwater—Quality—Management.
  I. Rifai, H. S. II. Newell, Charles J. III. Title.
  TD426.B44 1994.
    628.1'68—dc20   94-3073
                                                            CIP

Acquisitions editor: *Mike Hays*
Editorial/production supervision: *Kim Gueterman*
Cover design: *Jeannette Jacobs Design*
Production coordinator: *Alexis Heydt*

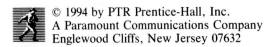 © 1994 by PTR Prentice-Hall, Inc.
A Paramount Communications Company
Englewood Cliffs, New Jersey 07632

Printed in the United States of America
10 9 8 7 6 5 4 3 2 1

ISBN 0-13-362592-3

Prentice-Hall International (UK) Limited, *London*
Prentice-Hall of Australia Pty. Limited , *Sydney*
Prentice-Hall Canada Inc., *Toronto*
Prentice-Hall Hispanoamericana, S.A., *Mexico*
Prentice-Hall of India Private, *New Delhi*
Prentice-Hall of Japan, Inc., *Tokyo*
Simon & Schuster Asia Pte. Ltd., *Singapore*
Editora Prentice-Hall do Brasil, Ltda., *Rio de Janeiro*

To Cindy, Eric, and Courtney
to my parents for their guidance
to all of my students and colleagues
PBB

To CJ, Imo, Steve, and Meaux
to my parents
HSR

To June, Wynne, Mark, Jon, and Phil
to Hanadi, for everything
CJN

CONTENTS

PREFACE

The 1970s ushered in a new decade of environmental awareness in response to major air pollution and water quality problems throughout the country. In 1970, one of the primary missions of the newly formed Environmental Protection Agency was to define, maintain, and protect the quality of the nation's surface waters and subsurface aquifers. The field of environmental engineering was in its infancy, and hydrologists, civil and environmental engineers, and other scientists were needed to provide the necessary expertise and engineering designs for water pollution control.

In the late 1970s, the discovery of hazardous wastes at sites such as Love Canal in New York and the Denver Arsenal in Colorado ushered in a new era in hazardous waste site problems. The 1980s could be called the hazardous waste decade. Numerous studies of active and abandoned waste sites and spills were conducted as required by the Resource Conservation and Recovery Act (RCRA) and the Superfund legislation administered by the EPA. All were designed to protect ground water quality. During this time, hydrogeologists, civil engineers, environmental engineers, and others were characterizing, evaluating, and remediating hazardous waste sites with respect to ground water contamination. More than 1500 hazardous waste sites have been placed on the National Priorities List and thousands of other sites still remain to be cleaned up. By 1985, leaking underground storage tanks became one of the most ubiquitous of all subsurface contamination issues.

Ground water hydrology has traditionally included the analysis of flow through porous media, infiltration into soils, flow prediction in various aquifer systems, and well mechanics for the analysis of aquifer characteristics (slug test, pump tests, tracer tests). Along with the maturing of environmental engineering and related ground water fields in the eighties, attention to hazardous waste problems has greatly expanded the scope and emphasis of traditional ground water investigations. Contaminant transport in the subsurface is now of paramount importance and encompasses physical, chemical, and biological mechanisms which affect rates of migration, degradation, and ultimate remediation. Scientists and engineers must now in turn be prepared to deal with problems in the unsaturated zone, NAPLs in the subsurface, and specialized remediation schemes for a mixture of complex chemicals in ground water.

Our book has been written to better address the scientific and engineering aspects of subsurface contaminant transport and remediation in ground water, with less emphasis on geological settings compared to earlier texts written primarily for hydrogeologists. The current text is a departure from past efforts in that it is written from both a theoretical and practical viewpoint with engineering methods and transport theory applied to hazardous waste site characterization, transport modeling, and remediation. For the first time, entire chapters on numerical methods and discussions of model applications to actual field sites exist alongside discussions of advection, adsorption, dispersion, chemical reaction, and biodegradation. Chapters on waste site characterization and remedial design are included along with detailed case studies that illustrate the various techniques currently being applied. This book is designed for hydrologists, civil and environmental engineers, hydrogeologists, and other scientists in the ground water field who are or will be involved in the evaluation and remediation of the nation's ground water. However, the field of ground water contamination is rapidly changing as new techniques are being researched at many field sites nationwide. We hope this book will provide the fundamentals for understanding and incorporating new remediation technology into the more traditional methods of the eighties.

The legal framework of ground water legislation under RCRA and Superfund has provided significant guidance and funding for many of the ground water studies which have been performed during the 1980s. These comprehensive legal instruments set into motion an entire industry devoted to the identification, characterization, and remediation of hazardous waste sites throughout the United States. As a result of 15 billion dollars allocated for remedial investigations and feasibility studies in the mid 1980s, thousands of engineers and scientists now form the core of the environmental remediation industry. During this time, college and university programs quickly added ground water flow and transport courses to their traditional fields of civil and environmental engineering and geology. And professional groups, such as the National Ground Water Association, saw their memberships grow by the thousands in response to the challenge of education and technology transfer. Our book was written in response to the tremendous demand in the environmental industry for a modern engineering approach to ground water contamination problems.

Organization of the Book

An engineering hydrologist today must be able to address processes of both ground water flow and issues relating to contaminant transport. This text is designed to assist engineering and science students, professional engineers, and hydrogeologists in dealing with four major issues in ground water: flow and contaminant transport processes, field investigation methods, flow and transport modeling, and ground water remediation. These are topics not generally covered in detail in earlier texts.

A review of ground water flow in the subsurface, including well mechanics, is required before one can make any progress toward explaining or predicting contamination processes. Chapter 2 provides the hydrologist with a working knowledge of methods that have been developed to predict rates of flow and directions of movement in ground water aquifer systems. Chapter 3 follows with coverage of well mechanics and a review of standard aquifer tests (pump tests, slug tests, tracer tests) for determining hydraulic conductivity in the field.

Sources of ground water contamination are widespread and include thousands of accidental spills, landfills, surface impoundments, underground storage tanks, above ground tanks, pipelines, injection wells, land application of wastes and pesticides, septic tanks, radioactive waste disposal, salt water intrusion, and acid mine drainage. These various sources are described in more detail in Chapter 4. The main contaminants of concern include petroleum hydrocarbons such as benzene, toluene, and xylene; chlorinated organics such as trichloroethylene (TCE); heavy metals such as lead, zinc, and chromium, and certain inorganic salts.

Sampling and monitoring methods (direct and indirect) have advanced significantly over the past decade with vast improvements in microelectronics and low level organic chemical analyses. Cone penetrometers and specialized multi-level samplers are routinely used as part of site investigations. Data collection and monitoring methods for ground water investigations are described in Chapter 5, authored by Robert S. Lee.

Chapters 6 through 8 introduces transport processes, theory, and equations, including mechanisms of advection, dispersion, adsorption, chemical reaction and biodegradation of organic contaminants. Chapter 9 presents a concise treatment of flow and transport in the unsaturated zone. Chapter 10 describes modern approaches to numerical modeling of ground water flow and transport systems, including a detailed treatment of finite difference methods.

Chapter 11 covers non-aqueous phase liquids (NAPLs) in the subsurface, an important topic which has received considerable attention in the past few years as it relates to the remediation of hazardous waste sites. NAPLs, which can provide a source of contamination for decades, are often associated with fuels and chlorinated organic contaminants. Chapter 12, authored by John A. Connor, applies much of the text's hydrologic and contaminant transport theory and addresses hydrogeologic site investigation. Chapter 13 covers ground water remediation and design. Several types of detailed case studies and investigations drawn from combined experiences around the United States for characterizing and remediating waste sites are included. This book presents the latest approaches to the design of remedial systems for hazardous waste sites.

Chapter 14, authored by James B. Blackburn, reviews the important legal measures relating to ground water contamination which have arisen due to legislation which has guided the EPA's mission to protect ground water quality in the U.S. federal legislation such as the Safe Drinking Water Act (SDWA, 1974), the Resource Conservation Recovery Act (RCRA, 1976), the Clean Water Act (1977), the Toxic Substances Control Act (TSCA, 1976), and the Comprehensive Environmental Response, Compensation, and Liability Act (CERCLA, 1980) all provide a complex and comprehensive group of laws to protect the quality of ground water. Together, these laws have created an entire industry devoted to the study and remediation of ground water.

Acknowledgments

Any book represents the evolution of ideas and concepts with input from a variety of colleagues, students, and friends over a long period of time. We would particularly like to thank reviewers of the original manuscript for their helpful guidance and numerous suggestions. Dr. Robert Borden provided valuable insight into the organization and content of the final book. Special thanks are due to personnel and researchers at the EPA Robert S. Kerr Environmental Research Lab in Ada, Oklahoma who supported and contributed to our research for more than a decade. We would like to thank Clint Hall, Marion Scalf, James McNabb, John Wilson, Carl Enfield, Steve Hutchins, Joe Williams, Dave Burden. Randall Ross, and Steve Acree. The National Center for Ground Water Research, headed by Dr. Herb Ward at Rice University, provided the mechanism for exchange of ideas. We would like to thank the Shell Foundation for support through the years as part of Dr. Bedient's Shell Distinguished Chair in Environmental Science. Without the Shell Chair, this text probably would never have been written. We appreciate the opportunity to present research and case studies performed together with the Monsanto Co., and acknowledge the contributions of Dale Wilson. Many Rice University students contributed to the text including Tameeza Asaria, Todd Fisher, Maged Hamed, and Howard Sweed. Karen Evans, Manar El-Beshry, and Anthony Holder developed and reviewed many of the homework problems. Paul LaWare, Carrie McDonald, and Kathy Moore completed most of the original figures for the text. Finally, we would like to thank Kathy Moore and Heather Dodge for typing and editing the original 900 page manuscript. We would like to thank Addison-Wesley for the use of portions of Chapter 8 in *Hydrology and Floodplain Analysis* (1992) by P. B. Bedient and W. C. Huber. We would also like to thank Technomic Publishing Co., Inc. for the use of portions of chapters 5, 6, and 7 in *Groundwater Remediation* (1992) edited by R. J. Charbeneau, P. B. Bedient, and R. C. Loehr.

Chapter 1 | INTRODUCTION TO GROUND WATER CONTAMINATION

1.1 THE HYDROLOGIC CYCLE

The science of **hydrology** includes the study of the storage and movement of water in streams and lakes on the surface of the earth and in the ground water aquifers in the subsurface. An **aquifer** is a geological unit that can store and supply significant quantities of water. Modern hydrology encompasses both flow and water-quality transport aspects of the water cycle. Figure 1.1 shows the various components of the hydrologic cycle, including both natural processes and engineered processes and transport pathways. Atmospheric water and solar energy provide the main inputs for the generation of precipitation, which falls over the land and oceans. Rainfall can either infiltrate into the soil system, percolate to deeper ground water, evaporate or transpire through vegetation back to the atmosphere, or run off to the nearest stream or river. Infiltrating water is the main source of recharge to the root zone and ground water aquifers below. Rivers can also recharge aquifers or can act as discharge points for aquifer outflows. The ocean is the ultimate receptor of surface and ground water contributions from surrounding land masses and provides the main source of water for evaporation back to the atmosphere.

Changes to the cycle caused by human activity have been recorded since the beginning of civilization. They include changes in infiltration patterns and evaporation due to land development, changes in runoff and evaporation patterns due to reservoir storage, increases in streamflow due to channelization and piping, and changes in ground water levels due to pumping of aquifers. Since the 1930s, hydrologists traditionally have spent much time and effort designing alterations to the natural hydrologic cycle for human use. Such alterations include providing surface or ground water supplies for industrial, agricultural, and municipal needs; providing water treatment for drinking water and the disposal of wastewater; meeting water supply needs through the building of dams and reservoirs or drilling of water-supply well fields; providing drainage and flood control via channelization and dams; and providing water quality and recreational benefits through the development and maintenance of reservoirs and stream corridors.

Considering the complexity of the hydrologic cycle shown in Figure 1.1, not all the transport pathways and storage elements can be measured easily, and some components can only be determined indirectly as unknowns in the overall hydrologic water-balance equation. Infiltration and evaporation are often computed as

1

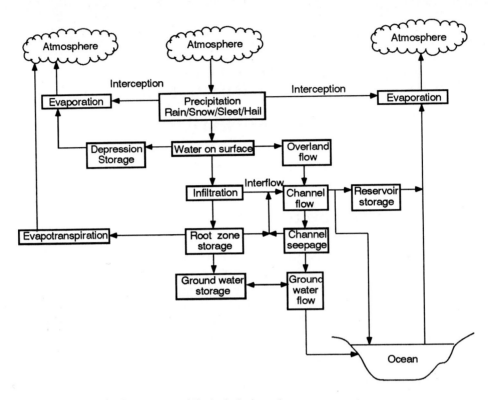

Figure 1.1 Components of the hydrologic cycle.

losses from the system and are not usually measured directly. Precipitation rates and stream levels can be directly measured by gages that have been located within a particular watershed being studied. Ground water levels and flow rates are measured from wells using flow meters and depth-to-water determinations. Overall water balances for a watershed or ground water basin can be computed if the preceding flow data are available. Computer methods have been developed in recent years to assist the hydrologist in watershed analysis and hydrologic design. Surface and ground water aspects of the hydrologic cycle are usually covered in modern hydrology texts such as Viessman, et al. (1989), Chow et al. (1988), Gupta (1989), and Bedient and Huber (1992).

1.2 GROUND WATER HYDROLOGY

Ground water is an important source of water supply for municipalities, agriculture, and industry. Figure 1.2 indicates the percentages of various types of ground water used in the United States. Agricultural irrigation use is clearly the largest category. Figure 1.3 depicts ground water use relative to total water use for each state in the continental United States; it indicates that western and midwestern

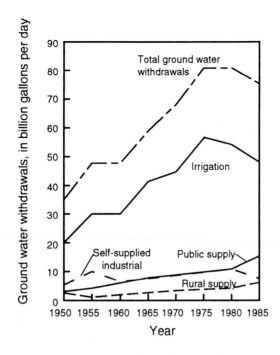

Figure 1.2 Trends in ground water use in the United States, 1950–1985.

areas are more dependent on ground water aquifers than those in the east and northeast.

Ground water hydrology has traditionally included the characterization of aquifers, application of Darcy's law for flow through porous media, infiltration into soils, and flow in shallow and deep aquifer systems (Chapter 2). More advanced topics include the mechanics of well flow in radial coordinates for single or multiple well systems. The prediction of flow in confined (under pressure) and unconfined (water table) aquifers is the starting point for understanding ground water mechanics. The **water table** in a shallow aquifer is defined as the level to which water will rise in a dug well under atmospheric conditions. Recently, more attention has been given to the connection between saturated and unsaturated zone processes near the water table as important aspects of modern ground water hydrology. Techniques for the analysis of aquifer characteristics using slug tests, pump tests, or tracer tests are still a major part of ground water investigations.

Ground water hydrology is of great importance because of the use of aquifer systems for water supply and because of the threat of contamination from leaking hazardous waste sites that occur at or below the ground surface. Properties of the porous media and subsurface geology govern both the rate and direction of ground water flow in any aquifer system. The injection or accidental spill of hazardous wastes into an aquifer or the pumping of the aquifer for water supply may alter the natural hydraulic flow patterns. For the hydrogeologist or civil or environmental engineer to obtain a full understanding of the mechanisms that lead to ground water contamination from spills or continuous leaks, it is necessary to first address properties of ground water flow, as covered in Chapters 2 and 3.

Geological aspects of ground water (hydrogeology) are of importance to un-

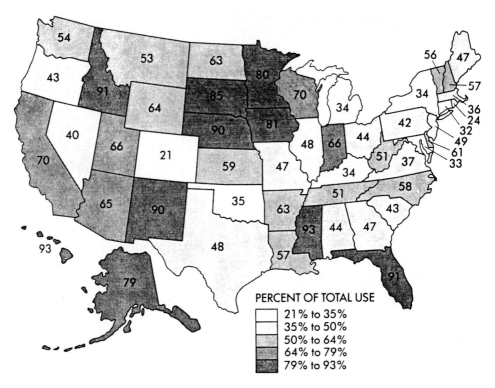

Figure 1.3 Ground water use as a percent of total water use, 1985. Source: Solley et al., 1988.

derstanding ground water flow and the fate and transport of contaminants in the subsurface. Regional geological aspects have been covered in detail in books by Freeze and Cherry (1979), Fetter (1988), and Domenico and Schwartz (1990) and will be addressed in this text on a limited basis. One useful generalization is the concept of ground water regions, which are geographical areas of similar occurrence of ground water. Useful comparisons can be made between areas of well-known hydrogeology and areas that are geologically similar but have not been as well studied. Meinzer (1923), considered the father of modern hydrogeology in the United States, proposed a classification system based on 21 different ground water provinces. Thomas (1952) revised the system based on 10 ground water regions, and Heath (1984) revised Thomas's system to include 15 different regions. Heath based his system on five features of ground water systems:

1. Components of the system and their arrangement
2. Nature of the water-bearing openings of the dominant aquifer or aquifers with respect to whether they are of primary or secondary origin
3. Mineral composition of the rock matrix of the dominant aquifers with respect to whether it is soluble or insoluble

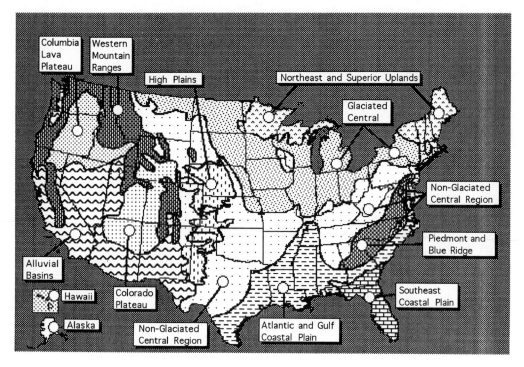

Figure 1.4 Ground water regions of the United States. Source: Newell et al., 1990.

4. Water storage and transmission characteristics of the dominant aquifer or aquifers

5. Nature and location of recharge and discharge areas

The 14 regions of the United States are shown in Figure 1.4 and are based on the DRASTIC system (Aller et al., 1987):

Each regional geologic category is unique and important to an understanding of the underlying stratigraphy, which may be affecting the transport of contaminants in the subsurface. In many cases, the first few shallow layers of sand or silty sand (5 to 50 m depth) may be the only zones of concern from a remediation standpoint. The text presents detailed methods and examples of hydrogeological site investigations and remediations at hazardous-waste sites. Thus, it is important for hydrologists and engineers to understand the relationship between the regional geological setting and the local, shallow stratigraphy surrounding a waste site.

1.3 GROUND WATER CONTAMINATION AND TRANSPORT

The occurrence of ground water contamination and the quality of ground water have become major issues since the discovery of numerous hazardous-waste sites in the late 1970s. Waste sites such as Love Canal in New York, the Denver Arsenal

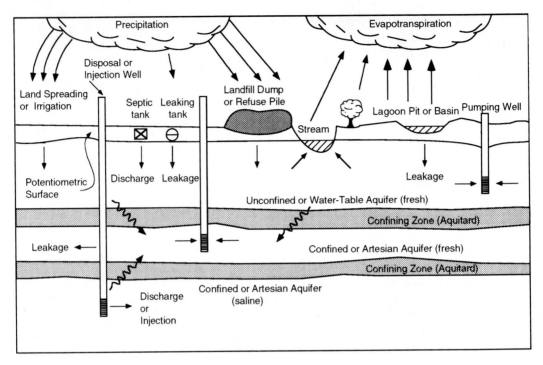

Figure 1.5 How waste-disposal practices contaminate the ground water system.

in Colorado, and Times Beach, Missouri, are three examples where hazardous wastes have created a serious ground water contamination problem for decades to come. Over 1500 sites nationwide are on the National Priority List from the Environmental Protection Agency (EPA). Sources of ground water contamination are widespread, as shown in Figure 1.5, and include thousands of accidental spills, landfills, surface impoundments, underground storage tanks, aboveground tanks, pipelines, injection wells, land application of wastes and pesticides, septic tanks, radioactive waste disposal, saltwater intrusion, and acid mine drainage. These various sources are described in more detail in Chapter 4. Figure 1.5 is a depiction of how waste-disposal practices can contaminate the ground water system in a general way. The main contaminants of concern include petroleum hydrocarbons, chlorinated organics such as trichloroethylene, heavy metals such as lead, zinc, and chromium, and inorganic salts.

In an effort to quantify ground water contamination at field sites, sampling and monitoring methods (direct and indirect) have advanced significantly over the years, contributing to better data-collection methods; these include cone penetrometers and multilevel sampling wells (Chapter 5). Chapter 6 develops and solves the equations that govern physical and chemical transport mechanisms in one and two dimensions. Chapters 7 and 8 describe some of the more important reactions that organic contaminants undergo in the subsurface, such as adsorption, chemical reaction, and aerobic and anaerobic biodegradation. Chapter 9 presents a concise treatment of flow and transport in the unsaturated zone. Chapter 10 reviews modern approaches to the numerical modeling of ground water flow and transport systems, including a detailed treatment of finite-difference methods along with applications to field settings.

Chapter 11 covers nonaqueous phase liquids (NAPLs) in the subsurface, an important topic that has received considerable attention in the past few years as it relates to the remediation of hazardous-waste sites. NAPLs, which can provide a source of contamination for decades, have been found at numerous waste sites as part of older pump-and-treat cleanup operations. Light (LNAPL) and dense (DNAPL) phase liquids present a considerably more complex problem in the sub-surface than soluble contaminant plumes and may create sources of slow release of contaminants for many years. Figure 1.6 presents the current conceptual model for what happens to a typical LNAPL hydrocarbon spill in the subsurface. It is now widely recognized that understanding transport and remediation of NAPLs may be the most important ground water problem of the 1990s.

Chapter 12 covers conceptual models for hydrogeologic site investigations. Chapter 13 addresses modern methods of ground water remediation, including containment, pump-and-treat, bioremediation, soil vapor extraction, and DNAPL issues. Major case studies from the authors' actual field experiences are used to highlight the various problems involved in site remediation. Finally, the driving force for the tremendous growth of the ground water remediation industry was the major environmental legislation (RCRA, 1976, and CERCLA, 1980) passed in

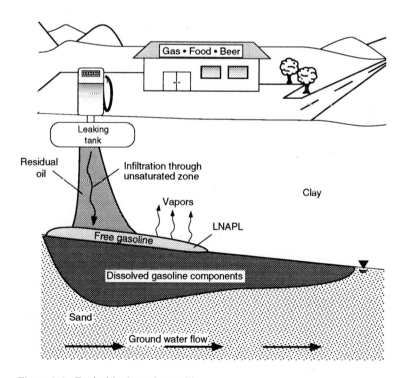

Figure 1.6 Typical hydrocarbon spill.

the late 1970s and 1980s. Chapter 14 covers the legal protection of ground water in detail.

1.4 EVOLUTION OF GROUND WATER INFORMATION

A number of classic textbooks in the ground water field have been written over the past 30 years. The field of ground water hydrology has expanded greatly since the first U.S. textbook by Tolman (1937). Todd's (1959) text, *Groundwater Hydrology,* stood as a classic in the field for many years and was updated with a new edition in 1980. DeWiest's text in 1965 and Davis and DeWiest in 1966 further advanced the subject with their books on geohydrology and hydrogeology. Bear's texts, written in 1972 and 1979, were departures from earlier approaches and emphasized the hydraulics of ground water, providing a very theoretical development of flow and transport in both the saturated and unsaturated zone for engineers, hydrologists, and hydrogeologists. In 1979, Freeze and Cherry's *Groundwater* quickly replaced others as the standard in the ground water field for more than a decade. Their chapters on transport, chemical properties, and contamination are still of great use today.

There has been an explosion of literature in the past decade, and there are numerous new sources of information and data in the ground water hydrology area. The U.S. Geological Survey (USGS) has primary responsibility for the collection of ground water data and evaluation of these data in terms of impacts on water supply, water quality, water depletion, and potential contamination. Studies performed by engineering consultants for the EPA during the remedial investigation or feasibility study of RCRA and Superfund sites also provide a useful description of applied methods in ground water. Other primary sources of information are state environmental and water resources agencies, the American Geophysical Union, and the National Ground Water Association. Journals such as *Water Resources Research, Ground Water, Journal of Hydrology*, the *ASCE Journal of Environmental Engineering,* and *Environmental Science and Technology* are major resources for exchange of information.

The current text is a departure from past efforts in that it is written from both a theoretical and engineering viewpoint, with hydrologic and transport theory applied to hazardous-waste site characterization, transport modeling, and remediation. For the first time, chapters on numerical methods and discussions of model applications to actual field sites exist alongside discussions of advection, adsorption, dispersion, chemical reaction, and biodegradation. Chapters on monitoring, hydrogeologic site characterization, and remediation are included, along with detailed case studies that illustrate the various techniques currently being applied. Several other textbooks on ground water and hydrogeology are available for more detailed treatment of specific hydrologic and transport processes (Bear, 1979; Freeze and Cherry, 1979; Todd, 1980; de Marsily, 1986; Fetter, 1988; Domenico and Schwartz, 1990; Fetter, 1993).

1.5 GROUND WATER REMEDIATION

The discovery of hazardous wastes at Love Canal in Niagara, New York, ushered in a new era in hazardous-waste problems. During the **hazardous-waste decade** of the 1980s, hydrologists, hydrogeologists, civil and environmental engineers, and other scientists were involved in characterizing, evaluating, and remediating hazardous-waste sites with respect to ground water contamination. The field of ground water has seen an explosion in the number of remedial investigation studies related to the thousands of abandoned and active hazardous-waste disposal sites and leaking tanks across the United States. One main objective of this text is to provide engineers and scientists with a modern treatment of the remediation methods currently being practiced, but with an eye toward the future. Emerging new technologies are rapidly coming into place as we learn more about the failures of the past. Earlier pump-and-treat systems that did not consider the presence of NAPLs have not cleaned aquifers to the required levels. Also, the EPA is now reconsidering the objective of remediating all aquifers to drinking-water standards. Chapters 11 and 13 address some of these issues in detail, but the reader is cautioned that ground water remediation is a rapidly changing and dynamic industry, and the general literature should be consulted for the latest methods.

REFERENCES

ALLER, L., BENNETT, T., LEHR, J., PETTY, R. J., and HACKETT, G., 1987, "DRASTIC: A Standardized System for Evaluating Ground Water Pollution Potential Using Hydrogeologic Settings," EPA-600/2-87-035. U.S. Environmental Protection Agency, Washington, DC.

BEAR, J., 1972, *Dynamics of Fluids in Porous Media,* American Elsevier, New York.

BEAR, J., 1979, *Hydraulics of Groundwater,* McGraw-Hill, New York.

BEDIENT, P. B., and HUBER, W. C., 1992, *Hydrology and Floodplain Analysis,* 2nd ed., Addison-Wesley, Reading, MA.

CERCLA (Comprehensive Environmental Response, Compensation, and Liability Act or Superfund), 1980, 40 C.F.R. Part 300.

CHOW, V. T., MAIDMENT, D. R., and MAYS, L. W., 1988, *Applied Hydrology,* McGraw-Hill, New York.

DAVIS, S. N., and DeWIEST, R. J. M., 1966, *Hydrogeology,* Wiley, New York.

DE MARSILY, G., 1986, *Quantitative Hydrogeology,* Academic Press, Orlando, FL.

DeWIEST, R. J. M., 1965, *Geohydrology,* Wiley, New York.

DOMENICO, P. A., and SCHWARTZ, F. W., 1990, *Physical and Chemical Hydrogeology,* Wiley, New York.

FETTER, C. W., 1988, *Applied Hydrogeology,* Merrill Publishing Company, Columbus, OH.

FETTER, C. W., 1993, *Contaminant Hydrogeology,* Macmillan, New York.

FREEZE, R. A., and CHERRY, J .A., 1979, *Groundwater*, Prentice Hall, Englewood Cliffs, NJ.

GUPTA, R. S., 1989, *Hydrology and Hydraulic Systems*, Prentice Hall, Englewood Cliffs, NJ.

HEATH, R. C., 1984, *Ground Water Regions of the United States*, U.S. Geological Survey Water Supply Paper 2242.

MEINZER, O. E., 1923, *Outline of Groundwater in Hydrology with Definitions*, U.S. Geological Survey Water Supply Paper 494.

NEWELL, C. J., HAASBEEK, J. F., and BEDIENT, P. B., 1990, "OASIS: A Graphical Decision Support System for Ground Water Contaminant Modeling," *Ground Water* 28:224–234.

RCRA (Resource Conservation and Recovery Act), 1976, 40 C.F.R. Parts 260–271.

SOLLEY, W. B., MERK, C. F., and PIERCE, R. R., 1988, "Estimated Water Use in the United States in 1985," *U.S. Geological Survey Circular 1004.*

THOMAS, H. E., 1952, "Ground Water Regions of the United States—Their Storage Facilities," in *The Physical and Economic Foundation of National Resources*, vol. 3, U.S. 83rd Congress, House Interior and Insular Affairs Committee, Washington, DC.

TODD, D. K., 1959, *Groundwater Hydrology*, 1st ed., Wiley, New York.

TODD, D. K., 1980, *Groundwater Hydrology*, 2nd ed., Wiley, New York.

TOLMAN, C. F., 1937, *Groundwater*, McGraw-Hill, New York.

VIESSMAN, W., JR., LEWIS, G. L., and KNAPP, J. W., 1989, *Introduction to Hydrology*, 3rd ed., Harper & Row, New York.

Chapter 2 | GROUND WATER HYDROLOGY

2.1 INTRODUCTION

This chapter is devoted to the properties of ground water, including the definition of aquifer systems and the parameters that can be used to characterize them. Porosity, hydraulic conductivity, storage coefficient, and hydraulic gradient are all important in determining the rate and direction of ground water flow. Governing equations of ground water flow are introduced and solved for simple flow systems in confined and unconfined aquifers. Flow nets are useful representations of streamlines and equipotential lines in two dimensions and provide a graphical picture of ground water heads and gradients of flow. The chapter ends with an introduction to the unsaturated zone, which generally lies above the water table up to the soil root zone.

2.2 PROPERTIES OF GROUND WATER

Vertical Distribution of Ground Water

Ground water can be characterized according to vertical distribution as defined by Todd (1980). Figure 2.1 indicates the divisions of subsurface water. The **soil-water zone,** which extends from the ground surface down through the major **root zone,** varies with soil type and vegetation. The amount of water present in the soil-water zone depends primarily on recent exposure to rainfall and infiltration. **Hygroscopic water** remains absorbed to the surface of soil grains, while gravitational water drains through the soil under the influence of gravity.

  Capillary water is held by surface tension forces just above the **water table,** which is defined as the level to which water will rise in a well drilled into the saturated zone. The **vadose zone** extends from the lower edge of the soil-water zone to the upper limit of the capillary zone (see Figure 2.1). Thickness may vary from zero for high-water-table conditions to more than 100 m in arid regions of the country, such as Arizona or New Mexico. Vadose zone water is held in place by hygroscopic and capillary forces, and infiltrating water passes downward toward the water table as gravitational flow, subject to retardation by these forces.

  The capillary zone, or fringe, extends from the water table up to the limit

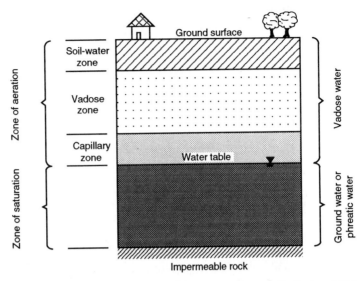

Figure 2.1 Vertical zones of subsurface water. Source: Bedient and Huber, 1992.

of **capillary rise,** which varies inversely with the pore size of the soil and directly with the surface tension. Capillary rise can range from 2.5 cm for fine gravel to more than 200 cm for silt (Lohman, 1972). Just above the water table, almost all pores contain capillary water, and the water content decreases with height depending on the type of soil. A typical soil-moisture curve is shown in Figure 2.2, which shows that the amount of moisture in the vadose zone generally decreases with vertical distance above the water table. Soil-moisture curves vary with soil type and with the wetting cycle; more details on the unsaturated zone are contained in Section 2.7 and Chapter 9.

In the zone of saturation, which occurs beneath the water table, the **porosity**

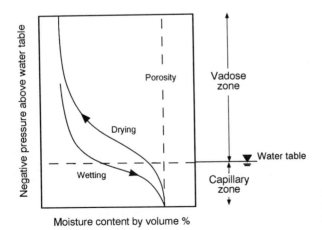

Figure 2.2 Typical soil-moisture relationship. Source: Bedient and Huber, 1992.

is a direct measure of the water contained per unit volume, expressed as the ratio of the volume of voids to the total volume. Only a portion of the water can be removed from the saturated zone by drainage or by pumping from a well. **Specific yield** is defined as the volume of water released from an unconfined aquifer per unit surface area per unit head decline in the water table. Fine-grained materials yield little water, whereas coarse-grained materials provide significant water and thus serve as aquifers. In general, specific yields for unconsolidated formations fall in the range from 7% to 25%.

Aquifers

An **aquifer** can be defined as a formation that contains sufficient saturated permeable material to yield significant quantities of water to wells and springs. Aquifers are generally areally extensive and may be overlain or underlain by a confining bed, defined as a relatively impermeable material. Figure 2.3 shows some typical examples of confined aquifers. An **aquiclude** is a relatively impermeable confining unit, such as clay, and an **aquitard** is a poorly permeable stratum, such as a sandy clay unit, that may leak water to adjacent sand aquifers.

Aquifers can be characterized by the porosity of a rock or soil, expressed as the ratio of the volume of voids V_ν to the total volume V. Porosity may also be expressed by

$$n = \frac{V_\nu}{V} = 1 - \frac{\rho_b}{\rho_m} \qquad (2.1)$$

where ρ_m is the density of the grains and ρ_b is the **bulk density,** defined as the oven-dried mass of the sample divided by its original volume. Table 2.1 shows a range of porosities for a number of aquifer materials. Figure 2.4 shows the theoretical porosity associated with cubic packing of spheres of equal diameter and the effect of adding smaller grains into the void space.

Unconsolidated geologic materials are normally classified according to their size and distribution. Soil classification based on particle size is shown in Table 2.2. Particle sizes are measured by mechanically sieving grain sizes larger than 0.05 mm and measuring rates of settlement for smaller particles in suspension. A typical particle-size distribution graph is shown in Figure 2.5. The value of the **uniformity coefficient,** defined as D_{60}/d_{10}, indicates the relative sorting of the material. A uniform material such as fine sand has a low coefficient, whereas a well-graded material such as alluvium has a high coefficient (Figure 2.5).

The texture of a soil is defined by the relative proportions of sand, silt, and clay present in the particle-size analysis and can be expressed most easily on a triangle of soil textures (Figure 2.6). For example, a soil with 30% clay, 60% silt, and 10% sand is referred to as a silty clay loam.

Most aquifers can be considered as undergound storage reservoirs that receive recharge from rainfall or from an artificial source. Water flows out of an aquifer due to gravity or to pumping from wells. Aquifers may be classified as **unconfined** or **confined,** depending on the existence of a water table. A leaky

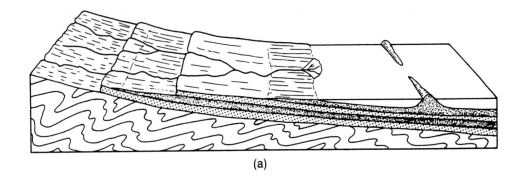

(a)

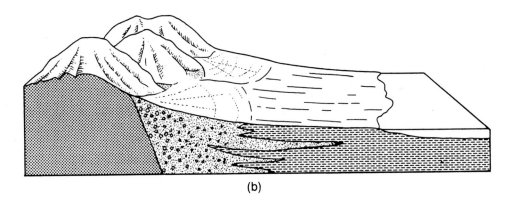

(b)

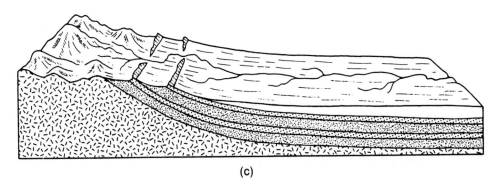

(c)

Figure 2.3 Confined aquifers created when aquifers are overlain by confining beds. (a) Confined aquifers created by alternating aquifers and confining units deposited on a regional dip. (b) Confined aquifers created by deposition of alternating layers of permeable sand and gravel and impermeable silts and clays deposited in intermontane basins. (c) Confined aquifer created by upwarping of beds by intrusions. Source: C. W. Fetter, *Applied Hydrology*, 2/E. Reprinted with the permission of Macmillan College Publishing © 1988.

TABLE 2.1 Representative Values of Porosity

Material	Porosity (%)	Material	Porosity (%)
Gravel, coarse	28[a]	Loess	49
Gravel, medium	32[a]	Peat	92
Gravel, fine	34[a]	Schist	38
Sand, coarse	39	Siltstone	35
Sand, medium	39	Claystone	43
Sand, fine	43	Shale	6
Silt	46	Till, predominantly silt	34
Clay	42	Till, predominantly sand	31
Sandstone, fine grained	33	Tuff	41
Sandstone, medium grained	37	Basalt	17
Limestone	30	Gabbro, weathered	43
Dolomite	26	Granite, weathered	45
Dune sand	45		

Source: Morris and Johnson, 1967.

[a] These values are for repacked samples; all others are undisturbed.

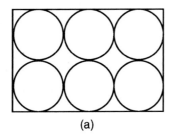 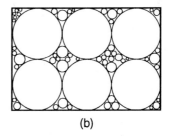

      (a)   (b)

Figure 2.4 (a) Cubic packing of spheres of equal diameter with a porosity of 47.65%. (b) Cubic packing of spheres with void spaces occupied by grains of smaller diameter, resulting in a much lower overall porosity.

TABLE 2.2 Soil Classification Based on Particle Size

Material	Particle size (mm)
Clay	<0.004
Silt	0.004–0.062
Very fine sand	0.062–0.125
Fine sand	0.125–0.25
Medium sand	0.25–0.5
Coarse sand	0.5–0.1
Very coarse sand	1.0–2.0
Very fine gravel	2.0–4.0
Fine gravel	4.0–8.0
Medium gravel	8.0–16.0
Coarse gravel	16.0–32.0
Very coarse gravel	32.0–64.0

Source: Morris and Johnson, 1967.

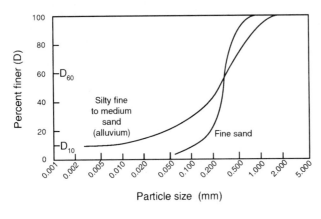

Figure 2.5 Particle-size distribution for two geologic samples. Source: Bedient and Huber, 1992.

confined aquifer represents a stratum that allows water to flow through the confining zone.

Figure 2.7 is a vertical cross section illustrating unconfined and confined aquifers. In an unconfined aquifer, a water table exists and rises and falls with changes in the volume of water stored. In a **perched water table,** an unconfined water body sits on top of a clay lens, separated from the main aquifer deeper below.

Confined aquifers, called **artesian aquifers,** occur where ground water is confined by a relatively impermeable stratum, or confining unit, and water is under pressure greater than atmospheric. If a well penetrates such an aquifer, the water

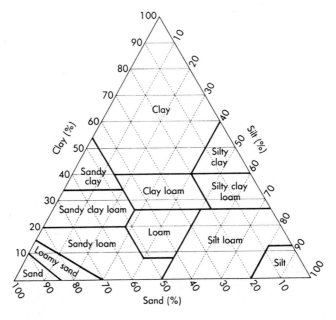

Figure 2.6 Triangle representation of soil textures for describing various combinations of sand, silt, and clay. Source: Bedient and Huber, 1992.

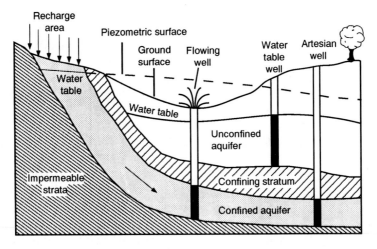

Figure 2.7 Schematic cross section of unconfined and confined aquifers. Black areas indicate screened zones. Source: Bedient and Huber, 1992.

level will rise above the bottom of the confining unit. If the water level rises above the land surface, a flowing well or spring results, which is referred to as an **artesian well** or spring.

A recharge area supplies water to a confined aquifer, and such an aquifer can convey water from the recharge area to locations of natural or artificial discharge. The **piezometric surface** (or **potentiometric surface**) of a confined aquifer is the hydrostatic pressure level of water in the aquifer, defined by the water level that occurs in a lined penetrating well. It should be noted that a confined aquifer can become unconfined when the piezometric surface falls below the bottom of the upper confining bed. Contour maps and profiles can be prepared for the water-table elevation for an unconfined aquifer or the piezometric surface for a confined aquifer. These **equipotential lines** are described in more detail in Section 2.6. Once determined from a series of wells in an aquifer, orthogonal lines can be drawn to indicate the general direction of ground water flow in the direction of decreasing head.

A parameter of some importance relates to the water-yielding capacity of an aquifer. The **storage coefficient** S is defined as the volume of water that an aquifer releases from or takes into storage per unit surface area per unit change in piezometric head. For a confined aquifer, values of S fall in the range from 0.00005 to 0.005, indicating that large pressure changes produce small changes in the storage volume. For unconfined aquifers, a change in storage volume is expressed simply by the product of the volume of aquifer lying between the water table at the beginning and end of a period of head change and the average specific yield of the formation. Thus, the storage coefficient for an unconfined aquifer is approximately equal to the specific yield, typically in the range from 0.07 to 0.25.

A vertical hole dug into the earth and penetrating an aquifer is referred to as a **well,** and the metal or plastic pipe that extends from the surface to the screened zone is called the **casing.** Wells are used for pumping, recharge, disposal, and water-level observation. Often the portion of the well hole that is open to the aquifer is **screened** to prevent aquifer material from entering the well. The annular space around the screen is often filled with sand or gravel to minimize hydraulic resistance to flow. The annulus is usually cemented with a material such as bentonite clay up to the surface to protect against contamination of the well from surface leakage. For more detail on well construction methods, see Chapter 5.

2.3 GROUND WATER MOVEMENT

Darcy's Law

The movement of ground water is well understood by hydraulic principles reported in 1856 by Henri Darcy, who investigated the flow of water through beds of permeable sand. Darcy discovered one of the most important laws in hydrology: *the flow rate through porous media is proportional to the head loss and inversely proportional to the length of the flow path.* Darcy's law serves as the basis for present-day knowledge of ground water flow and well hydraulics.

Figure 2.8 depicts the experimental setup for determining head loss through a sand column, with **piezometers** located a distance L apart. Total energy for this system can be expressed by the Bernoulli equation

$$\frac{p_1}{\gamma} + \frac{v_1^2}{2g} + z_1 = \frac{p_2}{\gamma} + \frac{v_2^2}{2g} + z_2 + h_1 \qquad (2.2)$$

where p = pressure z = elevation
       γ = specific weight of water h_1 = head loss
       v = velocity

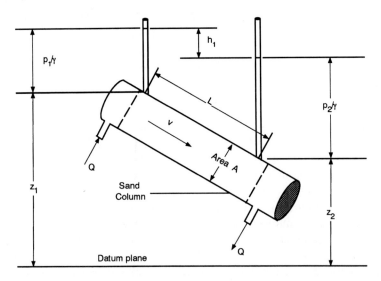

Figure 2.8 Head loss through a sand column. Source: Bedient and Huber, 1992.

Because velocities are very small in porous media, velocity heads may be neglected, allowing head loss to be expressed as

$$h_1 = \left(\frac{p_1}{\gamma} + z_1\right) - \left(\frac{p_2}{\gamma} + z_2\right) \qquad (2.3)$$

It follows that the head loss is independent of the inclination of the column. Darcy related flow rate to head loss and length of column through a proportionality constant referred to as K, the **hydraulic conductivity,** a measure of the **permeability** of the porous media. Darcy's law can be stated

$$V = -\frac{Q}{A} = -K\frac{dh}{dL} \qquad (2.4)$$

The negative sign indicates that flow of water is in the direction of decreasing head. The Darcy velocity that results from Eq. (2.4) is an average discharge velocity through the entire cross section of the column. The actual flow is limited to the pore space only, so the seepage velocity V_s is equal to the Darcy velocity divided by porosity:

$$V_s = \frac{Q}{nA} \qquad (2.5)$$

Thus, actual velocities are usually much higher (by a factor of 3) than the Darcy velocities.

It should be pointed out that Darcy's law applies to laminar flow in porous media, and experiments indicate that Darcy's law is valid for Reynolds number ($R = \rho Vd/\mu$) less than 1 and perhaps as high as 10. This represents an upper limit to the validity of Darcy's law, which turns out to be applicable in most ground water systems. Deviations can occur near pumped wells and in fractured aquifer systems with large openings.

Hydraulic Conductivity

The hydraulic conductivity of a soil or rock depends on a variety of physical factors and is an indication of an aquifer's ability to transmit water. Thus, sand aquifers have K values many orders of magnitude larger than clay units. Table 2.3 indicates representative values of hydraulic conductivity for a variety of materials. As can be seen, K can vary over many orders of magnitude in an aquifer that may contain different types of material. Thus, velocities and flow rates can also vary over the same range, as expressed by Darcy's law.

The term **transmissivity** is often used in ground water hydraulics as applied to confined aquifers. It is defined as the product of K and the **saturated thickness** of the aquifer b. Hydraulic conductivity is usually expressed in m/day (ft/day), and transmissivity T is expressed in m²/day (ft²/day). An older unit for T that is

TABLE 2.3 Representative Values of Hydraulic Conductivity

Unconsolidated sedimentary deposits	Hydraulic conductivity (cm/sec)
Gravel	3.0 to 3×10^{-2}
Coarse sand	6×10^{-1} to 9×10^{-5}
Medium sand	6×10^{-2} to 9×10^{-5}
Fine sand	2×10^{-2} to 2×10^{-5}
Silt, loess	2×10^{-3} to 1×10^{-7}
Till	2×10^{-4} to 1×10^{-10}
Clay	5×10^{-7} to 1×10^{-9}
Unweathered marine clay	2×10^{-7} to 8×10^{-11}
Sedimentary rocks	
Karst limestone	2 to 1×10^{-4}
Limestone and dolomite	6×10^{-4} to 1×10^{-7}
Sandstone	6×10^{-4} to 3×10^{-8}
Shale	2×10^{-7} to 1×10^{-11}
Crystalline rocks	
Permeable basalt	2 to 4×10^{-5}
Fractured igneous and metamorphic	3×10^{-2} to 8×10^{-7}
Basalt	4×10^{-5} to 2×10^{-9}
Unfractured igneous and metamorphic	2×10^{-8} to 3×10^{-12}
Weathered granite	3×10^{-4} to 5×10^{-3}

Source: Bedient and Huber, 1992.

Note: 1 cm/sec = 1×10^{-2} m/sec = 1.04×10^{3} darcys.

still reported in some applications is gal/day/ft. The **intrinsic permeability** of a rock or soil is a property of the medium only, independent of fluid properties. Intrinsic permeability k can be related to K by

$$K = k \frac{\rho g}{\mu} \tag{2.6}$$

where μ = dynamic viscosity
  ρ = fluid density
  g = gravitational constant

Intrinsic permeability k has units of m^2 or Darcys, equal to 0.987 μm^2; k is often used in the petroleum industry, whereas K is used in ground water hydrology for categorizing aquifer systems.

Determination of Hydraulic Conductivity

Hydraulic conductivity in saturated zones can be determined by a number of techniques in the laboratory as well as in the field. **Constant head** and **falling head permeameters** are used in the laboratory for measuring K and are described in more detail later. In the field, **pump tests, slug tests,** and **tracer tests** are available for determination of K. These tests are described in more detail in Chapter 3 under the general heading of well hydraulics.

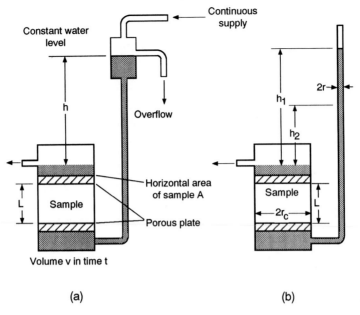

Figure 2.9 Permeameters for measuring hydraulic conductivity of geologic samples. (a) Constant head. (b) Falling head.   Source: Bedient and Huber, 1992.

A permeameter (Figure 2.9) is used in the laboratory to measure K by maintaining flow through a small column of material and measuring flow rate and head loss. For a constant-head permeameter, Darcy's law can be directly applied to find K, where V is volume flowing in time t through a sample of area A and length L, with constant head h:

$$K = \frac{VL}{Ath} \tag{2.7}$$

The falling-head permeameter test consists of measuring the rate of fall of the water level in the tube or column and noting that

$$Q = \pi r^2 \frac{dh}{dt} \tag{2.8}$$

Darcy's law can be written for the sample as

$$Q = \pi r_c^2 K \frac{h}{L} \tag{2.9}$$

After equating and integrating,

$$K = \frac{r^2 L}{r_c^2 t} \ln \frac{h_1}{h_2} \tag{2.10}$$

where L, r, and r_c are as shown in Figure 2.9 and t is the time interval for water to fall from h_1 to h_2.

Auger hole tests involve the measurement of a change in water level after rapid removal or addition of a volume of water to a bore hole. The value of K is applicable in the immediate vicinity of the hole and is generally useful for shallow water-table conditions. The **slug test** for shallow wells operates in much the same fashion, with a measurement of decline or recovery of the water level in the well through time.

The pump test involves the constant removal of water from a single well and observations of water-level declines at several adjacent wells. In this way, an integrated K value for a portion of the aquifer is obtained. Field methods generally yield significantly different values of K than corresponding laboratory tests performed on cores removed from the aquifer. Thus, field tests are preferable for the accurate determination of aquifer parameters (Chapter 3).

Anisotropic Aquifers

Most real geologic systems tend to have variations in one or more directions due to the processes of deposition and layering that can occur. In the typical field situation in alluvial deposits, we find the hydraulic conductivity in the vertical direction K_z to be less than the value in the horizontal direction K_x. For the case of a two-layered aquifer of different K in each layer and different thicknesses, we can apply Darcy's law to horizontal flow to show

$$K_x = \frac{K_1 z_1 + K_2 z_2}{z_1 + z_2} \tag{2.11}$$

or, in general,

$$K_x = \frac{\sum K_i z_i}{\sum z_i} \tag{2.12}$$

where $K_i = K$ in layer i
$\quad\quad z_i$ = thickness of layer i

For the case of vertical flow through two layers, q_z is the same flow per unit horizontal area in each layer:

$$dh_1 + dh_2 = \left(\frac{z_1}{K_1} + \frac{z_2}{K_2}\right) q_z \tag{2.13}$$

but $$dh_1 + dh_2 = \frac{z_1 + z_2}{K_z} q_z \tag{2.14}$$

where K_z is the hydraulic conductivity for the entire system. Equating Eqs. (2.13) and (2.14), we have

$$K_z = \frac{z_1 + z_2}{(z_1/K_1) + (z_2/K_2)} \tag{2.15}$$

or, in general,

$$K_z = \frac{\sum z_i}{\sum z_i/K_i} \tag{2.16}$$

Ratios of K_x/K_z usually fall in the range from 2 to 10 for alluvium, with values up to 100 where clay layers exist. In actual application to layered systems, it is usually necessary to apply ground water flow models that can properly handle complex geologic strata through numerical simulation. Various modeling techniques are described in Chapter 10.

Flow Nets

Darcy's law was originally derived in one dimension, but because many ground water problems are really two or three dimensional, methods are available for the determination of flow rate and direction. A specified set of **streamlines** and **equipotential lines** can be constructed for a given set of boundary conditions to form a **flow net** (Figure 2.10) in two dimensions.

Equipotential lines are prepared based on observed water levels in wells penetrating an **isotropic** aquifer. Flow lines are then drawn orthogonally to indicate the direction of flow. For the flow net of Figure 2.10, the hydraulic gradient i is given by

$$i = \frac{dh}{ds} \tag{2.17}$$

and constant flow q per unit thickness between two adjacent flow lines is

$$q = K \frac{dh}{ds} dm \tag{2.18}$$

If we assume $ds = dm$ for a square net, then for n squares between two flow lines over which total head is divided ($h = H/n$) and for m divided flow channels,

$$Q' = mq = \frac{KmH}{n} \tag{2.19}$$

where Q' = flow per unit width
K = hydraulic conductivity of the aquifer
m = number of flow channels
n = number of squares over the direction of flow
H = total head loss in direction of flow

Flow nets are useful graphical methods to display streamlines and equipotential lines. Since no flow can cross an impermeable boundary, streamlines must parallel it. Also, streamlines are usually horizontal through high K material and vertical through low K material because of refraction of lines across a boundary between different K media. It can be shown (Figure 2.11) that

$$\frac{K_1}{K_2} = \frac{\tan \theta_1}{\tan \theta_2} \tag{2.20}$$

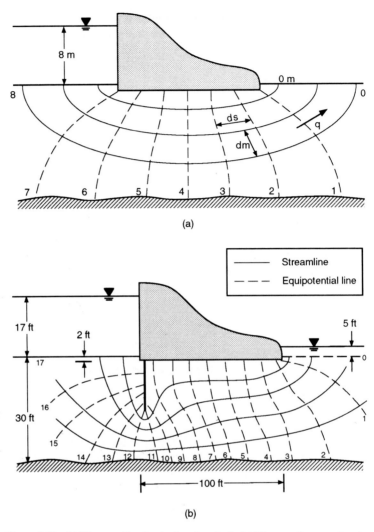

Figure 2.10 (a) Flow net for simple dam (section). (b) Flow net for complex dam (section). Source: Bedient and Huber, 1992.

Flow nets for seepage through a layered system may not be orthogonal and are shown in Figure 2.12. In Figure 2.12a, most of the flow is in the lower unit, and in part b it is in the upper unit. Such bending of streamlines becomes important in determining regional flow patterns. Figure 2.13 indicates possible local and regional patterns for a homogeneous isotropic aquifer that has a water table with a sinusoidal distribution (hilly terrain).

Flow nets can be used to evaluate the effects of pumping on ground water levels and directions of flow. Figure 2.14 depicts the contour map resulting from heavy pumping near Houston, Texas, over a period of years. Equipotential lines and streamlines are discussed further in Section 2.6.

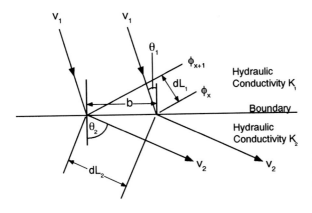

Figure 2.11 Refraction of flow lines across a boundary between media of different hydraulic conductivities. Source: Bedient and Huber, 1992.

Example 2.1 FLOW NET COMPUTATION

Compute the total flow seeping under the dam in Figure 2.10a, where width is 20 m and $K = 10^{-5}$ m/sec. Equation (2.19) is used to provide flow per unit width. From the figure $m = 4$, $n = 8$, $H = 8$ m, and K is as given.

Solution

$$Q' = \frac{(10^{-5} \text{ m/sec})(4)(8 \text{ m})}{8}$$

$$= 4 \times 10^{-5} \text{ m}^2/\text{sec} = 3.46 \text{ m}^2/\text{day}$$

Total flow $Q = 3.46$ m^2/day (20 m) $= 69.1$ m^3/day

2.4 GENERAL FLOW EQUATIONS

The governing flow equations for ground water are derived in most of the standard texts in the field (Bear, 1979; Freeze and Cherry, 1979; Todd, 1980; Domenico and Schwartz, 1990). The equation of continuity from fluid mechanics is combined with Darcy's law in three dimensions to yield a partial differential equation of flow in porous media, as shown in the next section. Both steady-state and transient flow equations can be derived. Mathematical solutions for specific boundary conditions are well known for the governing ground water flow equation. For complex boundaries and heterogeneous systems, numerical computer solutions must be used (Chapters 9 and 10).

Steady-state Saturated Flow

Consider a unit volume of porous media (Figure 2.15) called an elemental control volume. The law of conservation of mass requires that

Mass in − mass out = change in storage per time

For steady-state conditions, the right side is zero, and the equation of continuity becomes (Figure 2.15)

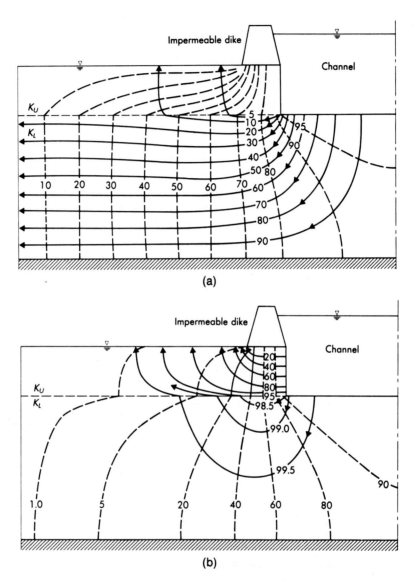

Figure 2.12 Flow nets for seepage from one side of a channel through two different anisotropic two-layer systems. (a) $K_u/K_l = 1/50$. (b) $K_u/K_l = 50$. The anisotropy ratio for all layers is $K_x/K_\chi = 10$.   Source: Todd and Bear, 1961.

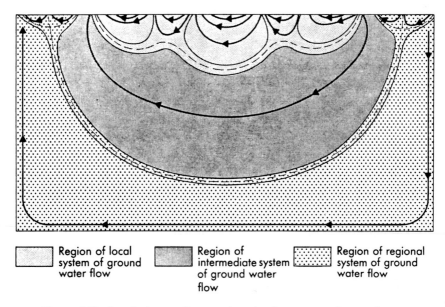

Region of local system of ground water flow

Region of intermediate system of ground water flow

Region of regional system of ground water flow

Figure 2.13 Local, intermediate, and regional systems of ground water flow. Source: Bedient and Huber, 1992.

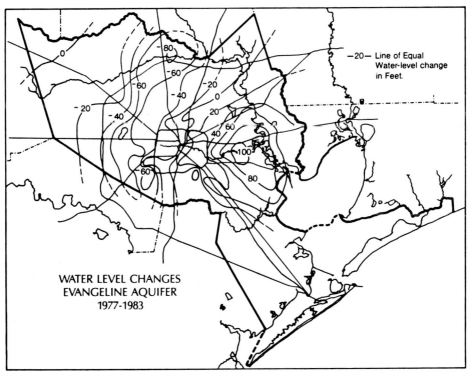

−20− Line of Equal Water-level change in Feet.

WATER LEVEL CHANGES
EVANGELINE AQUIFER
1977-1983

Figure 2.14 Drawdown in Harris and Galveston counties, Texas. Source: Bedient and Huber, 1992.

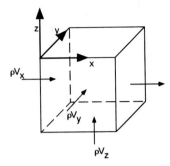

Figure 2.15 Elemental control volume.

$$-\frac{\partial}{\partial x}(\rho V_x) - \frac{\partial}{\partial y}(\rho V_y) - \frac{\partial}{\partial z}(\rho V_z) = 0 \qquad (2.21)$$

The units of ρV are mass/area/time, as required. For an incompressible fluid, $\rho(x, y, z)$ = constant, and ρ can be divided out of Eq. (2.21). Substitution of Darcy's law for V_x, V_y, and V_z yields

$$\frac{\partial}{\partial x}\left(K_x \frac{\partial h}{\partial x}\right) + \frac{\partial}{\partial y}\left(K_y \frac{\partial h}{\partial y}\right) + \frac{\partial}{\partial z}\left(K_z \frac{\partial h}{\partial z}\right) = 0 \qquad (2.22)$$

For an isotropic, homogeneous medium, $K_x = K_y = K_z = K$, which can be divided out of the equation to yield

$$\frac{\partial^2 h}{\partial x^2} + \frac{\partial^2 h}{\partial y^2} + \frac{\partial^2 h}{\partial z^2} = 0 \qquad (2.23)$$

Equation (2.23) is called Laplace's equation and is one of the best-understood partial differential equations. The solution is $h = h(x, y, z)$, the hydraulic head at any point in the flow domain. In two dimensions, the solution is equivalent to the graphical flow nets described in Section 2.3. If there were no variation of h with z, then the equation would reduce to two terms on the left side of Eq. (2.23).

Transient Saturated Flow

The transient equation of continuity for a confined aquifer becomes

$$-\frac{\partial}{\partial x}(\rho V_x) - \frac{\partial}{\partial y}(\rho V_y) - \frac{\partial}{\partial z}(\rho V_z) = \frac{\partial}{\partial t}(\rho n) = n\frac{\partial \rho}{\partial t} + \rho\frac{\partial n}{\partial t} \qquad (2.24)$$

The first term on the right side of Eq. (2.24) is the mass rate of water produced by an expansion of water under a change in ρ. The second term is the mass rate of water produced by compaction of the porous media (change in n). The first term relates to the compressibility of the fluid β and the second term to the aquifer compressibility α.

Compressibility and effective stress are discussed in detail in Freeze and Cherry (1979) and Domenico and Schwartz (1990) and will only be briefly reviewed here. According to Terzaghi (1925), the total stress acting on a plane in a saturated porous media is due to the sum of the weight of overlying rock and fluid pressure. The portion of the total stress not borne by the fluid is the effective stress σ_e. Since total stress can be considered constant in most problems, the change in effective stress is equal to the negative of the pressure change in the media, which is related to head change by $dp = \rho g\,dh$. Thus, a decrease in hydraulic head or pressure results in an increase in effective stress, since $d\sigma_e = -dp$.

The compressibility of water β implies that a change in volume occurs for a given change in stress or pressure and is defined as $(-dV/V)/dp$, where dV is volume change of a given mass of water under a pressure change of dp. The compressibility is approximately constant at 4.4×10^{-10} m^2/N for water at usual ground water temperatures.

The compressibility of the porous media or aquifer, α, is related to vertical consolidation for a given change in effective stress, or $\alpha = (db/b)/d\sigma_e$, where b is the vertical dimension. From laboratory studies, α is a function of the applied stress and depends on previous loading history. Clays respond differently than sands in this regard, and compaction of clays is largely irreversible for a reduced pressure in the aquifer compared to the response in sands. Land surface subsidence is a good example of aquifer compressibility on a regional scale where clays have been depressured over time.

Freeze and Cherry (1979) indicate that a change in head will produce a change in ρ and n in Eq. (2.24), and the volume of water produced for a unit head decline is S_s, the specific storage. Theoretically, we can show that specific storage is related to aquifer compressibility and the compressibility of water by

$$S_s = \rho g(\alpha + n\beta) \tag{2.25}$$

and the mass rate of water produced [right side of Eq. (2.24)] is $S_s(\partial h/\partial t)$. Equation (2.24) becomes, after substituting Eq. (2.25) and Darcy's law,

$$\frac{\partial}{\partial x}\left(K_x \frac{\partial h}{\partial x}\right) + \frac{\partial}{\partial y}\left(K_y \frac{\partial h}{\partial y}\right) + \frac{\partial}{\partial z}\left(K_z \frac{\partial h}{\partial z}\right) = S_s \frac{\partial h}{\partial t} \tag{2.26}$$

For homogeneous and isotropic media,

$$\frac{\partial^2 h}{\partial x^2} + \frac{\partial^2 h}{\partial y^2} + \frac{\partial^2 h}{\partial z^2} = \frac{S_s}{K} \frac{\partial h}{\partial t} \tag{2.27}$$

For the special case of a horizontal confined aquifer of thickness b,

$$S = S_s b, \quad \text{where } S \text{ is the storativity or storage coefficient}$$

$$T = Kb$$

and

$$\nabla^2 h = \frac{S}{T} \frac{\partial h}{\partial t} \quad \text{in two dimensions} \tag{2.28}$$

Solution of Eqs. (2.28) requires knowledge of S and T to produce $h(x, y)$ over the flow domain. The classical development of Eq. (2.28) was first advanced by Jacob (1940), along with considerations of storage concepts. More advanced treatments consider the problems of a fixed elemental control volume in a deforming media (Cooper, 1966), but these are unnecessary considerations for most practical problems.

2.5 DUPUIT EQUATION

For the case of unconfined ground water flow, Dupuit (1863) developed a theory that allows for a simple solution based on several important assumptions:

1. The water table or free surface is only slightly inclined.
2. Streamlines may be considered horizontal and equipotential lines vertical.
3. Slopes of the free surface and hydraulic gradient are equal.

Figure 2.16 shows a graphical example of Dupuit's assumptions for essentially one-dimensional flow. The free surface from $x = 0$ to $x = L$ can be derived by considering Darcy's law and the governing one-dimensional equation. The Dupuit approach neglects any vertical components of flow and reduces a complex two-dimensional problem to a one-dimensional problem that can be easily solved. For steady-state flow, the discharge through the system must be constant and requires the free surface to be a parabola. Example 2.2 shows the derivation of the Dupuit equations.

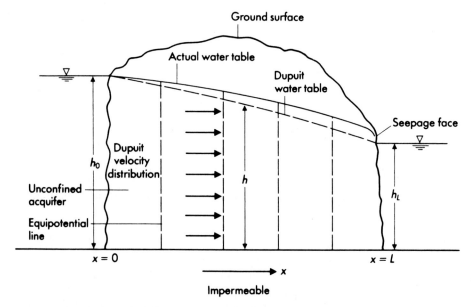

Figure 2.16 Steady flow in an unconfined aquifer between two water bodies with vertical boundaries. Source: Bedient and Huber, 1992.

Example 2.2 DUPUIT EQUATION

Derive the equation for one-dimensional flow in an unconfined aquifer using the Dupuit assumptions (Figure 2.16).

Solution. Darcy's law gives the one-dimensional flow per unit width as

$$q = -Kh \frac{dh}{dx}$$

where h and x are as defined in Figure 2.16. At steady state, the rate of change of q with distance is zero, or

$$\frac{d}{dx}\left(-Kh \frac{dh}{dx} \right) = 0$$

$$-\frac{K}{2} \frac{d^2h^2}{dx^2} = 0$$

or

$$\frac{d^2h^2}{dx^2} = 0$$

Integration yields

$$h^2 = ax + b$$

where a and b are constants. Setting the boundary condition $h = h_0$ at $x = 0$,

$$b = h_0^2$$

Differentiation of $h^2 = ax + b$ gives

$$a = 2h \frac{dh}{dx}$$

From Darcy's law,

$$h \frac{dh}{dx} = -\frac{q}{K}$$

so, by substitution,

$$h^2 = h_0^2 - \frac{2qx}{K}$$

Setting $h = h_L$ at $x = L$ and neglecting flow across the seepage face yields

$$h_L^2 = h_0^2 - \frac{2qL}{K}$$

Rearrangement gives

$$q = \frac{K}{2L} (h_0^2 - h_L^2), \qquad \text{Dupuit equation}$$

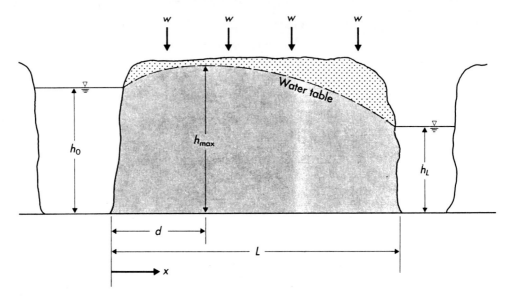

Figure 2.17 Dupuit parabola with recharge. Source: Bedient and Huber, 1992.

Then the general equation for the shape of the parabola is

$$h^2 = h_0^2 - \frac{x}{L}(h_0^2 - h_L^2), \qquad \text{Dupuit parabola}$$

The derivation of the Dupuit equations in Example 2.2 does not consider recharge to the aquifer. For the case of a system with recharge, the **Dupuit parabola** will take the mounded shape shown in Figure 2.17. The point where $h = h_{max}$ is known as the **water divide.** At the water divide, $q = 0$ since the gradient is zero. Example 2.3 derives the Dupuit equation for recharge and illustrates the use of the water-divide concept.

Example 2.3   DUPUIT EQUATION WITH RECHARGE

   (a) Derive the general Dupuit equation with the effect of recharge.

   (b) Two rivers located 1000 m apart fully penetrate an aquifer (Figure 2.17). The aquifer has a K value of 0.5 m/day. The region receives an average rainfall of 15 cm/yr and evaporation is about 10 cm/yr. Assume that the water elevation in river 1 is 20 m and the water elevation in river 2 is 18 m. Using the equation derived in part (a), determine the location and height of the water divide.

   (c) What is the daily discharge per meter of width into each river?

Solution

   (a) Designating recharge intensity as W, it can be seen that

$$\frac{dq}{dx} = W$$

From Darcy's law for one-dimensional flow, the flow per unit width is

$$q = -Kh \frac{dh}{dx}$$

Substituting the second equation into the first yields

$$-\frac{K}{2} \frac{d^2 h^2}{dx^2} = W$$

or

$$\frac{d^2 h^2}{dx^2} = -\frac{2W}{K}$$

Integration gives

$$h^2 = -\frac{W x^2}{K} + ax + b$$

where a and b are constants. The boundary condition $h = h_0$ at $x = 0$ gives

$$b = h_0^2$$

and the boundary condition $h = h_L$ at $x = L$ gives

$$a = \frac{h_L^2 - h_0^2}{L} + \frac{WL}{K}$$

Substitution of a and b into the previous equation for h^2 yields

$$\boxed{h^2 = h_0^2 + \frac{h_L^2 - h_0^2}{L} x + \frac{Wx}{K}(L - x), \qquad \text{Dupuit parabola}}$$

This equation will give the shape of the Dupuit parabola shown in Figure 2.17. If $W = 0$, this equation will reduce to the parabolic equation found in Example 2.2. Differentiation of the parabolic equation gives

$$2h \frac{dh}{dx} = \frac{h_L^2 - h_0^2}{L} + \frac{W}{K}(L - 2x)$$

But Darcy's law gives

$$h \frac{dh}{dx} = -\frac{q}{K}$$

so

$$-2\frac{q}{K} = \frac{1}{L}(h_L^2 - h_0^2) + \frac{W}{K}(L - 2x)$$

Simplifying,

$$\boxed{q = \frac{K}{2L}(h_0^2 - h_L^2) + W\left(x - \frac{L}{2}\right)}$$

(b) Given

$$L = 100 \text{ m}$$

$$K = 0.5 \text{ m/day}$$

$$h_0 = 20 \text{ m}$$

$$h_L = 18 \text{ m}$$

$$W = 5 \text{ cm/yr} = 1.369 \times 10^{-4} \text{ m/day}$$

At $x = d$, $q = 0$ (see Figure 2.17),

$$0 = \frac{K}{2L} (h_0^2 - h_L^2) + W \left(d - \frac{L}{2} \right)$$

$$d = \frac{L}{2} - \frac{K}{2WL} (h_0^2 - h_L^2)$$

$$= \frac{1000 \text{ m}}{2} - \frac{(0.5 \text{ m/day})(20^2 \text{ m}^2 - 18^2 \text{ m}^2)}{(2)(1.369 \times 10^{-4} \text{ m/day})(1000 \text{ m})}$$

$$= 500 \text{ m} - 138.8 \text{ m}$$

$$\boxed{d = 361.2 \text{ m}}$$

At $x = d$, $h = h_{max}$,

$$h_{max}^2 = h_0^2 + \frac{h_L^2 - h_0^2}{L} d + \frac{Wd}{K} (L - d)$$

$$= (20 \text{ m})^2 + \frac{(18^2 \text{ m}^2 - 20^2 \text{ m}^2)}{1000 \text{ m}} 361.2 \text{ m}$$

$$+ \frac{(1.369 \times 10^{-4} \text{ m/day})(361.2 \text{ m})}{0.5 \text{ m/day}} (1000 \text{ m} - 361.2 \text{ m})$$

$$= 400 \text{ m}^2 - 27.5 \text{ m}^2 + 63.2 \text{ m}^2$$

$$= 435.7 \text{ m}^2$$

$$\boxed{h_{max} = 20.9 \text{ m}}$$

Thus, the water divide is located 361.2 m from the edge of river 1 and is 20.9 m high.

(c) For discharge into river 1, set $x = 0$ m:

$$q = \frac{K}{2L} (h_0^2 - h_L^2) + W \left(0 - \frac{L}{2} \right)$$

$$= \frac{0.5 \text{ m/day}}{(2)(1000 \text{ m})} (20^2 \text{ m}^2 - 18^2 \text{ m}^2)$$

$$+ (1.369 \times 10^{-4} \text{ m/day})(- 1000 \text{ m/2})$$

$$= -0.0495 \text{ m}^2/\text{day}$$

The negative sign indicates that flow is in the opposite direction from the x direction. Therefore,

$$q = 0.0495 \text{ m}^2/\text{day into river 1}$$

For discharge into river 2, set $x = L = 1000$ m:

$$q = \frac{K}{2L} (h_0^2 - h_L^2) + W \left(1000 \text{ m} - \frac{L}{2} \right)$$

$$= \frac{0.5 \text{ m/day}}{(2)(1000 \text{ m})} (20^2 \text{ m}^2 - 18^2 \text{ m}^2)$$

$$+ (1.369 \times 10^{-4} \text{ m/day})(1000 \text{ m} - 1000 \text{ m/2})$$

$$q = 0.08745 \text{ m}^2/\text{day into river 2}$$

2.6 STREAMLINES AND EQUIPOTENTIAL LINES

The formal mathematical definition of the flow net can be derived using the equation of continuity for steady, incompressible, isotropic flow in two dimensions. The concept of velocity potential and a stream function was presented concisely by DeWiest (1965) in a classic text on geohydrology. The continuity equation states

$$\frac{\partial u}{\partial x} + \frac{\partial v}{\partial y} = 0$$

The governing steady-state flow equation is

$$\nabla^2 h = \frac{\partial^2 h}{\partial x^2} + \frac{\partial^2 h}{\partial y^2} = 0 \tag{2.29}$$

The **velocity potential** ϕ is a scalar function and can be written

$$\phi(x, y) = -K(z + p/\gamma) + c \tag{2.30}$$

where K and c are assumed constant. From Darcy's law in two dimensions,

$$u = \frac{\partial \phi}{\partial x}, \qquad v = \frac{\partial \phi}{\partial y} \tag{2.31}$$

Using Eq. (2.29) in two dimensions, we have

$$\nabla^2 \phi = \frac{\partial^2 \phi}{\partial x^2} + \frac{\partial^2 \phi}{\partial y^2} = 0 \tag{2.32}$$

where $\phi(x, y)$ = constant represents a family of equipotential curves on a two-dimensional surface. It can be shown that the **stream function** $\Psi(x, y)$ = constant is orthogonal to $\phi(x, y)$ = constant and that both satisfy the equation of continuity and Laplace's equation. The stream function $\Psi(x, y)$ is defined by

$$u = \frac{\partial \Psi}{\partial y}, \qquad v = -\frac{\partial \Psi}{\partial x} \qquad (2.33)$$

Combining Eqs. (2.31) and (2.33), the Cauchy–Riemann equations become

$$\frac{\partial \phi}{\partial x} = \frac{\partial \Psi}{\partial y}, \qquad \frac{\partial \phi}{\partial y} = -\frac{\partial \Psi}{\partial x} \qquad (2.34)$$

It can be shown that Ψ also satisfies Laplace's equation:

$$\nabla^2 \Psi = \frac{\partial^2 \Psi}{\partial x^2} + \frac{\partial^2 \Psi}{\partial y^2} = 0 \qquad (2.35)$$

Example 2.4 STREAMLINES AND EQUIPOTENTIAL LINES

Prove that Ψ and ϕ are orthogonal for isotropic flow, given Figure 2.18, where V is a velocity vector tangent to Ψ_2. Show that flow between two streamlines is constant.

Solution. We can write for the slope of the streamline

$$\frac{v}{u} = \frac{dy}{dx} = \tan \alpha$$

and

$$v \, dx - u \, dy = 0$$

Since

$$u = \frac{d\Psi}{dy}, \qquad v = -\frac{d\Psi}{dx}$$

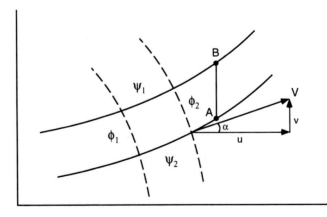

Figure 2.18 Streamlines and equipotential lines. Source: Bedient and Huber, 1992.

then

$$\frac{d\Psi}{dx} dx + \frac{d\Psi}{dy} dy = 0$$

or $d\Psi(x, y) = 0$. The total differential equals zero, and $\Psi(x, y) =$ constant, as required by the stream function. The ϕ_1 and ϕ_2 lines represent equipotential lines and can be represented by the total differential

$$d\phi = \frac{\partial\phi}{\partial x} dx + \frac{\partial\phi}{\partial y} dy = 0$$

Substituting for $\partial\phi/\partial x$ and $\partial\phi/\partial y$ from Darcy's law produces

$$u\ dx + v\ dy = 0$$

and

$$\frac{dy}{dx} = -\frac{u}{v}, \qquad \text{the slope of the equipotential line}$$

Thus, since the two slopes are negative inverses, equipotential lines are normal to streamlines. The system of orthogonal lines forms a flow net. Consider the flow crossing a vertical section AB between streamlines Ψ_1 and Ψ_2. The discharge across the section is designated Q, and it is apparent from fluid mechanics that

$$Q = \int_{\Psi_2}^{\Psi_1} u\ dy$$

or

$$Q = \int_{\Psi_2}^{\Psi_1} d\Psi$$

or

$$Q = \Psi_1 - \Psi_2$$

Thus, the flow between streamlines is constant and the spacing between streamlines reveals the relative magnitude of flow velocities between them.

Once either streamlines or equipotential lines are determined in a domain, the other can be evaluated from the Cauchy–Riemann equations (Eq. 2.34). Thus,

$$\Psi = \int \left(\frac{\partial\phi}{\partial x} dy - \frac{\partial\phi}{\partial y} dx \right)$$

and

$$\phi = \int \left(\frac{\partial\Psi}{\partial y} dx - \frac{\partial\Psi}{\partial x} dy \right)$$

Example 2.5   FLOW FIELD CALCULATION

A flow field is defined by $u = 2x$ and $v = -2y$. Find the stream function and potential function for this flow and sketch the flow net.

Solution. For this example, define

$$d\Psi = -v\ dx + u\ dy$$

$$d\phi = -u\ dx - v\ dy$$

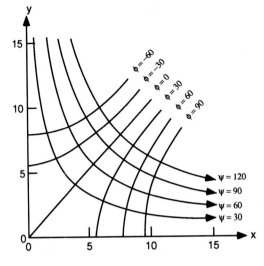

Figure 2.19 Flow net for flow in a corner.

Continuity requires that

$$\frac{\partial u}{\partial x} + \frac{\partial v}{\partial y} = 2 - 2 = 0$$

The stream function becomes

$$d\Psi = -v \, dx + u \, dy = 2y \, dx + 2x \, dy$$

or

$$\Psi = 2xy + C_1$$

where C_1 is a constant. The velocity potential ϕ exists if flow is irrotational and $(\partial v/\partial x) - (\partial u/\partial y) = 0$, which is satisfied in this case. Thus,

$$d\phi = -u \, dx - v \, dy = -2x \, dx + 2y \, dy$$

$$\phi = -(x^2 - y^2) + C_2$$

We can sketch the flow field by substituting values of Ψ into the expression $\Psi = 2xy$. For $\Psi = 60$, $x = 30/y$, and for $\phi = 60$, we have $x = \pm \sqrt{y^2 - 60}$. Figure 2.19 shows the flow field for the case of flow in a corner.

2.7 UNSATURATED FLOW AND THE WATER TABLE

Hydraulic conductivity $K(\theta)$ in the unsaturated zone above the water table relates velocity and hydraulic gradient in Darcy's law. **Moisture content** θ is defined as the ratio of the volume of water to the total volume of a unit of porous media. The **water table** defines the boundary between the unsaturated and saturated zones and is defined by the surface on which the fluid pressure P is exactly atmospheric, or $P = 0$. Hence, the total hydraulic head $h = \psi + z$, where $\psi = P/\rho g$, the pressure head. For saturated ground water flow, θ equals the **porosity** of the sample n, defined as the ratio of volume of voids to total volume of sample; for unsaturated flow above a water table, $\theta < n$.

Darcy's law is used with the unsaturated value for K and can be written

$$v = -K(\theta)\frac{\partial h}{\partial z} \tag{2.36}$$

where v = Darcy velocity
z = depth below surface
h = potential or head = $z + \psi$
ψ = tension or suction
$K(\theta)$ = unsaturated hydraulic conductivity
θ = volumetric moisture content

To complicate the analysis of unsaturated flow, the moisture content θ and the hydraulic conductivity K are functions of the capillary suction ψ. Also, it has been observed experimentally that the $\theta-\psi$ relationships differ significantly for different types of soil. Figure 2.2 shows the characteristic drying and wetting curves that occur in soils that are draining water or receiving infiltration of water. The nonlinear nature of these curves greatly complicates analyses in the unsaturated zone.

Near the water table a capillary fringe can occur where ψ is a small negative pressure corresponding to the air entry pressure. This capillary zone is small for sandy soils, but can be up to 2 m for fine-grained soils. By definition, pressure head is negative (under tension) at all points above the water table and is positive for points below the water table. The value of ψ is greater than zero in the saturated zone below the water table and equals zero at the water table. Soil physicists refer to $\psi < 0$ as the tension head or capillary suction head, which can be measured by an instrument called a tensiometer.

To summarize the properties of the unsaturated zone as compared to the saturated zone, Freeze and Cherry (1979) state that:

For the unsaturated zone (vadose zone):

1. It occurs above the water table and above the capillary fringe.
2. The soil pores are only partially filled with water; the moisture content θ is less than the porosity n.
3. The fluid pressure p is less than atmospheric; the pressure head ψ is less than zero.
4. The hydraulic head h must be measured with a tensiometer.
5. The hydraulic conductivity K and the moisture content θ are both functions of the pressure head ψ.

More details on the unsaturated zone can be found in Chapter 9 and in Freeze and Cherry (1979) and Domenico and Schwartz (1990).

SUMMARY

Chapter 2 presents mechanisms of ground water flow in the subsurface. Aquifer characteristics such as hydraulic conductivity, hydraulic gradient, and porosity are defined and used in governing equations of flow. Both steady-state and transient saturated flow equations are derived for confined and unconfined aquifers. Flow net theory is derived and several applications are presented for regional and local ground water problems. The chapter ends with a brief introduction to the unsaturated zone.

REFERENCES

BEAR, J., 1979, *Hydraulics of Groundwater*, McGraw-Hill, New York.

BEDIENT, P. B., and HUBER, W. C., 1992, *Hydrology and Floodplain Analysis,* 2nd ed., Addison-Wesley, Reading, MA.

COOPER, H. H., JR., 1966, "The Equation of Groundwater Flow in Fixed and Deforming Coordinates," *J. Geophys. Res.* 71:4785–4790.

DEWIEST, R. J. M., 1965, *Geohydrology,* Wiley, New York.

DOMENICO, P. A., and SCHWARTZ, F. W., 1990, *Physical and Chemical Hydrogeology,* Wiley, New York.

DUPUIT, J., 1863, *Études théoriques et pratiques sur le mouvement des eaux dans les canaux découverts et à travers les terrains perméables.* Dunod, Paris.

FETTER, C. W., 1988, *Applied Hydrogeology,* 2nd ed., Merrill Publishing Co., Columbus, OH, p. 103.

FREEZE, R. A., and CHERRY, J. A., 1979, *Groundwater,* Prentice Hall, Englewood Cliffs, NJ.

JACOB, C. E., 1940, "On the Flow of Water in an Elastic Artesian Aquifer," *Trans. Amer. Geophys. Union* 2:574–586.

LOHMAN, S. W., 1972, *Ground-water Hydraulics,* U.S. Geological Survey Prof. Paper 708.

MORRIS, D. A., and JOHNSON, A. I., 1967, *Summary of Hydrologic and Physical Properties of Rock and Soil Materials, as Analyzed by the Hydrologic Laboratory of the U.S. Geological Survey,* USGS Water Supply Paper 1839-D.

TERZAGHI, K., 1925. *Erdbaumechanic auf Bodenphysikalischer Grundlage.* Franz Deuticke, Vienna.

TODD, D.K., 1980, *Groundwater Hydrology,* 2nd ed., Wiley, New York.

TODD, D. K., and BEAR, J., 1961, "Seepage through Layered Anisotropic Porous Media," *J. Hydraulic Div. ASCE* 87:(HY3)31–57.

Chapter 3

GROUND WATER FLOW AND WELL MECHANICS

3.1 STEADY-STATE WELL HYDRAULICS

The case of steady flow to a well implies that the variation of head occurs only in space and not in time. The governing equations of flow presented in Section 2.4 can be solved for pumping wells in unconfined or confined aquifers under steady or unsteady conditions. Boundary conditions must be kept relatively simple, and aquifers must be assumed to be homogeneous and isotropic in each layer. More complex geometries can be handled by numerical simulation models in two or three dimensions (Chapter 10).

Steady One-dimensional Flow

For the case of ground water flow in the x direction in a confined aquifer, the governing equation becomes

$$\frac{d^2h}{dx^2} = 0 \tag{3.1}$$

and has the solution

$$h = \frac{-vx}{K} + h_0 \tag{3.2}$$

where $h = h_0$ at $x = 0$ and $dh/dx = -v/K$, according to Darcy's law. This states that head varies linearly with flow in the x direction.

The case of steady one-dimensional flow in an unconfined aquifer was presented in Section 2.5 using Dupuit's assumptions. The resulting variation of head with x is called the Dupuit parabola and represents the approximate shape of the water table for relatively flat slopes. In the presence of steep slopes near wells, the Dupuit approximation may be in error, and more sophisticated computer methods should be used.

Steady Radial Flow to a Well: Confined

The **drawdown curve** or **cone of depression** varies with distance from a pumping well in a confined aquifer (Figure 3.1). The flow is assumed two-dimensional for a completely penetrating well in a homogeneous, isotropic aquifer of unlimited

41

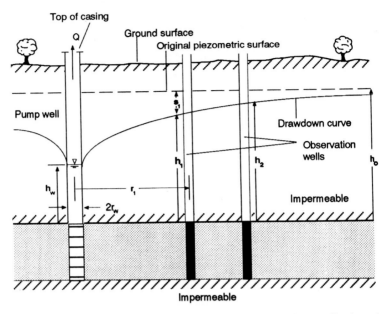

Figure 3.1 Radial flow to a well penetrating an extensive confined aquifer. Source: Bedient and Huber, 1992.

extent. For horizontal flow, these assumptions apply, and Q at any r equals, from Darcy's law,

$$Q = -2\pi r b K \frac{dh}{dr} \tag{3.3}$$

for steady radial flow to a well. Integrating after separation of variables, with $h = h_w$ at $r = r_w$ at the well, yields

$$Q = 2\pi K b \frac{h - h_w}{\ln(r/r_w)} \tag{3.4}$$

Equation (3.4) shows that h increases indefinitely with increasing r, yet the maximum head is h_0 for Figure 3.1. Near the well, the relationship holds and can be rearranged to yield an estimate for transmissivity T:

$$T = Kb = \frac{Q}{2\pi(h_2 - h_1)} \ln \frac{r_2}{r_1} \tag{3.5}$$

by observing heads h_1 and h_2 at two adjacent observation wells located at r_1 and r_2, respectively, from the pumping well. In practice, it is often necessary to use unsteady-state analyses because of the long times required to reach steady state.

Example 3.1   DETERMINATION OF K AND T IN A CONFINED AQUIFER

A well is constructed to pump water from a confined aquifer. Two observation wells, OW1 and OW2, are constructed at distances of 100 m and 1000 m, respectively. Water is pumped from the pumping well at a rate of 0.2 m³/min. At steady state, drawdown s' is observed as 2 m in OW2 and 8 m in OW1. Determine the hydraulic conductivity K and transmissivity T if the aquifer is 20 m thick.

Solution. Given

$$Q = 0.2 \text{ m}^3/\text{min}$$

$$r_2 = 1000 \text{ m}$$

$$r_1 = 100 \text{ m}$$

$$s_2' = 2 \text{ m}$$

$$s_1' = 8 \text{ m}$$

$$b = 20 \text{ m}$$

Equation (3.5) gives

$$T = Kb = \frac{Q}{2\pi(h_2 - h_1)} \ln\frac{r_2}{r_1}$$

Knowing that $s_1' = h_0 - h_1$ and $s_2' = h_0 - h_2$, we have

$$T = Kb = \frac{Q}{2\pi(s_1' - s_2')} \ln\frac{r_2}{r_1}$$

$$= \frac{0.2 \text{ m}^3/\text{min}}{2\pi(8\text{m} - 2\text{m})} \ln\frac{1000 \text{ m}}{100 \text{ m}}$$

$$\boxed{T = 0.0122 \text{ m}^2/\text{min} = 2.04 \text{ cm}^2/\text{sec}}$$

Then

$$K = \frac{T}{b}$$

$$= \frac{2.04 \text{ cm}^2/\text{sec}}{20 \text{ m}(100 \text{ cm}/1 \text{ m})}$$

$$\boxed{K = 1.02 \times 10^{-3} \text{ cm/sec}}$$

Steady Radial Flow to a Well: Unconfined

Applying Darcy's law for radial flow in an unconfined, homogeneous, isotropic, and horizontal aquifer and using Dupuit's assumptions (Figure 3.2),

$$Q = -2\pi r K h \frac{dh}{dr} \tag{3.6}$$

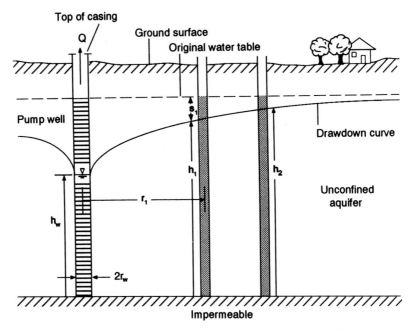

Figure 3.2 Radial flow to a well penetrating an unconfined aquifer. Source: Bedient and Huber, 1992.

Integrating, as before,

$$Q = \pi K \frac{h_2^2 - h_1^2}{\ln(r_2/r_1)} \qquad (3.7)$$

Solving for K,

$$K = \frac{Q}{\pi(h_2^2 - h_1^2)} \ln \frac{r_2}{r_1} \qquad (3.8)$$

where heads h_1 and h_2 are observed at adjacent wells located distances r_1 and r_2 from the pumping well, respectively.

Example 3.2 DETERMINATION OF K IN AN UNCONFINED AQUIFER

A fully penetrating well discharges 75 gpm from an unconfined aquifer. The original water table was recorded as 35 ft. After a long time period, the water table was recorded as 20 ft × mean sea level (MSL) in an observation well located 75 ft away and 34 ft MSL at an observation well located 2000 ft away. Determine the hydraulic conductivity of this aquifer in feet per second.

Solution. Given

$$Q = 75 \text{ gpm}$$
$$r_2 = 2000 \text{ ft}$$
$$r_1 = 75 \text{ ft}$$
$$h_2 = 34 \text{ ft}$$
$$h_1 = 20 \text{ ft}$$

Equation (3.8) gives

$$K = \frac{Q}{\pi(h_2^2 - h_1^2)} \ln \frac{r_2}{r_1}$$

$$= \frac{75 \text{ gpm}(0.134 \text{ ft}^3/\text{gal})(1 \text{ min}/60 \text{ sec})}{\pi(34^2 \text{ ft}^2 - 20^2 \text{ ft}^2)} \ln \frac{2000 \text{ ft}}{75 \text{ ft}}$$

$$\boxed{K = 2.32 \times 10^{-4} \text{ ft/sec}}$$

Well in a Uniform Flow Field

A typical problem in well mechanics involves the case of a well pumping from a uniform flow field (Figure 3.3). A vertical section and plan view indicate the sloping piezometric surface and the resulting flow net. The ground water divide between

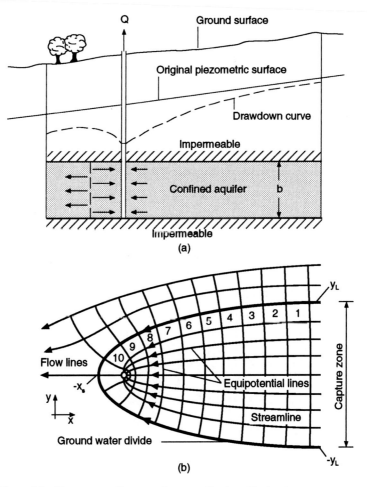

Figure 3.3 Flow to a well penetrating a confined aquifer having a sloping-plane piezometric surface. (a) Vertical section. (b) Plan view. Source: Bedient and Huber, 1992.

the region that flows to the well and the region flowing by the well can be found from

$$-\frac{y}{x} = \tan \frac{2\pi K b i}{Q} y \qquad (3.9)$$

Equation (3.9) results from the superposition of radial and one-dimensional flow field solutions, where i is the original piezometric slope (gradient). It can be shown that

$$y_L = \pm \frac{Q}{2Kbi} \qquad (3.10)$$

as $x \to \infty$, and the stagnation point (no flow) occurs at

$$x_s = -\frac{Q}{2\pi K b i}, \qquad y = 0 \qquad (3.11)$$

Equations (3.10) and (3.11) may be applied to unconfined aquifers for cases of relatively small drawdowns, where b is replaced by h_0, the average saturated aquifer thickness. An important application of the well in a uniform flow field involves the evaluation of pollution sources and impacts on downgradient well fields and the potential for pumping and capturing a plume as it migrates downgradient. Chapter 13 addresses capture zone methods in more detail.

Multiple-well Systems

For multiple wells with drawdowns that overlap, the principle of superposition can be used. Drawdown at any point in the area of influence of several pumping wells is equal to the sum of drawdowns from each well in confined or unconfined aquifers (Figure 3.4).

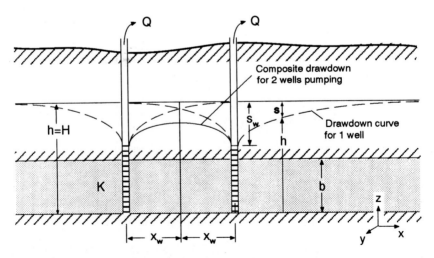

Figure 3.4 Individual and composite drawdown curves for two wells in a line. Source: Bedient and Huber, 1992.

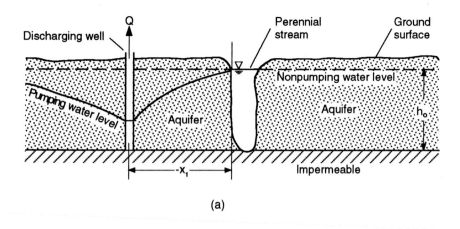

(a)

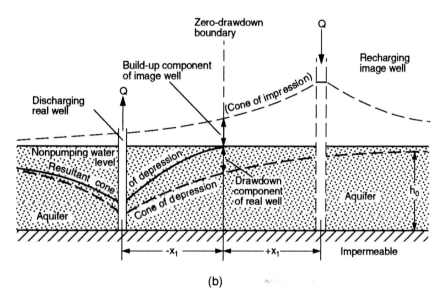

(b)

Figure 3.5 Sectional views. (a) Discharging well near a perennial stream. (b) Equivalent hydraulic system in an aquifer of infinite areal extent. Aquifer thickness h_0 should be very large compared with resultant drawdown near real well.

The same principle applies for well flow near a boundary. Figure 3.5 shows the case for a well pumping near a fixed head stream and near an impermeable boundary. **Image wells** placed on the other side of the boundary at a distance a can be used to represent the equivalent hydraulic condition. In one case the image well is recharging at the same rate Q, and in another case it is pumping at rate Q. The summation of drawdowns from the original pumping well and the image well provides a correct boundary condition at distance a from the well. Thus, the use of image wells allows an aquifer of finite extent to be transformed into an infinite aquifer so that closed-form solution methods can be applied. Figure 3.6

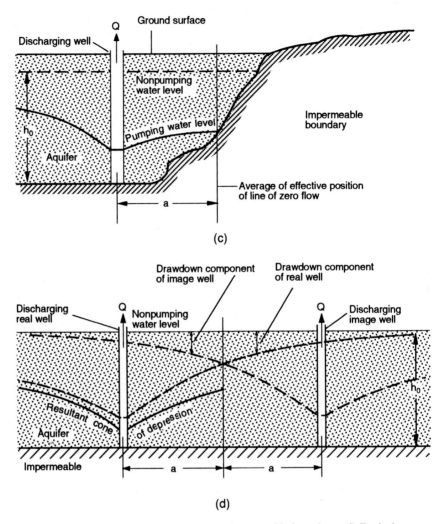

Figure 3.5 (c) Discharging well near an impermeable boundary. (d) Equivalent hydraulic system in an aquifer of infinite areal extent. Aquifer thickness h_0 should be very large compared with resultant drawdown near real well. Source: Ferris et al., 1962.

shows a flow net for a pumping well and a recharging image well and indicates a line of constant head between the two wells. The steady-state drawdown s' at any point (x, y) is given by

$$s' = \frac{Q}{4\pi T} \ln \frac{(x + x_w)^2 + (y + y_w)^2}{(x - x_w)^2 + (y - y_w)^2}$$
(3.12)

where $\pm x_w$, $y_w = 0$ are the locations of the recharge and discharge wells for the case shown in Figure 3.6.

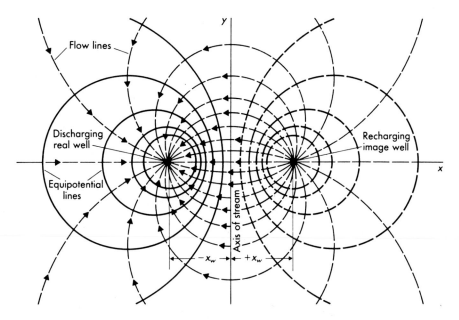

Figure 3.6 Flow net for a discharging real well and a recharging image well. Source: Ferris et al., 1962.

3.2 UNSTEADY WELL HYDRAULICS

The Theis Method of Solution

As a well penetrating a confined aquifer of infinite extent is pumped at a constant rate, a drawdown occurs radially extending from the well. The rate of decline of head times the storage coefficient summed over the area of influence equals the discharge. The rate of decline decreases continuously as the area of influence expands. The governing ground water flow equation [Eq. (2.27)] in plane polar coordinates is

$$\frac{\partial^2 h}{\partial r^2} + \frac{1}{r}\frac{\partial h}{\partial r} = \frac{S}{T}\frac{\partial h}{\partial t} \tag{3.13}$$

where h = head
       r = radial distance
       S = storage coefficient
       T = transmissivity

Theis (1935) obtained a solution for Eq. (3.13) by assuming that the well is a mathematical sink of constant strength and by using boundary conditions $h = h_0$ for $t = 0$ and $h \rightarrow h_0$ as $r \rightarrow \infty$ for $t \geq 0$:

$$s' = \frac{Q}{4\pi T}\int_u^\infty \frac{e^{-u}\,du}{u} = \frac{Q}{4\pi T}\,W(u) \tag{3.14}$$

where s' is drawdown, Q is discharge at the well, and

$$u = \frac{r^2 S}{4Tt} \tag{3.15}$$

Equation (3.14) is known as the nonequilibrium or Theis equation. The integral is written as $W(u)$ and is known as the exponential integral, or well function, which can be expanded as a series:

$$W(u) = -0.5772 - \ln(u) + u - \frac{u^2}{2 \cdot 2!} + \frac{u^3}{3 \cdot 3!} - \frac{u^4}{4 \cdot 4!} + \cdots \tag{3.16}$$

The equation can be used to obtain aquifer constants S and T by means of pumping tests at fully penetrating wells. It is widely used because a value of S can be determined, only one observation well and a relatively short pumping period are required, and large portions of the flow field can be sampled with one test.

The assumptions inherent in the Theis equation should be included since they are often overlooked:

1. The aquifer is homogeneous, isotropic, uniformly thick, and of infinite areal extent.
2. Prior to pumping, the piezometric surface is horizontal.
3. The fully penetrating well is pumped at a constant rate.
4. Flow is horizontal within the aquifer.
5. Storage within the well can be neglected.
6. Water removed from storage responds instantaneously with a declining head.

These assumptions are seldom completely satisfied for a field problem, but the method still provides one of the most useful and accurate techniques for aquifer characterization. The complete Theis solution requires the graphical solution of two equations with four unknowns:

$$s' = \frac{Q}{4\pi T} W(u) \tag{3.17}$$

$$\frac{r^2}{t} = \frac{4T}{S} u \tag{3.18}$$

The relation between $W(u)$ and u must be the same as that between s' and r^2/t because all other terms are constants in the equations. Theis suggested a solution based on graphical superposition. Example 3.3 indicates how a plot of $W(u)$ versus u, called a **type curve**, is superimposed over observed time–drawdown data while keeping the coordinate axes parallel. The two plots are adjusted until a position is found by trial, such that most of the observed data fall on a segment of the type curve. Any convenient point is selected, and values of $W(u)$, u, s', and r^2/t are used in Eqs. (3.17) and (3.18) to determine S and T (see Figure 3.7).

It is also possible to use Theis's solution for the case where several wells are sampled for drawdown simultaneously near a pumped well. Distance–drawdown data are then fitted to the type curve similarly to the method just outlined.

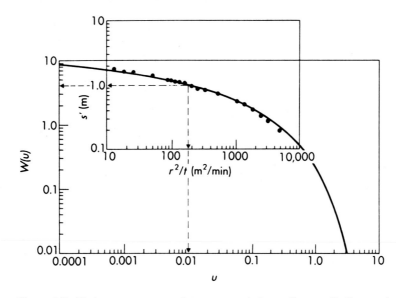

Figure 3.7 Theis curve compared to measured data. Source: Bedient and Huber, 1992.

Example 3.3   DETERMINATION OF *T* AND *S* BY THE THEIS METHOD

A fully penetrating well in a 25-m-thick confined aquifer is pumped at a rate of 0.2 m³/sec for 1000 min. Drawdown is recorded versus time at an observation well located 100 m away. Compute the transmissivity and storativity using the Theis method. (See Tables 3.1 and 3.2.)

Solution. A plot of s' versus r^2/t is made on log–log paper. This is superimposed on a plot of $W(u)$ versus u, which is also on log–log paper. A point is chosen at some convenient point on the matched curve, and values for $s'r^2/t$, $W(u)$, and u are read (see Figure 3.7 and the accompanying table of values). From the plot,

TABLE 3.1

Time (min)	s' (m)	Time (min)	s' (m)
1	0.11	60	1.02
2	0.20	70	1.05
3	0.28	80	1.08
4	0.34	90	1.11
6	0.44	100	1.15
8	0.50	200	1.35
10	0.54	400	1.55
20	0.71	600	1.61
30	0.82	800	1.75
40	0.85	1000	1.80
50	0.92		

TABLE 3.2   Values of the Function $W(u)$ for Various Values of u

u	$W(u)$	u	$W(u)$	u	$W(u)$	u	$W(u)$
1×10^{-10}	22.45	7×10^{-8}	15.90	4×10^{-5}	9.55	1×10^{-2}	4.04
2	21.76	8	15.76	5	9.33	2	3.35
3	21.35	9	15.65	6	9.14	3	2.96
4	21.06	1×10^{-7}	15.54	7	8.99	4	2.68
5	20.84	2	14.85	8	8.86	5	2.47
6	20.66	3	14.44	9	8.74	6	2.30
7	20.50	4	14.15	1×10^{-4}	8.63	7	2.15
8	20.37	5	13.93	2	7.94	8	2.03
9	20.25	6	13.75	3	7.53	9	1.92
1×10^{-9}	20.15	7	13.60	4	7.25	1×10^{-1}	1.823
2	19.45	8	13.46	5	7.02	2	1.223
3	19.05	9	13.34	6	6.84	3	0.906
4	18.76	1×10^{-6}	13.24	7	6.69	4	0.702
5	18.54	2	12.55	8	6.55	5	0.560
6	18.35	3	12.14	9	6.44	6	0.454
7	18.20	4	11.85	1×10^{-3}	6.33	7	0.374
8	18.07	5	11.63	2	5.64	8	0.311
9	17.95	6	11.45	3	5.23	9	0.260
1×10^{-8}	17.84	7	11.29	4	4.95	1×10^{0}	0.219
2	17.15	8	11.16	5	4.73	2	0.049
3	16.74	9	11.04	6	4.54	3	0.013
4	16.46	1×10^{-5}	10.94	7	4.39	4	0.004
5	16.23	2	10.24	8	4.26	5	0.001
6	16.05	3	9.84	9	4.14		

$$\frac{r^2}{t} = 180 \text{ m}^2/\text{min}$$

$$s' = 1.0 \text{ m}$$

$$u = 0.01$$

$$W(u) = 4.0$$

Equation (3.17) gives

$$s' = \frac{Q}{4\pi T} W(u)$$

$$T = \frac{QW(u)}{4\pi s'}$$

$$= \frac{0.2 \text{ m}^3/\text{sec } (4.0)}{4\pi (1.0 \text{ m})}$$

$$\boxed{T = 6.37 \times 10^{-2} \text{ m}^2/\text{sec}}$$

Equation (3.18) gives

$$\frac{r^2}{t} = \frac{4Tu}{S}$$

$$S = \frac{4Tu}{r^2/t}$$

$$= \frac{4(6.37 \times 10^{-2} \text{ m}^2/\text{sec})(0.01)}{180 \text{ m}^2/\text{min}(1 \text{ min}/60 \text{ sec})}$$

$$\boxed{S = 8.49 \times 10^{-4}}$$

Cooper–Jacob Method of Solution

Cooper and Jacob (1946) noted that for small values of r and large values of t the parameter u in Eq. (3.16) becomes very small, so the infinite series can be approximated by

$$s' = \frac{Q}{4\pi T}\left(-0.5772 - \ln\frac{r^2 S}{4Tt}\right) \tag{3.19}$$

Further rearrangement and conversion to decimal logarithms yields

$$s' = \frac{2.30Q}{4\pi T} \log \frac{2.25Tt}{r^2 S} \tag{3.20}$$

Thus, a plot of drawdown s' versus the logarithm of t forms a straight line, as shown in Figure 3.8. A projection of the line to $s' = 0$, where $t = t_0$, yields

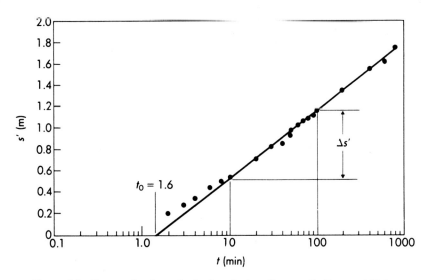

Figure 3.8 Cooper–Jacob method of analysis. Source: Bedient and Huber, 1992.

$$0 = \frac{2.3Q}{4\pi T} \log \frac{2.25Tt_0}{r^2 S} \tag{3.21}$$

and it follows that, since log (1) = 0,

$$S = \frac{2.25Tt_0}{r^2} \tag{3.22}$$

Finally, by replacing s' by $\Delta s'$, where $\Delta s'$ is the drawdown difference of data per log cycle of t, Eq. (3.20) becomes

$$T = \frac{2.3Q}{4\pi \Delta s'} \tag{3.23}$$

The Cooper–Jacob method first solves for T with Eq. (3.23) and then for S with Eq. (3.22) and is applicable for small values of u (less than 0.01). Calculations with the Theis method were presented in Example 3.3, and the Cooper–Jacob method is used in Example 3.4.

Example 3.4 DETERMINATION OF T AND S BY THE COOPER–JACOB METHOD

Using the data given in Example 3.3, determine the transmissivity and storativity of the 25-m-thick confined aquifer using the Cooper–Jacob method.

Solution. Values of s' and t are plotted on semilog paper with the t axis logarithmic (see Figure 3.8). A line is fitted through the later time periods and is projected back to a point where $s' = 0$. This point determines t_0. $\Delta s'$ is measured over 1 log cycle of t. From the plot,

$$t_0 = 1.6 \text{ min}$$

$$\Delta s' = 0.65 \text{ m}$$

Equation (3.23) gives

$$T = \frac{2.3Q}{4\pi \Delta s'}$$

$$= \frac{2.3(0.2 \text{ m}^3/\text{sec})}{4\pi(0.65 \text{ m})}$$

$$\boxed{T = 5.63 \times 10^{-2} \text{ m}^2/\text{sec}}$$

Equation (3.22) gives

$$S = \frac{2.25Tt_0}{r^2}$$

$$= \frac{2.25(5.63 \times 10^{-2} \text{ m}^2/\text{sec})(1.6 \text{ min})(60 \text{ sec}/1 \text{ min})}{100^2 \text{ m}^2}$$

$$\boxed{S = 1.22 \times 10^{-3}}$$

Slug Tests

Slug tests involve the use of a single well for the determination of aquifer formation constants. Rather than pumping the well for a period of time, as described in the preceding examples, a volume of water is suddenly removed or added to the well casing, and observations of recovery or drawdown are noted through time. By careful evaluation of the drawdown curve and knowledge of the well screen geometry, it is possible to derive K or T for an aquifer.

Typical procedure for a slug test requires the use of a rod of slightly smaller diameter than the well casing or a pump to evacuate the well casing. The simplest slug test method in a piezometer was published by Hvorslev (1951), who used the recovery of water level over time to calculate hydraulic conductivity of the porous media. Hvorslev's method relates the flow $q(t)$ at the piezometer at any time to the hydraulic conductivity and the unrecovered head distance, $H_0 - h$ in Figure 3.9, by

$$q(t) = \pi r^2 \frac{dh}{dt} = FK(H_0 - h) \qquad (3.24)$$

where F is a factor that depends on the shape and dimensions of the piezometer intake. If $q = q_0$ at $t = 0$, then $q(t)$ will decrease toward zero as time increases. Hvorslev defined the basic time lag

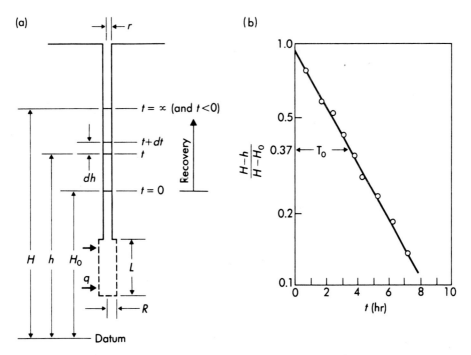

Figure 3.9 Hvorslev piezometer test. (a) Geometry. (b) Method of analysis. Source: Bedient and Huber, 1992.

$$T_0 = \frac{\pi r^2}{FK}$$

and solved Eq. (3.24) with initial conditions $h = H_0$ at $t = 0$. Thus,

$$\frac{H - h}{H - H_0} = e^{-t/T_0} \tag{3.25}$$

By plotting recovery $(H - h)/(H - H_0)$ versus time on semilog graph paper, we find that $t = T_0$, where recovery equals 0.37 (Figure 3.9). For piezometer intake length divided by radius (L/R) greater than 8, Hvorslev has evaluated the shape factor F and obtained an equation for K:

$$K = \frac{r^2 \ln(L/R)}{2LT_0} \tag{3.26}$$

Several other slug test methods have been developed by Cooper et al. (1967) and Papadopoulos et al. (1973) for confined aquifers. These methods are similar to Theis's in that a curve-matching procedure is used to obtain S and T for a given aquifer. A family of type curves $H(t)/H_0$ versus Tt/r_c^2 was published for five values of the variable α, defined as $(r_s^2/r_c^2)S$ in Figure 3.10. Papadopoulos et al. (1973)

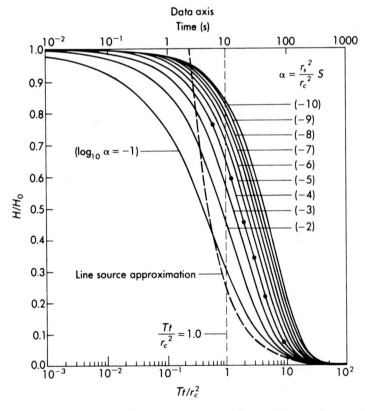

Figure 3.10 Papadopoulos slug test type curves. Source: Papadopoulos, et al., 1973. © 1973 American Geophysical Union.

added five additional values of α. The solution method is graphical and requires a semilogarithmic plot of measured $H(t)/H_0$ versus t, where H_0 is the assumed initial excess head. The data are then curve matched to the plotted type curves by horizontal translation until the best match is achieved (Figure 3.10), and a value of α is selected for a particular curve. The vertical time axis t, which overlays the vertical axis for $Tt/r_c^2 = 1.0$ is selected, and a value of T can then be found from $T = 1.0 r_s^2/t_1$. The value of S can be found from the definition of α. The method is representative of the formation only in the immediate vicinity of the test hole and should be used with caution.

The most commonly used method for determining hydraulic conductivity in ground water investigations is the Bouwer and Rice (1976) slug test. Although it was originally designed for unconfined aquifers, it can be used for confined or stratified aquifers if the top of the screen is some distance below the upper confining layer. The method is based on the following equation:

$$K = \frac{r_c^2 \ln(R_e/r_w)}{2L_e} \frac{1}{t} \ln \frac{y_0}{y_t} \tag{3.27}$$

where r_c = radius of casing
  y_0 = vertical difference between water level inside well and water level outside at $t = 0$
  y_t = vertical difference between water level inside well and water table outside (drawdown) at time t
  R_e = effective radial distance over which y is dissipated, and varying with well geometry
  r_w = radial distance of undisturbed portion of aquifer from centerline (usually thickness of gravel pack)
  L_e = length of screened, perforated, or otherwise open section of well
  t = time

In Eq. (3.27), y and t are the only variables. Thus, if a number of y and t measurements are taken, they can be plotted on semilogarithmic paper to give a straight line. The slope of the best-fitting straight line will provide a value for $\ln(y_0/y_t)/t$. All the other parameters in Eq. (3.27) are known from well geometry, and K can be calculated. A point to note is that drawdown on the ground water table becomes increasingly significant as the test progresses, and the points will begin to deviate from the straight line for large t and small y. Hence, only the straight-line portion of the data must be used in the calculation for K.

Example 3.5 SLUG TEST METHOD

A screened, cased well penetrates a confined aquifer. The casing radius is 5 cm and the screen is 1 m long. A gravel pack 2.5 cm wide surrounds the well, and a slug of water is injected that raises the water level by 0.28 m. The change in water level with time is as listed in the accompanying table. Given that R_e is 10 cm, calculate K for the aquifer (see Figure 3.11a).

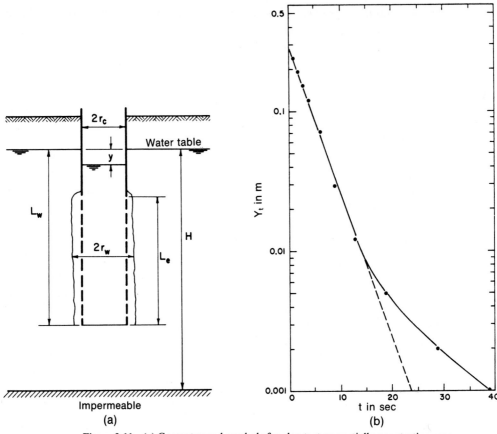

Figure 3.11 (a) Geometry and symbols for slug test on partially penetrating, partially screened well in unconfined aquifer with gravel pack and/or developed zone around screen. (b) Graph of log Y_t versus t for slug test on well in Salt River bed, 27th Avenue, Phoenix, Arizona. Source: Bouwer and Rice, 1976. © American Geophysical Union.

t (sec)	y_t (m)
1	0.24
2	0.19
3	0.16
4	0.13
6	0.07
9	0.03
13	0.013
19	0.005
29	0.002
40	0.001

Solution. Data for y versus t are plotted on semilog paper as shown in Figure 3.11b. The straight line from $y_0 = 0.28$ m to $y_t = 0.001$ m covers 2.4 log cycles. The time increment between the two points is 24 sec. To convert the log cycles to natural log, a factor of 2.3 is used. Thus, $1/t \ln(y_0/y_t) = (2.3 \times 2.4)/24 = 0.23$. Using this value in the Bouwer and Rice equation gives

$$K = \frac{(5 \text{ cm})^2 \ln(10 \text{ cm}/7.5 \text{ cm})}{2 \times 100 \text{ cm}} 0.23 \text{ sec}^{-1}$$

$$= 8.27 \times 10^{-3} \text{ cm/sec}$$

Example 3.5 indicates how slug tests are applied to field data. It should be noted that slug tests are often used at hazardous-waste sites since large volumes of contaminated water do not have to be dispersed, as in the case of a pump test. However, the pump test generally gives a better picture of overall hydraulic conductivity than the slug test at a site.

Radial Flow in a Leaky Aquifer

Leaky aquifers represent a unique and complex problem in well mechanics. When a leaky aquifer is pumped, as shown in Figure 3.12, water is withdrawn both from the lower aquifer and from the saturated portion of the overlying aquitard. By creating a lowered piezometric surface below the water table, ground water can migrate vertically downward and then move horizontally to the well. While steady-state conditions in a leaky system are possible, a more general nonequilibrium analysis for unsteady flow is more applicable and more often occurs in the field.

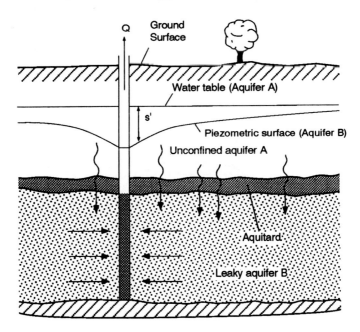

Figure 3.12 Well pumping from a leaky aquifer. Source: Bedient and Huber, 1992.

When pumping starts from a well in a leaky aquifer, drawdown of the piezometric surface can be given by

$$s' = \frac{Q}{4\pi T} W\left(u, \frac{r}{B}\right) \tag{3.28}$$

where the quantity r/B is given by

$$\frac{r}{B} = \frac{r}{\sqrt{T/(K'/b')}}$$

where T is transmissivity of the aquifer, K' is vertical hydraulic conductivity of the aquitard, and b' is thickness of the aquitard. Values of the function $W(u, r/B)$ have been tabulated by Hantush (1956) and have been used by Walton (1960) to prepare a family of type curves, shown in Figure 3.13. Equation (3.28) reduces to the Theis equation for $r/B = 0$. The method of solution for the leaky aquifer works in the same way as the Theis solution with a superposition of drawdown data on top of the leaky type curves. A curve of best fit is selected, and values of W, $1/u$, s', and t are found, which allows T and S to be determined. Finally, based on the value of r/B, it is possible to calculate K' and b'.

In general, leaky aquifers are much more difficult to deal with than confined or unconfined systems. But the method just described does provide a useful tool for evaluating leaky systems analytically. For more complex geologies and systems with lenses, a three-dimensional computer simulation may be employed to properly represent ground water flow. These types of models are described in detail in Chapter 10.

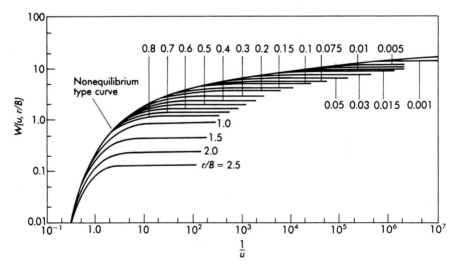

Figure 3.13 Type curves for analysis of pumping test data to evaluate storage coefficient and transmissivity of leaky aquifers. Source: Walton, 1960.

Example 3.6 APPLICATION OF THEIS EQUATION

A square excavation (375 m on a side) is to be dewatered by the installation of four wells at the corners. Point A is in the middle and point B is on one side equidistant from two of the wells. For an allowable pumping period of 24 hr determine the pumping rate required to produce a minimum drawdown of 4 m everywhere within the limits of the excavation. The confined aquifer has a transmissivity of 2×10^{-4} m^2/s and a storage factor of 7×10^{-5}.

$$ s = \frac{Q}{4\pi T} W(u) $$

$$ u = \frac{r^2 S}{4Tt} $$

Determine the required pumping rate for 4-m drawdown at A:

$$ d = \sqrt{2}\,\frac{375}{2} = 265 \text{ m} $$

$$ u = \frac{(265 \text{ m})^2 (7 \times 10^{-5})}{4(2 \times 10^{-4} \text{ m}^2/\text{sec})(24 \text{ hr}) (3600 \text{ sec/hr})} = 0.071 $$

$$ W(u) = 2.14 $$

and each well contributes 25%.

$$ Q = \frac{s}{4W(u)} 4\pi T = \frac{4 \text{ m}}{4(2.14)} 4(\pi)(2 \times 10^{-4} \text{ m}^2/\text{sec}) $$

$$ = 1.17 \times 10^{-3} \text{ m}^3/\text{sec} = 4.23 \text{ m}^3/\text{hr}, \qquad \text{for each well} $$

Determine the drawdown at B:

$$ u = \frac{(187.5 \text{ m})^2 (7 \times 10^{-5})}{4(2 \times 10^{-4} \text{ m}^2/\text{sec})(24 \text{ hr})(3600 \text{ sec/hr})} = 0.036 $$

$$ W(u) = 2.79 $$

and the drawdown is produced by four wells:

$$ s = \frac{2(1.17 \times 10^{-3} \text{ m}^3/\text{sec})}{4(\pi)(2 \times 10^{-4} \text{ m}^2/\text{sec})} 2.79 = 2.60 \text{ m} $$

$$ u = \frac{(419 \text{ m})^2 (7 \times 10^{-5})}{4(2 \times 10^{-4} \text{ m}^2/\text{sec})(24 \text{ hr})(3600 \text{ sec/hr})} = 0.18 \Rightarrow W(u) = 1.34 $$

$$ s = \frac{2(1.17 \times 10^{-3} \text{ m}^3/\text{sec})}{4(\pi)(2 \times 10^{-4} \text{ m}^2/\text{sec})} 1.34 = 1.25 \text{ m} $$

Total drawdown at $B = 2.60 + 1.25 = 3.85$ m.

Thus, the pumping rate needs to be increased in order to achieve 4-m drawdown at B and to determine the required pumping rate for the four wells:

$$\frac{2Q}{4\pi T} 2.79 + \frac{2Q}{4\pi T} 1.34 = 4$$

$$Q = \frac{(4 \text{ m})(4)(\pi)(2 \times 10^{-4} \text{ m}^2/\text{sec})}{8.26} = 1.21 \times 10^{-3} \text{ m}^3/\text{sec} = \boxed{4.38 \text{ m}^3/\text{hr}}$$

This automatically meets the criteria at point A in the middle of the square area.

SUMMARY

Chapter 3 has provided a review of well mechanics under both steady-state and transient conditions. The principle of superposition was applied for multiple well systems and image well problems. The Theis method of solution was derived and several examples are shown. Slug test methods for single wells are covered in detail. Well mechanics solutions are important in the application of pump tests and capture zones, which are discussed in more detail in Chapters 10 and 13.

REFERENCES

AMERICAN WATER WORKS ASSOCIATION, 1967, *AWWA Standard for Deep Wells*, AWWA-A100-66, Denver, CO.

BEDIENT, P. B., and HUBER, W. C., 1992, *Hydrology and Floodplain Analysis*, 2nd ed., Addison-Wesley, Reading, MA.

BOUWER, H., and RICE, R. C., 1976, "A Slug Test for Determining Hydraulic Conductivity of Unconfined Aquifers with Completely or Partially Penetrating Wells," *Water Resources Res.* 12:423–428.

CAMPBELL, M. D., and LEHR, J. H., 1973, *Water Well Technology*, McGraw-Hill, New York.

COOPER, H. H., JR., and JACOB, C. E., 1946, "A Generalized Graphical Method for Evaluating Formation Constants and Summarizing Well Field History," *Trans. Amer. Geophys. Union* 27:526–534.

COOPER, H. H., JR., BREDEHOEFT, J. D., and PAPADOPOULOS, I. S., 1967, "Response of a Finite-diameter Well to an Instantaneous Charge of Water," *Water Resources Res.* 3: 263–269.

FERRIS, J. G., KNOWLES, D. B., BROWNE, R. H., and STALLMAN, R. W., 1962, *Theory of Aquifer Tests*, U.S. Geological Survey Water Supply Paper 1536-E.

FREEZE, R. A., and CHERRY, J. A., 1979, *Groundwater*, Prentice Hall, Englewood Cliffs, NJ.

HANTUSH, M. S., 1956, "Analysis of Data from Pumping Tests in Leaky Aquifers," *J. Geophys. Res.* 69:4221–4235.

HVORSLEV, M. J., 1951, *Time Lag and Soil Permeability in Groundwater Observations*, U.S. Army Corps of Engineers Waterways Exp. Sta. Bull. 36, Vicksburg, MS.

PAPADOPOULOS, I. S., BREDEHOEFT, J. D., and COOPER, H. H., JR., 1973, "On the Analysis of Slug Test Data," *Water Resources Res*. 9:(4)1087–1089.

THEIS, C. V., 1935, "The Relation between the Lowering of the Piezometric Surface and the Rate and Duration of Discharge of a Well Using Ground-water Storage," *Trans. Amer. Geophys. Union* 16:519–524.

WALTON, W. C., 1960, *Leaky Artesian Aquifer Conditions in Illinois,* Illinois State Water Surv. Rept. Invest. 39, Urbana, IL.

SOURCES AND TYPES OF GROUND WATER CONTAMINATION

4.1 INTRODUCTION

Humans have been exposed to hazardous substances dating back to prehistoric times when they inhaled noxious gases from volcanoes and in cave dwellings. Pollution problems started in the industrial sector with the production of dyes and other organic chemicals developed from the coal tar industry in Germany during the 1800s. In the 1900s, the variety of chemicals and chemical wastes increased drastically from the production of steel and iron, lead batteries, and petroleum refining. During that time, radium and chromic wastes created serious problems as well. In the 1930s and 1940s, the World War II era ushered in massive production of chlorinated solvents, pesticides, polymers, plastics, paints, and wood preservatives. Very little was known about the environmental impacts of many of these chemical wastes until much later.

The Love Canal hazardous-waste site attracted major public attention in 1979 and heralded the hazardous-waste decade of the 1980s. The site in Niagara Falls, New York, had received 20,000 metric tons of chemical waste containing at least 80 different chemicals and was creating serious environmental impacts on nearby residents. By 1989, state and federal governments had spent $140 million to clean up the site and relocate the residents. Several other sites during the 1980s also received national attention, including Woburn, Massachusetts (tannery and glue-making chemicals dating back to 1850), the Stringfellow Acid Pits near Riverside, California, the Valley of the Drums in Kentucky, the Brio chemical waste site in Houston, Texas, and Times Beach, Missouri (the town was abandoned because of dioxin contamination).

This chapter describes most of the significant chemical threats to ground water quality from various sources of contamination. In a 1984 report, *Protecting the Nation's Groundwater from Contamination,* the Office of Technology Assessment (OTA, 1984) listed more than 30 different potential sources of contamination. Table 4.1 lists the major sources of ground water contamination and divides them into six major categories. Section 305(b) of the Federal Clean Water Act requires states to submit reports to the EPA on sources and types of ground water contamination. In 1988, the *National Water Quality Inventory—1988 Report to Congress* (USEPA, 1990) presented the data on the relative importance of various sources of contamination and various types of contaminants. State inventories showed

TABLE 4.1 Sources of Ground Water Contamination

Category I	Category II	Category III
Sources designed to discharge substances	Sources designed to store, treat, and/or dispose of substances; discharged through unplanned release	Sources designed to retain substances during transport or transmission
Subsurface percolation (e.g., septic tanks and cesspools) Injection wells Land application	Landfills Open dumps Surface impoundments Waste tailings Waste piles Materials stockpiles Aboveground storage tanks Underground storage tanks Radioactive disposal sites	Pipelines Materials transport and transfer

Category IV	Category V	Category VI
Sources discharging as consequence of other planned activities	Sources providing conduit or inducing discharge through altered flow patterns	Naturally occurring sources whose discharge is created and/or exacerbated by human activity
Irrigation practices Pesticide applications Fertilizer applications Animal feeding operations De-icing salts applications Urban runoff Percolation of atmospheric pollutants Mining and mine drainage	Production wells Other wells (nonwaste) Construction excavation	Ground water–surface water interactions Natural leaching Saltwater intrusion, brackish water

Source: Office of Technology Assessment, 1984.

that more than half the states and territories listed underground storage tanks, septic tanks, agricultural activities, municipal landfills, and abandoned hazardous-waste sites as major threats to ground water. Other sources that were listed include industrial landfills, injection wells, regulated hazardous-waste sites, land application, road salt, saltwater intrusion, and brine pits from oil and gas wells. The highest priority rankings were given to underground storage tanks, abandoned waste sites, agricultural activity, septic tanks, surface impoundments, and municipal landfills.

Figures 4.1 and 4.2 indicate the priority rankings of the sources and various contaminants as reported to Congress. Each section of this chapter discusses how the major sources of contamination may degrade ground water quality and provides the latest information about the scope of the problem. Figure 4.3 shows the various mechanisms of ground water contamination associated with some of the major sources, which are described next. A wide variety of materials has been

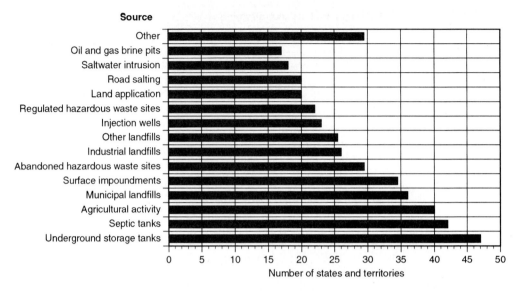

Figure 4.1 Frequency of various contamination sources considered by states and territories of the United States to be major threats to ground water quality. Source: National Water Quality Inventory, 1988 Report to Congress, Environmental Protection Agency, 1990.

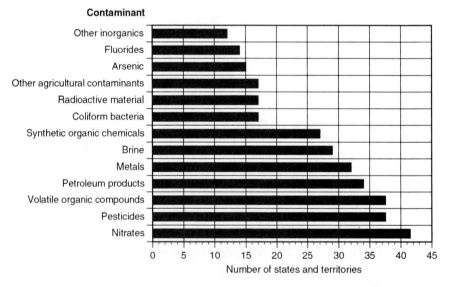

Figure 4.2 Frequency of various contaminants considered by states and territories of the United States to be a major threat to ground water quality. Source: National Water Quality Inventory, 1988 Report to Congress, Environmental Protection Agency, 1990.

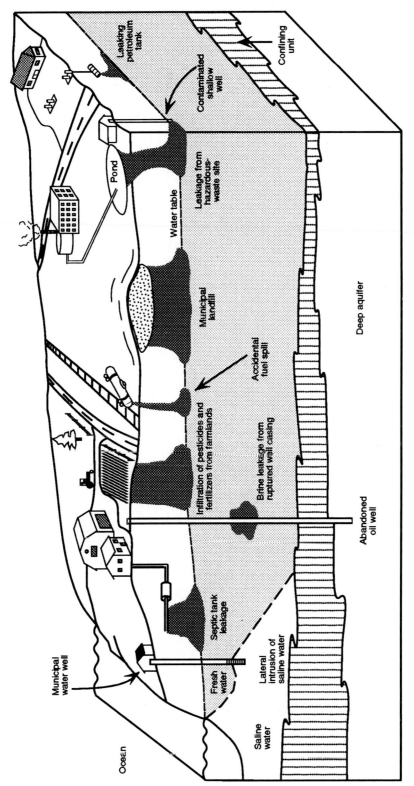

Figure 4.3 Mechanisms of ground water contamination. Source: Fetter, 1993.

67

TABLE 4.2 Environmental Protection Agency List of Priority Pollutants (Organic compounds are subdivided into four categories according to the method of analysis.)

Base-neutral extractables

Acenaphthene	Diethyl phthalate
Acenaphthylene	Dimethyl phthalate
Anthracene	2,4-Dinitrotoluene
Benzidine	2,6-Dinitrotoluene
Benzo[a]anthracene	Di-n-octyl phthalate
Benzo[b]fluoranthene	1,2-Diphenylhydrazine
Benzo[k]fluoranthene	Fluoranthene
Benzo[ghi]perylene	Fluorene
Benzo[a]pyrene	Hexachlorobenzene
Bis(2-chloroethoxy) methane	Hexachlorobutadiene
Bis(2-chloroethyl) ether	Hexachlorocyclopentadiene
Bis(2-chloroisopropyl) ether	Hexachloroethane
Bis(2-ethylhexyl) phthalate	Indeno[1,2,3-cd] pyrene
4-Bromophenyl phenyl ether	Isophorone
Butyl benzyl phthalate	Naphthalene
2-Chloronaphthalene	Nitrobenzene
4-Chlorophenyl phenyl ether	N-Nitrosodimethylamine
Chrysene	N-Nitrosodiphenylamine
Dibenzo[a,h] anthracene	N-Nitrosodi-n-propylamine
Di-n-butyl phthalate	Phenanthrene
1,2-Dichlorobenzene	Pyrene
1,3-Dichlorobenzene	2,3,7,8-Tetrachlorodibenzo-p-dioxin
1,4-Dichlorobenzene	1,2,4-Trichlorobenzene
3,3'-Dichlorobenzidine	

Acid extractables

p-Chloro-m-cresol	2-Nitrophenol
2-Chlorophenol	4-Nitrophenol
2,4-Dichlorophenol	Pentachlorophenol
2,4-Dimethylphenol	Phenol
4,6-Dinitro-o-cresol	2,4,6-Trichlorophenol
2-4-Dinitrophenol	Total phenols

Volatiles

Acrolein	1,2-Dichloroethane
Acrylonitrile	1,1-Dichloroethylene
Benzene	trans-1,2-Dichloroethylene
Bis(chloromethyl) ether	1,2-Dichloropropane
Bromodichloromethane	cis-1,3-Dichloropropene
Bromoform	trans-1,3-Dichloropropene
Bromomethane	Ethylbenzene
Carbon tetrachloride	Methylene chloride
Chlorobenzene	1,1,2,2-Tetrachloroethane
Chloroethane	1,1,2,2-Tetrachloroethene
2-Chloroethyl vinyl ether	Toluene
Chloroform	1,1,1-Trichloroethane
Chloromethane	1,1,2-Trichloroethane
Dibromochloromethane	Trichloroethylene
Dichlorodifluoromethane	Trichlorofluoromethane
1,1-Dichloroethane	Vinyl chloride

Pesticides

Aldrin	Dieldrin	PCB-1016[a]
α-BHC	α-Endosulfan	PCB-1221[a]
β-BHC	β-Endosulfan	PCB-1232[a]
γ-BHC	Endosulfan sulfate	PCB-1242[a]
δ-BHC	Endrin	PCB-1248[a]
Chlordane	Endrin aldehyde	PCB-1254[a]
4,4'-DDD	Heptachlor	PCB-1260[a]
4,4'-DDE	Heptachlor epoxide	Toxaphene
4,4'-DDT		

Inorganics

Antimony	Chromium	Nickel
Arsenic	Copper	Selenium
Asbestos	Cyanide	Silver
Beryllium	Lead	Thallium
Cadmium	Mercury	Zinc

[a] Not pesticides.

TABLE 4.3 Typical Organic Compounds Found in Waste-Disposal Sites in the United States

Ground water contaminant	
Acetone	Methylene chloride
Benzene	Naphthalene
Bis(2-ethylhexyl)phthalate	Phenol
Chlorobenzene	Tetrachloroethylene
Chloroethane	Toluene
Chloroform	1,2-trans-Dichloroethane
1,1-Dichloroethane	1,1,1-Trichloroethane
1,2-Dichloroethane	Trichloroethylene
Di-*n*-butyl phthalate	Vinyl chloride
Ethyl benzene	Xylene

identified as contaminants in ground water. These include inorganic compounds such as nutrients and trace metals, synthetic organic chemicals, radioactive contaminants, and pathogens. Table 4.2 provides a list of major pollutants according to the Environmental Protection Agency. Most of the materials will dissolve in water to some degree depending on their solubility.

Large quantities of organic compounds are manufactured and used by industry, agriculture, and municipalities. They have created the greatest potential for ground water contamination, as described later in this chapter. One such group is the soluble aromatic hydrocarbons associated with petroleum fuels or lubricants. The group includes benzene, toluene, ethyl benzene, and various xylene isomers (BTEX) often associated with petroleum spills. Chlorinated hydrocarbons such as trichloroethylene (TCE) and tetrachloroethylene (PCE) have been used as degreasers, solvents, paints, dry cleaning fluids, and dye stuffs. Table 4.3 lists some of the more abundant organic compounds found in ground water; examples of their uses are described in Section 4.11.

The inorganic compounds occur in nature and may come from natural sources as well as human activities. Metals from mining, industry, waste water, agriculture, and fossil fuels can present serious problems in ground water. Table 4.4 lists some of the more important trace metals occurring in ground water.

TABLE 4.4 Examples of Trace Metals Occurring in Ground Water

Aluminum	Copper	Selenium
Antimony	Gold	Silver
Arsenic	Iron	Strontium
Barium	Lead	Thallium
Beryllium	Lithium	Tin
Boron	Manganese	Titanium
Cadmium	Mercury	Uranium
Chromium	Molybdenum	Vanadium
Cobalt	Nickel	Zinc

4.2 UNDERGROUND STORAGE TANKS

Underground tanks are ubiquitous. While most often associated with gasoline service stations, these tanks are also used by industry, agriculture, government agencies, and homes as storage facilities for gasoline, oil, hazardous chemicals, and chemical waste products. The Office of Technology Assessment (OTA, 1984) estimates the number of tanks, both abandoned and in use, at 2.5 million. A recent EPA survey found that 47 states suffered ground water contamination from faulty underground tanks.

Many tanks installed in the 1950s and 1960s are still in use today or have been abandoned or forgotten. Undergound tanks can leak due to internal or external corrosion of the metal. Leaks can occur through holes in the tank or in associated piping. In a recent survey of motor fuel storage tanks, the EPA found that 35% of the estimated 800,000 such tanks leaked. Steel tanks are being replaced by fiberglass tanks, but faulty piping and subsequent leaks still occur. Figure 4.4 shows a typical double-wall tank and leak-detection system, a possible solution to the problems resulting from leaking tanks. Obviously, such systems are more expensive than older tanks, and they have yet to be tested over time.

The state of Texas alone is spending millions per year for investigation and cleanup of leaking underground storage tanks estimated at more than 5000 in number. The remediation of underground storage tank plumes has become a major focus of hydrogeologic assessments in the United States. One of the most studied underground storage tank incidents in the United States was a fuel spill at the U.S. Coast Guard Station at Traverse City, Michigan. The spill of aviation gas and jet fuel resulted in a plume of contamination more than 1 mile long and 500 feet wide, which polluted about 100 shallow municipal water wells. The site has been the subject of extensive evaluation, and more details are provided in Chapters 8 and 13.

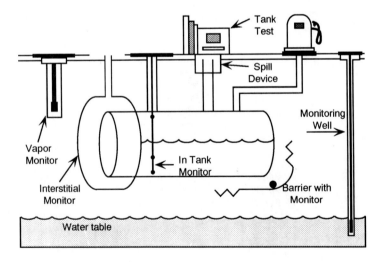

Figure 4.4   Leak-detection alternatives.   Source: Jorgensen, 1989.

4.3 LANDFILLS

Landfills today may be built with elaborate leak-prevention systems, but most, particularly the older ones, are simply large holes in the ground filled with waste and covered with dirt. Originally designed to reduce air pollution and unsightly trash that accompanied open dumping and burning, landfills became the disposal method for every conceivable type of waste. However, many were poorly designed and are leaking liquids or leachates, which are contaminating surrounding shallow ground water. Table 4.8 lists the main constituents of landfill leachate. According to EPA reports, there are approximately 2395 open dumps and 24,000 to 36,000 closed or abandoned landfills in the United States. The EPA estimates that 12,000 to 18,000 municipal landfills may contain hazardous waste. In addition, there are an estimated 75,000 on-site industrial landfills.

Materials placed in landfills include garbage, trash, debris, sludge, incinerator ash, foundry waste, and hazardous substances. Liquid hazardous wastes can no longer be legally disposed of in municipal landfills.

Many older landfills were located based on convenience, rather than hydrogeologic study, and consequently have been situated in environmentally sensitive marshlands, abandoned mines, gravel and sand pits, and sink holes. The disposal technology simply involved filling the hole with liquid and solid wastes, compacting with a bulldozer, and then covering with a layer of soil. As rainwater infiltrates the landfill, leaching of contaminants into the ground water can occur (Figure 4.3).

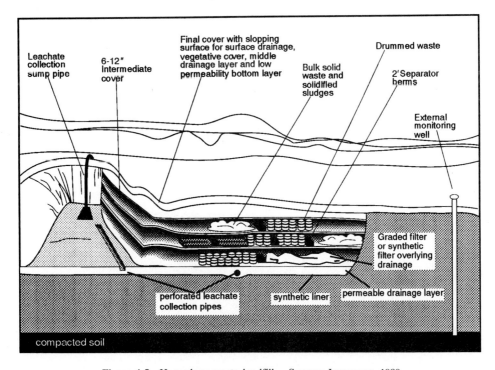

Figure 4.5 Hazardous waste landfill. Source: Jorgensen, 1989.

Modern landfills have liner systems and leachate collection systems to control the migration of contaminants so that they can be transported to a treatment plant. Extensive engineering and hydrogeologic designs are required for permission to establish municipal and industrial landfills today. But the landfills of the 1950s through the 1970s contain no such liners or leachate collection systems. Hazardous-waste landfills are now regulated under the Resource Conservation and Recovery Act (Chapter 14). Figure 4.5 depicts a modern hazardous-waste landfill with leachate collection system.

4.4 SURFACE IMPOUNDMENTS

Surface impoundments are often called pits, ponds, or lagoons. Ranging in size from a few square feet to several thousand acres, surface impoundments serve as disposal or temporary storage sites for hazardous and nonhazardous wastes. They are designed to accept purely liquid wastes or mixed solids and liquids that separate in the impoundment. Chemical wastes in the impoundment are either treated and discharged to the environment or are allowed to infiltrate the soil or evaporate to the atmosphere. Prior to the passage of RCRA, liquid hazardous wastes were also discharged into pits which may have been unlined or lined with clay or other liner membranes.

Surface impoundments are commonly used by municipal wastewater and sewage-treatment operations for settling of solids, biological oxidation, and chemical treatment. Surface impoundments are also used by animal feedlots and farms and by many industries, including oil and gas, mining, paper, and chemical operations. Water from surface impoundments may be discharged to streams and lakes. Many of these surface impoundments have been found to leak (Figure 4.6) and create large contaminated zones in the subsurface. The most famous case is the Rocky Mountain Arsenal near Denver, which discharged nerve gas and pesticides into unlined evaporation ponds from 1942 until 1956. Contamination of nearby

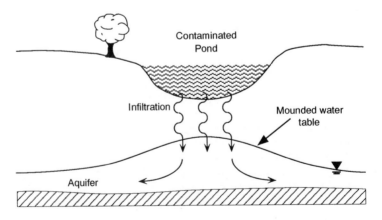

Figure 4.6 Pond leak and associated water table mound.

wells was detected in the early 1950s when irrigated crops died, and ground water contamination extended over an 8-mile region. The ground water under the Rocky Mountain Arsenal has been found more recently to contain many synthetic organic contaminants associated with the manufacture of nerve gas and pesticides (Konikow and Thompson, 1984). It is estimated that the cleanup of contaminated soil and ground water at the arsenal will ultimately cost more than $1 billion.

In 1982, the EPA identified over 180,000 waste impoundments including 37,000 municipal, 19,400 agricultural, 27,912 industrial, 25,000 mining, and 65,688 brine pits for oil and gas (EPA, 1985). Of the industrial sites evaluated, 95% were within 1 mile of drinking-water wells, 70% were unlined, and 50% were on top of aquifers.

4.5 WASTE-DISPOSAL INJECTION WELLS

Injection wells are used to discharge liquid hazardous waste, brine, agricultural and urban runoff, municipal sewage, aquifer recharge water, and fluids used in solution mining and oil recovery into the subsurface. Every year in the United States millions of tons of toxic, hazardous, radioactive, and other liquid wastes are dumped directly into the earth through many thousands of waste-disposal wells. This practice, most commonly utilized by the chemical, petroleum, metals, minerals, aerospace, and wood-preserving industries, has contaminated ground water in at least 20 states.

Injection wells can cause ground water contamination if the fluid enters a drinking-water aquifer due to poor well design, faulty construction, or inadequate understanding of the geology. Wastewater can migrate vertically upward into a drinking-water aquifer through cracks, fault zones, or abandoned well casings. Figure 4.7 shows a typical deep-well injection of liquid waste. Normally, such wells are designed to have pressure gages and monitoring wells to detect any leak or fracture problems with the injection. Injection wells are now regulated under the Underground Injection Control Program of the Safe Drinking Water Act, and the 1984 RCRA amendments prohibit the underground injection of certain hazardous wastes. Probably, the injection wells that pose the greatest threat to ground water include agricultural wells, septic-system wells, brine-injection wells, and deep wells for hazardous waste. An additional concern regarding injection wells is that wastes that have been disposed of years earlier may migrate into drinking-water aquifers due to fractures and faults in abandoned casings (Figure 4.8).

4.6 SEPTIC SYSTEMS

Approximately 22 million septic systems are operating in the United States today, and about one-half million new systems are installed every year. These systems serve nearly 30 percent of the nation's population.

Septic systems generally are composed of a septic tank and a drain field into which effluent flows from the tank (Figure 4.9). Within the tank, physical pro-

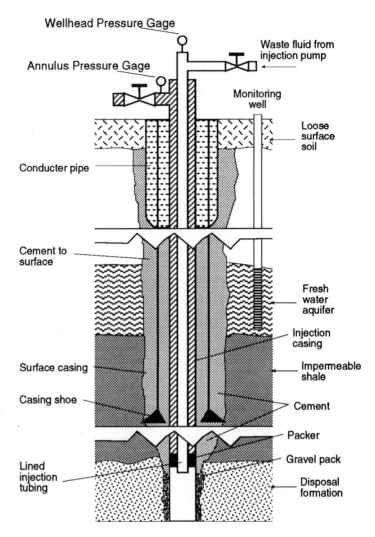

Figure 4.7 Deep-well injection of liquid waste.

cesses separate the inflow into sludge (which accumulates on the bottom of the tank), wastewater, and scum (which forms on top of the wastewater). Once a tank reaches a certain percentage of its capacity, the sludge and scum, called septage, must be pumped out so that the tank will continue to function properly.

Serious system failures are usually quite evident because wastes will surface and flood the drainage field (not only causing an odor, but also exposing people to pathogenic bacteria and viruses). Unfortunately, we cannot see or smell contaminants from underground systems that leach into aquifers. Years may pass before contamination emanating from poorly designed systems is detected.

Septic systems discharge a variety of organic and inorganic compounds, including biochemical oxygen demand, chemical oxygen demand, total suspended

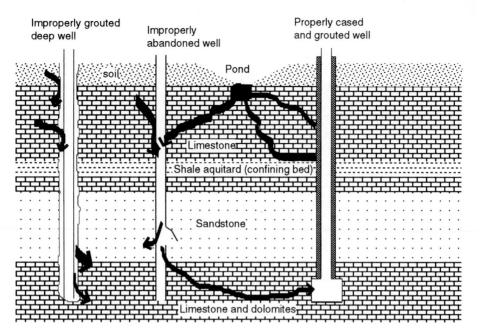

Figure 4.8 Aquifer contamination through improperly constructed or abandoned wells.

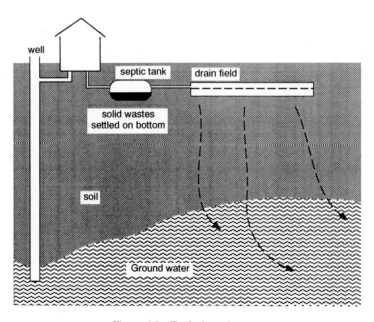

Figure 4.9 Typical septic system.

solids, fecal coliforms, nitrates and nitrites, ammonia, and phosphorus. Synthetic organic chemicals such as trichloroethylene (TCE), benzene, and methylene chloride may be discharged to the subsurface.

Commercial and industrial septic systems present unique and potentially more severe problems to ground water contamination than domestic systems due to the hazardous nature of the wastes disposed of in these systems. Chemicals including nitrates, heavy metals, such as lead, copper, and zinc, and certain synthetic organic chemicals, such as toluene, trichloroethylene, chloroform, and tetrachloroethylene, are dumped into such systems. The EPA has identified several commercially used septic systems as sources of chemical contamination at sites around the nation designed for cleanup under the federal Superfund law. Additionally, many small businesses, including laundries, hardware stores, restaurants, service stations, and laboratories, contaminate ground water through commercial septic systems.

4.7 AGRICULTURAL WASTES

Pesticides were first identified in ground water less than 10 years ago, but now over 25 states report ground water contaminated by pesticides. Recent limited ground water monitoring efforts are only beginning to tell the story of decades of sometimes indiscriminate pesticide use. Pesticides have been widely used for many purposes, such as herbicides, insecticides, fungicides, and defoliants. There are 50,000 different pesticide products in the United States composed of 600 active ingredients. They are used on golf courses, lawns and gardens, roadsides, parks, home foundations, and wood products. They can contaminate ground water through migration through the soil to the water table. Many in use today are biodegradable to some extent. More than 65% of pesticides are applied by aerial spraying and pose a special problem. (Rachel Carson's *Silent Spring,* published in 1962, is a classic book that exposed the serious problem of pesticide use in the United States.)

Fertilizers from agriculture can provide a major source of elevated nutrient levels to the subsurface. Nitrogen, potassium, and phosphorus are the three basic fertilizers, but nitrogen represents over half of the total used and is the most likely to leach. Phosphorus is not very mobile and does not pose a significant threat to ground water. The use of nitrogen on U.S. agricultural lands increased 38% from 1975 to 1981, bringing the total to over 10 million metric tons. In a recent U.S. Geological Survey study, 20% of the sample had a nitrate concentration of over 3 mg/L, and 6% had a nitrate concentration exceeding EPA's 10-mg/L limit for drinking water. Nitrates represent the most frequently reported contaminant considered a major threat to ground water quality according to the *National Water Quality Inventory* (EPA, 1990).

The production of millions of tons of manure annually contaminates underlying aquifers with nitrogen, bacteria, viruses, hormones, and salts. Although ground water can be contaminated by relatively small livestock operations if they are located above porous soils, the most obvious threat stems from animal feedlots, where dense livestock populations are confined to small areas. Facilities that treat or dispose of animal wastes likewise pose a threat to ground water.

Modern irrigation practices can lead to salt contamination and high levels of total dissolved solids (TDS) in underlying aquifers. Irrigation water contains small quantities of salts, which, because they are not transpired by crops or evaporated from soil, build up within the soil and eventually leach into ground water. Irrigation return flows, which eventually reach rivers and streams, may also contribute to ground water contamination, especially in arid areas. In arid and semi-arid areas of the country, excess irrigation water is applied to rid the root zone of potentially crop-devastating salt buildup. Although it may maintain crop productivity, this practice degrades underlying ground water supplies.

4.8 LAND APPLICATION

Land application is a treatment and disposal method also called land treatment and land farming. The practice involves spreading waste sludges and wastewater generated by public treatment works, industrial operations such as paper, pulp, and textile mills, tanneries and canneries, livestock farms, and oil and gas exploration and extraction operations. Wastewater is applied primarily by a spray irrigation system, whereas sludge from wastewater plants is generally applied to soil as a fertilizer. Oily wastes from refining operations are often land farmed in soil to be broken down by soil microbes. If properly designed and operated, land application recycles nutrients and waters to the soil and aquifer.

Over 20 states reported land application as a major threat to ground water. Contamination occurs when heavy metals, toxic chemicals, nitrogen, and pathogens leach to underlying aquifers. This occurs if the sludge or wastewater has not received adequate pretreatment or if the depth to ground water has not been properly considered. In some cases the hazardous materials do not degrade in the subsurface. For example, 40% of the state of California's hazardous wastes are treated by land-farming practices. The land application of hazardous wastes has received major attention from the EPA in recent years and is no longer an approved technology in many aquifer settings.

4.9 RADIOACTIVE CONTAMINANTS

The massive production of radioactive isotopes by weapons and nuclear reactors since World War II has been accompanied by increasing concern about environmental and health effects. Radionuclides are fission products of heavy nuclei of such elements as uranium and plutonium. The ultimate disposal of radioactive wastes has caused major controversy regarding the widespread use of nuclear power.

Radionuclides emit ionizing radiation in the form of alpha particles, beta particles, and gamma rays. Gamma rays are the most damaging and are similar to x-rays, although more energetic. The decay of a specific radionuclide follows

TABLE 4.5   Radionuclides in Water

Radionuclide	Half-life
Naturally occurring and from cosmic reactions	
Carbon 14	5730 years
Silicon 32	~300 years
Potassium 40	~1.4×10^9 years
Naturally occurring from ^{238}U series	
Radium 226	1620 years
Lead 210	21 years
Thorium 230	75,200 years
Thorium 234	24 days
From reactor and weapons fission	
Strontium 90	28 years
Iodine 131	8 days
Cesium 137	30 years
Barium 140	13 days
Zirconium 95	65 days
Cerium 141	33 days
Strontium 89	51 days
Ruthenium 103	40 days
Krypton 85	10.3 years
Cobalt 60	5.25 years
Manganese 54	310 days
Iron 55	2.7 years
Plutonium 239	24,300 years

a first-order decay law that can be expressed as $C = C_0 e^{-\lambda t}$, where C is the activity at time t, C_0 is the initial activity at time 0, and λ, the decay coefficient, is related to the half-life by $(t_{1/2}) = 0.693/\lambda$. The half-life is defined as the time during which 50% of a given number of radioactive atoms will decay. First-order decay is described in more detail in Chapter 6. Table 4.5 summarizes the major natural and artificial radionuclides typically encountered in water and their associated half-lives.

The nuclear industry is currently the main generator of radioactive contaminants. Potential sources occur in the mining and milling of uranium, fuel fabrication, power-plant operation, fuel reprocessing, and waste disposal. The disposal of civilian radioactive wastes and uranium mill tailings is licensed under the Nuclear Regulatory Commission. High-level radioactive wastes from nuclear power plants are currently in temporary storage, but will eventually go into an underground repository, such as the one planned for Yucca Mountain, Nevada. Low-level wastes and medical wastes are currently buried in shallow landfills.

Unless radioactive wastes are properly handled in well-designed sites, the potential for migration to ground water exists. Enormous contamination problems currently exist at a number of Department of Energy (DOE) facilities, including Oakridge, Tennessee, the Hanford Reactor in Washington State, the Savannah River in Georgia, and the Idaho National Engineering Lab. These and other associated nuclear sites are the subject of massive investigations and remediation studies. The health hazards associated with radiation leaks are well known, but the risks are difficult to assess at low levels of exposure. The true impact of radioactive waste disposal may not be known for years.

4.10 OTHER SOURCES OF CONTAMINATION

According to the Citizen's Clearinghouse for Hazardous Waste, the U.S. military branches may be the largest generators of hazardous waste in the country, producing over 1 billion pounds per year, more than the top five civilian chemical companies combined. Numerous spills, leaks, and landfills have been discovered on military bases throughout the country and are the subject of intense investigation and remediation efforts. The U.S. Air Force alone estimates more than 4300 waste sites and spills on more than 100 of their bases. Many of these military sites are currently on the EPA national priority list as Superfund sites. One of these Air Force sites in Tucson, Arizona, created a TCE and chromium plume of contamination that extends 6 miles in length and a half-mile in width. The water-supply wells for the city of Tucson are in the downgradient direction of the plume. The site is currently being remediated with a pump-and-treat system (Chapter 10) designed to withdraw 3700 gal/min of water from the aquifer located 150 ft below the surface.

The construction techniques, products, and by-products of mining operations are serious threats to the quality and quantity of nearby aquifers. Surface and underground mining may disrupt natural ground water flow patterns and create the potential for acid mine drainage to seep from the mine. Millions of acres of U.S. land have been mined for coal, copper, uranium, and other minerals. Inactive and abandoned mines, as well as active mines, can be steady and serious sources of contamination; there are an estimated 67,000 inactive or abandoned mines in the United States.

4.11 CLASSIFICATION OF ORGANIC COMPOUNDS

In the past 15 years, organic compounds in ground water have come to be recognized as a major threat to human health. This section will introduce some of the major classes and some of the most important compounds found in ground water. More detailed coverage on organic compounds can be found in Freeze and Cherry (1979), Pankow (1991), Manahan (1991), and Fetter (1993). Organic chemistry deals with the chemistry of carbon compounds.

Carbon is a unique element in that it forms four covalent bonds and is capable of bonding to other carbon atoms, even forming double and triple bonds. It is this characteristic of carbon that gives rise to the possibility of great diversity in the physical and chemical properties of organic compounds. The simplest organic compounds are hydrocarbons consisting of carbon and hydrogen alone. The traditional approach to classifying organic compounds involves defining functional groups, which include a simple combination of two or more of the following atoms: C, H, O, S, N, and P. Domenico and Schwartz (1990) present a condensed scheme of classification consisting of 16 major classes (Table 4.6), which is a useful approach for organizing organic contaminants in ground water. Other elements such as O, N, S, P, H, and Cl can bond with carbon at any of four locations. Hydrocarbons can be divided into aromatics, which contain a benzene ring, and aliphatics, which do not contain a benzene ring.

Aliphatic Hydrocarbons

The aliphatic hydrocarbons with more than one carbon atom can be classified as alkanes (single bonds), alkenes (double bonds), or alkynes (triple bonds). Straight-chain alkanes include methane (CH_4), ethene (CH_3), propane, butane, pentane, and hexane. These compounds are known as saturated hydrocarbons or paraffins. Alkanes can also have branched chains, creating isomers with the same formula but with different properties. Alkanes conform to the general formula C_nH_{2n+2}, where n is the number of carbon atoms. Cycloalkanes are characterized by a ring structure that contains single C–C bonds such as cyclopropane and cyclohexane.

Alkenes have a carbon–carbon double bond with the general formula C_nH_{2n} and include such compounds as ethene and propene. Alkenes are unsaturated hydrocarbons or olefins. Ethene is also referred to as ethylene. If functional groups are present, then their position is indicated by the carbon atom to which they are bonded, and structural isomers can exist.

Aromatic Hydrocarbons

Aromatic hydrocarbons are compounds with a molecular structure based on that of benzene, C_6H_6. These compounds are a major constituent of petroleum and related products. Typical benzene-related compounds are shown in Figure 4.10. The benzene molecule consists of six carbon and six hydrogen atoms in a cyclical form. The ring in the center represents a delocalized cloud of electrons. The carbon atoms in benzene are also capable of bonding to functional groups, and isomerism is possible. More than one functional group may result in *ortho-, meta-,* or *para-*isomers (see the xylene isomers in Figure 4.10).

When several of the benzene rings are joined together, polycyclic aromatic hydrocarbons (PAH) are formed such as naphthalene (two benzene rings), phenanthrene (three benzene rings), and benzo[*a*]pyrene (five benzene rings). Figure 4.11 shows several of the important PAHs in ground water. These compounds are found in petroleum products, asphalt, coal tar, and creosote and result from the incomplete combustion of fossil fuels. If the benzene ring is joined to another group, it may be referred to as a functional group, phenyl, and in combination with chlorine these compounds are called polychlorinated biphenyls (PCBs). They are extemely resistant to chemical, thermal, or biological degradation and tend to persist in the environment. PCBs associated with the energy industry comprise some of the most serious contaminants in ground water.

Phenols are characterized by an aromatic ring with an attached hydroxyl group. They originate in ground water mostly as contaminants from industrial wastes or biocides. Phenol is a common ground water contaminant due to its many

TABLE 4.6   Classification of Organic Compounds

1. Miscellaneous nonvolatile compounds

2. Halogenated hydrocarbons

   Aliphatic   Aromatic

   Trichloroethylene   Chlorobenzene

3. Amino acids

   Basic structure   Aspartic acid

4. Phosphorus compounds

   Basic structure   Malathion

5. Organometallic compounds

   Tetraethyllead

81

TABLE 4.6 *(continued)*

6. Carboxylic acid

Basic structure

$$R{-}\overset{\displaystyle O}{\overset{\|}{C}}{-}OH$$

Acetic acid

$$CH_3{-}\overset{\displaystyle O}{\overset{\|}{C}}{-}OH$$

7. Phenols

Basic structure

OH

Cresol

CH₃

CH

8. Amines

Basic structure

Aliphatic   Aromatic

N—
|
CH₃

NH₂

Dimethylamine

CH₃
|
CH₃—N—H

9. Ketones

Basic structure

$$R{-}\overset{\displaystyle O}{\overset{\|}{C}}{-}R'$$

Acetone

$$CH_3{-}\overset{\displaystyle O}{\overset{\|}{C}}{-}CH_3$$

10. Aldehydes

Basic structure

$$R{-}\overset{\displaystyle O}{\overset{\|}{C}}{-}H$$

Formaldehyde

$$H{-}\overset{\displaystyle O}{\overset{\|}{C}}{-}H$$

11. Alcohols

Basic structure

R—OH

Methanol

CH₃—OH

12. Esters

Basic structure

$$R{-}\overset{\displaystyle O}{\overset{\|}{C}}{-}OR'$$

Vinyl acetate

$$H_2C{=}CH{-}O{-}\underset{\displaystyle O}{\underset{\|}{C}}{-}CH_3$$

TABLE 4.6 *(continued)*

13. Ethers

Basic structure

C—O—C

1,4-Dioxane

14. Polynuclear aromatic hydrocarbons

Phenanthrene

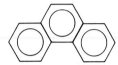

15. Aromatic hydrocarbons

Basic structure

Toluene

CH₃

16. Alkane, alkene, and alkyne hydrocarbons

Ethane

H H
| |
H—C—C—H
| |
H H

Ethene

Ethyne

H—C≡C—H

R = aliphatic backbone

Source: Domenico and Schwartz, 1990.

industrial uses and can also occur naturally with decomposing organic matter. Cresols have a methyl group attached to the benzene ring of toluene and are used in the coal tar refining industry. Chlorophenols are used as wood and leather preservatives and as antimildew agents. Phenol, in general, has acute toxicological effects on the central nervous system and can result in death after even a short exposure. Pentachlorophenol has been found at a number of creosote wood preserving sites and represents a highly toxic and nonbiodegradable compound.

Benzene is a carcinogen, and inhaled benzene is readily absorbed by blood

Name	Structure	Molecular weight	Solubility in water (mg/L)	Soil-water partition coefficient
Benzene		78.11	1780	97
Toluene	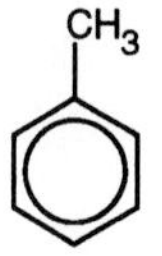	92.1	500	242
Xylene, ortho	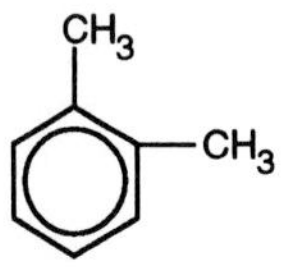	106.17	170	363
Xylene, meta	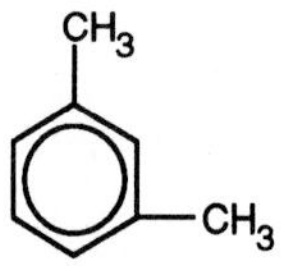	106.17	173	182
Xylene, para		106.17	200	331
Ethyl benzene	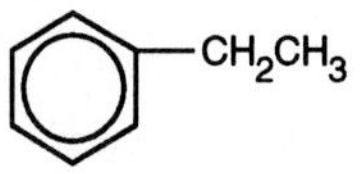	106.17	150	622

Figure 4.10 Benzene-related compounds.

and is strongly taken up by fatty tissues. Benzene can be converted to phenol by an oxidation reaction in the liver, which is responsible for the unique toxicity of benzene, which involves damage to bone marrow. Benzene is also a skin irritant and can affect the central nervous system. Toluene is classified as moderately toxic and is much less toxic than benzene because it is readily excreted from the body (Manahan, 1991).

Halogenated Hydrocarbons

Halogenated hydrocarbons are one of the largest and most important groups of contaminants found in ground water. This group consists of both aliphatic and aromatic subclasses and is characterized by the presence of one or more halogen atoms (Cl, Br, F). Included in the aliphatic subclass group are solvents such as methylene chloride, chloroform, carbon tetrachloride, TCE, and vinyl chloride

Name	Structure	Molecular Weight	Solubility in Water (mg/L)	Soil-Water Partition Coefficient
Naphthalene	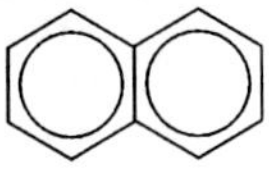	128.16	31.7	1300
Acenaphthene	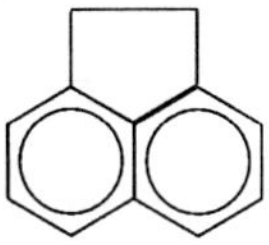	154.21	7.4	2580
Ancenaphthylene	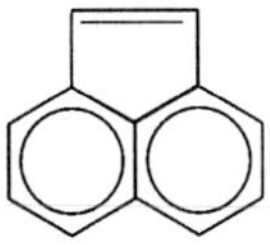	152.2	3.93	3814
Fluorene	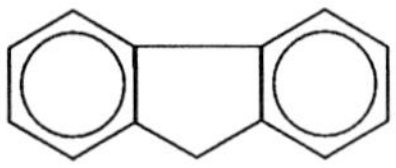	166.2	1.98	5835
Fluoranthene	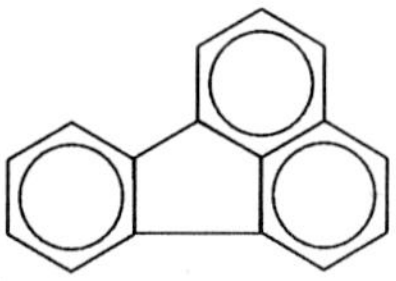	202	0.275	19,000
Phenanthrene	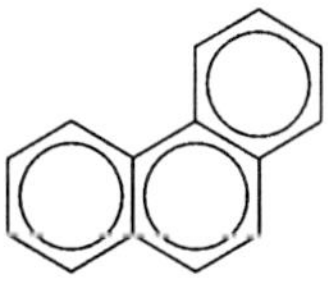	178.23	1.29	23,000
Anthracene	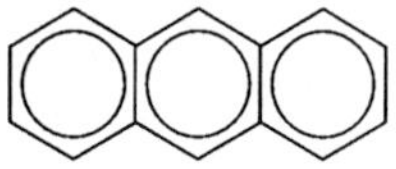	178.23	0.073	26,000

Figure 4.11 Structure and properties of some aromatic and polycyclic aromatic hydrocarbons.

and pesticides such as aldrin and dieldrin. The aromatic subclass includes pesticides such as DDD, DDE, and various PCB compounds. Many of these compounds in the halogenated group are used as solvents, insecticides, and metal degreasers and are degradation products from other chlorinated compounds in the production of plastics (Figure 4.12 and Table 4.3).

The specific toxicity of halogenated hydrocarbons varies with the compound, but most affect the central nervous system. Carbon tetrachloride, for example, is a systemic poison that affects the nervous system, the intestinal tract, liver, and kidneys. Over the years the Food and Drug Administration (FDA) compiled a grim record of toxic effects and eventually banned its household use in 1970. Vinyl chloride has been used widely in the production of polyvinyl chloride (PVC)

Name	Structure	Uses and Other Sources
Trichloromethane (chloroform)	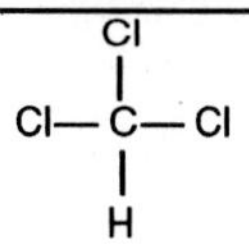 Cl–C(Cl)(Cl)–H	Liquid used in manufacture of anesthetics, pharmaceuticals, fluorocarbon refrigerants, and plastics. Used as solvent and insecticide. Formed from methane when chlorinating drinking water.
Vinyl chloride (chloroethene)	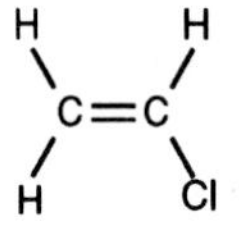 H₂C=CHCl	Gas used in the manufacture of polyvinyl chloride. End product of microbial degradation of chlorinated ethenes.
Chloroethane	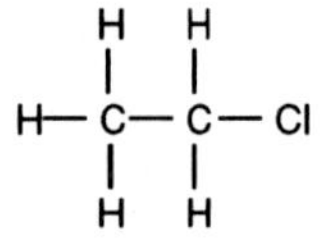 H–CH₂–CH₂–Cl	Liquid used to manufacture tetraethyl lead. Degradation product of chlorinated ethanes.
1,2-Dichloroethane	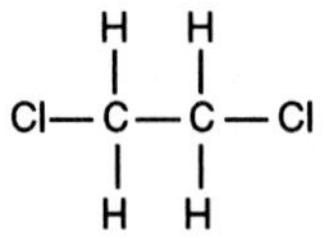 Cl–CH₂–CH₂–Cl	Liquid used to manufacture vinyl chloride. Degradation product of trichloroethane.
Trichloroethene (Trichloroethylene)	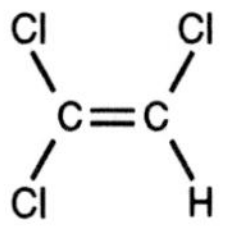 ClCH=CCl₂	Solvent used in dry cleaning and metal degreasing. Organic synthesis. Degradation product of tetrachloroethene.
Tetrachloroethene (perchloroethene) (perchloroethylene)	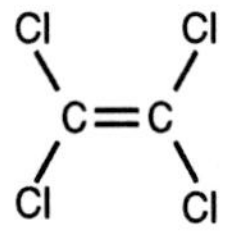 Cl₂C=CCl₂	Solvent used in dry cleaning and metal degreasing. Used to remove soot from industrial boilers. Used in manufacture of paint removers and printing inks.
1,2-Dibromo-3-chloropropane (DBCP)	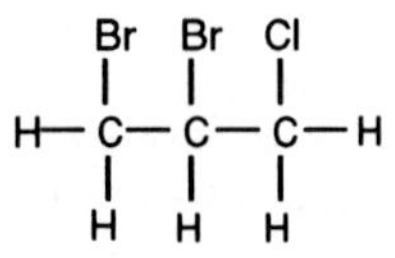 H–CHBr–CHBr–CHCl–H	Soil fumigant to kill nematodes. Intermediate in organic synthesis.
o-Dichlorobenzene (1,2-dichlorobenzene)	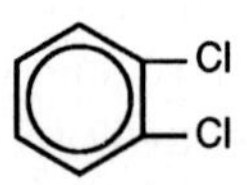 C₆H₄Cl₂	Chemical intermediate. Solvent. Fumigant and insecticide. Used for industrial odor control. Found in sewage from odor-control chemicals used in toilets.

Figure 4.12 Chlorinated organic found in hazardous waste. Source: Fetter, 1993.

materials, and exposure can affect the central nervous system, respiratory system, liver, blood, and lymph systems. Most notably, vinyl chloride is a carcinogen. The dichlorobenzenes are irritants that affect the respiratory system, liver, skin, and eyes through inhalation or contact. Polychlorinated biphenyls (PCBs) have been widely used in the electrical industry as hydraulic fluids in transformers. They represent extremely persistent environmental pollutants with a strong tendency to undergo bioaccumulation in lipid tissue (Manahan, 1991). The MOTCO Superfund site in Texas is an example of a site contaminated with halogenated hydrocarbons; it is described in Chapter 13.

Other Organic Compounds

Table 4.6 indicates other classes of organic compounds. Those most important in ground water contamination include the phosphorus-based compounds (pesticides), ketones, aldehydes, and alcohols. Alcohols have one or more OH groups, are miscible with water, and have the potential for significant mobility; however, they are readily biodegraded. Esters are the result of the combination of an alcohol with a carboxylic acid. Esters are used in flavorings, perfumes, solvents, and paints. One important class of esters is the phthalates, which are used to improve the flexibility of various plastics. Ethers have an oxygen atom strongly bonded between two carbon atoms and thus have relatively low toxicities. Organic compounds that contain nitrogen are common in industry and the manufacturing of explosives. TNT (2,4,6-trinitrotoluene) has been reported as a soil contaminant in areas where munitions are manufactured. More details on organic compounds and their degradation and reactions can be found in Pankow (1991), Fetter (1993), Manahan (1991), and Schwarzenbach et al., 1993. Chapters 7 and 8 present important chemical reactions, adsorption, and biodegradation mechanisms that affect organic contaminants in the subsurface.

4.12 INORGANIC COMPOUNDS IN GROUND WATER

The quality of water is a direct result of the reactions that occur between it and other compounds that it may come into contact with. In ground water, chemistry and chemical processes are important primarily because ground water is in contact with soil and rocks that contain a variety of minerals. In addition, the carbon and nitrogen cycles contribute to the quality of ground water. For instance, precipitation may come in contact with high levels of carbon dioxide in the atmosphere and become acidic. There is potential for this acidic water to infiltrate to ground water and dissolve minerals as it encounters them. Because of the processes that affect it, ground water naturally contains dissolved inorganic ions. A list of the inorganic constituents of ground water is presented in Table 4.7, and Table 4.8 lists the constituents of landfill leachate and is an indication of the major inorganic contaminants in ground water. Chemical processes of particular importance to ground water constituents and contaminants include solubility product, pH, and oxidation–reduction reactions (Chapter 7). More detailed information may be

TABLE 4.7 Dissolved Constituents in Ground Water Classified According to Relative Abundance

Major constituents (greater than 5 mg/L)

Bicarbonate	Silicon
Calcium	Sodium
Chloride	Sulfate
Magnesium	Carbonic acid

Minor constituents (0.01–10.0 mg/L)

Boron	Nitrate
Carbonate	Potassium
Fluoride	Strontium
Iron	

Trace constituents (less than 0.1 mg/L)

Aluminum	Molybdenum
Antimony	Nickel
Arsenic	Niobium
Barium	Phosphate
Beryllium	Platinum
Bismuth	Radium
Bromide	Rubidium
Cadmium	Ruthenium
Cerium	Scandium
Cesium	Selenium
Chromium	Silver
Cobalt	Thallium
Copper	Thorium
Gallium	Tin
Germanium	Titanium
Gold	Tungsten
Indium	Uranium
Iodide	Vanadium
Lanthanum	Ytterbium
Lead	Yttrium
Lithium	Zinc
Manganese	Zirconium

Organic compounds (shallow)

Humic acid	Tannins
Fulvic acid	Lignins
Carbohydrates	Hydrocarbons
Amino acids	

Organic compounds (deep)

Acetate	Propionate

found in Freeze and Cherry (1979), Domenico and Schwartz (1990), and Fetter (1993) or in environmental chemistry texts such as Manahan (1991) or Pankow (1991).

Of the inorganic contaminants in ground water, those of greatest concern are nitrates, ammonia, and trace metals. Nitrates in ground water originate from nitrate sources on land and are associated with fertilizers and the disposal of sewage waste. Feedlots are also a major source of nitrate in ground water, espe-

TABLE 4.8 Representative Ranges for Various Inorganic Constituents in Leachate from Sanitary Landfills

Parameter	Representative Range (mg/L)
K^+	200–1000
Na^+	200–1200
Ca^{2+}	100–3000
Mg^+	100–1500
Cl^-	300–3000
SO_4^{2-}	10–1000
Alkalinity	500–10,000
Fe (total)	1–1000
Mn	0.01–100
Cu	<10
Ni	0.01–1
Zn	0.1–100
Pb	<5
Hg	<0.2
NO_3^-	0.1–10
NH_4^+	10–1000
P as PO_4	1–100
Organic nitrogen	10–1000
Total dissolved organic carbon	200–30,000
COD (chemical oxidation demand)	1000–90,000
Total dissolved solids	5000–40,000
pH	4–8

cially in rural areas. Nitrate concentrations are not limited by solubility constraints, resulting in high mobility of nitrates in ground water.

Arsenic, cadmium, chromium, lead, zinc, and mercury are metal pollutants of major concern in ground water. Most of them arise from industrial practices and discharges from mining, metal plating, plumbing, coal, gasoline, and pesticide-related industries. Many of these metals are very toxic to humans, especially cadmium, lead, and mercury. Cadmium and zinc are common water and sediment pollutants in areas associated with industrial installations. A major source of lead is from leaded gasoline and lead piping. Mercury is associated with discarded batteries, laboratory products, and lawn fungicides. Arsenic is produced through phosphate mining and is a by-product of copper, gold, and lead refining.

These metals are of concern in ground water due to their unique acid–base and solubility characteristics in aerobic systems. Metals occur as cations in ground water of low pH and have a greater mobility in acidic waters. Mobility tends to decrease as the solid phase is approached. Mobility of metals is also increased by complexation of ions, and nearly all the trace metals in ground water are influenced by redox conditions, especially when complexation occurs. Heavy metals are particularly toxic in their chemically combined forms, and some, notably mercury, are toxic in the elemental form.

Chromium plumes have been identified and sampled at a number of industrial facilities where metal plating was a predominant activity. A classic plume on Long Island, New York, was reported by Wilson and Miller (1978). More recently the U.S. Air Force Hughes Plant 44 Site in Tucson, Arizona, has a major chromium plume migrating more than 150 feet below the surface in a sand and gravel aquifer.

SUMMARY

Chapter 4 has reviewed the main sources and types of ground water contamination. In particular, underground storage tanks, septic tanks, agricultural activities, municipal landfills, and abandoned hazardous-waste sites are considered major threats to ground water. These are described in some detail in the chapter. Specific organic compounds are addressed briefly in Section 4.11; other major references on organic compounds include Manahan (1991), Pankow (1991), Schwarzenbach et al. (1993), and Fetter (1993). Section 4.12 briefly discusses inorganic compounds in ground water.

REFERENCES

DOMENICO, P. A., and SCHWARTZ, F. W., 1990, *Physical and Chemical Hydrogeology,* Wiley, New York.

FETTER, C. W., 1988, *Applied Hydrogeology,* Merrill Publishing Co., Columbus, OH.

FETTER, C. W., 1993, *Contaminant Hydrogeology,* Macmillan, New York.

FREEZE, R. A., and CHERRY, J. A., 1979, *Groundwater,* Prentice Hall, Englewood Cliffs, NJ.

JORGENSEN, E. P., ed., 1989, *The Poisoned Well,* Chapter 3, Island Press, Washington, DC.

KONIKOW, L. F., and THOMPSON, D. W., 1984, "Groundwater Contamination and Aquifer Reclamation at the Rocky Mountain Arsenal, Colorado," *Studies in Geophysics, Groundwater Contamination,* National Academy Press, Washington, DC.

MANAHAN, S. E., 1991, *Environmental Chemistry,* 5th ed., Lewis Publishers, Chelsea, MI.

OFFICE OF TECHNOLOGY ASSESSMENT (OTA), 1984, *Protecting the Nation's Groundwater from Contamination,* Report

PANKOW, J. F., 1991, *Aquatic Chemistry Concepts,* Lewis Publishers, Chelsea, MI.

SCHWARZENBACH, R. P., GSCHWEND, P. M., and IMBODEN, D. M., 1993, *Environmental Organic Chemistry,* Wiley, New York.

U.S. ENVIRONMENTAL PROTECTION AGENCY, 1985, Office of Groundwater Protection, *Overview of State Groundwater Program Summaries,* Washington, DC.

U.S. ENVIRONMENTAL PROTECTION AGENCY, 1990, the *National Water Quality Inventory—1988 Report to Congress,* Washington, DC.

WILSON, J. D., and MILLER, P. J., "Two-dimensional Plume in Uniform Ground-water Flow," *J. Hydraulic Division,* April 1978.

Chapter 5 | DATA-COLLECTION METHODS

Robert S. Lee

5.1 INTRODUCTION

This chapter presents an overview of the data-collection methods most commonly used in conducting hydrogeologic site investigations. In general, the methods employed were originally developed for the geotechnical, water well, and petroleum industries, but have been refined to accommodate the special requirements associated with defining the extent of contaminant plumes.

The design of a site investigation that provides the required technical information in a cost-effective manner is presented in Chapter 12. By way of preview, two general statements concerning project objectives and economic constraints should be made.

First, three broad classes of data, geologic, hydrologic, and chemical, must be collected during the investigation to facilitate the development of a conceptual model describing the occurrence and movement of contaminants within a hydrogeologic setting. Defining the site's geologic setting, its hydrologic properties, and the lateral and vertical extent of the contaminants present constitutes three steps in the investigation. However, there is significant overlap in the acquisition of the three classes of data. For example, a soil boring drilled to characterize the geology of the site may also provide qualitative hydrologic data, such as whether the aquifer underlying the site is confined or unconfined and if there are soil samples for laboratory analysis of contaminant concentration. The boring may subsequently be converted to a monitoring well to permit collection of ground water samples and quantitative hydrologic data.

Second, information concerning the nature of the geologic materials underlying the site, the occurrence of ground water, and the presence or absence of contaminants in the ground water can be obtained by both direct observation, that is, collection and analysis of soil and ground water samples and by indirect means such as geophysical measurements and soil vapor surveys. Use of such indirect methods of assessing site conditions can reduce the cost of the investigation and in some cases can also replace subjective description with more readily comparable numerical data. However, such data are subject to interpretation and may not be considered definitive of site conditions by regulatory agencies. At a minimum, such indirect measurements should be calibrated against direct data obtained from soil and ground water samples collected on site.

The following sections describe methods for the acquisition of geologic, hydrologic, and soil and ground water quality (chemical) data. The emphasis is on the more conventional techniques, which employ direct observation and measurement of soil and ground water samples. Techniques employing indirect measurement of soil and water properties are treated more briefly, but their value should not be discounted. Use of indirect sampling methods in environmental assessments is an evolving field, and improvements in data quality and cost effectiveness are likely to make such techniques increasingly common in the future.

Project Safety

Safety considerations should figure prominently in all site investigations. As specified by OSHA regulations (29 CFR §1910.120), special training is required for all workers on hazardous-waste sites, and a site-specific health and safety plan must be prepared for all projects. All drilling locations should be cleared for the presence of underground utilities prior to rig mobilization.

5.2 GEOLOGIC DATA ACQUISITION

The essential geologic data required in all hydrogeologic site investigations is a description of the principal stratigraphic units underlying the site, including their thickness, lateral continuity, and water-bearing properties. This is most commonly assessed by examination of soil or rock samples collected from core borings. On sites underlain by unconsolidated materials and where drilling depths are fairly shallow (100 ft or less), the collection of core samples from soil borings is a cost-effective method of collecting geologic data. In areas underlain by consolidated rock or where site conditions require investigation at greater depths, the collection of core samples is more costly and may be augmented or replaced by cone penetrometer testing or borehole geophysical methods.

Drilling Methods

The most common drilling methods used in the environmental industry are described in this section. Selection of the drilling method will depend on such factors as drilling depth, nature of the geologic formations under investigation, and the specific purpose of the boring, for example, lithologic sampling, soil sampling for chemical analysis, and/or well installation. This summary is only a brief introduction to the subject. Driscoll (1986) presents detailed descriptions of the drilling techniques generally used in the water-well industry. Scalf et al. (1981) includes a description of drilling techniques that focuses on environmental applications.

In conducting environmental site investigations, the introduction of foreign materials into the borehole is generally undesirable due to the possibility of reaction with the geologic media or ground water. Even when a potable water source is used in the preparation of drilling fluids, chlorine may react with naturally occurring organic material in the soil to produce detectable concentrations of trihalomethane compounds (such as chloroform and dichlorobromomethane) in the ground water. Therefore, drilling is frequently performed "dry," that is, without the use of drilling fluids or "mud." Drilling dry also allows identification of the depth of the first occurrence of ground water. When ground water is first encountered, drilling operations should be halted long enough to observe whether or not water rises in the borehole, to determine if the ground water is confined or unconfined. Dry drilling is most commonly performed using either solid-flight or hollow-stem augers, as described in the following paragraphs.

Solid-flight Auger Drilling

Solid-flight augers consist of sections of solid rod with a continuous ramp of upward spiraling "flights" welded around it (see Figure 5.1). The auger sections are pinned together as drilling proceeds. As the drill stem is rotated, cutting teeth

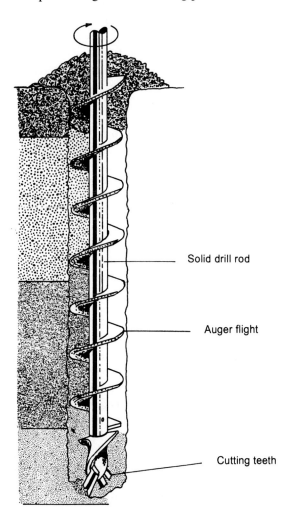

Solid drill rod

Auger flight

Cutting teeth

Figure 5.1 Solid-flight auger drilling equipment.   Source: Scalf et al., 1981.

on the lead auger dig into the formation and the loosened soil rides up the flights to the ground surface. The string of auger sections is tripped out of the hole at each soil sample interval, and a Shelby tube or split-spoon sampler, described later, is pushed using the rig hydraulic system or driven with the rig hammer into the underlying undrilled formation. The drill string is then tripped back into the hole, and drilling proceeds to the next sample point.

Solid-flight augers are effective in cohesive soils above the saturated zone. Because of the tendency for borehole walls to collapse, solid-flight augers are not useful in loose soils or below the water table and are not generally suitable for monitoring well installation.

Hollow-stem Auger Drilling

Hollow-stem augers are another type of flight auger, but instead of being welded to a solid rod, the flights are welded around a hollow pipe (Figure 5.2). The lead auger is fitted with cutting bits located around the circumference of its base. During drilling, a center rod equipped with a pilot bit is lowered inside the auger, and the center rod and hollow stem are rotated together.

As drilling proceeds, loosened soil rides up the outer ramp of auger flights, as with the solid-flight auger. A plug, positioned above the bit, prevents soil from traveling up the inside of the hollow stem. When the sample interval is reached, the center rod, plug, and bit are removed, and the soil sampling tool is lowered inside the hollow stem, which remains in the borehole to prevent collapse of the walls. The sampling tool, generally a split-spoon or Shelby tube sampler, is then pushed or driven into the soil ahead of the auger. If a monitoring well is to be installed, the well screen and riser can also be lowered within the hollow stem, which provides a temporary casing during well installation, as described in the following section.

Under the most favorable conditions, hollow-stem augers can be used to depths exceeding 150 ft. However, at depths below about 50 ft, drilling and soil sampling slow down considerably. Auger drilling is not applicable to hard rock formations. At greater depths, in harder formations, and in flowing sand conditions, wet rotary drilling is a more effective method.

Wet Rotary Drilling

During wet rotary drilling, a dense drilling fluid or "mud" is pumped down a hollow drill pipe and through holes or jets located above the teeth of the drill bit (see Figure 5.3). As the fluid is pumped down the hole, drill cuttings are circulated up the borehole to the surface. The drilling mud also cools the drill bit, exerts hydrostatic pressure on the formation, and forms a thin coating or "mudcake" on the borehole wall, which keeps the unconsolidated aquifer material from collapsing and closing the borehole during soil sampling and well installation. In environmental applications, the drilling mud is most commonly prepared from powdered, additive-free bentonite (a mixture of dense clay minerals of volcanic origin) and potable water. During soil sample collection, the drill stem is removed

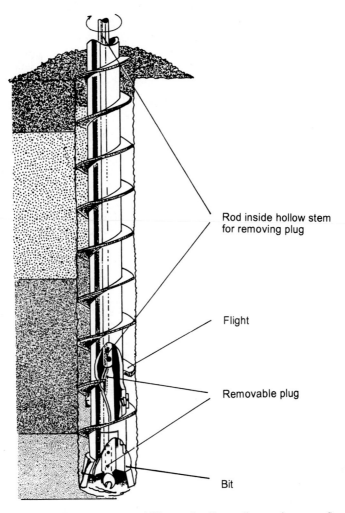

Rod inside hollow stem
for removing plug

Flight

Removable plug

Bit

Figure 5.2 Hollow-stem auger drilling and soil sampling equipment. Source: Scalf et al., 1981.

from the hole, and the drill bit is replaced with either a split-spoon or Shelby tube sampler.

Wet rotary drilling is effective in environments where auger drilling is not practical, including hard, consolidated formations, such as well-cemented sandstones or shales, and very loose, flowing sand formations. In general, drilling and soil sample collection by wet rotary drilling are faster than in a hollow-stem operation, particularly as drilling depth increases. For soil sampling, a smaller borehole can be drilled than with a hollow-stem auger, generating a smaller volume of cuttings, which can be an important consideration if cuttings must be drummed and disposed of as hazardous waste.

On the other hand, because drilling fluid is present in the borehole from the start of the hole, the depth of the first occurrence of ground water cannot be

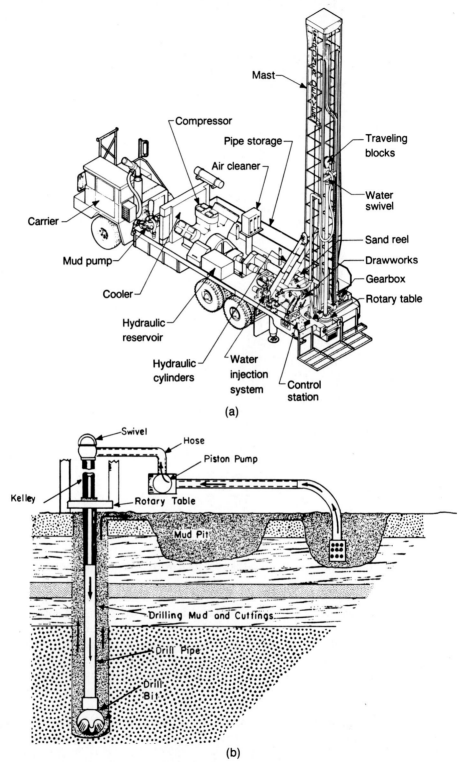

Figure 5.3 Wet rotary drilling equipment. Sources: (a) Driscoll, 1986, (b) Scalf et al., 1981.

determined during wet rotary drilling. Also, some contact between soil samples and drilling fluid is unavoidable. If a monitoring well is installed in a wet rotary drilled hole, more extensive well development, as described in Section 5.3, is required to remove the introduced fluids. Some drill rigs are equipped for both wet and dry drilling methods and can be switched from auger drilling above the water table to wet rotary drilling below it.

Air Rotary Drilling

Formations such as vesicular basalts and highly fissured or cavernous limestones are frequently drilled by air rotary methods because of the potential for loss of drilling mud circulation in these rock formations. Air rotary drilling operates on the same principle as wet rotary drilling. Direct circulation of compressed air down the drill string and through the bit raises cuttings to the surface in small-diameter boreholes without introducing water from the surface. As formation water is blown out the borehole along with the cuttings, identification of the uppermost water-bearing zone is possible.

Hand Auger Soil Borings

Hand auger soil borings are frequently used to collect soil samples from the shallowest portion of the subsurface, from depths of less than 10 ft. Hand auger borings can be advanced to somewhat deeper depths (approximately 20 ft), but with depth the method becomes increasingly inefficient. Still, it may be the most practical method at some locations, such as where the presence of overhead or belowground structures may prevent access by a truck-mounted drill rig. Hand auger borings are not generally used for the installation of monitoring wells, but under certain circumstances they may be acceptable for temporary piezometers or ground water sampling points.

Soil Trenches

Soil trenches dug with a backhoe or other excavation equipment are sometimes used for soil sampling and geologic description in the shallow subsurface. Trenches can be particularly useful in investigating landfills, where the presence of rubble frequently hampers drilling operations. Trenches deeper than 3 ft with a slope greater than 1:1 should not be entered unless shored. If contaminants are present, proper evaluation of air-quality conditions should be made and appropriate respiratory protection should be worn.

Sample Collection

Soil Sample Collection

In most environmental applications, soil samples are collected from cohesive clay soils using a 2.5-ft-long by 3-in.-diameter thin-walled steel tube known as a Shelby tube sampler (see Figure 5.4). The sample tube is pushed into and extracted

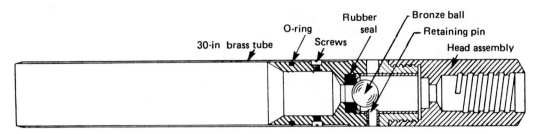

Figure 5.4 Shelby tube soil sampler. Source: Hunt, 1984.

from the soil using the hydraulic system on the drill rig. Once the sample is retrieved to the surface, it is extruded, also using the hydraulic system, and examined. Such soil samples are frequently referred to as "undisturbed."

In coarser-grained, less cohesive soils, such as sands and clay-poor silts, a split-spoon sampler is used. Split-spoon samplers are constructed of two half-cylinders held together by threaded fittings at the top and base to form a 1.5-ft-long by 1.5-in.-diameter tube (see Figure 5.5). The split-spoon sampler is driven into the soil by repeated blows from a rig-mounted hammer and then retrieved to the surface by the rig hydraulic system. A core catcher inserted at the base of the sampler prevents unconsolidated materials from falling out of the sampler as it is removed from the borehole. Upon retrieval, the threaded ends are removed and the two half-cylinders separated to reveal the core for examination.

Due to the repeated impact as the sampler is driven to depth, split-spoon samples are disturbed samples. Original sedimentary structures such as cross-bedding are not generally preserved, and any apparent bedding structures are probably artifacts of the sampling process. The upper portion of the sample may contain materials that accumulated on the borehole floor, and these should be discarded.

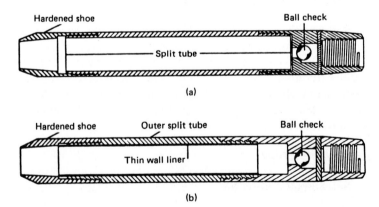

Figure 5.5 (a) Split-spoon soil sampler. (b) Split-spoon soil sampler with liner. Source: Hunt, 1984.

Rock Core Sample Collection

Samples are collected from formations of consolidated rock using rock coring barrels, which vary in size and design depending on the particular drilling conditions. Hunt (1984) provides a summary of the various designs and their applications. Generally, the core barrel is fitted at its base with a ring-shaped carbide or diamond-tipped bit that cuts the core during drilling. The coring device is rotated under pressure while drilling fluid is circulated down the drill string, through the bit, and up the annulus. The bit cuts a circular groove, leaving a column of intact rock standing within the core barrel. When the full length of the core barrel has been drilled, the core is broken from the formation and lodged in the barrel by the core lifter, and the core barrel is retrieved to the surface.

The time required to sample a borehole in a hard rock formation can be reduced by the use of a wire-line core barrel. In this application, a small-diameter core is diverted into a tube that is brought to the surface by means of a retrieving "spear" lowered on a wire line within the drill pipe, eliminating the need to trip the drill string out of the hole to retrieve the core barrel.

Core Logging

Logging of core samples for the purposes of hydrogeologic site investigations includes a description of both the geologic characteristics and the visual evidence of contamination, if present. Of primary importance from a hydrogeologic standpoint are the characteristics that influence the water-bearing capacity of the soil or rock, including grain size and secondary porosity features. Evidence of contamination may include chemical staining or odor. Core samples collected during drilling may be used for field screening or laboratory analysis, and special handling of the samples may be required, as described in Section 5.4.

Soil Core Logging

Numerous soil classification schemes have been developed for various purposes. The system most commonly used in the environmental industry is the Unified Soil Classification System (USCS) (see Figure 5.6 and Hunt, 1984). The USCS places most soils into one of two major divisions: coarse-grained soils, including gravelly and sandy soils, and fine-grained soils, including silts and clays. Highly organic soils, such as peat and humus, form a third, smaller division. In general, unconsolidated aquifers are composed of gravels, sands, and low-plasticity silts, while clays and clay-rich or highly compacted silts form aquitards or aquicludes.

In logging cores using the USCS, soils are classified on the basis of the major constituent: gravel, sand, silt, or clay. Modifiers referring to secondary constituents are used when that constituent accounts for 10% or more of the sample. Thus, a silty sand or clayey sand (SM or SC in USCS shorthand) contains greater than 50% sand and 10% or more silt or clay, respectively. If the silt or

Major Divisions			Graph symbol	Letter symbol	Typical descriptions
Coarse-grained soils More than 50% of material is larger than no. 200 sieve size	Gravel and gravelly soil More than 50% of coarse fraction **retained** on a no. 4 sieve	Clean gravels (little or no fines)		GW	Well-graded gravel-sand mixtures, little or no fines
				GP	Poorly graded gravels, gravel-sand mixtures, little or no fines
		Gravels with fines (appreciable amount of fines)		GM	Silty gravels, gravel-sand-silt mixtures
				GC	Clayey gravels, gravel-sand-clay mixtures
	Sand and sandy soils More than 50% of coarse fraction **passing** a no. 4 sieve	Clean sand (little or no fines)		SW	Well-graded sands, gravelly sands, little or no fines
				SP	Poorly graded sands, gravelly sands, little or no fines
		Sands with fines (appreciable amount of fines)		SM	Silty sands, sand-silt mixtures
				SC	Clayey sands, sand-clay mixtures
Fine-grained soils More than 50% of material is smaller than no. 200 sieve size	Silts and clays	Liquid limit **less** than 50%		ML	Inorganic silts and very fine sands, rock flour silty or clayey fine sands or clayey silts with slight plasticity
				CL	Inorganic clays of low to medium plasticity, gravelly clays, sandy clays, silty clays, lean clays
				OL	Organic silts and organic silty clays or low plasticity
	Silts and clays	Liquid limit **greater** than 50%		MH	Inorganic silts, micaceous or diatomaceous fine sand or silty soils
				CH	Inorganic clays or high plasticity, fat clays
				OH	Organic clays of medium to high plasticity, organic silts
Highly organic soils				PT	Peat, humus, swamp soils with high organic contents

Figure 5.6 Unified soil classification system.

clay content is less than 10%, the sand is either a well-graded or poorly graded sand (SW or SP), with minor or trace silt or clay. Under the USCS, clay soils are divided into high- and low-plasticity varieties based on their liquid and plastic limit values, which are determined by laboratory testing (see Hunt, 1984).

Although the precise quantification of grain size distribution and liquid and plastic limits are determined by laboratory testing, field classification of soils with borderline compositions must be based on judgment. As a rule of thumb, high-plasticity clays can be rolled between the hands to form elongate strings, while low-plasticity clays will crumble when rolled. If a soil sample that appears to be either a silty clay or a clayey silt yields free ground water, it is most likely a clayey silt.

Secondary porosity features in fine-grained soils can transmit fluids at much higher rates than through the soil formation as a whole. Slickenside fractures, horizons of woody or organic material, root zones, burrows, calcareous zones, and desiccation features frequently provide avenues of contaminant migration through low-permeability soils into underlying aquifers.

Other descriptive features such as color should be noted to the extent that they can be used to distinguish strata of similar composition from one another and may sometimes provide consistent stratigraphic markers, such as a gray clay overlying a red clay.

Rock Classification and Logging

Rocks are broadly classified in terms of their origin as igneous (those formed from a molten fluid, or magma), metamorphic (those formed by recrystallization of a preexisting rock subjected to heat and/or pressure), and sedimentary (those formed from deposition and consolidation of soil or rock particles or as chemical precipitates from mineral saturated water). Core samples should be logged in terms of their mineralogic composition, as well as their water-bearing properties (see Hunt, 1984).

Sedimentary rocks formed by accumulation and consolidation of soil or rock fragments, ranging in size from fine powders to house-sized boulders, are referred to as clastic or detrital rocks. The most commonly encountered water-bearing varieties include sandstones and conglomerates (rocks composed of grains of variable sizes, for example, gravel and sand). The porosity in these rocks may be primary (intergranular porosity, as encountered in well-sorted, poorly cemented sandstones) or secondary (fracture porosity, as encountered in well-cemented, jointed rocks), or both. Rocks formed by chemical precipitation from a mineral-saturated water body include carbonate rocks (limestones and dolomites) and the less common evaporites (halites, gypsum, and anhydrites). Porosity in these rocks is usually secondary, occurring as fractures or solution cavities. The water-bearing capacity of sedimentary rocks can also be influenced by the degree of weathering.

Surveying and Completion of Core Borings

Soil boring locations should be surveyed, either prior to drilling or upon completion, in order to permit construction of accurate geologic maps and cross sections. Locations may be surveyed relative to a site grid or to some easily identified permanent feature. The elevations of the ground surface should be surveyed to the nearest 0.1 ft relative to a common datum, such as mean sea level, or an arbitrary datum established by a site bench mark. Ground surface elevations provide a reference datum for structural geologic cross sections.

Core borings should be sealed upon completion to prevent migration of fluids from the surface down the borehole. This is especially critical when contamination is present at the surface and the soil boring has been advanced to the ground water. The grout seal may consist of neat cement, a cement–bentonite mixture, or specially developed sealing materials such as Volclay. Some state regulatory agencies have specifications for grout composition and density.

To ensure distribution of the grout over the full depth of the boring, grout should be placed using the tremie method. A length of pipe is lowered within a few feet of the base of the borehole, and the grout is pumped downhole under pressure. As the grout fills the hole, ground water is displaced to the surface. To prevent flow of contaminated ground water on the ground surface, it may be necessary to catch the return flow in a portable mud pit or other vessel. When grout is seen flowing from the hole, placement of the seal is complete, although it may settle during curing and require topping off.

Cone Penetrometer Testing

The cone penetrometer test (or CPT) can be useful for defining stratigraphy to depths of up to 100 ft (rarely more) in fine-grained soils and unconsolidated sands. In cone penetrometer testing, electronic strain gages mounted in a steel, cone-shaped probe are pushed at a constant rate into the subsurface by a truck-mounted hydraulic system (see Figure 5.7).

The gages measure the resistance encountered at the tip of the cone tool and the friction encountered along its side as the tool is advanced, producing a

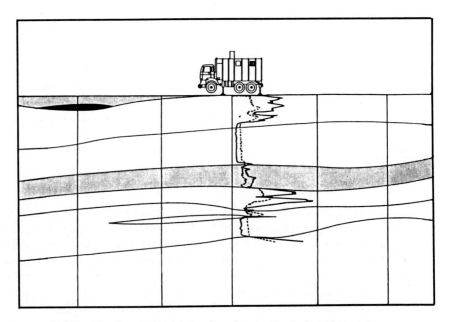

Figure 5.7 Cone penetrometer rig. Source: Fugro Geosciences, Inc.

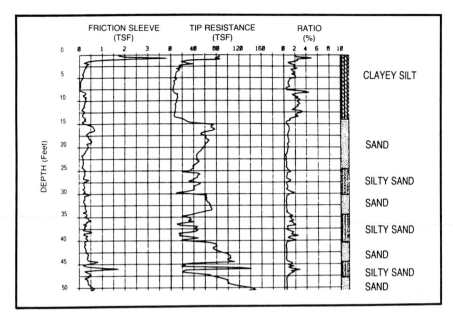

Figure 5.8 Log of cone penetrometer test.   Source: Fugro Geosciences, Inc.

continuous log of these soil properties with depth (see Figure 5.8). Empirical relationships for soils in various regions of the country have been developed to allow interpretation of soil type based on the resulting set of curves. High tip resistance and low friction indicate coarser-grained soils; lower tip resistance and higher friction indicate finer-grained soils. It is advisable to locate at least one CPT immediately adjacent to a continuously sampled soil boring to allow correlation of the data to local soil conditions.

The major advantage of the cone penetrometer is the speed and cost of data collection. At a rate of 2 cm/sec, a continuous log to a depth of 50 ft can be obtained in less than 13 min, whereas a continuously sampled soil boring to the same depth can take several hours. Footage rates for cone testing can be as little as half those of auger drilling, and there are no soil cuttings to dispose of.

Geophysical Methods

Subsurface geological conditions can also be evaluated using a variety of geophysical methods. Geophysical methods are broadly divided into surface methods and borehole methods. In surface geophysical surveys, measurements are collected at regularly spaced intervals along a traverse or on a grid. Geologic strata are differentiated by measuring the contrasting responses of differing geologic materials to physical forces, such as electricity, magnetism, or seismic energy. Borehole techniques utilize a variety of probes or sondes that measure various physical properties of the soil or rock or contrasts between the drilling fluid and the fluids in the formation.

Geophysical methods can be substituted for conventional drilling and core sampling to varying degrees and can reduce the cost of evaluating subsurface geological conditions, particularly when drilling conditions are difficult and required drilling depths are deep. Techniques employed in geophysical data acquisition and interpretation range from very simple to extremely sophisticated. Interpretations may be qualitative or quantitative, depending on the particular technique. For best results, geophysical data should be correlated to direct geologic data, that is, core samples. Regulatory agencies generally require at least some verification of these methods with actual geological data.

The main limitations of geophysical methods arise from the fact that identical responses can be caused by a variety of conditions, and a unique and definitive interpretation of the data is not generally possible. Because the data acquired by the simpler methods are not often definitive, these techniques are frequently not a satisfactory substitute for drilling. The problem of interpretation is increased by lateral and vertical heterogeneities in the geologic system, and more sophisticated techniques may be required. The relatively high costs associated with the more sophisticated techniques limit their ability to compete cost-wise with drilling and sampling at shallow depths.

TABLE 5.1 Summary of Major Geophysical Techniques for Environmental Applications

Type of geophysical survey	How method works	Detectors	Comments
Electrical surveys	A.C. current is introduced into ground; current and potential are measured at specific distances away from source.	Presence of conducting fluids, porosity	
Magnetometer surveys	Strength of magnetic field is measured at various locations on site.	Presence of different rock types: igneous rocks show response, while sedimentary rocks do not	Can also be used to detect buried metallic objects.
Seismic surveys	Seismic waves are generated by energy source (hammer or explosive charge) and measured with geophones at various locations on site.	Differences in density and elasticity of soil or rock types	Pressure-wave (P-wave) methods are more common. Shear-wave (S-wave) may provide more resolution for shallow unconsolidated units.
Borehole methods	Measuring device ("sonde") is lowered downhole to measure various properties.	Electric log: properties of fluids in borehole Spontaneous Potential: Salinity contrasts between borehole fluids and formation fluids Resistance: Different types soil or rock column Gamma: Differences in radiation between clay/shales and sand	Combined sondes can be employed to get several type of data.

Zohdy et al. (1984) and Keys and MacCary (1971) provide guidelines for the applicability, acquisition, and interpretation of surface and borehole geophysical data, respectively. Table 5.1 summarizes the major geophysical techniques that are most frequently employed for ground water investigations.

Determining subsurface geology from borehole geophysical logs requires considerable interpretation. To illustrate this, Figure 5.9 presents a comparison of borehole log responses to a stratigraphic column from the Texas Gulf Coast. To provide definitive, quantitative data, a large suite of borehole logging techniques must be run and interpreted. Such analysis is generally not cost effective compared with drilling and sampling. Therefore, in environmental applications, borehole logging is usually limited to qualitative evaluation for correlation purposes. Borehole geophysical logs should always be compared directly to at least one immediately adjacent, continuously sampled core boring.

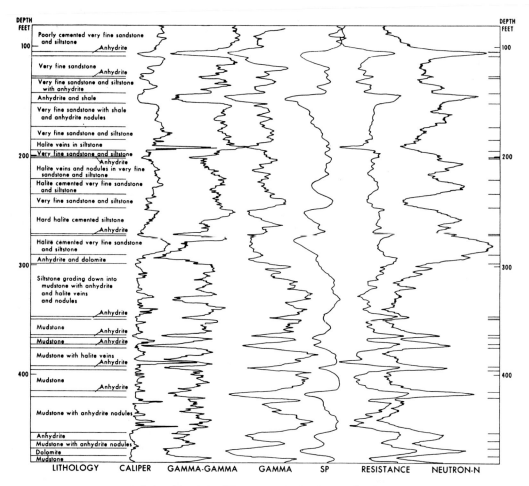

Figure 5.9 Relationship of six different geophysical logs to lithology, Upper Brazos River Basin, Texas.   Source: Keys and MacCary, 1976.

5.3 HYDROLOGIC DATA ACQUISITION

Assessment of the direction and rate of ground water flow beneath a site requires the following hydrologic data: lateral hydraulic gradient, hydraulic conductivity, and effective porosity. Of these, hydraulic gradient and conductivity are obtained by field measurement. Effective porosity (the connected pore spaces through which ground water flows) is not measured directly and is generally an estimated value (see Chapter 2). This section describes methods for field determination of hydraulic gradient and hydraulic conductivity. Because such determinations are most commonly made from measurements in piezometers and monitoring wells, we begin with a description of monitoring-well construction.

Monitoring-well Construction

The monitoring well is the primary source of hydrologic and ground water-quality data used in hydrogeologic site assessments. Most of the special requirements for

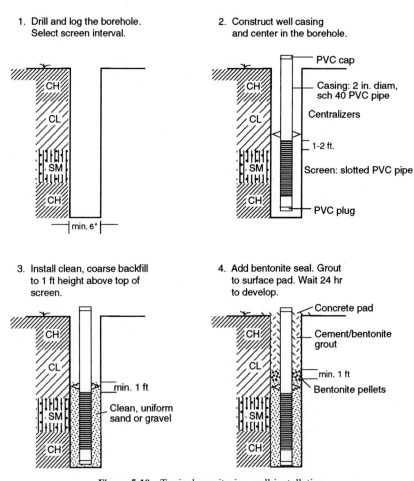

Figure 5.10 Typical monitoring-well installation.

monitoring-well construction are due to its use in the collection of ground water-quality data. For collection of hydrologic data, a piece of slotted pipe stuck into a borehole would be sufficient in most instances, but because the wells serve a dual purpose, careful attention must be paid to the materials used and the methods of construction. Many state environmental regulatory agencies have very particular construction specifications and require that well installation be performed by licensed drillers.

The essential elements of a monitoring well are the well screen and riser, the filter pack, and the seal (see Figure 5.10). Well screens, usually sections of slotted pipe, allow water to flow from the formation into the well while screening out most soil particles. The riser, sometimes referred to as casing, is the solid-walled pipe that connects the well screen with the surface. Silt traps or sumps, consisting of a short section of riser beneath the well screen, are frequently installed to prevent sediment passing through the screen section from accumulating in the screened interval and clogging it. A bottom cap or plug at the base of the screen or pump prevents the flow of sediment into the bottom of the well. As described later, the filter pack, or gravel pack, creates a zone of enhanced permeability around the well, thus aiding inflow of ground water. Above the gravel pack a seal composed of bentonite, cement, or other low-permeability material prevents fluids from above the screened formation interval (including percolating rainwater) from entering the well. A concrete surface pad is installed to protect the grout seat from weathering. The well is protected from damage and tampering by a locking protective cover and, where appropriate, guard posts.

Well Materials

Monitoring-well screens and risers are most frequently constructed of 2- or 4-in.-diameter schedule 40 PVC. Two and 4-inch diameter wells are usually installed within 6- and 10-in.-diameter boreholes, respectively. The pipe joints should be threaded, since PVC glues contain volatile organic compounds and their use is generally not permissible.

PVC reacts with some contaminants and so is not always a suitable material for monitoring-well construction. Contact with some nonaqueous-phase liquids (NAPLs) or high concentrations of some contaminants can attack PVC, potentially compromising ground water samples or even damaging the well. Stainless steel is frequently used under such conditions, at significantly greater expense. Information on the compatibility of various well materials with common contaminants can be found in Driscoll (1986).

Screen Interval Selection

The purpose of the well and the hydrogeologic setting determine the selections of the well screen interval. In confined aquifers, well screens are most often positioned with the top of the screen 1 to 2 ft above the top of the aquifer. In unconfined aquifers, the top of the screen is generally placed 1 to 2 ft above the highest expected water-table elevation. In deeper wells, centralizers are frequently placed above the screen section to prevent it from contacting the borehole wall.

Placement of the screen across the top of the water-bearing zone permits detection of floating NAPLs. If the presence of dense nonaqueous-phase liquids (DNAPLs) is under investigation, the well screen should be positioned to intersect the base of the aquifer (see Chapter 11). Screen length should likewise be chosen based on the objectives of the investigation. Long screen sections will yield water-level elevations and water samples representing an average of conditions across their length, while shorter screens yield more depth-specific data. However, monitoring wells are rarely designed with more than 15 to 20 ft of screen to minimize the potential for cross-contamination and vertical movement of contaminants down through the well. The top of the well should also be fitted with a threaded or slip-on cap.

Installation Procedures

When wells are drilled using hollow-stem augers, the well screen and riser are lowered within the hollow stem. For rotary-drilled wells, the drilling mud should be thinned by dilution with potable water to the extent practical prior to well installation. The screen and riser sections are screwed together and lowered in the borehole. Final placement should be based on careful measurement.

Once the screen and riser have been positioned, a filter pack, or gravel pack, is installed within the annulus around the well screen (see Figure 5.10). The filter pack creates a zone of enhanced permeability around the well screen and aids in the collection of ground water samples free of fine sediment.

Usually, the gravel pack is composed of graded, preferably well-rounded pure silica sand. Blasting sand and other general-use sands may contain minerals that adsorb dissolved metals, potentially compromising the integrity of the ground water sample. The grain size interval of the filter pack material should be selected based on the analysis of aquifer grain size distribution as described by Driscoll (1986). The filter pack is generally placed from the base of the well to 1 to 2 ft above the top of the screen. In wells drilled by hollow stem, the filter pack material is frequently poured down the inside of the augers. The auger sections are pulled from the hole one at a time as the annulus is filled with sand. In deeper wells and wells drilled by wet rotary, to ensure that the gravel pack is placed uniformly over the length of the screen interval, it is frequently placed using the tremie method. The filter material is washed down a pipe lowered to the base of the well.

Following placement of the filter pack, the well is sealed to the ground surface to prevent migration of fluids from the surface or water-bearing zones above the screened interval down the borehole. A 1- to 2-ft-thick layer of bentonite in granular or pellet form is usually placed atop the filter pack. Once hydrated, the bentonite swells to form a seal and protect the filter pack from invasion by the grout, which completes the seal to the ground surface. Grout may be composed of neat Portland cement, a mixture of cement and powdered bentonite, or other specialty materials such as Volclay. Grout is frequently placed using the tremie method to ensure even placement up the borehole. Monitoring wells are usually completed at the surface with a locking steel protective casing and concrete surface pad.

Survey Specifications

The elevations of the ground surface and top of well casings should be surveyed at all well locations relative to a common datum, such as mean sea level, or an arbitrary datum established by a site bench mark. The top of casing elevations should be surveyed to the nearest 0.01 ft. For ground-surface elevations, the nearest 0.1 ft is usually sufficient. Top of casing elevations are required to convert depth to water measurements to static water-level elevations for the construction of potentiometric surface maps. The actual point of measurement should be marked on the top of the well casing to allow collection of accurate water-level data.

Protective Casing Requirements

To establish the vertical extent of ground water contamination, it is frequently necessary to drill monitoring wells through a contaminated upper zone into an uncontaminated lower zone. This may mean drilling through the upper portion to the lower portion of a single aquifer or through a contaminated upper aquifer into an uncontaminated lower aquifer. In such cases, it is necessary to first install isolation casing to prevent entrainment of contaminants from the upper zone to the lower zone during drilling.

A relatively large diameter hole is first drilled through the zone to be isolated, and an appropriately sized length of riser pipe is grouted in place in the hole. For example, to accommodate a 2-in.-diameter well, which is typically installed within a 6-in.-diameter borehole, an 8-in.-diameter isolation casing will need to be installed. This will require the drilling of at least a 10-in. and preferably a 12-in.-diameter borehole.

The isolation casing should be fitted at the bottom with a watertight cap or plug to prevent contaminants from entering the casing, and the casing is filled with water to overcome buoyancy and allow setting at the correct depth. The grout should be placed in the annular space by the tremie method and allowed to set up for 24 hr before drilling proceeds. During that time, sufficient weight, generally supplied by the drill rig, must be kept on the casing to prevent it from rising in the hole. Once the grout has set up, the water is pumped from the casing and drilling proceeds. Naturally, the bottom cap must be composed of a drillable material such as PVC.

Well Development Procedures

Prior to collection of ground water samples, monitoring wells must be developed by the removal of some minimum quantity of water. Some work plans specify the volume to be removed. Others require that development continue until the pH of the well discharge has stabilized. If the well has been installed in a low-permeability aquifer using a dry drilling method, bailing out three to ten casing volumes may be sufficient to permit collection of representative ground water samples. If fluids have been introduced during drilling, larger volumes of water must be removed. Wells that are to be equipped with pumps, whether for ground water sampling or control of ground water flow gradients, require development sufficient to remove a large fraction of the fine materials from the filter pack and the formation adjacent to the borehole to prevent clogging of the pump and well screen.

Typical development consists of a combination of pumping and surging. Surging the well, by running a close-fitting cylinder up and down the inside of the well over the screened interval, causes a back-flushing action in the gravel pack around the well, loosening fine sediment. Pumping from the well (preferably at a rate somewhat higher than the expected normal pumping rate) pulls fine sediment through the well screen into the well, where it can be pumped to the surface.

Static Water-level Surveys

Following well development and a stabilization period of sufficient length to allow water levels within the well to return to static conditions, static water-level surveys are conducted to permit estimation of ground water flow gradients. The water level in each site well should be measured to the nearest 0.01 ft using an electric water-level sounding device or other appropriate instrument. Static water level is obtained by subtracting the depth to water from the top of casing elevation. The water level probe is rinsed with deionized water prior to each measurement to prevent cross-contamination between wells.

Ideally, static water-level surveys represent aquifer conditions at one instant in time. Since "static" water levels within wells can change in response to such factors as changes in barometric pressure or tidal influence, measurements should be made in as short a time as possible. It is advisable, particularly on sites with large numbers of wells that may require several hours to survey, to remeasure the first well measured at the end of the survey to detect possible changes in the potentiometric surface during the period of the survey. If significant changes have occurred, most often the potential error in the elevation data is simply noted. In some cases, a correction factor might be applied to the water-level data. If more than one instrument is to be used in the survey, a common well should be measured simultaneously using each instrument to confirm that all instruments give the same reading.

If a floating free-phase layer is present within the well, the top of the hydrocarbon layer as well as the water level should be measured and the thickness of the floating hydrocarbon layer calculated. Care must be taken during measurement to disturb the floating layer as little as possible. Due to buoyancy contrasts, a correction factor must be applied to the thickness of the hydrocarbon layer to provide a corrected water depth. For gasoline, a density correction of 0.75 is often applied. Thus, if the depth from the top of the well casing to the top of gasoline layer is 10.00 ft, and the depth to the top of the water is 12.00 ft, the corrected depth to water is $12.00 - (2.00 \times 0.75) = 10.50$ ft. This value is then subtracted from the top of the casing elevation to obtain the corrected static water-level elevation. It should be noted that depending on site conditions, the thickness of a "free-product" layer in a well may not necessarily reflect an equivalent accumulation in the adjacent formation.

Upon completion of the survey, static water-level elevations relative to a common datum are plotted on a scaled site map. Lines of equal water-level elevation are drawn. Contours should be spaced evenly between data points and dashed where inferred, for example, beyond the polygon defined by the outermost measuring points. In the absence of interferences due to active ground water recharge and discharge features and lateral variation in aquifer permeability, most ground water flow gradients are nearly planar features. If the resulting contour map indicates highly irregular flow pattern and no hydrologic reason for it, mismeasurement should be suspected. It may be necessary to repeat the static water-level survey or check the top of the casing elevation survey.

Aquifer Tests

Single-well Slug Tests

Single-well slug tests are the most common tool for the estimation of hydraulic conductivity in hydrogeologic site assessments. Two major varieties, rising-head tests and falling-head tests, can be used. Because falling-head tests necessitate the addition of water to the well and are more difficult to perform and analyze, rising-head slug tests, described here, are more commonly performed.

The rising-head test is conducted by first measuring the static water level in the well and then lowering a rod of known displacement inside the well to just below the static water level (see Figure 5.11). When the water level has reequilibrated, the rod is pulled rapidly from the well, causing a sudden drop in the water level. The return of the water level to static conditions is then monitored. Monitoring the water level can be done by hand in low-permeability systems. Higher yield systems may recover too quickly to permit manual collection of the most critical early data and may require the use of pressure transducers placed in the well and monitored with an electronic data logger.

The resultant change in head over time is plotted on semilog paper, and the curve is analyzed according to one of several methods, depending on aquifer and well conditions. The method of analysis will depend on such factors as whether the aquifer is confined or unconfined and what percentage of the saturated interval is screened in the well. Analytical methods for slug tests are described in more detail in Chapter 3.

Slug tests evaluate only the portion of the aquifer immediately surrounding the tested well. Therefore, tests should be performed at a representative selection of site wells to provide a more accurate estimate of hydraulic conductivity for the aquifer.

Pump Tests and Well Performance Tests

While slug tests can provide reasonable estimates of hydraulic conductivity at little cost, they test only a small portion of the aquifer and can be misleading if used exclusively in the detailed design of ground water remediation systems.

1. Measure depth to water (DTW) in well.

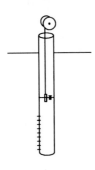

2. Lower displacement rod below water surface. Monitor DTW until static water level returns to initial reading.

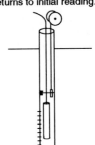

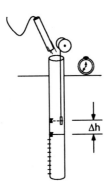

3. Rapidly pull the displacement rod from the well, start stopwatch, and measure DTW at close time intervals (i.e., every 15 sec to 2 min, every 30 sec to 5 min, every 1 min to 10 min, etc.) until water level has risen at least 90% of the distance back to the initial level.

4. Plot water level recovery versus time on semi log paper. Calculate aquifer type and well configuration.

Figure 5.11 Rising-head slug test procedure.

Constant-rate aquifer pumping tests are used to provide estimates of aquifer transmissivity and hydraulic conductivity, which reflect conditions over a larger portion of the aquifer. They also provide the means of calculation of the aquifer storage coefficient, which slug tests do not. Because these tests are relatively expensive to conduct, they are usually deferred until the detailed design phase of a ground water remediation system.

Stepped-rate well performance tests provide an estimate of the specific capacity of a pumping well from which transmissivity can be estimated. A small-scale well performance test can be conducted during well development at relatively little extra cost. A well stepped-rate test, also referred to as a step-drawdown test, is generally conducted before a constant-rate aquifer pump test to allow selection of the optimal pumping rate for the pump test.

During a stepped-rate test, the well is pumped at a constant rate until the water level in the well stabilizes. The pumping rate is then increased and the water level observed until it stabilizes again. For each pumping rate, a specific capacity for the well is calculated by dividing the well discharge by the amount of drawdown that occurs (Q/s). Transmissivity (gpd/ft) can then be estimated from the empirical relationship $Q/s = T/2000$ for a confined aquifer or $Q/s = T/1500$ for an unconfined aquifer.

Unlike slug tests or well performance tests, constant-rate aquifer pumping tests provide a measurement of aquifer characteristics beyond the immediate vicinity of a single well. A well is pumped at a constant rate for a fixed period, generally 24 hr for a confined aquifer to several days for an unconfined formation. Drawdown of water levels is observed in the pumping well and one or more observation wells. Upon completion of the pumping period, the pump is turned off and the recovery of water levels is monitored in the pumping and observation wells.

Analysis of constant-rate aquifer pumping tests is performed by plotting water-level drawdown over elapsed pumping time in individual wells and drawdown versus distance from the pumping well in two or more observation wells at a given time. The slopes of the resulting curves are used to calculate aquifer transmissivity and storativity. Well efficiency and radius of influence can also be calculated. Each of these values is necessary for the design of an efficient ground water remediation system. The more common methods of analysis are summarized in Chapter 3.

5.4 ACQUISITION OF SOIL AND GROUND WATER QUALITY DATA

Chemical analysis of soil and ground water samples is required to assess the nature of the contaminant and to delineate its lateral and vertical extent. This section describes procedures for the collection and handling of soil and ground water samples. The following discussion is intended only as a general guide. The project work plan should specify sampling and analysis protocol based on project goals and applicable state regulations or guidelines.

General Sampling Procedures

To assure collection of data that accurately represent site conditions, proper protocol must be followed during sample collection and handling. In addition to providing design basis information for site cleanup, the data must also be of a quality acceptable to regulatory agencies. Data may also be admitted as evidence in legal proceedings, and their integrity must withstand the scrutiny of opposing legal counsel.

Measures must be taken to prevent the loss of contaminant mass from the sample, for instance, by volatilization or biodegradation. Samples should be collected in appropriate sample containers (frequently specified by the analytical method) and appropriately preserved. Equally important is preventing the introduction of contaminants into the sample from some other source (cross-contamination). Sampling equipment should be thoroughly cleansed between sample locations. Where possible, locations should be sampled in order of increasing contaminant concentration to prevent cross-contamination between locations.

Most analyses have specified maximum holding times between sample collection and analysis. Samples should be transported to the laboratory as expeditiously as possible, generally within 24 hr of collection. The shipment of samples must be tracked with and accompanied by a chain-of-custody form that includes the signatures and affiliations of the personnel relinquishing and receiving the samples, as well as the date and time of the transfer. Essential sample collection data (sample identification number, date and time of collection) and the specifications of the laboratory program must also be included.

The same collection and handling protocol described next should be regarded as a representative minimum standard procedure for the collection of representative samples. Certain analytical methods, state regulations, or the particular needs of the project may impose additional requirements on the sample-collection process.

Soil Sample-collection and Handling Procedures

Soil samples should always be collected using clean sample tools composed of nonreactive materials, such as stainless-steel hand augers or trowels when collection is done by hand or thin-walled Shelby tube samplers or split spoons when collection is performed during drilling operations. Tools should be thoroughly cleansed with steam, pressured hot water, or an appropriate detergent solution and clean-water rinse prior to sample collection and between sampling locations.

Soil samples should be collected in clean glass jars (unless otherwise specified by laboratory procedure) with tight-fitting lids and identified with an appropriate label bearing the unique identification number of the sample. The outer surface of soil cores should be trimmed using a clean knife prior to shipment to the laboratory to ensure that soils that have been in contact with formation fluids above the sample point or with the inside wall of the sampler are not tested.

Ground Water Sample Collection and Handling

Prior to collection of ground water samples, static water-level elevations in all relevant wells are measured as described previously. The volume of water in casing storage in the well is then calculated using the well depth, inside casing diameter, and depth to water, to determine the required purge volume.

Wells should also be checked for the presence of free-phase hydrocarbon layers using an electric hydrocarbon interface probe or transparent bailer. The thickness of the hydrocarbon layer at each location should be measured to the nearest 0.01 ft. Ground water samples from wells containing a free-product layer may not accurately reflect the dissolved-phase concentration and are frequently not analyzed.

Prior to sampling, wells are purged using a clean bailer or pump to ensure collection of samples representative of formation conditions. Removal of three to five well casing volumes is the generally accepted minimum purge. More stringent protocol may require periodic measurement of pH and purging until pH values cease to change. Samples should be collected within 24 hr of well purging. Collection of contaminated purge water for proper treatment or disposal is frequently required.

To minimize the potential for cross-contamination, monitoring wells should be sampled in order of increasing contaminant concentration, to the extent that this can be determined prior to sampling. The pumps or bailers used for well purging and sampling should be of materials such as stainless steel or Teflon that will not react with the contaminants in the sample. Clean, chemical-resistant gloves should be worn and changed between samples.

Bailers and pumps should be scrubbed in an appropriate detergent solution and thoroughly rinsed in clean water, with a final rinse in deionized water, prior to use in each well. The string or rope used to lower the bailer should be changed between sampling each well, and care should be taken to prevent contact between the ground and all sampling and purging equipment. The use of dedicated sampling equipment, that is, a pump or bailer for each well, further reduces the potential for cross-contamination, as well as the time and effort required for decontamination procedures.

Sampling Quality-control Measures

Field blank samples are often prepared to provide a quality-control check on field sampling procedures. This is most often done at the end of the sampling episode, but can be done at any point in the program, depending on the specific requirements of the sampling program. The bailer should be subjected to exactly the same decontamination procedures used prior to every other sampling point. A sample of deionized water is then decanted from the bailer or run through the sampling pump and submitted for laboratory analysis.

The sampling and analysis program may specify that sample kits packed by the lab may contain a trip blank. This is a blank sample prepared in the laboratory, which accompanies the other samples through the analytical process. It provides a more direct quality-control check on the laboratory than does a field blank, because it is not affected by the sample-collection process.

Ground water samples collected for analysis of volatile organic constituents must be collected free of head space (air bubbles) to prevent volatile constituents from coming out of solution. Note, too, that excessive fine sediment content can interfere with some analyses, resulting in unacceptably high detection limits. Acids used in the preservation of samples collected for heavy metal analyses can leach naturally occurring metals from sediments in the sample, yielding false ground water sample results. It may be necessary to filter such samples prior to the addition of preservatives.

Temporary Ground Water Sampling Points

Ground water samples are most commonly collected from monitoring wells. As permanent installations, monitoring wells are necessary to permit resampling and evaluation of changing site conditions. However, the installation, completion, and development of monitoring wells is fairly expensive. To reduce the costs of a ground water plume delineation program, the use of temporary ground water sampling points is becoming increasingly common. These may be installed using a drill rig or cone penetrometer rig and provide ground water samples in a fraction of the time required to install a well. Following plume delineation, a relatively small number of permanent wells can be installed to confirm plume boundaries and facilitate future monitoring.

Field Testing Methods

To aid in the preliminary assessment of the limits of the contaminated zone and to facilitate selection of samples for laboratory analysis, a variety of field testing methods is available for soil and ground water. Some methods can provide quantitative data that are roughly repeatable in the laboratory, although at higher detection limits. Others, while yielding a concentration value, for practical purposes provide merely qualitative results, indicating the presence of the constituent. In either case, confirmation of contaminant concentrations by an approved laboratory method is generally required. Some service companies provide mobile laboratories that follow EPA-approved laboratory Quality Assurance/Quality Control (QA/QC) procedures and can provide results on site within a few hours of collection.

Soil Gas Surveys

Soil gas surveys are a means of evaluating the presence of volatile organic constituents present in near-surface soils and ground water without actually collecting soil or ground water samples. A probe mounted on hollow pipe is pushed into the soil to the desired depth. Vapors are extracted from the probe by means of a vacuum pump and analyzed at the surface by a detection instrument, such as an organic vapor analyzer or portable gas chromatograph (GC).

Because the results of vapor analyses are affected by a wide range of variables and are frequently not acceptable to regulatory agencies as definitive of site conditions, soil gas surveys are most commonly used as a qualitative tool for assessing site conditions prior to a drilling and sampling program.

Soil Sample Headspace Vapor Analyses

An organic vapor analyzer or portable GC can also be used to measure the organic vapors in a soil sample jar headspace. Samples collected for this purpose must be sealed tightly in a partially filled jar and either agitated or allowed to sit for a period to permit constituents to volatilize. If laboratory analysis of volatile organics is also to be performed, the sample should be split and a portion properly stored for shipment to the laboratory.

The results of such headspace analyses depend on variables such as soil type, the volatility of the constituents present, the presence of NAPLs, the temperature of the sample, residence time in the sample container, and the volume of soil versus headspace in the container. Accordingly, they should be viewed as qualitative, providing a relative measure of the constituents present in the samples.

Colorimetric Soil Tests

Contaminant-specific test methods are also available for hydrocarbon compounds in soils. Like laboratory analysis of soil samples, these methods consist of an extraction procedure and analysis of a liquid extract sample. The cost of the test kit and the time required to perform the extraction and analytical procedures must be weighed against the cost of a more definitive laboratory analysis to ascertain whether such tests are worth conducting.

Ground Water Field Measurements

Ground water field measurements usually include such general water-quality parameters as temperature, specific conductivity, and pH. Redox potential (Eh) and dissolved oxygen content are also frequently measured at the time of sample collection. Such data are used in assessing the potential for biodegradation and for the design of treatment systems. Water that has contacted field measurement probes or containers used in conjunction with them should not be included in the sample collected for laboratory analysis.

Ground Water Colorimetric Tests

Contaminant-specific field test kits for ground water are available for a much wider variety of constituents than for soil. Because the extraction procedure is not required, ground water field testing is simpler and faster to perform and may be more reliable than field testing of soil samples. Used in conjunction with a temporary ground water sampling installation, such as those described previously, the use of these methods can result in significant savings to the project budget by preventing the installation of an unnecessary number of monitoring wells.

Field screening tests generally do not have detection limits sufficiently low to provide definitive ground water plume delineation. The quality control and repeatability of the data are also inadequate for regulatory purposes in most instances. Therefore, results are generally confirmed by laboratory analysis of duplicate samples.

SUMMARY

This chapter summarizes field methods for acquisition of geologic, hydrologic, and soil and ground water quality data commonly used in hydrogeologic site investigations. Topics include drilling methods; description of geologic materials; installation of ground water monitoring wells; aquifer permeability testing; and proper handling of soil and ground water samples for contaminant analysis. Emphasis is on the more conventional and commonly used techniques. Less common methods, such as geophysical assessment of stratigraphy and field tests for ground water quality, are also presented.

REFERENCES

BOUWER, HERMAN, 1989, "The Bouwer Rice Slug Test—An Update," *Groundwater* 27(3): 304–309.

BOUWER, HERMAN, and RICE, R. C., 1976, "A Slug Test for Determining Hydraulic Conductivity of Unconfined Aquifers with Completely or Partially Penetrating Wells," *Water Resources Res.* 12:423–428.

DRISCOLL, F. G., 1986, *Groundwater and Wells,* Johnson Division, St. Paul, MN.

FUGRO GEOSCIENCES INC., from vendor brochure, 1993, Houston, TX.

HUNT, R. E., 1984, *Geotechnical Engineering Investigation Manual,* McGraw-Hill, New York.

HVORSLEV, M. J., 1951, "Time Log and Soil Permeability in Groundwater Observations," *U.S. Army Corps of Engineers Waterways Experimental Station Bulletin 36,* Vicksburg, MS.

KEYS, W. S., and MACCARY, L. M., 1976, *Application of Borehold Geophysics to Water Resources Investigations,* U.S. Government Printing Office, Washington, DC.

SCALF, M. R., MCNABB, J. F., DUNLAP, W. J., COSBY, R. L., and FRYBERGER, J., 1981, *Manual of Groundwater Sampling Procedures,* National Water Well Association, Worthington, OH.

U.S. DEPARTMENT OF THE INTERIOR, BUREAU OF RECLAMATION, 1985, *Groundwater Manual,* U.S. Government Printing Office, Washington, DC.

ZOHDY, A. A. R., EATON, G. P., and MAYBEY, D. R., 1984, *Application of Surface Geophysics to Groundwater Investigations,* U.S. Government Printing Office, Washington, DC.

Chapter 6 | CONTAMINANT TRANSPORT MECHANISMS

6.1 INTRODUCTION

This chapter explores transport concepts relating to the migration and fate of contaminants from hazardous-waste sites dissolved in ground water. The discussion will include case studies with applications for environmental engineers, hydrologists, hydrogeologists, and others in the ground water area. The approach considers the advective and dispersive transport of solutes dissolved in ground water, which may undergo simple first-order decay or linear adsorption. Much effort has been devoted to solving the governing partial differential equations over the past two decades. Both one-dimensional (soil column) and two-dimensional (landfill plume) problems have generally been addressed. Mathematical derivations of the governing mass transport equations are reviewed in Section 6.4 and have been presented earlier by Ogato (1970), Bear (1972, 1979), and Freeze and Cherry (1979). This chapter deals primarily with analytical and semianalytical solutions, whereas Chapter 10 presents numerical approaches in detail.

Several reviews of solute transport modeling have been written at an introductory level, such as those by Mercer and Faust (1981), Wang and Anderson (1982), and Fetter (1993). Freeze and Cherry (1979) in their classic text offer clear descriptions of many transport mechanisms and derive many of the governing transport equations. Domenico and Schwartz (1990) present a number of analytical equations and solutions in their recent text on physical and chemical hydrogeology.

The concepts presented in this chapter serve as a synthesis of existing theory highlighted by examples and case studies for understanding and predicting migration and transport of solutes in ground water from hazardous-waste sources. The main transport processes of concern in ground water include advection, diffusion, dispersion, adsorption, biodegradation, and chemical reaction. **Advection** is the movement of contaminants along with flowing ground water at the seepage velocity in porous media. **Diffusion** is a molecular mass-transport process in which solutes move from areas of higher concentration to areas of lower concentration. **Dispersion** is a mixing process caused by velocity variations in the porous media. Dispersion causes sharp fronts to spread out and results in the dilution of the solute at the advancing edge of the contaminant front (Figure 6.1). **Adsorption,** the partitioning of organic contaminants from the soluble phase onto the soil ma-

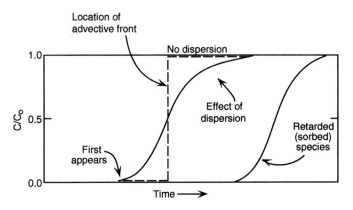

Figure 6.1 Breakthrough curves in one dimension showing effects of dispersion and retardation.

trix, results in retarded fronts (Chapter 7). **Biodegradation** represents the transformation of certain organics to simple CO_2 and water in the presence of microbes in the subsurface. Biodegradation kinetics and other chemical reactions in ground water are covered in Chapter 8. Transport mechanisms in the unsaturated zone are described in Chapter 9. Recent attention also includes a consideration of non-aqueous-phase liquids (NAPLs), which can be major sources of contamination in the subsurface. These are defined as separate phase products, such as gasoline, which can float just above the water table. Chlorinated solvents such as TCE, which are denser than water, can sink to the bottom of an aquifer and cause a source of contamination for many years (Chapter 11).

Advection, diffusion, dispersion, adsorption, and biodegradation have been analyzed in some detail under both laboratory and field conditions. Soil columns have been set up in the laboratory and loaded with numerous organic compounds similar to those found at hazardous-waste sites. Dispersion coefficients and rates of adsorption (retardation) and biodegradation have been evaluated with some success in these columns. Figure 6.1 shows the typical response (breakthrough curve) of a soil column loaded with a tracer such as chloride or retarding organic at one end and monitored at the outlet as a function of time.

Field efforts in the mid 1980s attempted to quantify dispersion, adsorption, and biodegradation mechanisms. These processes are the focus of current research efforts at many hazardous-waste research sites. One of the most successful field sites for tracer studies has been the Borden landfill site in Canada, which resulted in a series of important papers on dispersion and adsorption processes measured during a 2-year field experiment (Mackay et al., 1986; Roberts et al., 1986). LeBlanc et al. (1991) describe a natural gradient tracer experiment performed at Otis Air Base on Cape Cod, Massachusetts, where more than 30,000 samples were analyzed over a 2-year period. Detailed comparisons with the results from the Borden landfill indicate many similarities (Section 6.10).

Borden and Bedient (1986), Rifai et al. (1988), and Berry-Spark and Barker (1987) reported some success on measuring and modeling the biodegradation of contaminant plumes associated with naphthalene and benzene-related compounds in ground water. Several of these field projects are described in detail in later sections of the text.

The incorporation of the preceding transport mechanisms into ground water models for the prediction and evaluation of waste sites has been described in many references over the past two decades. Some of the earliest efforts are presented in Bear (1972), Bredehoeft and Pinder (1973), Fried (1975), Anderson (1979), Bear (1979), and Freeze and Cherry (1979). More recently, Domenico and Schwartz (1990), Mercer (1990), Anderson and Woessner (1992), and Fetter (1993) provide discussion of some of the more complex flow and transport issues facing the ground water analyst.

6.2 ADVECTION PROCESSES

Advection represents the movement of a contaminant with the flowing ground water according to the seepage velocity in the pore space, which was defined in Eq. (2.5).

$$v_x = \frac{K}{n} \frac{dh}{dL}$$

It is important to realize that seepage velocity equals the average linear velocity of a contaminant in porous media and is the correct velocity for use in governing solute transport equations. As defined earlier, the average linear velocity equals the Darcy velocity divided by effective porosity, n, associated with the pore space through which water can actually flow. The average linear velocity, or seepage velocity, is less than the microscopic velocities of water molecules moving along individual flow paths, due to tortuosity. The one-dimensional mass flux (F_x) due to advection equals the product of velocity and concentration (C) of solute, or $F_x = v_x n C$.

For certain cases in the field, an advective model provides a useful estimate of contaminant transport. Some models include the concept of arrival time by integration along known streamlines (Nelson, 1977). Steamline models are used to solve for arrival times of particles that move along the streamlines at specified velocities, usually in a two-dimensional flow net (Section 6.8). Others have set up an induced flow field through injection or pumping and evaluated breakthrough curves by numerical integration along flow lines. Dispersion is not directly considered in these models, but results from the variation of velocity and arrival times in the flow field (Charbeneau, 1981, 1982). In cases where pumping of ground water dominates the flow field, it may be useful to neglect dispersion processes without loss of accuracy. In Chapter 10, it will be shown how advection is treated numerically by particle mover models, which represent the velocity of individual particles moving according to the local velocity field in the x or y direction.

6.3 DIFFUSION AND DISPERSION PROCESSES

Diffusion is a molecular-scale process that causes spreading due to concentration gradients and random motion. Diffusion causes a solute in water to move from an area of higher concentration to an area of lower concentration. Diffusive transport can occur in the absence of velocity. Mass transport in the subsurface due to diffusion in one dimension can be described by Fick's first law of diffusion,

$$f_x = -D_d \frac{dC}{dx} \tag{6.1}$$

where f_x = mass flux ($M/L^2/T$)
$\quad D_d$ = diffusion coefficient (L^2/T)
$\quad dC/dx$ = concentration gradient ($M/L^3/L$)

Diffusion is usually only a factor in the case of very low velocities, such as in a tight soil or clay liner, or in the case of mass transport involving very long time periods. Typical values of D_d are relatively constant and range from 1×10^{-9} to 2×10^{-9} m²/sec at 25°C. Typical dispersion coefficients in ground water are several orders of magnitude larger and tend to dominate the spreading process when velocities are present.

A soil column similar to the one used by Darcy can be used to introduce the concept of advective–dispersive transport. The column is loaded continuously with tracer at relative concentration of $C/C_0 = 1$ across the entire cross section. Figure 6.1 shows the resulting concentration versus time response measured at $x = L$, called the breakthrough curve. The step function loading is begun at $t = 0$, and the breakthrough curve develops due to dispersion processes that create a zone of mixing between the displacing fluid and the fluid being displaced. The advective front (center of mass) moves at the average linear velocity (or seepage velocity) and $C/C_0 = 0.5$ at that point, as shown in Figure 6.1. Velocity variations within the soil column cause some of the mass to leave the column in advance of the advective front and some lags behind, producing a dispersed breakthrough curve in the direction of flow. The size of the mixing zone increases as the advective front moves farther from the source, but at some distance behind the advective front the source concentration C_0 is encountered, and it remains at steady state in this region. If there were no dispersion, the shape of the breakthrough curve would be identical to the step input function.

Dispersion is caused by heterogeneities in the medium that create variations in flow velocities and flow paths. These variations can occur due to friction within a single pore channel, to velocity differences from one channel to another, or to variable path lengths. Laboratory column studies indicate dispersion is a function of average linear velocity and a factor called dispersivity, α. Dispersivity in a soil column is on the order of centimeters, while values in field studies may be on the order of one to thousands of meters. Figure 6.2 shows the factors causing longitudinal dispersion (D_x) of a contaminant in the porous media.

Mass transport due to dispersion can also occur normal to the direction of flow. This transverse dispersion, D_y, is caused by diverging flow paths in the

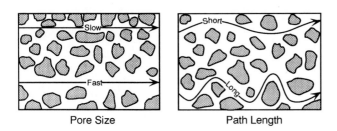

Pore Size

Path Length

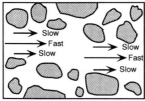

Friction in Pore

Figure 6.2 Factors causing longitudinal dispersion.

porous media that cause mass to spread laterally from the main direction of flow. In most cases involving a two-dimensional plume of contamination, D_y is much less than D_x, and the shape of the plume tends to be elongated in the direction of flow.

Freeze and Cherry (1979) defined hydrodynamic dispersion as the process in which solutes spread out and are diluted compared to simple advection alone. It is defined as the sum of molecular diffusion and mechanical dispersion, where much dispersion is caused by local variations in velocity around some mean velocity of flow.

Dispersion in two dimensions causes spreading in the longitudinal (x) and transverse directions (y) both ahead of and lateral to the advective front. Many typical contaminant plumes in ground water are represented by two-dimensional advective–dispersive mechanisms. Figure 6.3 depicts the normal shape of such a plume in two dimensions compared to advection alone. Longitudinal dispersion causes spreading and decreases concentrations near the frontal portions of the plume. The main difference with the one-dimensional soil column results described

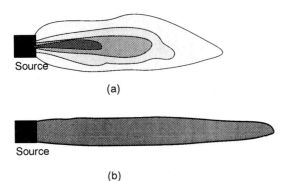

Figure 6.3 (a) Advection and dispersion. (b) Advection only.

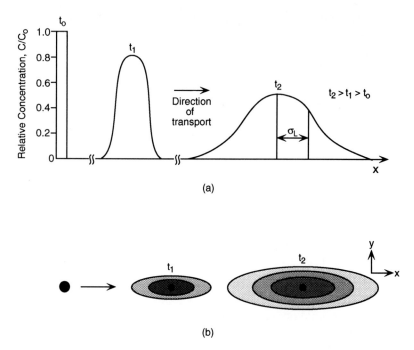

Figure 6.4 Instantaneous (pulse) source. The shapes can be represented by Gaussian (normal) distributions. (a) One dimension. (b) Two dimensions.

is that two-dimensional plumes have transverse spreading, which reduces concentrations everywhere behind the advective front.

For the case of an instantaneous or pulse source (such as a sudden release or spill of contaminant into the subsurface), the variation in concentration with time and space is shown in Figure 6.4. It can be seen that the concentration distributions start out as sharp fronts and are smoothed out as concentrations are diluted through dispersive mixing. The center of mass is advected at the average linear velocity. Concentrations can be described by Gaussian or normal distributions in either one-, two-, or three-dimensional geometries. Equations for prediction of pulse source concentrations as a function of space and time are presented in Sections 6.5 and 6.7. From the characteristics of the normal curves in Figure 6.4, we can show that dispersion cofficients can be related to the variance of the distributions: $D_x = \sigma_x^2/2t$ and $D_y + \sigma_y^2/2t$. Thus, by measuring the spread or variance in a plume, we can estimate a value for D_x and D_y at a field site. Dispersion measurements are discussed in more detail in Sections 6.9 and 6.10.

6.4 MASS-TRANSPORT EQUATIONS

Dispersive flux in a flow field with average linear velocity components v_x and v_y can be written in terms of the statistical fluctuations of velocity about the average, v_{x*} and v_{y*}. For the case of a uniform flow field where $v_x = \bar{v}_x$, a constant, and

$v_y = 0$, dispersive flux is assumed proportional to the concentration gradient in the x direction:

$$f_{x*} = nCv_{x*} = -nD_x \frac{\partial C}{\partial x} \qquad (6.2)$$

where D_x is the longitudinal dispersion coefficient, n is effective porosity, and C is concentration of contaminant tracer. Similarly, the dispersive flux f_{y*} is assumed proportional to the concentration gradient in the y direction:

$$f_{y*} = nCv_{y*} = -nD_y \frac{\partial C}{\partial y} \qquad (6.3)$$

For a simple uniform flow field with average linear (seepage) velocity $\bar{v}_x$,

$$D_x = \alpha_x \bar{v}_x \qquad (6.4)$$

$$D_y = \alpha_y \bar{v}_x \qquad (6.5)$$

where α_x and α_y are the longitudinal and transverse dispersivities, respectively.

Dispersivity values have usually been set constant in transport models, but studies by Smith and Schwartz (1980) and Gelhar et al. (1979) indicate that dispersivity depends on the distribution of heterogeneities and the scale of the field problem. Many investigators during the 1980s worked on the complex problem of estimating dispersivity from field tracer studies and pump tests, and both statistical and deterministic models have been postulated (Anderson, 1979; Freeze and Cherry, 1979; Mackay et al., 1986; Gelhar and Axness, 1983; Freyberg, 1986; Dagan, 1986, 1987, 1988; Neuman et al., 1987).

It should be noted that dispersion coefficients become more complex in a nonuniform flow field characterized by v_x and v_y. The dispersion coefficient relates the mass flux vector to the gradient of concentration and can be represented as a second-rank tensor (Bear, 1979). Through careful definition of the coordinate system, relationships can be developed among D_x, D_y, and D_z and the components of the tensor (D) (Wang and Anderson, 1982).

Derivation of the Advection–Dispersion Equation for Solute Transport

The equation governing transport in ground water is a statement of the law of the conservation of mass. The derivation is based on those of Ogata (1970) and Bear (1972) and is presented in Freeze and Cherry (1979). It will be assumed that the porous medium is homogeneous, isotropic, and saturated; it is further assumed that the flow is steady state and that Darcy's law applies. The flow is described by the average linear velocity or seepage velocity, which transports the dissolved substance by advection. If advection were the only transport mechanism operating, conservative solutes would move according to plug flow concepts. In reality, there is an additional mixing process, hydrodynamic dispersion, which is caused by velocity variations within each pore channel and from one channel to another.

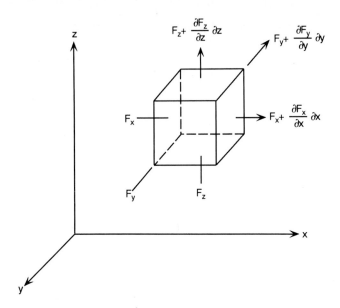

Figure 6.5 Mass balance in a cubic element in space.

Hydrodynamic dispersion is used to account for the additional transport (spreading) caused by fluctuations in the velocity field.

To establish the mathematical statement of the conservation of mass, the solute flux into and out of a representative elemental volume in the porous medium will be considered (Figure 6.5). In Cartesian coordinates, the specific discharge v has components v_x, v_y, v_z, and the average linear velocity $\bar{v} = v/n$ has components $\bar{v}_x$, $\bar{v}_y$, $\bar{v}_z$. The rate of advective transport is equal to $\bar{v}$. The concentration of the solute C is defined as the mass of solute per unit volume of solution. The mass of solute per unit volume of porous media is therefore nC. For a homogeneous medium, the effective porosity n is a constant, and $\partial(nC)/\partial x = n(\partial C/\partial x)$. The mass of solute transported in the x direction by the two mechanisms of solute transport can be represented as

$$\text{Transport by advection} = \bar{v}_x nC \; dA$$

$$\text{Transport by dispersion} = nD_x \frac{\partial C}{\partial x} \; dA$$

(6.6)

where D_x is the hydrodynamic dispersion coefficient in the x direction and dA is the elemental cross-sectional area of the cubic element. The dispersion coefficient D_x is related to the dispersivity a_x and the diffusion coefficient D_d by

$$D_x = a_x \bar{v}_x + D_d \qquad (6.7)$$

The form of the dispersive component embodied in Eq. (6.6) is analogous to Fick's first law.

If F_x represents the total mass of solute per unit cross-sectional area transported in the x direction per unit time, then

$$F_x = \bar{v}_x nC - nD_x \frac{\partial C}{\partial x} \qquad (6.8)$$

The negative sign before the dispersive term indicates that the contaminant moves toward the area of lower concentration. Similarly, expressions for flux in the other two directions can be written as

$$F_y = \bar{v}_y nC - nD_y \frac{\partial C}{\partial y} \qquad (6.9)$$

$$F_z = \bar{v}_z nC - nD_z \frac{\partial C}{\partial z} \qquad (6.10)$$

The total amount of solute entering the fluid element (Figure 6.5) is

$$F_x \, dz \, dy + F_y \, dz \, dx + F_z \, dx \, dy$$

The total amount leaving the representative fluid element is

$$\left(F_x + \frac{\partial F_x}{\partial x} \, dx \right) dz \, dy + \left(F_y + \frac{\partial F_y}{\partial y} \, dy \right) dz \, dx + \left(F_z + \frac{\partial F_z}{\partial z} \, dz \right) dx \, dy$$

The difference in the amount entering and leaving the fluid element is

$$\left(\frac{\partial F_x}{\partial x} + \frac{\partial F_y}{\partial y} + \frac{\partial F_z}{\partial z} \right) dx \, dy \, dz$$

Because the dissolved tracer is assumed to be conservative (nonreactive), the difference between the flux into the element and the flux out of the element equals the amount of dissolved substance accumulated in the element. The rate of mass change in the element can be represented by

$$-n \frac{\partial C}{\partial t} \, dx \, dy \, dz$$

The complete conservation of mass expression, therefore, becomes

$$\frac{\partial F_x}{\partial x} + \frac{\partial F_y}{\partial y} + \frac{\partial F_z}{\partial z} = -n \frac{\partial C}{\partial t} \qquad (6.11)$$

Substitution of expressions for F_x, F_y, and F_z into (6.11) and cancellation of n from both sides of the equation yields

$$\left[\frac{\partial}{\partial x} \left(D_x \frac{\partial C}{\partial x} \right) + \frac{\partial}{\partial y} \left(D_y \frac{\partial C}{\partial y} \right) + \frac{\partial}{\partial z} \left(D_z \frac{\partial C}{\partial z} \right) \right]$$

$$- \left[\frac{\partial}{\partial x} (\bar{v}_x C) + \frac{\partial}{\partial y} (\bar{v}_y C) + \frac{\partial}{\partial z} (\bar{v}_z C) \right] = \frac{\partial C}{\partial t} \qquad (6.12)$$

In a homogeneous medium in which the velocity is steady and uniform (that is, it does not vary through time or space) and the dispersion coefficients D_x, D_y, and D_z do not vary through space (but $D_x \neq D_y \neq D_z$, in general), Eq. (6.12) becomes

$$\left[D_x \frac{\partial^2 C}{\partial x^2} + D_y \frac{\partial^2 C}{\partial y^2} + D_z \frac{\partial^2 C}{\partial z^2} \right]$$

$$- \left[\bar{v}_x \frac{\partial C}{\partial x} + \bar{v}_y \frac{\partial C}{\partial y} + \bar{v}_z \frac{\partial C}{\partial z} \right] = \frac{\partial C}{\partial t} \quad (6.13)$$

In two dimensions, the governing equation for a one-dimensional velocity becomes

$$\left[D_x \frac{\partial^2 C}{\partial x^2} + D_y \frac{\partial^2 C}{\partial y^2} \right] - \left[\bar{v}_x \frac{\partial C}{\partial x} \right] = \frac{\partial C}{\partial t} \quad (6.14)$$

In one dimension, such as for a soil column, the governing equation reduces to the familiar advective–dispersive equation, which can be solved using Laplace transforms:

$$D_x \frac{\partial^2 C}{\partial x^2} - \bar{v}_x \frac{\partial C}{\partial x} = \frac{\partial C}{\partial t} \quad (6.15)$$

A number of analytical solutions exist for Eqs. (6.13), (6.14), and (6.15) under various simplifying assumptions, and some of these are presented in Section 6.5.

6.5 ONE-DIMENSIONAL MODELS

The governing mass transport Eq. (6.13) is difficult to solve in field cases of practical interest due to boundary irregularities and variations in aquifer characteristics, so numerical methods must generally be employed. There are, however, a limited number of relatively simple, one-dimensional problems for which analytical solutions exist. Several of these cases are presented next in order to gain insights into the effect of advection, dispersion, and adsorption on the overall patterns produced.

The simplifying assumptions include the following: (1) the tracer is ideal, with constant density and viscosity; (2) the fluid is incompressible; (3) the medium is homogeneous and isotropic; and (4) only saturated flow is considered. For the case of a nonreactive tracer in one-dimensional flow in the $+x$ direction, Eq. (6.15) applies.

$$D_x \frac{\partial^2 C}{\partial x^2} - v_x \frac{\partial C}{\partial x} = \frac{\partial C}{\partial t} \quad (6.16)$$

where D_x is the coefficient of hydrodynamic dispersion and v_x is the average seepage velocity. Note that the bar over the velocity terms is dropped for convenience.

Several different solutions can be derived for Eq. (6.16) depending on initial and boundary conditions and whether the tracer input is a slug input or a continuous release (Figures 6.1 and 6.4a). Initial conditions ($t = 0$) in a soil column are usually set to zero [$C(x, 0) = 0$] or to some constant background concentration.

Boundary conditions must be specified at the two ends of the one-dimensional column. For a continuous source load at $x = 0$, the concentration is set to $C(0, t) = C_0$ for $t > 0$. The concentration at the other boundary, $x = \infty$, is set to zero $C(\infty, t) = 0$ for $t > 0$.

Case 6.1 Continuous Source in One Dimension. For an infinite column with background concentration of zero and input tracer concentration C_0 at $-\infty \le x \le 0$ for $t \ge 0$, Bear (1961) solves the problem using the Laplace transform at $x = L$, the length of the column; and erfc is the complementary error function.

$$\frac{C(x, t)}{C_0} = \frac{1}{2} \left(\text{erfc} \left[\frac{L - v_x t}{2\sqrt{D_x t}} \right] + \exp \left(\frac{v_x L}{D_x} \right) \text{erfc} \left[\frac{L + v_x t}{2\sqrt{D_x t}} \right] \right) \tag{6.17}$$

where $\text{erfc}(x) = 1 - \text{erf}(x) = 1 - (2/\sqrt{\pi}) \int_0^x e^{-u^2} du$.

The center of mass ($C/C_0 = 0.5$) of the breakthrough curve travels with the average linear velocity v_x and corresponds to the point where $x = v_x t$. Note that the second term on the right side of Eq. (6.17) can generally be neglected for most practical problems. The error functions erf (β) and erfc (β) are tabulated in Table 6.1.

Case 6.2 Instantaneous Source in One Dimension. The corresponding solution can be derived for the injection of a tracer pulse (instantaneous input) at $x = 0$ with background concentration equal to zero in the column. As the slug moves downstream with v_x in the $+x$ direction, it spreads out according to

$$C(x, t) = \frac{M}{(4\pi D_x t)^{1/2}} \exp \left[-\frac{(x - v_x t)^2}{4 D_x t} \right] \tag{6.18}$$

where M is the injected mass per unit cross-sectional area. Figure 6.4 shows the resulting Gaussian distribution of concentration for the instantaneous pulse source in one dimension.

Plots comparing the shapes of Eqs. (6.17) and (6.18) are shown in Figures 6.1 and 6.4, respectively, at $x = L$ at the end of a soil column. The differences between instantaneous (spike) source and continuous-source transport problems are obvious in one dimension. The continuous source produces a response or breakthrough curve that starts at a low concentration and eventually levels off to the initial input concentration C_0 as a function of time. The spike source produces a normal or Gaussian distribution that continues to decrease in maximum concentration due to spreading out with time in the direction of flow. For the spike source, the amount of mass under each curve is identical if the tracer is conservative.

Case 6.3 Adsorption Effects. While there exist many reactions that can alter contaminant concentrations in ground water, adsorption onto the soil matrix appears to be one of the dominant mechanisms. Adsorption is covered in detail in Chapter 7. The concept of the isotherm is used to relate the amount of contaminant

TABLE 6.1   Values of erf (β) and erfc (β)
for Positive Values of β

β	erf (β)	erfc (β)
0	0	1.0
0.05	0.056372	0.943628
0.1	0.112463	0.887537
0.15	0.167996	0.832004
0.2	0.222703	0.777297
0.25	0.276326	0.723674
0.3	0.328627	0.671373
0.35	0.379382	0.620618
0.4	0.428392	0.571608
0.45	0.475482	0.524518
0.5	0.520500	0.479500
0.55	0.563323	0.436677
0.6	0.603856	0.396144
0.65	0.642029	0.357971
0.7	0.677801	0.322199
0.75	0.711156	0.288844
0.8	0.742101	0.257899
0.85	0.770668	0.229332
0.9	0.796908	0.203092
0.95	0.820891	0.179109
1.0	0.842701	0.157299
1.1	0.880205	0.119795
1.2	0.910314	0.089686
1.3	0.934008	0.065992
1.4	0.952285	0.047715
1.5	0.966105	0.033895
1.6	0.976348	0.023652
1.7	0.983790	0.016210
1.8	0.989091	0.010909
1.9	0.992790	0.007210
2.0	0.995322	0.004678
2.1	0.997021	0.002979
2.2	0.998137	0.001863
2.3	0.998857	0.001143
2.4	0.999311	0.000689
2.5	0.999593	0.000407
2.6	0.999764	0.000236
2.7	0.999866	0.000134
2.8	0.999925	0.000075
2.9	0.999959	0.000041
3.0	0.999978	0.000022

Source: Domenico and Schwartz, 1990.

adsorbed by the solids S to the concentration in solution, C. One of the most commonly used isotherms is the Freundlich isotherm,

$$S = K_d C^b \tag{6.19}$$

where S is the mass of solute adsorbed per unit bulk dry mass of porous media,

K_d is the distribution coefficient, and b is an experimentally derived coefficient. If $b = 1$, Eq. (6.19) is known as the linear isotherm and is incorporated into the one-dimensional advective–dispersive equation in the following way:

$$D_x \frac{\partial^2 C}{\partial x^2} - v_x \frac{\partial C}{\partial x} - \frac{\rho_b}{n} \frac{\partial S}{\partial t} = \frac{\partial C}{\partial t} \tag{6.20}$$

where ρb is the bulk dry mass density, n is porosity, and

$$- \frac{\rho_b}{n} \frac{\partial S}{\partial t} = \frac{\rho_b}{n} \frac{dS}{dC} \frac{\partial C}{\partial t}$$

For the case of the linear isotherm, $(dS/dC) = K_d$, and

$$D_x \frac{\partial^2 C}{\partial x^2} - v_x \frac{\partial C}{\partial x} = \frac{\partial C}{\partial t} \left(1 + \frac{\rho_b}{n} K_d \right)$$

or, finally,

$$\frac{D_x}{R} \frac{\partial^2 C}{\partial x^2} - \frac{v_x}{R} \frac{\partial C}{\partial x} = \frac{\partial C}{\partial t} \tag{6.21}$$

where $R = [1 + (\rho_b/n)K_d]$ = retardation factor, which has the effect of retarding the adsorbed species relative to the advective velocity of the ground water (Figure 6.1). The retardation factor is equivalent to the ratio of velocity of the sorbing contaminant and the ground water and ranges from 1 to several thousand.

The use of the distribution coefficient assumes that partitioning reactions between solute and soil are very fast relative to the rate of ground water flow. Thus, it is possible for nonequilibrium fronts to occur that appear to migrate faster than retarded fronts, which are at equilibrium. These complexities involve other rate kinetic factors beyond the scope of simple models discussed in this chapter (see Chapter 7).

An interesting case study is that of a semi-infinite column for which the column ($x > 0$) is initially at $C = 0$ and is connected to a contaminant source containing tracer at $C = C_0$ ($x = 0$). The tracer continuously moves down the column at seepage velocity v_x in the $+x$ direction. It is assumed that $C = 0$ at $x = \infty$. For the case of linear adsorption, Eq. (6.21) is used, where R = retardation factor ≥ 1 as described earlier for adsorption. The solution in this case becomes (Bear, 1972)

$$C(x, t) = \frac{C_0}{2} \left[\text{erfc} \left(\frac{Rx - v_x t}{2\sqrt{RD_x t}} \right) + \exp \left(\frac{v_x x}{D_x} \right) \text{erfc} \left(\frac{Rx + v_x t}{2\sqrt{RD_x t}} \right) \right] \tag{6.22}$$

Ogata and Banks (1961) showed that the second term in Eq. (6.22) can be neglected where $D_x/v_x x < 0.002$; this condition produces an error of less than 3%. Equation (6.22) then reduces to Eq. (6.17) with $R = 1$.

Equation (6.17) or (6.22) can be used in laboratory studies to determine dispersion coefficients for nonreactive and adsorbing species. Figure 6.1 shows several graphs depicting the effect of dispersion and adsorption on the observed breakthrough curve for a continuous source. Retarded fronts can be derived from conservative fronts by adjusting v_x/R and D_x/R in one dimension. Typical values of R for organics often encountered in field sites range from 2 to 10 (Roberts et al., 1986). Larger values of D_x tend to spread out the fronts, while large values of R tend to slow the velocity of the center of mass ($C/C_0 = 0.5$) and reduce D_x by a factor of $1/R$. Conceptually, in one-dimension, these transport mechanisms are relatively well understood for laboratory-scale experiments.

Case 6.4 Transport (One Dimensional) with First-order Decay. One example of mass transport that includes a simple kinetic reaction is the first-order decay of a solute. This could be caused by radioactive decay, biodegradation, or hydroloysis, and it is generally represented in the transport equation by adding the term $-\lambda C$, where λ is the first-order decay rate in units of t^{-1}. For example, Eq. (6.16) would become

$$D_x \frac{\partial^2 C}{\partial x^2} - v_x \frac{\partial C}{\partial x} - \lambda C = \frac{\partial C}{\partial t} \qquad (6.23)$$

The resulting solution to the equation for a pulse source is given by Eq. (6.18), but multiplied by the factor $e^{-\lambda t}$. In general, the effect of the decay coefficient is to reduce mass and concentration as a function of time. This concept is described in more detail for hydrolysis and biodegradation in Chapters 7 and 8.

Example 6.1 THE ONE-DIMENSIONAL APPLICATION OF TRANSPORT EQUATION

(a) An underground tank leaches an organic (benzene) continuously into a one-dimensional aquifer having a hydraulic conductivity of 2.15 m/day, an effective porosity of 0.1, and a hydraulic gradient of 0.04 m/m. Assuming an initial concentration of 1000 mg/L and longitudinal dispersity of 7.5 m, find the time taken for the contaminant concentration to reach 100 mg/L at $L = 750$ m. Neglect any other degradation processes.

Solution. Seepage velocity v_x is calculated from Darcy's law:

$$v_x = \frac{K\,\Delta h/\Delta x}{n} = \frac{2.15 \times 0.04}{0.1} = 0.86 \text{ m/day}$$

$$D_x = \alpha_x v_x = 7.5 \times 0.86 = 6.45 \text{ m}^2/\text{day}$$

Using Eq. (6.17) and ignoring the second term,

$$C(L, t) = \frac{C_0}{2} \text{ erfc } \frac{L - v_x t}{2\sqrt{D_x t}}$$

$$100 = 500 \text{ erfc } \frac{750 - 0.86t}{2\sqrt{6.45t}}$$

$$\frac{750 - 0.86t}{2\sqrt{6.45t}} = 0.9064$$

By trial and error,

$$\boxed{t = 728 \text{ days} = 1.99 \text{ yr}}$$

(b) One of the drums stored at the site breaks and suddenly releases 1 kg of decaying

^{137}Cs (half-life of 33 years) over the cross section of flow, which is estimated to be 10 m². What is the concentration (in mg/m³) at $L = 100$ m 90 days later?

Solution. The decay rate is calculated as

$$\lambda = \frac{\ln 2}{(33 \times 365)} = 5.755 \times 10^{-5} \text{ day}^{-1}$$

Using Eq. (6.18) and incorporating radioactive decay,

$$C(x, t) = \left\{ \frac{M}{\sqrt{4\pi D_x t}} \exp\left[-\frac{(x - v_x t)^2}{4 D_x t} \right] \right\} \exp(-\lambda t)$$

$$= \frac{10^6 \text{ mg}/10 \text{ m}^2}{\sqrt{4\pi(6.45)(90)}} \exp\left\{ -\frac{[100 - (0.86)(90)]^2}{4 \times 6.45 \times 90} \right\} \exp(-5.755 \times 10^{-5} \times 90)$$

$$\boxed{C(x, t) = 934.79 \text{ mg/m}^3 = 0.935 \text{ mg/L}}$$

6.6 GOVERNING FLOW AND TRANSPORT EQUATIONS IN TWO DIMENSIONS

The differential equation for simulating ground water flow in two dimensions is usually written [see Eq. (2.26) and (2.28)]

$$\frac{\partial}{\partial x}\left(T_x \frac{\partial h}{\partial x} \right) + \frac{\partial}{\partial y}\left(T_y \frac{\partial h}{\partial y} \right) = S \frac{\partial h}{\partial t} + W \tag{6.24}$$

where $T_x = K_x b$ = transmissivity in the x direction (L²/T)
$T_y = K_x b$ = transmissivity in the y direction (L²/T)
b = aquifer thickness (L)
S = storage coefficient
W = source or sink term (L/T)
h = hydraulic head (L)

The governing flow equation in two dimensions must be solved before the transport equation can be solved. Chapter 10 describes numerical methods that can be used to solve Eq. (6.24) for an actual field site. The resulting head distribution $h(x, y)$ can then be used to determine gradients and seepage velocities in two dimensions.

The governing transport equation in two dimensions is usually written

$$\frac{\partial}{\partial x}\left(D_x \frac{\partial C}{\partial x} \right) + \frac{\partial}{\partial y}\left(D_y \frac{\partial C}{\partial y} \right) - \frac{\partial}{\partial x}(Cv_x) - \frac{\partial}{\partial y}(Cv_y) - \frac{C_0 W}{nb} + \sum R_k = \frac{\partial C}{\partial t} \tag{6.25}$$

where C = concentration of solute (M/L³)
v_x, v_y = seepage velocity (L/T) averaged in the vertical direction
D_x, D_y = coefficient of dispersion (L²/T) in x and y directions

C_0 = solute concentration in source or sink fluid (M/L^3)
R_k = rate of addition or removal of solute ($\pm$) (M/L^3T)
n = effective porosity
W = source or sink term

Equation (6.25) can only be solved analytically under the most simplifying conditions where velocities are constant, dispersion coefficients are constant, and source terms are simple functions. There are difficulties in attempting to use the mass-transport equation to describe an actual field site in two dimensions, since dispersivities in the x and y directions are difficult to estimate from tracer tests due to the presence of spatial heterogeneities and other reactions in the porous media. Estimation of hydraulic conductivity and associated velocities can be very difficult due to the presence of field heterogeneities that are often unknown. The source or sink concentrations that drive the model are usually assumed constant in time, but may actually have varied significantly. A particularly serious problem appears to be the reaction term, which may represent adsorption, ion exchange, or biodegradation. The assumption of equilibrium conditions and the selection of rate coefficients are both subject to some error and may create difficult prediction problems at many field sites. In addition, fuels and solvents can create non-aqueous-phase liquids (NAPLs) in the subsurface, which are difficult to measure and can provide sources of soluble contaminants for years to come (Chapter 11).

Despite all the above-mentioned problems, mass-transport models still offer the most reliable approach to the organization of field data, prediction of plume migration, and ultimate management of waste-disposal problems. Although a complete three-dimensional scenario with all rate coefficients included will probably never be achieved, presently existing one- and two-dimensional solute transport models have much to offer in simplifying and providing insight into complex ground water problems. Chapter 10 presents numerical methods for the solution of two- and three-dimensional flow and solute transport problems.

6.7 TWO-DIMENSIONAL MODELING

Two-dimensional modeling is generally the most useful tool for analyzing and predicting solute transport in a field situation. This type of model is considerably more versatile than the one-dimensional models previously discussed. Horizontal two-dimensional modeling can be used where (1) no vertical velocity components exist, or (2) observation wells or monitoring points are fully screened and provide averaged data in two dimensions for a uniformly thick aquifer. Nonuniformities in the flow field may be due to variations in topography and permeability, or natural or artificial sinks such as wells and springs, or recharge points such as injection wells or leakage from lagoons. Probably the most common reason for using a horizontal two-dimensional analysis is that the monitoring wells are typically distributed about the horizontal plane. Most wells are fully penetrating, and therefore any samples collected represent an approximate vertical average in that portion of the aquifer.

Sampling in the vertical direction may be warranted where there are significant differences between the contaminant density and water density or where detailed information is available on vertical variations in aquifer hydraulic conductivity or porosity. Profile two-dimensional models that represent a vertical slice of the aquifer are available for such cases. A full three-dimensional picture is only rarely available for a field site, and requires the use of numerous wells in the *x, y* plane screened at vertical intervals for discrete sampling. The classic studies performed at the Borden Landfill in Canada provide excellent examples of a well-monitored field test in three dimensions (Mackay et al., 1986).

To study the variation in contaminants in two dimensions, it is first necessary to solve the governing equations [Eqs. (6.24) and (6.25)] in the *x* and *y* directions. The techniques available for developing a solution include both analytical, semi-analytical, and sophisticated numerical methods that have been developed over the past 25 years. Analytical methods basically provide a closed-form solution for relatively simple boundary and initial conditions. Semianalytical methods usually result in an integral type of solution that must be evaluated by a simple numerical integration procedure. Numerical models are briefly discussed in Section 6.8 and in Chapters 9 and 10.

Analytical Models

Analytical models are developed by solving the transport equation for certain simplified boundary and initial conditions. Numerous analytical solutions are available in the literature (Bear, 1979; Hunt, 1978; Wilson and Miller, 1978; Cleary and Ungs, 1978; Shen, 1976; Galya, 1987; Javandel et al., 1984) for pulse and continuous contaminant sources, with boundary conditions ranging from no flow to constant heads. Processes that may be included in these models include advection, dispersion, adsorption, and first-order decay (biological or radioactive). Analytical solutions generally require constant parameters, simple geometries, and boundary conditions, but do provide useful insights to many ground water contaminant problems at two-dimensional field sites.

Case 6.5   Wilson and Miller 1978 Model.   One of the first two-dimensional analytical models was that developed by Wilson and Miller (1978). It is one of the simplest to use and can account for lateral and transverse dispersion, adsorption, and first-order decay in a uniform flow field. Concentration C at any point in the *x, y* plane can be predicted by solving Eq. (6.14) for an instantaneous spike source or for continuous injection. First-order decay is added to the equation by adding a term, $-\lambda C$, to the left side of Eq. (6.14). The first-order decay rate is λ $(1/T)$. Velocity in the *y* direction is assumed to be zero, and the *x* axis is oriented in the direction of flow. Contaminants are assumed to be injected uniformly throughout the vertical axis. The flow velocity must be obtained from a flow model or from detailed field monitoring. The solution becomes

$$C(x, y, t) = \frac{m'}{4\pi nt \sqrt{D_x D_y}} \exp\left[-\frac{(x - vt)^2}{4D_x t} - \frac{y^2}{4D_y t} - \lambda t \right] \qquad (6.26)$$

$$C(x, y, t) = \frac{f'_m \exp(x/B)}{4\pi n \sqrt{D_x D_y}} W\left(u, \frac{r}{B} \right) \qquad (6.27)$$

where m' = injected contaminant mass per vertical unit aquifer (M/L)

f'_m = continuous rate of contaminant injection per vertical unit aquifer (M/LT)

D_x = longitudinal dispersion coefficient (L²/T)

D_y = transverse dispersion coefficient (L²/T)

n = porosity

t = time since start of injection (T)

λ = decay coefficient (l/T)

v = seepage velocity (L/T)

$B = 2D_x/v$

$\gamma = 1 + \dfrac{2B\lambda}{v}$

$u = \dfrac{r^2}{4\gamma D_x t}$

$r = \left[\left(x^2 + \dfrac{D_x y^2}{D_y} \right) \gamma \right]^{1/2}$

$W(u, r/B)$ is the leaky well function described in papers by Hantush (1956) and Wilson and Miller (1978) and tabulated in many ground water texts (Bear, 1979; Freeze and Cherry, 1979). Note that the first-order decay is incorporated into Eq. (6.26). When analyzing the effects of adsorption, the retardation coefficient R is used to redefine v, D_x, and D_y for the sorbing contaminant:

$$v' = \frac{v}{R}, \qquad D'_x = \frac{D_x}{R}, \qquad D'_y = \frac{D_y}{R}$$

Equation (6.26) was verified by estimating the size and shape of a plume of chromium contamination and comparing with previous numerical modeling and field monitoring (Wilson and Miller, 1978). In general, the predicted contaminant plume compared very well with the measured extent of contamination. Some deviation from the prediction did occur below a small stream that caused the natural flow field to deviate from the idealized uniform flow field used in the analytical model.

Case 6.6 Steady-state 2-D Model.   Bear (1972) provides a solution to Eq. (6.14) for the condition where steady-state conditions have been reached and the plume has been stabilized. The solution requires use of K_0, defined as the modified Bessel function of second kind and zero order. Q (L^2/T) equals the rate at which a tracer of concentration C_0 is being injected.

$$C(x, y) = \frac{C_0 Q}{2\pi(D_x D_y)^{1/2}} \exp\left(\frac{v_x x}{2D_x}\right) K_0 \left\{\frac{v_x^2}{4D_x}\left(\frac{x^2}{D_x} + \frac{y^2}{D_y}\right)^{1/2}\right\} \qquad (6.28)$$

Case 6.7 Pulse Source Models. If a pulse of contaminant is injected over the full thickness of a two-dimensional homogeneous aquifer, it will move in the direction of flow and spread out with time. Figure 6.6 represents the theoretical pattern of contamination at various points in time. A line source model from De Josselin De Jong (1958) assumes that the spill occurs as a line source (well) loaded vertically into a two-dimensional flow field (x, y). If a tracer with concentration C_0 is injected over an area A at a point (x_0, y_0), the concentration at any point (x, y) at time t after the injection is given by the following equation. Average linear velocity v_x, longitudinal dispersion D_x, and transverse dispersion D_y are all assumed to be constant in this equation (see Example 6.2).

$$C(x, y, t) = \frac{C_0 A}{4\pi t(D_x D_y)^{1/2}} \exp\left\{-\frac{((x - x_0) - v_x t)^2}{4D_x t} - \frac{(y - y_0)^2}{4D_y t}\right\}$$

$$(6.29)$$

The corollary equation in three dimensions from a point source was derived by Baetsle (1969) and can sometimes be used to represent the sudden release from a single source or tank in the subsurface. Baetsle's model gives the following equation:

$$C(x, y, z, t) = \frac{C_0 V_0}{8(\pi t)^{3/2}(D_x D_y D_z)^{1/2}} \exp\left[-\frac{(x - vt)^2}{4D_x t} - \frac{y^2}{4D_y t} - \frac{z^2}{4D_z t} - \lambda t\right]$$

$$(6.30)$$

where C_0 is the original concentration; V_0 is the original volume so that $C_0 V_0$ is the mass involved in the spill; D_x, D_y, and D_z are the coefficients of hydrodynamic dispersion; v is the velocity of the contaminant; and λ is the first-order decay constant for a radioactive substance. For a nonradioactive substance, the term λt is ignored.

With an idealized three-dimensional point source spill, spreading occurs in

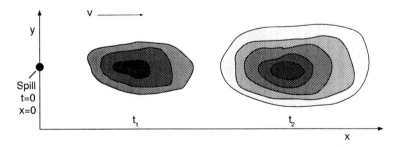

Figure 6.6 Plan view of plume instant point source in time (two dimensions).

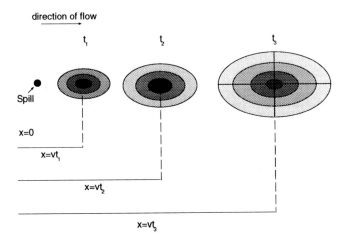

direction of flow

t₁   t₂   t₃

Spill

x=0

x=vt₁

x=vt₂

x=vt₃

Figure 6.7 Plan view of plume developed from an instantaneous point source at three different times (three dimensions).

the direction of flow, and the peak or maximum concentration occurs at the center of the Gaussian plume where $y = z = 0$ and $x = vt$ (Figure 6.7):

$$C_{max} = \frac{C_0 V_0 e^{-\lambda t}}{8(\pi t)^{3/2} (D_x D_y D_z)^{1/2}}$$

The dimensions of the plume are related to

$$\sigma_x = (2D_x t)^{1/2}, \qquad \sigma_y = (2D_y t)^{1/2}, \qquad \sigma_z = (2D_z t)^{1/2}$$

where σ is the standard deviation and $3\sigma_x$, $3\sigma_y$, and $3\sigma_z$ represent three standard deviations away from the mean within which 99.7% of the contaminant mass is contained.

Example 6.2 PULSE SOURCE IN TWO DIMENSIONS

A tank holding chloride at a concentration of 10,000 mg/L accidentally leaks over an area of 10 m² into an aquifer. Assuming that chloride is a completely conservative tracer, $D_x = 1$ m²/day, $D_y = 0.1$ m²/day, and the seepage velocity is 1.0 m/day, calculate the following:

(a) Time required for the center of mass of the plume to reach a point 75 m away
(b) Peak concentration at that point
(c) x and y dimensions of the plume at that point

Solution
(a) Time required to reach 75 m:

$$t = \frac{R_t x}{V_w} = \frac{1 \, (75 \text{ m})}{1 \text{ m/day}} = 75 \text{ days}$$

(b) Peak concentration at 75 m:

$$C_{\max} = \frac{C_0 A}{4\pi t (D_x D_y)^{1/2}}$$

$$= \frac{10^4 \text{ mg/L} \times 10 \text{ m}^2}{4\pi \times 75 \text{ m} \times (1 \text{ m}^2/\text{day} \times 0.1 \text{ m}^2/\text{day})^{1/2}}$$

$$= 335.7 \text{ mg/L}$$

(c) Plume dimensions:

$$\sigma_x = (2D_x t)^{1/2} = (2 \times 1 \times 75)^{1/2} \text{ m} = 12.25 \text{ m}$$

$$x \text{ dimension} = 3\sigma_x = 36.7 \text{ m}$$

$$\sigma_y = (2D_y t)^{1/2} = (2 \times 0.1 \times 75)^{1/2} = 3.87 \text{ m}$$

$$y \text{ dimension} = 3\sigma_y = 11.6 \text{ m}$$

6.8 SEMIANALYTICAL METHODS

Semianalytical methods usually involve the computation of streamlines and equi-potential lines over a flow domain. Numerical integration is useful for determining travel times along path lines. The basis for the numerical determination of front positions in space is the knowledge of the velocity at every instant for all fluid particles located on an advancing front. The displacement of each particle in a time step Δt is given by $\Delta s = v \, \Delta t$. Bear (1979) describes the method in detail in evaluating the movement of a body of injected water in uniform flow. For the case of steady flow in a two-dimensional flow field, the movement of a front is obtained by displacing fluid particles along the streamlines. If ds denotes an ele-ment of length along a streamline, the time required for fluid particle to travel a length s along a streamline is given by

$$t = \int_0^{s(t)} \frac{ds}{v_s(s)} = -\frac{n}{K} \int_0^{s(t)} \frac{ds}{\partial\phi/\partial s} \tag{6.31}$$

where the aquifer is homogeneous and isotropic, and v_s is the velocity component tangent to the streamline.

Of particular interest is the case of an aquifer with a number of injection and pumping wells, each having its own positive or negative rate of production. For example, consider the case of steady uniform flow in a confined aquifer in which m pumping and recharging wells are operating. For this case, the analytical solu-tion in terms of piezometric head ϕ (Bear and Verruijt, 1987) is

$$\phi = -\frac{n}{K}(xV_{0x} + yV_{0y}) + \sum_{j=1}^{m} \frac{Q_j}{2\pi KB} \ln \frac{r_j}{R} \tag{6.32}$$

where V_{0x} and V_{0y} are the components of the background uniform flow. Q_j is the production of well number j and r_j is the distance from the point at which ϕ is being determined to that well. Using Darcy's law for V_x and V_y and dividing by porosity, we obtain

$$V_x = V_{0x} - \sum \frac{Q_j}{2\pi nB} \frac{x - x_j}{r_j^2}, \qquad V_y = V_{0y} - \sum \frac{Q_j}{2\pi nB} \frac{y - y_j}{r_j^2} \tag{6.33}$$

These equations give the components of the velocity vector at any point of the flow field, where the flow rates and the well locations can be varied. Equation (6.33) can then be used to determine travel times along streamlines in the flow field using Eq. (6.31). Note that a limitation of this method is that effects of diffusion and dispersion are not considered. However, in advection-dominated systems where pumping or injection wells are present, the method is quite satisfactory.

Ground water generally moves slowly and can be assumed at steady state for many ground water flow problems of interest. Two very useful tools for summarizing the variation of a contaminant in space and time are location and arrival time distributions. In a series of classic papers, Nelson (1978) introduced the concept of arrival distribution and presented a technique for developing those distributions based on determining the travel time along individual flow lines. Variation in the arrival distribution is the result of convergence and divergence of flow lines. Nelson's procedure does not account for dispersion directly, but the arrival distribution concept is useful because it uses the actual flow field to predict transport processes. Unlike analytical methods, semianalytical approaches allow the solving of two-dimensional flow problems with an arbitrary number of sources and sinks.

In Figure 6.8a from Nelson (1978), the flow components include a pond (contaminant source), an injection well (Q_2), and a pump well (Q_1). The hydraulic conductivity is K_0 and the effective porosity, in this case, is P. Other terms are defined in Figure 6.8a. All these sources and sinks are superimposed on a background flow due to a hydraulic gradient. The calculated distribution of path lines and time contours based on a particle mover concept is shown in Figure 6.8b. The time contours illustrate the progress of the contaminant fronts. Contaminants arrive at the pumped well after about 9 yr, and they arrive at the river between 15 and 21 yr. Nelson summarizes arrival times at the river in terms of location/arrival time distribution, which is shown in Figure 6.8c.

Case 6.8   Horizontal Plane Source Model.   Galya (1987) developed the horizontal plane source (HPS) model for the geometry shown in Figure 6.9a. The model assumes one-dimensional velocity, but allows for three-dimensional dispersion in the x, y, and z direction. The governing equations are solved for a continuous release, although a finite source can be handled through linear superposition in time. The model's geometry is useful to analyze ponds or pits that leach contaminants from the surface down to the ground water table. The source term is treated in a unique way in order to maintain mass balances in the solution.

A model was derived through the use of Green's functions to simulate three-dimensional contaminant transport from an HPS. This analytical model incorporates retardation and decay and can simulate varying source emission rates. Uses of the model include simulations of contaminant transport from landfills, waste lagoons, land-treatment facilities, and areas of pesticide application. Comparison between HPS and point-source solutions indicates that for such simulations the HPS model will provide more accurate results than the point-source solution, particularly near the source. Representative model applications indicate the model's sensitivity to variations in retardation, decay, and temporal period of source emission.

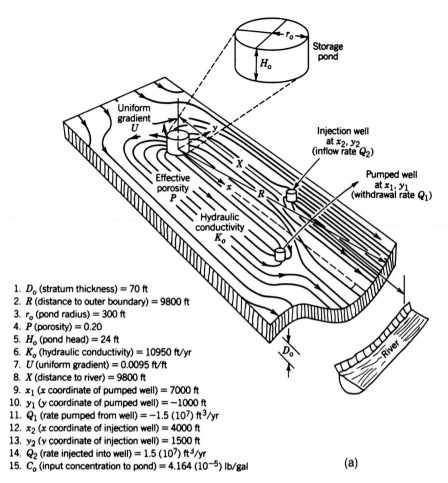

1. D_o (stratum thickness) = 70 ft
2. R (distance to outer boundary) = 9800 ft
3. r_o (pond radius) = 300 ft
4. P (porosity) = 0.20
5. H_o (pond head) = 24 ft
6. K_o (hydraulic conductivity) = 10950 ft/yr
7. U (uniform gradient) = 0.0095 ft/ft
8. X (distance to river) = 9800 ft
9. x_1 (x coordinate of pumped well) = 7000 ft
10. y_1 (y coordinate of pumped well) = −1000 ft
11. Q_1 (rate pumped from well) = −1.5 (10^7) ft^3/yr
12. x_2 (x coordinate of injection well) = 4000 ft
13. y_2 (y coordinate of injection well) = 1500 ft
14. Q_2 (rate injected into well) = 1.5 (10^7) ft^3/yr
15. C_o (input concentration to pond) = 4.164 (10^{-5}) lb/gal

(a)

Figure 6.8 (a) Example of how simple flow components including a uniform, regional flow field, a recharging well, a discharging well, and a pond (source) can be represented in a semianalytical model. The list of 15 parameters gives the data required as input to the calculation.

The governing equation to be solved is

$$\frac{\partial C}{\partial t} + \frac{u}{R}\frac{\partial C}{\partial x} = \frac{D_x}{R}\frac{\partial^2 C}{\partial x^2} + \frac{D_y}{R}\frac{\partial^2 C}{\partial y^2} + \frac{D_z}{R}\frac{\partial^2 C}{\partial z^2} - \lambda C + \frac{M_s}{\theta} \qquad (6.34)$$

where u = pore velocity in the x direction
$\quad\quad D_i$ = dispersion coefficient in i direction = $a_i u$
$\quad\quad a_i$ = dispersivity in the i direction
$\quad\quad \lambda$ = first-order decay coefficient
$\quad\quad R$ = retardation factor
$\quad\quad \theta$ = porosity
$\quad\quad M_s$ = mass flux of contaminant source

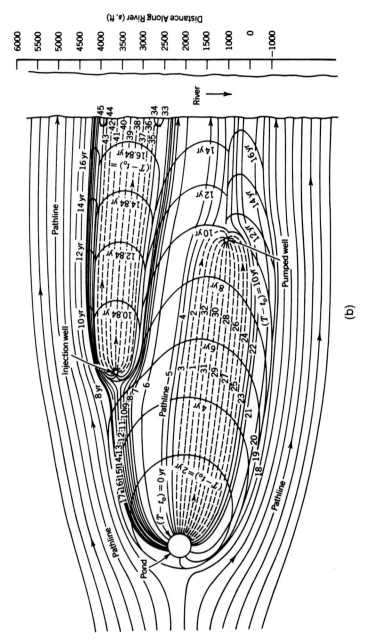

Figure 6.8 (b) Calculated distribution of path lines and time contours for the example data of part (a).

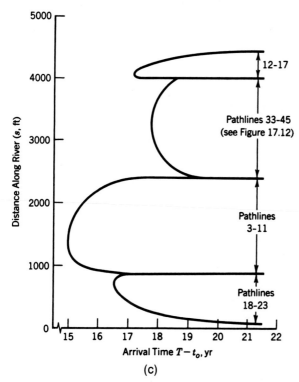

12-17

Pathlines 33-45
(see Figure 17.12)

Pathlines
3-11

Pathlines
18-23

Distance Along River (s, ft)

Arrival Time $T - t_o$, yr

(c)

Figure 6.8 (c) Location/arrival time distribution for the example problem. Source: Nelson, 1978. © American Geophysical Union.

The initial condition on the problem is $C = 0$ at $t = 0$:

Boundary conditions: $C = 0$ $y = \pm\infty$

$$D_z \frac{\partial C}{\partial Z} = 0 \quad z = 0$$

$$D_z \frac{\partial C}{\partial Z} = 0 \quad z = H$$

$$C = 0 \quad z = \infty$$

where H is the aquifer thickness. Galya obtained a solution of the following form utilizing Green's functions:

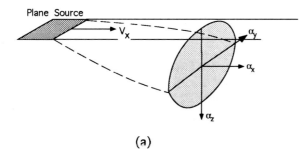

Plane Source

V_x

α_y

α_x

α_z

(a)

Figure 6.9 (a) Galya geometry for the HPS model.

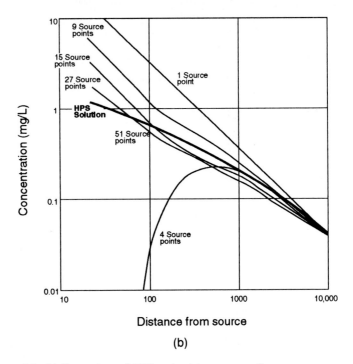

Figure 6.9 (b) Comparison of HPS and point-source results.

$$C = \frac{1}{\theta R} X_0(x,\, t)Y_0(y,\, t)Z_0(z,\, t)T(t) \qquad (6.35)$$

where T is the degradation function [$\exp(-\lambda t)$ for first-order degradation] and X_0, Y_0, and Z_0 are Green's functions for transport in the x, y, and z directions, defined as

$$X_0 = \frac{1}{2L}\left(\operatorname{erf}\frac{(L/2) + x - x_s - (ut/R)}{\sqrt{4D_x t/R}} + \operatorname{erf}\frac{(L/2) - x + x_s + (ut/R)}{\sqrt{4D_x t/R}} \right) \qquad (6.36)$$

$$Y_0 = \frac{1}{2B}\left(\operatorname{erf}\frac{(B/2) + y - y_s}{\sqrt{4D_y t/R}} + \operatorname{erf}\frac{(B/2) - y + y_s}{\sqrt{4D_y t/R}} \right) \qquad (6.37)$$

$$Z_0 = \frac{1}{H}\left(1 + 2 \sum_{n=1}^{\infty} \exp\left(-\frac{n^2\pi^2 D_z t}{H^2 R} \right) \cos\frac{n\pi z}{H} \cos\frac{n\pi z_s}{H} \right) \qquad (6.38)$$

where x = distance in x direction to calculation point
x_s = distance in x direction to center of line source
y = distance in y direction to calculation point
y_s = distance in y direction to center of line source
z = depth of calculation point
z_s = depth of source

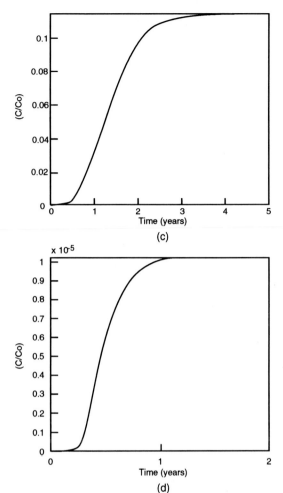

Figure 6.9 (c) Breakthrough curve for chloride at well location.
(d) Breakthrough curve for ethylbenzene at well location.

L = source length
B = source width

A series of simulations were run for comparison to point-source results and for sensitivity analyses, and the HPS model appears to be a useful tool. Figure 6.9b shows a comparison of results from the HPS model and various point-source results according to Galya (1987). The code is available from Galya (1987).

Example 6.3 HPS EXAMPLE

A landfill of dimensions 20 m by 60 m has been leaching chloride and ethylbenzene into an underlying ground water aquifer for 5 yr. Assume aquifer thickness of 30.0 m. The seepage velocity is 40.0 m/yr, soil porosity is 0.38, concentration of chloride and ethylbenzene in the leachate is 100 mg/L, and the infiltration rate is estimated to be 0.8 m/yr. Assume that the maximum longitudinal dispersivity is 5 m and that the transverse and vertical dispersivities are 10 times smaller than the longitudinal dispersivity. Assume that site conditions favor the aerobic biodegradation of ethylbenzene, that the natural biodegradation process can be modeled using first-order kinetics, with a decay rate of 10.4/yr, and that ethylbenzene adsorbs to the soil matrix. The retardation factor is 1.4.

(a) What are the ethylbenzene and chloride concentrations at a well located 50.0 m downgradient from the landfill and screened 1.0 m below the water table after 5 yr?

(b) Plot the breakthrough curve for both the chloride and ethylbenzene at the well location.

Solution

(a) An input file is prepared with the following values (definitions of the parameters are given in Galya, 1987):

Variable	Value (ethylbenzene)	Value (chloride)
NUMPT	1	1
NINT	2000	2000
NUMST	1 (continuous source)	1 (continuous source)
NUMSRS	1	1
CS	100.0 mg/L	100.0 mg/L
INF	0.8 m/yr	0.8 m/yr
TS	5 yr	5 yr
XS	0.0 m	0.0 m
YS	0.0 m	0.0 m
ZS	0.0 m	0.0 m
L	20.0 m	20.0 m
B	60.0 m	60.0 m
X	50.0 m	50.0 m
Y	0.0 m	0.0 m
Z	1.0 m	1.0 m
U	40.0 m/yr	40.0 m/yr
ALPHX	5.0 m	5.0 m
ALPHY	0.5 m	0.5 m
ALPHZ	0.5 m	0.5 m
H	30.0 m	30.0 m
T	5.0 yr	5.0 yr
R	1.4	1.0
LAMB	10.4 yr^{-1}	0.0
POR	0.38	0.38

Carrying out the calculations using HPS results in
    Ethylbenzene concentration at the well = 0.52322E-03 mg/L, or 0.523 µg/L
    Chloride concentration at the well = 11.52 mg/L
Note the effect of adsorption and biodegradation on dramatically reducing the ethylbenzene concentration at the well location.

(b) Chloride and ethylbenzene breakthrough curves are shown in Figure 6.9c and d.

The semianalytical approach has been used by Javandel et al. (1984) to show

how concentration versus time data for a single well can be mapped to other observation points to estimate spatial distribution in concentration. The computer program RESSQ calculates two-dimensional contaminant transport by advection and adsorption in a steady-state flow field. Recharge wells and ponds act as sources and pumping wells act as sinks. RESSQ calculates the streamline pattern in the aquifer, the location of contaminant fronts around sources at various times, and the variation of contaminant concentration with time at sinks. RESSQ was developed at the Lawrence Berkeley Laboratory based on a solution procedure by Gringarten and Sauty (1975). Javandel et al. (1984) present a detailed discussion, listing, and users guide for the semianalytical computer program RESSQ.

6.9 TESTS FOR DISPERSIVITY

To apply any one- or two-dimensional transport model properly to an actual site, it is first necessary to define the input parameters. For modeling a nonadsorbing solute such as chloride, the parameters of primary concern are water-table elevation, hydraulic conductivity K, and dispersivity α. For small project sites, accurate delineation of the water table depends on the number of wells that can be drilled. Porosity can easily be determined by laboratory testing of a few samples. Although this limited sampling may result in errors in porosity of 10% to 20%, these errors will probably not be significant compared to the problems inherent in measuring hydraulic conductivity and dispersivity. Measurement of hydraulic conductivity was described in Chapters 3 and 5.

Laboratory Tests for Dispersivity

Soil column studies are often reported in terms of pore volumes, where one pore volume represents the volume of water that will fill the voids completely along the column length. The total number of pore volumes U during a particular column test is the total discharge divided by a single pore volume.

$$U = \frac{v_x n A t}{A L n} = \frac{v_x t}{L}$$

(6.39)

where v_x is seepage velocity, A is the cross-sectional area, L is the length of the column, and n is porosity.

By rearranging Eq. (6.17) in terms of pore volumes and neglecting the second term on the right side, we obtain a useful relation between C/C_0 and the pore volume function $(U - 1)/U^{0.5}$. By plotting C/C_0 versus the pore volume function on normal probability paper, Pickens and Grisak (1981) demonstrated that data plot as a straight line, and the slope of the line is related to D_x, the longitudinal dispersion coefficient. The value of D_x can be found from

$$D_x = \frac{v_x L}{8} [J(0.84) - J(0.16)]^2$$

(6.40)

where $J(0.84)$ = value of the pore volume function when $C/C_0 = 0.84$
  $J(0.16)$ = value of the pore volume function when $C/C_0 = 0.16$

Because $D_x = \alpha_x v_x + D_d$,

$$\alpha_x = \frac{D_x - D_d}{v_x} \tag{6.41}$$

The authors found a dispersivity of 0.033 cm for the column test.

Example 6.4 DISPERSION IN SAND COLUMNS

Pickens and Grisak (1981) studied dispersion in sand columns. They used sand with a mean grain size of 0.2 mm, porosity of 0.36, and uniformity coefficient of 2.3. The column had a length of 30 cm and diameter of 4.45 cm. Chloride (tracer) at a concentration of 200 mg/L was run through the column at a rate of 5.12×10^{-3} mL/ sec. Average linear velocity was 9.26×10^{-4} cm/sec. The concentration of chloride in the effluent was measured as a function of U, and C/C_0 was plotted as a function of $(U - 1)/U^{1/2}$ on probability paper as shown in Figure 6.10a.

At 25°C, the molecular diffusion coefficient for chloride in water is 2.03×10^{-5} cm^2/sec. From this, Pickens and Grisak estimated the effective diffusion coefficient to be 1.02×10^{-5} cm^2/sec. Hydrodynamic dispersion coefficients are based on the slope of the straight line. In the case of this run, a hydrodynamic dispersivity of 4.05 $\times 10^{-5}$ cm^2/sec was obtained.

Finally, the dispersivity for the run was calculated using Eq. (6.41) as follows:

$$\alpha_x = \frac{D_x - D_d}{V_x}$$

$$= \frac{4.05 \times 10^{-5} \text{ cm}^2/\text{sec} - 1.02 \times 10^{-5} \text{ cm}^2/\text{sec}}{9.26 \times 10^{-4} \text{ cm/sec}} = 0.033 \text{ cm}$$

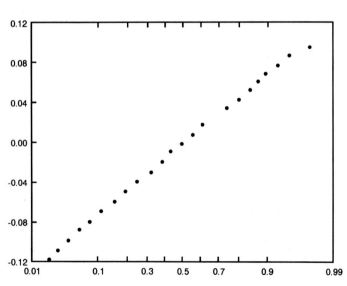

Figure 6.10 (a) Plot of $(U - 1)/U^{1/2}$ versus C/C_0 on probability paper for determination of dispersion in a laboratory sand column.

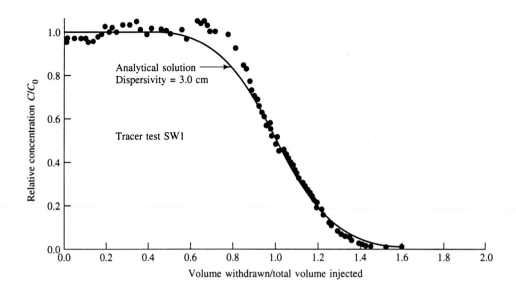

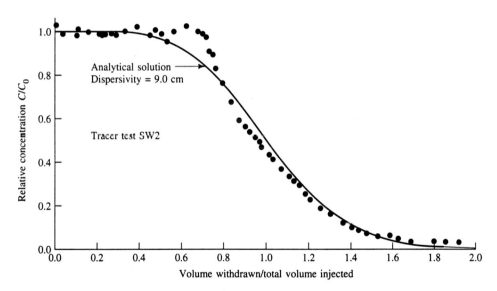

Figure 6.10 (b) Comparison of measured C/C_0 values for a single-well injection–withdrawal test versus an analytical solution. Source: Pickens and Grisak, 1981.

Single-well Tracer Test

Pickens et al. (1981) report one of the very few field experiments for the in situ determination of dispersive and adsorptive properties of a well-defined sandy aquifer system. The aquifer was 8.2 m thick with $K = 1.4 \times 10^{-2}$ cm/sec and porosity of 0.38. The technique involves the use of a radial injection dual-tracer test with

[131]I as the nonreactive tracer and [85]Sr as the reactive tracer. Tracer migration was monitored at various radial distances and depths with multilevel point sampling devices. In the analysis of curves, nonequilibrium adsorption effects were incorporated into the dispersion terms of the solute transport equation, rather than introducing a separate kinetic term.

Various curves were computed for different values of dispersivity, and the curves were best fit to the field data collected at the well. Effective dispersivity values (α) obtained for [85]Sr ranged from 0.7 to 3.3 cm (mean of 1.9 cm), typically a factor of 2 to 5 larger than those obtained for [131]I (range of 0.4 to 1.5 cm with a mean of 0.8 cm). The K_d values obtained by various analyses of the breakthrough curves ranged from 2.6 to 4.5 mL/g; these were within a factor of 4 of the mean values of K_d for [85]Sr based on a separate analysis of sediment cores in another part of the aquifer. The withdrawal phase for the well showed essentially full tracer recoveries for both compounds. The comparison of measured C/C_0 values for a single-well tracer test versus an analytical solution is shown in Figure 6.10b. The two curves are based on longitudinal dispersivities of 3.0 and 9.0 cm, respectively. Thus, the usefulness of the dual-tracer injection tests has been demonstrated along with the existence of nonequilibrium conditions at a field site.

Scale Effect of Dispersion

Dispersivity α is defined to be the characteristic mixing length and is a measure of the spreading of the contaminant. Unfortunately, this parameter has little physical significance and varies with the scale of the problem and sample method. Laboratory values range from 1 to 10 cm, whereas field studies often yield values from less than 1 m to over 1000 m.

Many investigators have recognized the scale effect of dispersion (Fried, 1975; Pickens and Grisak, 1981; Gelhar, et al., 1985). A study by Gelhar et al. (1985) shows the scale dependency of longitudinal dispersivity at 55 actual field sites around the world (Figure 6.11). The data on the Gelhar graph from many sites around the world indicate a general increase in longitudinal dispersivity with scale, although the relation may not be linear. At the Borden landfill study described next, Mackay et al. (1986) determined that there was a small scale effect.

Since dispersion can be caused by slight differences in fluid velocity within a pore, between pores, and because of different flow paths, the effect of different hydraulic conductivities in an aquifer can also cause spreading due to dispersion. The natural variation of hydraulic conductivity and porosity in a vertical and horizontal sense can play a key role in the scale effect of dispersion. As the flow path gets longer, ground water can encounter greater and greater variations in the aquifer. The deviations from the average increase along with the mechanical dispersion. It is logical to assume that dispersivity will approach some upper limit at long traveled distances and large travel times (Gelhar and Axness, 1983; Dagan, 1986, 1988). Due to a lack of field data, explanations of dispersion still represent an active research area.

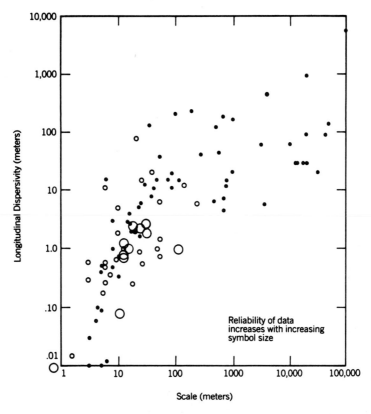

Figure 6.11 Scale of observation versus longitudinal dispersibility: reliability classification. Source; Gelhar et al., 1985.

6.10 NATURAL GRADIENT FIELD TESTS FOR DISPERSION

Borden Landfill Natural Gradient Test

One of the most extensively monitored field tracer tests in history was performed by Cherry's research group at the Canadian Forces Base Borden (Sudicky et al., 1983; Mackay et al., 1986; Freyberg, 1986). An area near the landfill, shown in Figure 6.12a, was monitored extensively with multilevel sampling devices. The aquifer is about 20 m thick and thins to about 10 m in the direction of ground water flow, with an average hydraulic conductivity of 1.16×10^{-2} cm/sec. In August 1982, a natural gradient field experiment was begun by injecting 12 m^3 into the subsurface of the sand aquifer. The average linear ground water velocity was computed to be 29.6 m/yr. The ground water flow direction is indicated in Figure 6.12b.

The concept was simply to inject a known amount of two conservative tracers and five volatile organic compounds for one day into a shallow, sand aquifer and monitor for about 2 years at over 5000 observation points in three dimensions

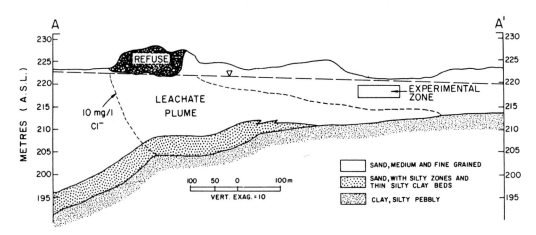

Figure 6.12 (a) Approximate vertical geometry of aquifer along section AA'. Rectangle illustrates the vertical zone in which the experiment was conducted, which is above the landfill leachate plume (denoted by 10 mg/L chloride isopleth from 1979 data). Source: Mackay et al., 1986.

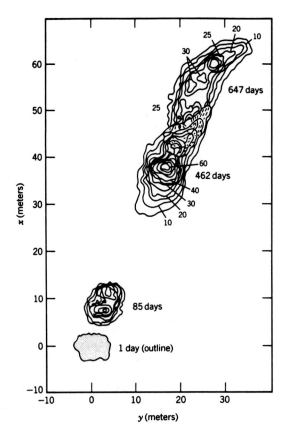

Figure 6.12 (b) Vertically averaged concentration distribution of Cl⁻ at various times after injection. Source: Mackay et al., 1986.

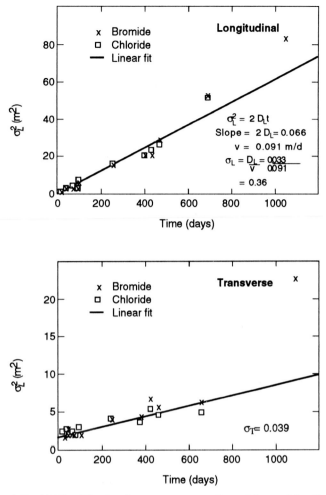

Figure 6.12 (c) Fit of Borden data to a constant dispersivity model. Source: Freyberg, 1986. Copyright by American Geophysical Union.

(each well had up to 18 vertical screened intervals for sampling). Figure 6.12b shows the vertically averaged Cl^- distribution at four different times, and the various plumes were analyzed statistically by Freyberg (1986) to evaluate advection and dispersion in the natural gradient test. The mean ground water velocity was measured to be 0.091 m/day. It is clear from the gradual plume elongation depicted in Figure 6.12b that longitudinal dispersion was occurring and was much larger than transverse dispersion.

Freyberg (1986) computed the center of mass of the chloride plume in three dimensions as well as the variances in the longitudinal and transverse directions. The data were further analyzed by relating variance, time, and dispersivity. The dispersivities were first assumed to be constants with time and then were shown to vary with time, indicating scale-dependent behavior. Freyberg (1986) calculated average linear values of α_x and α_y to be 0.36 and 0.039 m, respectively, but indicated that α_x may be a function of time. Assuming scale-dependent behavior, a value of 0.43 m was computed and extrapolated to an asymptotic value of 0.49 m. Figure 6.12c shows the fit of the Borden data to a constant dispersivity model. The reader is referred to the complete set of six articles on the Borden natural gradient test in the December 1986 issue of *Water Resources Research*.

Roberts et al. (1986) studied adsorption by analyzing arrival times of Cl^- compared to organics such as carbon tetrachloride at the wells at the Borden site. Chapter 7 describes the sorption results in more detail at the Borden site.

At an actual waste site, experimental determination of α_x and α_y is often impractical due to the very long flow times. In these cases the most common approach is to run the transport model for a range of dispersivities and then adjust them until the predicted plume matches observed concentration data. Based on the Borden test data, a typical starting range for longitudinal dispersivity (α_x) in a sandy aquifer is 1 to 10 m. Transverse dispersivity typically ranges from 10% to 30% of α_x. The procedure of fitting dispersivities to predict observed concentration patterns is admittedly crude, but can lead to useful descriptions of contaminant transport.

Natural Gradient Tracer Test: Cape Cod, Massachusetts

This natural gradient tracer experiment was designed to provide a well-defined set of initial conditions associated with contaminant release. An extensive and systematic monitoring program was undertaken so that a detailed description of the resulting contaminant plume could be established and maintained over an extended period of time as it moved and dispersed within the aquifer. The collected data were then used to evaluate the validity of currently accepted theories used to predict solute movement and behavior in the saturated zone. Results were compared to those that were found at the Borden Landfill Experiment.

Description of the Site and Experiment

The site of the experiment was a sand and gravel aquifer located on Cape Cod, Massachusetts (Figure 6.13), near Otis Air Base. The nature of the site and the experiment were originally described by LeBlanc et al. (1991). The site was composed of approximately 100 m of unconsolidated sediments that overlay an impermeable bedrock strata. The upper 30 m of the site consisted of horizontally stratified sand and gravel deposits. Beneath these deposits the sediments consisted of fine-grained sand and silt. These deposits formed an unconfined aquifer with a mean water-table elevation located approximately 5 m beneath the ground surface. Several small-scale tracer tests conducted at the site revealed an overall effective porosity of 0.39. The magnitude of the horizontal hydraulic gradient was found to vary from 0.0014 to 0.0018, while a vertical gradient was nondetectable. The average hydraulic conductivity was estimated from an in situ flow test to be 110 m/day. More detailed analyses conducted with a borehole flowmeter in conjunction with permeameter testing of core samples showed the hydraulic conductivity

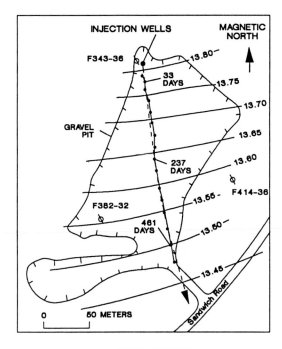

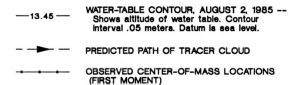

Figure 6.13 Horizontal trajectory of the center of mass of the bromide plume at the Cape Cod site. Source: Garabedian et al., 1991.

to vary by about 1 order of magnitude across the site due to the interbedded lens structure of the aquifer. Based on the values quoted for effective porosity, hydraulic gradient, and hydraulic conductivity, the mean ground water velocity at the site was calculated to be 0.4 m/day.

Solute injection at the Cape Cod site was accomplished by the placement of 7.6 m^3 of tracer solution into the saturated zone through three injection wells at the rate of 7.6 L/min (2.5 L/min/well). This slow injection rate was selected to minimize spreading of the plume during injection. Both nonreactive and reactive tracers were used. Bromide was selected as the nonreactive tracer due to the fact that chloride was present at relatively high natural concentrations in the aquifer. The reactive tracers used were lithium, molybdate, and fluoride. A summary of the solution composition is presented in Table 6.2.

The sampling array was composed of 656 multilevel samplers. The samplers were arranged as shown in Figure 6.14a. A typical vertical section through the sampling well grid (Figure 6.14b) shows the vertical distribution of sampling points. There were a total of 9840 individual sampling points at the site. The focus of sample collection was on obtaining synoptic ("snapshot") data; thus, sampling

TABLE 6.2   Injected Solutes at the Cape Code Site

Solute	Injected concentration (mg/L)	Injected mass (g)	Background concentration (mg/L)
Bromide	640	4900	<0.10
Lithium	78	590	<0.01
Molybdate	80	610	<0.02
Flouride	50	380	<0.20

Source: LeBlanc et al., 1991.

episodes took place on roughly monthly intervals over a 3-yr period (1985–1988). The data collected at the site included approximately 30,000 bromide analyses, 33,000 lithium analyses, and 38,000 molybdate analyses. Fluoride was abandoned as a tracer early in the test due to the fact that it was so strongly sorbed that trace concentrations remaining in solution became indistinguishable from background levels.

Overview of Plume Behavior

Significant features of the resulting contaminant cloud observed at the Cape Cod site include a predominant spreading in the direction of ground water flow, minimal spreading in the horizontal transverse direction, and essentially no spreading in the vertical direction. These features combined to create a narrow plume that was aligned in the direction of ground water flow. Figure 6.15 provides a comparison of the bromide, lithium, and molybdate plumes at various times following injection. Note that the movement of the reactive species, lithium and molybdate, is significantly retarded in comparison with the conservative bromide.

Inspection of the vertical concentration profiles for bromide recorded at the site reveals minimal vertical spreading (Figure 6.16). However, also of interest is the downward movement of the plume with horizontal travel distance. This vertical movement was attributed to two effects: (1) vertical flow associated with areal recharge and (2) sinking of the plume due to density differences between the tracer solution and the natural ground water (LeBlanc et al., 1991).

Experimental Observations

As a conservative tracer, the total mass of bromide in solution should be constant with time and should be equal to the mass injected. Evaluation of the mass of bromide contained in each plume for each of 16 separate synoptic data sets yielded the plot of total mass versus time data. The trend of effectively constant mass over time may be inferred. The differences between the calculated mass and the injected mass may be attributed to measurement and estimation error (Garabedian et al., 1991).

A graphical representation of the horizontal trajectory of the center of mass of the tracer plume is given in Figure 6.13. The movement of the center of mass is quite linear and closely corresponds to the direction predicted by the water-

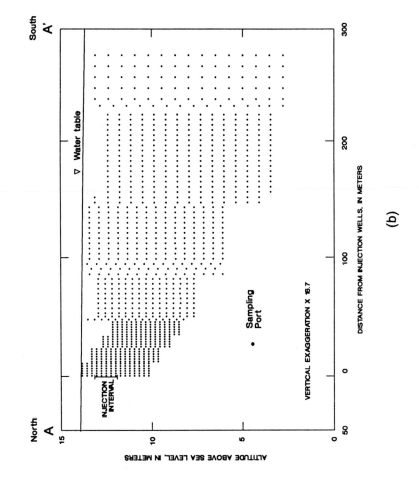

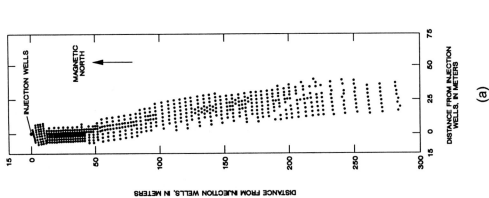

Figure 6.14 Locations of multilevel samplers and injection wells at the Cape Cod Site. (a) Plan view. (b) Vertical distribution of sampling points along representative section. Source: LeBlanc et al., 1991. © American Geophysical Union.

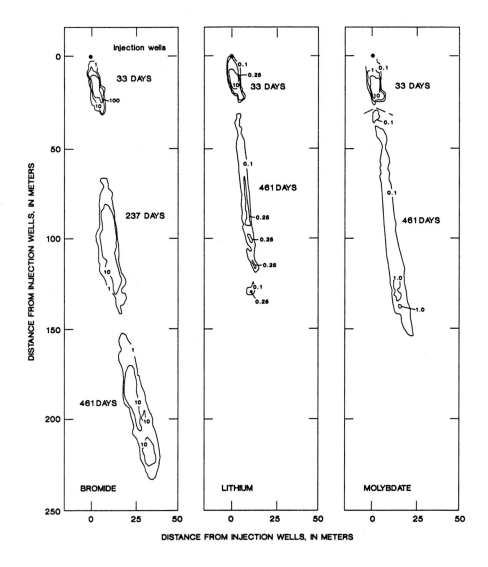

Figure 6.15 Areal distribution of maximum concentrations of bromide, lithium, and molybdate at various times after injection at the Cape Cod site. Source: LeBlanc et al., 1991. © American Geophysical Union.

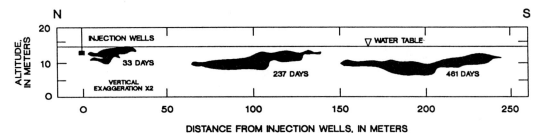

Figure 6.16 Vertical location of bromide plume at 33, 237, and 461 days after injection at the Cape Cod site. Source: LeBlanc et al., 1991. © American Geophysical Union.

table gradient. Although some vertical displacement of the center of mass was observed, as noted previously, this movement was considered inconsequential by the researchers in comparison with the large horizontal displacement, and so a strictly horizontal transport velocity could be assumed. A plot of horizontal displacement of the plume center of mass versus time (Figure 6.17) reveals a highly linear trend indicating that the transport velocity in the aquifer remained constant over time. The observed ground water flow velocity at the site as given by the slope of the line fitted through these data was computed as 0.42 m/day. This value is consistent with that calculated using the average measured porosity, hydraulic gradient, and hydraulic conductivity.

A theoretical approach to describing dispersivity in the ground water environment as a function of time has been developed by Dagan (1982, 1984). Dagan's theory suggests that limiting values of dispersivity will be reached for large values of time. Typically, this large time value corresponds to a plume that has reached a scale of spreading that is significantly larger than the scale of hydraulic conductivity variability (Garabedian et al., 1991).

To study this idea of an asymptotic dispersivity, the Cape Cod study was designed to incorporate a large (280 m) plume travel distance. Data from the experiment showed a constant value of longitudinal dispersivity develop after a short period of nonlinear growth. The data also indicated a positive-valued transverse dispersivity that was constant with time. The horizontal and transverse dispersivity values were determined to be 0.96 and 0.018 m, respectively. The vertical dispersivity was calculated to be 0.0015 m. While the asymptotic nature of the longitudinal dispersivity was clear, the nonzero value of transverse dispersivity was not in agreement with the theory of Dagan (1984). We may note from these results, however, that the typical assumption that the longitudinal dispersivity is an order of magnitude larger than the transverse dispersivity was supported by the data.

Significant Findings and Comparisons

A large-scale natural gradient tracer experiment was performed in a shallow, unconfined sand and gravel aquifer. The purpose of this experiment was to assess the validity of the often made simplifying assumptions used in the application of theoretical transport models to field situations where extensive data are not avail-

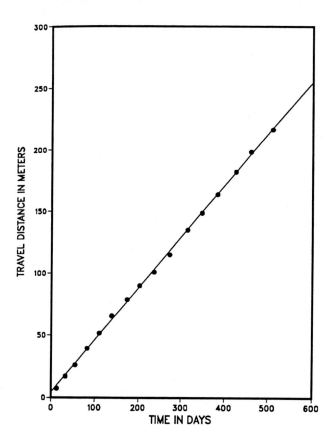

Figure 6.17 Horizontal displacement of the center of mass of the bromide plume at the Cape Cod site. Source: Garabedian et al., 1991. © American Geophysical Union.

able. Similarities between the results of the two tests at Borden and Cape Cod include the following:

1. The advective velocity and trajectory of a solute can be adequately predicted using conventionally collected data (that is, porosity, hydraulic gradient, and hydraulic conductivity estimated from any of several techniques).

2. The relative magnitudes of dispersion follow the trend of longitudinal dispersion > transverse dispersion > vertical dispersion and transverse dispersion is an order of magnitude less than longitudinal dispersion.

3. A limiting value of dispersivity is reached at large values of time (travel distance) after an early period of nonlinear growth.

4. A retardation mechanism functions to limit the movement of reactive solutes relative to nonreactive solutes.

Differences between the two tests indicate a much greater hydraulic conductivity and seepage velocity at Cape Cod (0.42 m/day) versus Borden (0.091 m/day). Longitudinal dispersivity was 0.96 m at Cape Cod versus 0.43 m at Borden, and the ratio of longitudinal to transverse dispersivity was much greater at Cape

Cod (50 to 1) compared to at Borden (11 to 1). Garabedian et al. (1991) present more details on the comparison of the two tests.

SUMMARY

Basic concepts in ground water transport processes include problems in one and two dimensions with advection, dispersion, adsorption, and biodegradation. The one-dimensional solutions to the advective dispersion equation allow simple predictions of breakthrough with adsorption or decay. Analytical expressions allow reasonable estimates for travel time and spreading of contaminant fronts in one-dimensional soil columns. Two-dimensional models can be solved analytically for simple boundary conditions, but numerical models are generally required for actual field problems. Semianalytical methods such as the HPS model from Galya (1987) can provide useful estimates of concentrations from simple source areas. Other models such as RESSQ can be used to calculate streamline patterns and contaminant transport in the presence of recharge wells and ponds.

Tests for dispersivity in the laboratory and in the field are reviewed. There are serious problems associated with parameter estimation at field sites, including estimation of hydraulic conductivity, dispersivity, and adsorption coefficients. Extrapolation of results from laboratory column experiments to field sites has proved to be very difficult; therefore, methods that can be directly applied to the field are being developed by researchers. Results from the extensive natural gradient tracer studies performed at the Borden landfill in Canada and at Cape Cod, Massachusetts, are briefly described. These two sites represent the most comprehensive data bases on advective–dispersive transport that are available in the general literature.

REFERENCES

ANDERSON, M. P., 1979, "Using Models to Simulate the Movement of Contaminants through Ground Water Flow Systems," *CRC Critical Rev. Environ. Control,* Chemical Rubber Co. 9:96.

ANDERSON, M. P., and WOESSNER, W. W., 1992, *Applied Groundwater Modeling,* Academic Press, San Diego, CA.

BAETSLE, L. H., 1969, "Migration of Radionuclides in Porous Media," *Progress in Nuclear Energy Series XII, Health Physics,* A. M. F. Duhamel, ed., Pergamon Press, Elmsford, NY, pp. 707–730.

BEAR, J., 1961, "Some Experiments on Dispersion," *J. Geophys. Res.* 66(8):2455–67.

BEAR, J., 1972, *Dynamics of Fluids in Porous Media.* American Elsevier, New York.

BEAR, J., 1979, *Hydraulics of Ground Water,* McGraw-Hill, New York.

BEAR, J., and VERRUIJT, A., 1987, *Modeling Groundwater Flow and Pollution,* D. Reidel, Dordrecht, Holland.

BERRY-SPARK, K., and BARKER, J. F., 1987, "Nitrate Remediation of Gasoline Contaminated Ground Waters: Results of a Controlled Field Experiment," in *Proc., Petroleum Hydrocarbons and Organic Chemicals in Ground Water,* National Water Well Association/American Petroleum Institute, pp. 3–10.

BORDEN, R. C., and BEDIENT, P. B., 1986, "Transport of Dissolved Hydrocarbons Influenced by Reaeration and Oxygen Limited Biodegradation: 1. Theoretical Development," *Water Resources Res.* 22:1973–1982.

BREDEHOEFT, J. D., and PINDER, G. F., 1973, "Mass Transport in Flowing Groundwater," *Water Resources Res.* 9:192–210.

CHARBENEAU, R. J., 1981, "Groundwater Contaminant Transport with Adsorption and Ion Exchange Chemistry: Method of Characteristics for the Case without Dispersion," *Water Resources Res.* 17(3):705–713.

CHARBENEAU, R. J., 1982, "Calculation of Pollutant Removal during Groundwater Restoration with Adsorption and Ion Exchange," *Water Resources Res.* 18(4):1117–1125.

CLEARY, R. W., and UNGS, M. J., 1978, "Groundwater Pollution and Hydrology, Mathematical Models and Computer Programs," *Rep. 78-WR-15,* Water Resources Program, Princeton University, Princeton, NJ.

COOPER, H. H., JR., BREDEHOEFT, J. D., and PAPADOPULOS, S. S., 1967, "Response of a Finite Diameter Well to an Instantaneous Charge of Water," *Water Resources Res.* 3(1): 263–269.

DAGAN, G., 1982, "Stochastic Modeling of Groundwater Flow by Unconditional and Conditional Probabilities, 2, The Solute Transport," *Water Resources Res.* 18(4):835–848.

DAGAN, G., 1984, "Solute Transport in Heterogeneous Porous Formations," *J. Fluid Mech.* 145:151–177.

DAGAN, G., 1986, "Statistical Theory of Ground Water Flow and Transport: Pore to Laboratory, Laboratory to Formation, and Formation to Regional Scale," *Water Resources Res.* 22(9):120S–134S.

DAGAN, G., 1987, "Theory of Solute Transport in Water," *Ann. Rev. Fluid Mechanics* 19:183–215.

DAGAN, G., 1988, "Time-dependent Macrodispersion for Solute Transport in Anistropic Heterogeneous Aquifers," *Water Resources Res.* 24(9):1491–1500.

DE JOSSELIN DE JONG, G., 1958, "Longitudinal and Transverse Diffusion in Granular Deposits," *Trans. Amer. Geophys. Union* 39(1):67.

DOMENICO, P. A., and SCHWARTZ, F. W., 1990, *Physical and Chemical Hydrogeology,* Wiley, New York.

FETTER, C. W., 1988, *Applied Hydrogeology,* Merrill Publishing Co., Columbus, OH.

FETTER, C. W., 1993, *Contaminant Hydrogeology,* Macmillan Publishing Co., New York.

FREEZE, R. A., 1975, "A Stochastic-conceptual Analysis of One-dimensional Groundwater Flow in a Nonuniform Homogeneous Media," *Water Resources Res.* 11(5):725–741.

FREEZE, R. A., and CHERRY, J. A., 1979, *Groundwater,* Prentice Hall, Englewood Cliffs, NJ.

FREYBERG, D. L., 1986, "A Natural Gradient Experiment on Solute Transport in a Sand Squifer, (2) Spatial Moments and the Advection and Dispersion of Nonreactive Tracers," *Water Resources Res.* 22(13):2031–2046.

FRIED, J. J., 1975, *Ground Water Pollution,* Elsevier, Amsterdam.

GALYA, D. P., 1987, "A Horizontal Plane Source Model for Ground-water Transport," *Ground Water,* 25(6):733–739.

GARABEDIAN, S. P., LEBLANC, D. R., GELHAR, L. W., and CELIA, M. A., 1991, "Large-scale Natural Gradient Tracer Test in Sand and Gravel, Cape Cod, Massachusetts, 2, Analysis of Spatial Moments for a Nonreactive Tracer," *Water Resources Res.* 27(5): 911–924.

GELHAR, L. W., and AXNESS, C. L., 1983, "Three-dimensional Stochastic Analysis of Macrodispersion in Aquifers," *Water Resources Res.* 19(1):161–180.

GELHAR, L. W., GUTJAHR, A. L., and NAFF, R. L., 1979, "Stochastic Analysis of Macro-dispersion in a Stratified Aquifer," *Water Resources Res.* 15(6):1387–1397.

GELHAR, L. W., MONTOGLOU, A., WELTY, C., and REHFELDT, K. R., 1985, "A Review of Field-scale Physical Solute Transport Processes in Saturated and Unsaturated Porous Media," *Final Proj. Rep. EPRI EA-4190,* Electric Power Research Institute, Palo Alto, CA.

GRINGARTEN, A. C., and SAUTY, J. P., 1975, "A Theoretical Study of Heat Extraction from Aquifers with Uniform Regional Flow," *J. Geophys. Res.* 80(35):4956–4962.

HANTUSH, M. S., 1956, "Analysis of Data from Pumping Tests in Leaky Aquifers," *J. Geophys. Res.* 69:4221–4235.

HUNT, B., 1978, "Dispersive Sources in Uniform Ground-water Flow," *J. Hydrol. Div. ASCE* 104:75–85.

HVORSLEV, M. J., 1951, "Time Lag and Soil Permeability in Groundwater Observations," *U.S. Army Corps Engrs. Waterways Exp. Sta. Bull. 36,* Vicksburg, MS.

JAVANDEL, I., DOUGHTY, C., and TSANG, C. F., 1984, *Groundwater Transport: Handbook of Mathematical Models,* American Geophysical Union, Water Resources Monograph 10, Washington, DC.

KEYS, W. S., and MACCARY, L. M., 1976, *Application of Borehole Geophysics to Water Resources Investigations,* 2nd ed., from *Techniques of Water Resources Investigations,* Chapter 1, Book 2, Collection of Environmental Data, U.S. Geological Survey, U.S. Government Printing Office, Washington, DC.

KONIKOW, L. F., and BREDEHEOFT, J. D., 1978, "Computer Model of Two-dimensional Solute Transport and Dispersion in Ground Water, Automated Data Processing and Computations," *Techniques of Water Resources Investigations of the U.S. Geological Survey,* Book 7, Chapter C2, Washington, DC.

LEBLANC, D. R., GARABEDIAN, S. P., HESS, K. M., GELHAR, L. W., QUADRI, R. D., STOLLENWERK, K. G., and WOOD, W. W., 1991, "Large-scale Natural Gradient Tracer Test in Sand and Gravel, Cape Cod, Massachusetts, 1, Experimental Design and Observed Tracer Movement," *Water Resources Res.* 27(5):895–910.

MACKAY, D. M., FREYBERG, D. L., ROBERTS, P. V., and CHERRY, J. A., 1986, "A Natural Gradient Experiment on Solute Transport in a Sand Aquifer, (1) Approach and Overview of Plume Movement," *Water Resources Res.* 22(13):2017–2029.

MERCER, J. W., and FAUST, C. R., 1981, *Ground-water Modeling,* 2nd ed. National Water Well Association, Columbus, OH.

MERCER, J. W., SKIPP, D. C., and GIFFIN, D., 1990, *Basics of Pump-and-Treat Ground-water Remediation Technology,* Robert S. Kerr Environmental Research Laboratory, U.S. Environmental Protection Agency, Ada, OK, EPA-600/8-90/003.

NELSON, R. W., 1977, "Evaluating the Environmental Consequences of Groundwater Contamination. 1. An Overview of Contaminant Arrival Distributions as General Evaluation Requirements," *Water Resources Res.* 14:409–415.

NELSON, R. W., 1978, "Evaluating the Environmental Consequences of Groundwater Contamination, Parts 1–4," *Water Resources Res.* 14(3):409–450.

NEUMAN, S. P., WINTER, C. L., and NEWMAN, C. N., 1987, "Stochastic Theory of Field-scale Fickina Dispersion in Anisotropic Porous Media," *Water Resources Res.* 23(3): 453–466.

OGATA, A., 1970, "Theory of Dispersion in a Granular Medium," U.S. Geological Survey Professional Paper 411-I.

OGATA, A., and BANKS, R. B., 1961, "A Solution of the Differential Equation of Longitudinal Dispersion in Porous Media," U.S. Geological Survey Professional Paper 411-A, U.S. Government Printing Office, Washington, DC.

PERLMUTTER, N. M., and LIEBER, M., 1970, "Dispersal of Plating Wastes and Sewage Contaminants in Groundwater and Surface Water, South Farmingdale–Massapequa Area, Nassau County, New York," *Water Supply Paper 1879-G,* U.S. Geological Survey, Washington, DC.

PICKENS, J. F., and GRISAK, G. E., 1981, "Scale-dependent Dispersion in a Stratified Granular Aquifer," *Water Resources Res.* 17(4):1191–1211.

PICKENS, J. F., JACKSON, R. E., INCH, K. J., and MERRITT, W. F., 1981, "Measurement of Distribution Coefficients Using a Radial Injection Duel-tracer Test," *Water Resources Res.* 17(3):529–44.

PINDER, G. F., 1973, "A Galerkin–Finite Element Simulation of Groundwater Contamination on Long Island, New York," *Water Resources Res.* 9:1657–1669.

PINDER, G. F., and GRAY, W. G., 1977, *Finite Element Simulation in Surface and Subsurface Hydrology,* Academic Press, New York.

PRICKETT, T. A., 1975, "Modeling Techniques for Groundwater Evaluation," *Advances in Hydroscience,* vol. 10, Academic Press, New York, pp. 1–143.

RIFAI, H. S., BEDIENT, P. B., WILSON, J. T., MILLER, K. M., and ARMSTRONG, J. M., 1988, "Biodegradation Modeling at a Jet Fuel Spill Site," *ASCE J. Environmental Engr. Div.* 114:1007–1019.

ROBERTS, P. V., GOLTZ, M. N., and MACKAY, D. M., 1986, "A Natural Gradient Experiment on Solute Transport in a Sand Aquifer, (4), Sorption of Organic Solutes and Its Influence on Mobility," *Water Resources Res.* 22(13):2047–2058.

SHEN, H. T., 1976, "Transient Dispersion in Uniform Porous Media Flow," *J. Hydrau. Div. ASCE* 102:707–716.

SMITH, L., and SCHWARTZ, F. W., 1980, "Mass Transport: 1. A Stochastic Analysis of Macroscopic Dispersion," *Water Resources Res.* 16:303–313.

SUDICKY, E. A., CHERRY, J. A., and FRIND, E. O., 1983, "Migration of Contaminants in Groundwater at a Landfill: A Case Study, 4, A Natural Gradient Dispersion Test," *J. Hydrology* 63(1/2):81–108.

WANG, H. F., and ANDERSON, M. P., 1982, *Introduction to Groundwater Modeling, Finite Difference and Finite Element Methods,* W. H. Freeman, San Francisco.

WILSON, J. L., and MILLER, P. J., 1978, "Two-dimensional Plume in Uniform Groundwater Flow," *J. Hydrau. Div. ASCE* 104:503–514.

Chapter
7

SORPTION AND OTHER CHEMICAL REACTIONS

7.1 INTRODUCTION

Many processes take place in ground water that are of interest when considering the fate and transport of contaminants in the subsurface. Contaminants can sorb onto the soil, volatilize, undergo chemical precipitation, be subjected to abiotic reactions and/or biodegradation, and be a part of oxidation–reduction sequences. Furthermore, radioactive compounds can decay. These processes and their impact on contaminant transport in soils and ground water are the main topic of this chapter.

Sorption causes contaminants to move more slowly than the flowing ground water; this effect is sometimes referred to as **retardation.** Sorption effects must be considered both in evaluating the potential for movement of the contaminants in an aquifer and in designing remediation activities at hazardous-waste sites. Radioactive decay, biodegradation, hydrolysis (an abiotic reaction), volatilization, precipitation, and oxidation–reduction affect the concentration of the contaminant within the plume, but may not necessarily have an impact on the rate of plume movement. Thus, to predict the movement of contaminants in the subsurface, it is necessary to have a fundamental understanding of and ability to quantify these interactions of contaminants in aquifers. Domenico and Schwartz (1990) and Fetter (1993) provide additional details on chemical processes in ground water.

7.2 THE CONCEPT OF SORPTION

Sorption can be defined as the interaction of a contaminant with a solid (Piwoni and Keeley, 1990). The process of sorption can be divided into **adsorption** and **absorption.** The former refers to an excess contaminant concentration at the surface of a solid, while the latter implies a more or less uniform penetration of the solid by a contaminant. In most environmental settings, we have little information concerning the specific nature of the interaction. The term sorption is used in a generic way to encompass both phenomena.

Two terms commonly used in discussing sorption theory are **sorbate** and **sorbent.** The sorbate is the contaminant that adheres to the sorbent, or sorbing material. In reference to ground water contamination, the sorbate will usually be an organic molecule, and the sorbent will be the soil or aquifer matrix. Another commonly used term is **partitioning,** which refers to the process by which a contaminant, originally in solution, distributes itself between the solution and the solid phase.

Sorption is determined experimentally by measuring the partitioning of a contaminant onto a particular sediment, soil, or rock type. A number of solutions containing various concentrations of the contaminant are well mixed with the solid, and the amount of contaminant removed is determined. The results of the experiment are plotted on a graph, known as an **isotherm,** that shows the concentration of the contaminant versus the amount sorbed on the solid. If the sorption process is rapid compared with the flow velocity, the dissolved contaminant will reach an equilibrium condition with the sorbed phase, and the process can be described using an **equilibrium isotherm.** If the sorption process is slow compared with the rate of fluid flow in the porous media, a **kinetic** model is needed. Only equilibrium models are discussed in this chapter; for information on kinetic models of sorption, the reader is referred to Fetter (1993). A detailed discussion of equilibrium sorption isotherms is presented in Section 7.4.

7.3 FACTORS INFLUENCING SORPTION

A number of factors control sorption. These include the chemical and physical characteristics of the contaminant, the composition of the surface of the solid, and the fluid media encompassing both. By gaining an understanding of these factors, conclusions can often be drawn about the impact of sorption on the movement and distribution of contaminants in the subsurface. Failure to account for sorption can result in significant underestimation of the mass of a contaminant at a site and of the time required for it to move from one point to another.

Contaminant Characteristics

The properties of a contaminant have a profound impact on its sorption behavior. Some of these include the following:

- Water solubility
- Polar–ionic character
- Octanol–water partition coefficient

Of the various parameters that affect the fate and transport of organic chemicals in the environment, water solubility is one of the most important. Highly soluble chemicals are easily and quickly distributed by the hydrologic cycle (Lyman et al., 1990). These chemicals tend to have relatively low adsorption coefficients for soils and sediments. The solubility of a chemical in water may be defined as the maximum amount of the chemical that will dissolve in pure water at a specified temperature. Above this concentration, two phases exist if the organic chemical is a liquid at the given temperature: a saturated aqueous solution and a liquid organic phase.

In discussing sorption, it might be useful to divide chemicals into three classes to define their polar–ionic character: (1) ionic or charged species, (2) uncharged polar species, and (3) uncharged nonpolar species. Most inorganic chemicals in aqueous solution will occur as ionic or charged species. Organic contaminants have representatives in all three of the sorption categories. Some examples of nonpolar species include trichloroethene (TCE), tetrachloroethene (PCE), the chlorinated benzenes, and the more soluble components of hydrocarbon fuels, such as benzene, toluene, and xylene. Pesticides and phenols, on the other hand, exist in solution as either charged or polar molecules.

The polar–ionic character of the contaminant affects sorption according to the following rule: for charged species, *opposites attract* and for uncharged species, *likes interact with likes*. Likes refers to the three categories of contaminants discussed and to the properties of the soil matrix presented in the next section.

The octanol–water partition coefficient (K_{ow}) represents the distribution of a chemical between octanol and water in contact with each other at equilibrium conditions:

$$K_{ow} = \frac{\text{concentration in octanol phase}}{\text{concentration in aqueous phase}}$$

K_{ow} has been measured in the laboratory for many chemicals and is a readily available parameter (see Table 7.1 for examples). Measured values of K_{ow} for organic chemicals have been found as low as 10^{-3} and as high as 10^7. The octanol–water partition coefficient is a key parameter in studies of the environmental fate of organic chemicals. It has been found to be related to water solubility and soil–sediment adsorption coefficients. In general, K_{ow} is a measure of the hydrophobicity of an organic compound. The more hydrophobic the contaminant is, the more likely it is to partition onto soils and to have a low solubility in water.

Soil Characteristics

Some of the most important characteristics of soil affecting the sorptive behavior of subsurface materials include the following:

- Mineralogy
- Permeability–porosity
- Texture
- Homogeneity
- Organic carbon content
- Surface charge
- Surface area

The combination of these characteristics describes the surfaces offered as

TABLE 7.1 Octanol–Water Partition Coefficient for Selected Chemicals

Compound	Observed log K_{ow}
Methylacetylene	0.94
Fluoroform	0.64
Isobutylene	2.34
Ethanol	-0.31
Dimethyl ether	0.10
Cyclohexane	3.44
Propane	2.36
2-Propanol	0.05
tert-Butylamine	0.40
2-Phenylethylamine	1.41
N-Phenylacetamide	1.16
p-Nitrophenol	1.91
Cyclohexene	2.86
1,2-Dichlorotetrafluoroethane	2.82
Hexachlorophene	3.93
1,2-Methylenedioxybenzene	2.08
2-Phenyl-1,3-indandione	2.90
Carbon tetrachloride	2.83
Dioxane	-0.42
2-Bromoacetic acid	0.41
2-Chloroethanol	0.03
Indene	2.92
Flourene	4.12
Anthracene	4.45
Pyrene	4.88
Phenyl vinyl ketone	1.88
Styrene	2.95
1-Phenyl-3-hydroxypropane	1.95
Methyl styryl ketone	2.07
1,1,2-Trichloroethylene	2.29
2-Methoxyanisole	2.08
Ethyl vinyl ether	1.04
Pyrazole	0.13
1,1-Difluoroethylene	1.24
1,2,3,4-Tetrahydroquinoline	2.29
2,2,2-Trifluoroethanol	0.41
2,2,2-Trifluoroacetamide	0.12
2,2,2-Trichloroethanol	1.35
2,2,2-Trichloroacetamide	1.04
Pyrimidine	-0.40
Glucose	-3.24
Cyclohexylamine	1.49
Neopentane	3.11
2-Methylpropane	2.76

Source: Lyman et al., 1990.

sorptive sites to contaminants in water passing through the subsurface matrix. For example, silts and clays have much higher surface areas than sand, usually carry a negative charge, and almost invariably associate with natural organic matter. Sandy materials offer little in the way of sorptive surfaces to passing contaminants. Silts and clays, on the other hand, and particularly those having substantial amounts of organic matter, provide a rich sorptive environment for all three categories of contaminants. Even the most porous and highly productive aquifers, composed of sands and gravels, usually have some fine-grained material, and a few percent of silts and clays can result in a substantial increase in the sorptive behavior of the aquifer material.

Fluid Media Characteristics

One of the most important characteristics of water that affect sorption is pH, for it dictates the chemical form and therefore the mobility of all contaminants susceptible to the gain or loss of a proton (Kan and Tomson, 1990). As an example, pentachlorophenol will primarily be an uncharged polar molecule in an aqueous solution whose pH is below about 4.7 and an anion when the pH is above that value, increasing its solubility from 14 to 90 mg/L.

Other characteristics of water that can influence the behavior of contaminants include the salt content and the dissolved organic carbon content. For example, dissolved organic matter at the relatively high concentrations found in many leachates has a significant effect on the mobility of most nonpolar organics.

7.4 SORPTION ISOTHERMS

The partitioning of solutes between liquid and solid phases in a porous medium as determined by laboratory experiments is commonly expressed in two-ordinate graphical form, where mass adsorbed per unit mass of dry solids (S) is plotted against the concentration of the constituent (C) in solution. These graphical relations of S versus C are known as **isotherms.**

Linear Sorption Isotherm

A linear adsorption isotherm (Figure 7.1) is described by the equation

$$S = K_d C^N \tag{7.1}$$

where S is the mass of solute sorbed per dry unit weight of solid (mg/kg), C is the concentration of the solute in solution in equilibrium with the mass of the solute sorbed onto the solid (mg/L), and K_d is the distribution coefficient (L/kg). Note that K_d is equal to the slope of the linear sorption isotherm.

Freundlich Sorption Isotherm

A more general equilibrium isotherm is the Freundlich isotherm shown graphically in Figure 7.2 and mathematically given by

$$S = K_d C^{1/N} \tag{7.2}$$

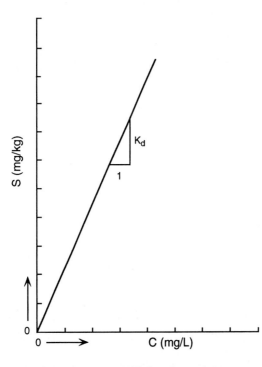

Figure 7.1   Linear sorption isotherm with S versus C plotting as a straight line.

where K_d is the Freundlich adsorption constant, N is the Freundlich exponent, C is the solution concentration (mg/L), and S is the adsorbed concentration (mg/kg).

Langmuir Sorption Isotherm

The Langmuir isotherm is based on the concept that a solid surface possesses a finite number of sorption sites. When all the sorption sites are filled, the surface will no longer sorb solute from solution. The Langmuir isotherm (Figure 7.3) is given by

$$S = \frac{\alpha\beta C}{1 + \alpha C} \tag{7.3}$$

where α is an absorption constant related to·the binding energy (L/mg), and β is the maximum amount of solute that can be absorbed by the solid (mg/kg).

7.5 HYDROPHOBIC THEORY FOR ORGANIC CONTAMINANTS

Many organic compounds dissolved in ground water adsorb onto solids according to the **hydrophobic theory.** These compounds have low solubilities in water, but can be dissolved in many nonpolar organic solvents. The hydrophobic theory advanced by Karickhoff et al. (1979) mainly proposes that the sorption of the

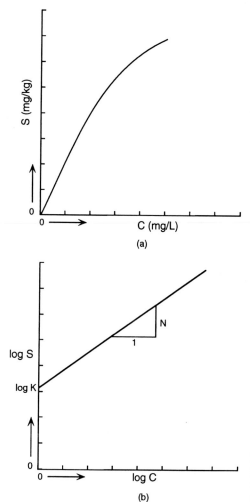

Figure 7.2 (a) Freundlich sorption isotherm plotted in terms of S versus C. (b) Freundlich sorption isotherm plotted in terms of $\log S$ versus $\log C$.

solute onto solids is almost exclusively onto the organic carbon fraction, f_{oc}, if it constitutes at least 1% of the soil or aquifer on a weight basis. Karickhoff et al. (1979) found a strong correlation between K_d and the organic carbon content of the sediment (Figure 7.4), and thus conveniently defined a term, K_{oc}, as

$$K_{oc} = \frac{K_d}{f_{oc}} \tag{7.4}$$

Karickhoff et al. (1979) reached these conclusions by investigating the sorption of polycyclic aromatics and chlorinated hydrocarbons, two hydrophobic organic families, on river and pond sediments. Special emphasis in these studies was placed on the sorption role of sediment particle size and organic matter content and on the correlation of sorption with sorbate aqueous solubility and octanol–water distribution coefficients.

Furthermore, Karickhoff et al. (1979) found a strong linear correlation be-

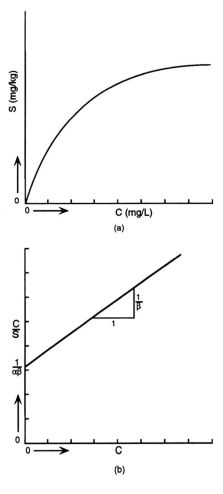

Figure 7.3 (a) Nonlinear Langmuir sorption isotherm. (b) Langmuir sorption isotherm plotted in terms of C/S versus C.

tween K_{oc} and K_{ow} and a rather poor linear correlation in K_{oc} relating to compound solubility (Figure 7.5). Linear least-square fitting of the K_{ow} and the K_{oc} data gave

$$K_{oc} = 0.63 K_{ow} \qquad (r^2 = 0.96) \qquad (7.5)$$

and

$$\log K_{oc} = 1.00 \log K_{ow} - 0.21 \qquad (r^2 = 1.00) \qquad (7.6)$$

In a subsequent paper, Karickhoff (1981) proposed

$$K_{oc} = 0.411 K_{ow} \qquad (r^2 = 0.994) \qquad (7.7)$$

Table 7.2 lists additional relationships that have been proposed by other researchers. The use of such relationships is predicated on the following (Karickhoff, 1984): (1) Sorption is primarily on the organic carbon in the soil or aquifer, (2) sorption is primarily hydrophobic, and (3) there is a linear relationship between sorption and the concentration of the solute.

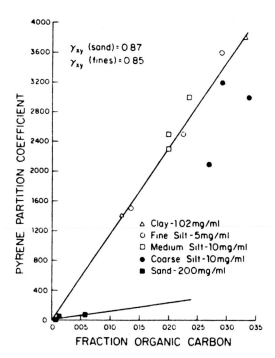

Figure 7.4 Pyrene partition coefficient as a function of sediment organic carbon. Source: Karickhoff et al., 1979. © American Geophysical Union.

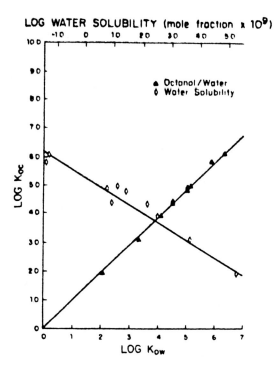

Figure 7.5 Sorption constant as a function of compound water solubility and octanol–water distribution coefficients. Source: Karickhoff et al., 1979. © American Geophysical Union.

TABLE 7.2 Regression Equations for the Estimation of K_{oc}

Equation[a]	No.[b]	$r^{2[c]}$	Chemical class represented	Reference
$\log K_{oc} = -0.55 \log S + 3.64$ (S in mg/L)	106	0.71	Wide variety, mostly pesticides	Kenaga and Goring (1978)
$\log K_{oc} = -0.54 \log S + 0.44$ (S in mole fraction)	10	0.94	Mostly aromatic or polynuclear aromatics; two chlorinated	Karickhoff et al. (1979)
$\log K_{oc} = -0.557 \log S + 4.277$ (S in μmoles/L)[d]	15	0.99	Chlorinated hydrocarbons	Chiou et al. (1979)
$\log K_{oc} = 0.544 \log K_{ow} + 1.377$	45	0.74	Wide variety, mostly pesticides	Kenaga and Goring (1978)
$\log K_{oc} = 0.937 \log K_{ow} - 0.006$	19	0.95	Aromatics, polynuclear aromatics, triazines, and dinitroaniline herbicides	Brown et al. (1981)
$\log K_{oc} = 1.00 \log K_{ow} - 0.21$	10	1.00	Mostly aromatic or polynuclear aromatics; two chlorinated	Karickhoff et al. (1979)
$\log K_{oc} = 0.94 \log K_{ow} + 0.02$	9	e	s-Triazines and dinitroaniline herbicides	Brown (1979)
$\log K_{oc} = 1.029 \log K_{ow} - 0.18$	13	0.91	Variety of insecticides, herbicides, and fungicides	Rao and Davidson (1980)
$\log K_{oc} = 0.524 \log K_{ow} + 0.855$	30	0.84	Substituted phenylureas and alkyl-N-phenylcarbamates	Briggs (1973)

[a] K_{oc} = soil (or sediment) adsorption coefficient; S = water solubility; K_{ow} = octanol–water partition coefficient

[b] No. = number of chemicals used to obtain

[c] r^2 = correlation coefficient for regression equation.

[d] Equation originally given in terms of K_{om}. The relationship $K_{om} = K_{oc}/1.724$ was used to rewrite the equation in terms of K_{oc}.

[e] Not available.

Source: Lyman et al., 1990.

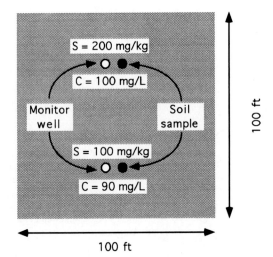

Figure 7.6 Aquifer system in Example 7.1.

Example 7.1 ESTIMATING SORBED MASS OF AN ORGANIC IN AQUIFER

For the system shown in Figure 7.6, calculate the mass of organics dissolved in the ground water and sorbed on the soil. Assume that the depth of the saturated zone is 10 ft, with a porosity of 0.25 and a solids density of 2.65 g/mL (1 gallon = 3.78 L and 1 ft^3 = 28.3 L).

Solution

$$\text{Volume of ground water in aquifer} = 100\,\text{ft} \times 100\,\text{ft} \times 10\,\text{ft}$$
$$\times (28.3\,\text{L}/1\,\text{ft}^3) \times 0.25$$
$$= 7.075 \times 10^5\,\text{L}$$

$$\text{Average concentration of benzene in ground water} = \frac{100 + 90}{2} = 95\,\text{mg/L}$$

$$\text{Mass of dissolved benzene} = 7.075 \times 10^5\,\text{L} \times 95\,\text{mg/L}$$
$$= 6.721 \times 10^7\,\text{mg}$$

$$\text{Mass of soil in aquifer} = 100\,\text{ft} \times 100\,\text{ft} \times 10\,\text{ft} \times 0.75$$
$$\times (28.3\,\text{L/ft}^3)$$
$$\times 2650\,\text{g/L} \times (1\,\text{kg}/1000\,\text{g})$$
$$= 5.625 \times 10^6\,\text{kg}$$

$$\text{Average soil concentration} = \frac{200 + 100}{2} = 150\,\text{mg/kg}$$

$$\text{Mass of benzene sorbed on soil} = 5.625 \times 10^6\,\text{kg} \times (150\,\text{mg/kg})$$
$$= 8.438 \times 10^8\,\text{mg}$$

7.6 SORPTION EFFECTS ON FATE AND TRANSPORT OF POLLUTANTS

The transfer by adsorption from the ground water to the solid part of the porous medium while flow occurs causes the advance rate of the contaminant front to be retarded. This concept is illustrated for one- and two-dimensional contaminant transport in Figure 7.7. In one dimension, the concentration profile of the retarded species travels behind the front of the conservative species. In two dimensions, the retarded plume is less spread out and exhibits higher concentrations than the nonreactive tracer plume.

When the partitioning of the contaminant can be described with a linear isotherm, the retardation of the front relative to the ground water flow can be expressed using a retardation factor, R:

$$R = \frac{v}{v_c} = 1 + \frac{\rho_b}{n} K_d \tag{7.8}$$

where v is the velocity of the ground water, v_c is the velocity of the $C/C_0 = 0.5$

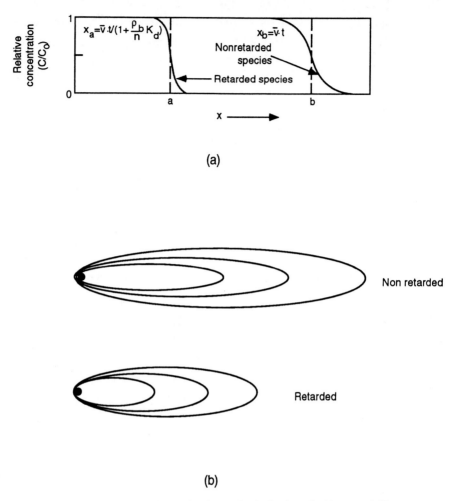

Figure 7.7 Advance of adsorbed and nonadsorbed solutes in (a) one and (b) two dimensions.

point on the concentration profile of the retarded constituent, ρ_b is the bulk mass density of the porous medium (the oven-dried mass of the sample divided by its field volume; values for this parameter range from 1.6 to 2.1 g/cm^3), K_d is the distribution coefficient, and n is the porosity.

A retardation factor of 10 implies that the contaminant plume moves 10 times slower than the ground water velocity. The determination of K_d is essential to the evaluation of the mobility of compounds in water. Generally, if $K_d = 0$, the relevant chemical species migrates at the same velocity as the water. If $K_d > 0$, the chemical species will be retarded or move at a slower velocity than the ground water. Distribution coefficients for reactive solutes range from values near zero to 10^3 mL/g or greater. For K_d values that are orders of magnitude larger than 1, the solute is essentially immobile. Integrating adsorption into the one-dimensional advection–dispersion equation at steady state was discussed in Chapter 6. See Section 6.5 for details.

Example 7.2   ESTIMATING THE RETARDATION FACTOR USING FIELD DATA

Using the data from Example 7.1, estimate the retardation factor for the system.

Solution. The distribution coefficient can be estimated from

$$K_d = \frac{200 - 100}{100 - 90} = 10 \text{ L/kg}$$

$$R = 1 + \frac{2.65}{0.25} \; 10 \text{ L/kg}$$

$$= 107$$

7.7 ESTIMATION OF SORPTION

Adsorption can be evaluated and predicted using eight methods (Olsen and Davis, 1990):

- Use of empirical field data
- Methods based on K_{ow}
- Methods based on water solubility
- Methods based on molecular structure
- Methods based on surface area
- Laboratory methods
- Field column devices and injection tests
- Methods based on plume location

Use of Empirical Field Data

This method is based on knowing the concentrations of the chemical in the ground water and in the soil samples in contact with the ground water. The partition coefficient, K_d, can be calculated using Eq. (7.1). The adsorbed concentration, S, should be based on analyses of a dry soil sample (that is, only the chemical adsorbed on the soil and not in the soil water should be considered).

The wet soil samples collected from the field site should be analyzed for moisture content, and the chemical concentration analyzed in the soil containing water should be corrected for the chemical actually contained in the water fraction of the soil. The concentration of the chemical in the soil moisture is usually assumed equal to the chemical concentration in the water taken from the same depth as the soil samples.

Methods Based on K_{ow}

K_{ow} has been measured in the laboratory for many chemicals. Because of this, many researchers have developed relationships between K_{oc} and K_{ow}. Table 7.2 summarizes most of these equations and lists the compounds that were used to develop the relationships.

Methods Based on Water Solubility

Water solubility has also been correlated with K_{oc}. Table 7.2 provides three equations that have been proposed. However, Olsen and Davis (1990) advise caution when using the equations in Table 7.2 because of the wide range of solubility values reported in the literature for some compounds.

Molecular Structure Methods

Estimates of K_{oc} may be made based on structure activity relationships (SAR). Three SAR methods can be used and the reader is referred to Olsen and Davis (1990) for a discussion of these methods.

Methods Based on Surface Area

These methods apply to systems where inorganic adsorption dominates adsorption; that is, the amount of organic carbon is below the critical level and the mineral surface area is high, allowing adsorption on the inorganic mineral surfaces. Two equations have been proposed for calculating K_d, as shown in Table 7.3.

Laboratory Methods

Laboratory studies can be done to estimate K_d. Two types of experiments can be performed: batch tests and column tests. Batch experiments are performed with contaminated water and noncontaminated soil from the site. The water and soil are mixed either at a constant water–soil ratio or as a geometrically progressing series of water–soil ratios. After the soil–water system has reached equilibrium, the water is analyzed to determine the decrease in concentration of the chemical

---

TABLE 7.3 Equations for Estimating K_d from Surface Area

1. $\log K_d = -0.061\,(SA) + 2.89$

2. $K_d = \dfrac{SA}{(K_{ow})^{0.16}}$

SA = surface area of soil (m^2/g)

Source: Olsen and Davis, 1990.

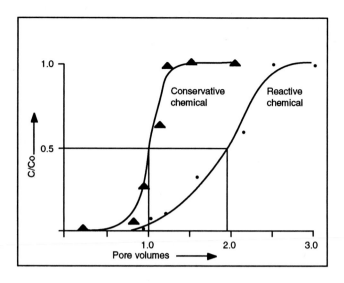

Figure 7.8 Typical results of column test (for adsorption). Source: Olsen and Davis, 1990.

of concern. This decrease is attributed to adsorption by the soil. Adsorption isotherms similar to those shown in Figures 7.1 through 7.3 are developed by plotting the equilibrium concentrations in solution versus the amount of the chemical on the soil. The data are usually interpreted using the Freundlich and Langmuir equations discussed in Section 7.4.

In column experiments, noncontaminated soil is packed in a column and contaminated water is introduced to the column. The concentration in the effluent is measured at selected intervals. The effluent data, expressed as a ratio of the influent concentration (C/C_o), are plotted versus time (measured as the number of pore volumes flushed into the column), as shown in Figure 7.8. A nonreactive tracer is usually introduced in the influent for comparison purposes. The velocity of the contaminant is compared to that of the nonreactive tracer, and the retardation coefficient is estimated using Eq. (7.8).

Field Column Devices and Injection Tests

A variety of field tests can be conducted to estimate sorption. Examples of field tests for sorption include tracer tests or actual injection of contaminants in the field. These tests accurately predict sorption at field sites. There are many excellent examples of injection field studies; the reader is referred to Chapters 6 and 13 for discussion of the Moffett site, the Borden landfill, Traverse City, and the Cape Cod sites. The data in Figure 7.9, for example, show the response at downgradient observation wells to the injection of reactive and nonreactive constituents at the Borden landfill.

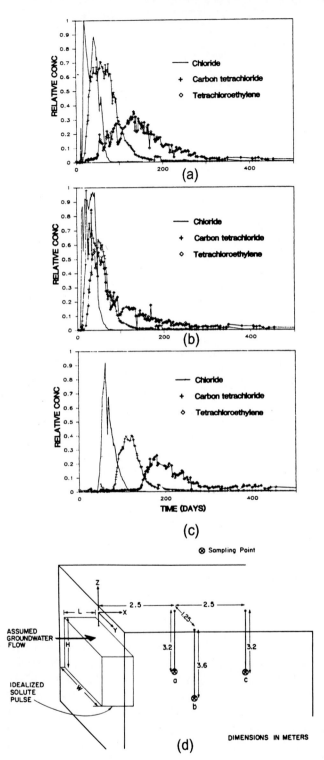

Figure 7.9 Field retardation data from the Borden landfill site.   Source: Roberts et al., 1986. © American Geophysical Union.

Methods Based on Plume Location

This method involves comparing the location of the plume relative to the location of a nonreactive chemical plume. Basically, the locations of the reactive and nonreactive contaminant chemical fronts relative to the source area are compared, and the retardation coefficient is calculated using

$$R = \frac{\text{distance from source of nonreactive plume}}{\text{distance from source of reactive plume}} \qquad (7.9)$$

In Eq. (7.9), the distances to the concentrations equal to 0.5 of the source concentration should be used.

As an alternative to using location data, retardation can be estimated using temporal data; that is, the retardation factor is determined using

$$R = \frac{\text{time for reactive chemical to reach a given point}}{\text{time for nonreactive chemical to reach a given point}} \qquad (7.10)$$

The temporal data approach was used at the Borden landfill site. The distances traveled by the injected chemicals are plotted for different points in time in Figure 7.10. The resulting retardation factors shown in Figure 7.11 indicate that higher retardation factors are achieved at locations farther from the source.

Example 7.3 SORPTION IN A TWO-DIMENSIONAL DISSOLVED PLUME

The USGS two-dimensional transport model (see Chapter 10) was used to simulate the migration of a contaminant in ground water for two scenarios: (1) no retardation of the contaminant, and (2) a retardation factor of 9. A hypothetical aquifer about

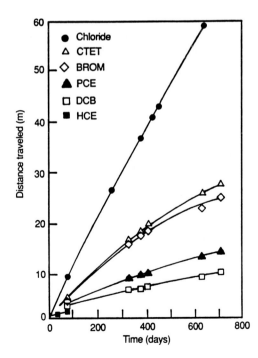

Figure 7.10 Distances traveled by plume centers of mass at the Borden landfill site. Source: Roberts et al., 1986. © American Geophysical Union.

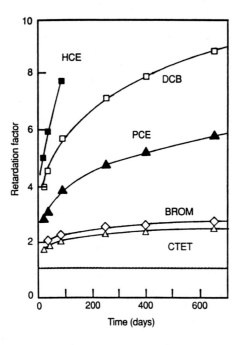

Figure 7.11 Estimated retardation factors from synoptic sampling data at the Borden landfill site. Source: Roberts et al., 1986. © American Geophysical Union.

25 ft thick with a transmissivity of 0.0025 ft²/sec was used in the analysis. A contaminant source leaking at a rate of 129 gal/day was also assumed. The source concentration equaled 100 mg/L.

The resulting plumes for the two scenarios after 8 yr are shown in Figure 7.12. For the no retardation case, the plume extends about 450 ft downgradient from the source and has a width of approximately 150 ft. The maximum concentration within the plume in this case is about 36 mg/L. On the other hand, the plume for the retardation scenario only extends about 50 ft and has a maximum concentration of 18 mg/ L. The retardation plume also has a total sorbed mass on the soil of about 130 kg, and only 10% of the total released contamination is present in the dissolved form.

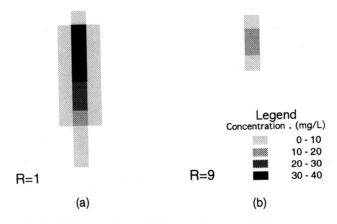

Figure 7.12 (a) Nonretarded and (b) retarded plumes for Example 7.3.

7.8 REDOX REACTIONS

A redox reaction involves electrons (Pankow, 1991). When a chemical species accepts electrons, it undergoes a reduction in its total charge. A specific example is the reduction of Fe^{3+} to Fe^{2+}:

$$Fe^{3+} + e^- = Fe^{2+} \qquad (7.11)$$

Similarly, when a chemical species loses an electron, it is oxidized:

$$Fe^{2+} = Fe^{3+} + e^- \qquad (7.12)$$

It is important to note that no free electrons result from a redox reaction because each complete reaction involves a pair of redox reactions (Morel, 1983). For example, the redox equation

$$O_2 + 4Fe^{2+} + 4H^+ = 2H_2O + 4Fe^{3+} \qquad (7.13)$$

consists of two half-redox reactions:

$$O_2 + 4H^+ + 4e^- = 2H_2O \qquad (7.14)$$

$$4Fe^{2+} = 4Fe^{3+} + 4e^- \qquad (7.15)$$

The conservation of mass equations can be written in terms of the concentration or activities of reactants and products and an appropriate equilibrium constant, K (Domenico and Schwartz, 1990). For example, the mass law expression for the half-reaction

$$\overset{\overrightarrow{\text{reduction}}}{Ox + ne^- = Red}_{\underset{\longleftarrow}{\text{oxidation}}} \qquad (7.16)$$

where Ox and Red refer to oxidants and reductants, respectively, can be written as

$$K = \frac{[\text{Red}]}{[\text{Ox}]\,[e^-]^n} \qquad (7.17)$$

The electron activity can be written by rearranging Eq. (7.17):

$$[e^-] = \left\{ \frac{[\text{Red}]}{[\text{Ox}]K} \right\}^{1/n} \qquad (7.18)$$

Another method of defining the electron activity is to use pe such that

$$pe = -\log[e^-] = \left(\frac{1}{n}\right)\left\{ \log K - \log \frac{[\text{Red}]}{[\text{Ox}]} \right\} \qquad (7.19)$$

The term $\log K$ is replaced with $pe°$ when $n = 1$, or the reaction is written in terms of a single electron such that

$$pe = pe° - \log \frac{[\text{Red}]}{[\text{Ox}]} \tag{7.20}$$

Large values of pe indicate low values of electron activity. Low values of electron activity favor the existence of electron-poor (oxidized) species. Small values of pe indicate high electron activity and thus correspond to electron-rich species (reduced). The subsurface environment can be divided up into redox regions as follows (for a pH of 7.0): $pe > 7$, oxic; $2 < pe < 7$, suboxic; and $pe < 2$, anoxic (Knox et al., 1993).

Another common method of characterizing redox conditions is in terms of E_H or the redox potential, which has units of volts. The redox potential is related to pe by the following equation:

$$E_H = \frac{2.3RT}{F} pe \tag{7.21}$$

where F is the Faraday constant defined as the electrical charge of 1 mole of electrons, with $2.3RT/F$ equal to 0.059 V at 25°C. In practice, however, E_H is measured using an electrochemical cell and therefore represents the electrode potential.

Many redox reactions are mediated by naturally occurring microorganisms. In a closed aqueous system containing organic material, say CH_2O, oxidation of organic matter is observed to occur first by reduction of O_2. This will be followed by reduction of NO_3^- and NO_2^-. The succession of these reactions follows the decreasing pe level. Reduction of MnO_2, if present, should occur at about the same pe levels as that of nitrate reduction, followed by reduction of $FeOOH(s)$ to Fe^{2+} (see Table 7.4 and Figure 7.13). When sufficiently negative pe levels have

TABLE 7.4   Reduction and Oxidation Reactions That May Be Combined to Result in Biologically Mediated Exergonic Processes (pH = 7)

Reduction	$pe°(W) =$ $\log K(W)$	Oxidation	$pe°(W) =$ $-\log K(W)$
(A) $\frac{1}{4}O_2(g) + H^+(W) + e = \frac{1}{2}H_2O$	+13.75	(L) $\frac{1}{4}CH_2O + \frac{1}{4}H_2O = \frac{1}{4}CO_2(g) + H^+(W) + e$	−8.20
(B) $\frac{1}{5}NO_3^- + \frac{6}{5}H^+(W) + e = \frac{1}{10}N_2(g) + \frac{3}{5}H_2O$	+12.65	(L-1) $\frac{1}{2}HCOO^- = \frac{1}{2}CO_2(g) + \frac{1}{2}H^+(W) + e$	−8.73
(C) $\frac{1}{2}MnO_2(s) + \frac{1}{2}HCO_3^- (10^{-3}) + \frac{3}{2}H^+(W) + e$ $= \frac{1}{2}MnCO_3(s) + H_2O$	+8.9	(L-2) $\frac{1}{2}CH_2O + \frac{1}{2}H_2O = \frac{1}{2}HCOO^- + \frac{3}{2}H^+(W) + e$	−7.68
(D) $\frac{1}{8}NO_3^- + \frac{5}{4}H^+(W) + e = \frac{1}{8}NH_4^+ + \frac{3}{8}H_2O$	+6.15	(L-3) $\frac{1}{2}CH_3OH = \frac{1}{2}CH_2O + H^+(W) + e$	−3.01
(E) $FeOOH(s) + HCO_3^- (10^{-3}) + 2H^+(W) +$ $e = FeCO_3(s) + 2H_2O$	−0.8	(L-4) $\frac{1}{2}CH_4(g) + \frac{1}{2}H_2O = \frac{1}{2}CH_3OH + H^+(W) + e$	+2.88
(F) $\frac{1}{2}CH_2O + H^+(W) + e = \frac{1}{2}CH_3OH$	−3.01	(M) $\frac{1}{8}HS^- + \frac{1}{2}H_2O = \frac{1}{8}SO_4^{2-} + \frac{9}{8}H^+(W) + e$	−3.75
(G) $\frac{1}{8}SO_4^{2-} + \frac{9}{8}H^+(W) + e = \frac{1}{8}HS^- + \frac{1}{2}H_2O$	−3.75	(N) $FeCO_3(s) + 2H_2O = FeOOH(s) + HCO_3^-$ $(10^{-3}) + 2H^+(W) + e$	−0.8
(H) $\frac{1}{8}CO_2(g) + H^+(W) + e = \frac{1}{8}CH_4(g) +$ $\frac{1}{4}H_2O$	−4.13	(O) $\frac{1}{8}NH_4^+ + \frac{3}{8}H_2O = \frac{1}{8}NO_3^- + \frac{5}{4}H^+(W) + e$	+6.16
(J) $\frac{1}{6}N_2 + \frac{4}{3}H^+(W) + e = \frac{1}{3}NH_4^+$	+4.68	(P) $\frac{1}{2}MnCO_3(s) + H_2O = \frac{1}{2}MnO_2(s) + \frac{1}{2}HCO_3^-$ $(10^{-3}) + \frac{3}{2}H^+(W) + e$	8.9

Source: Stumm and Morgan, 1981.

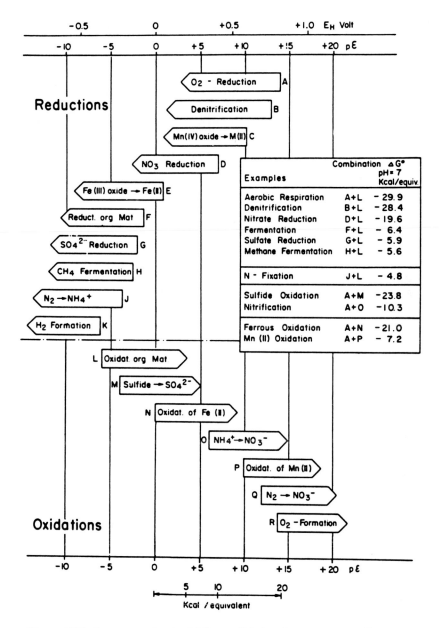

Figure 7.13 Sequence of microbially mediated redox processes. Source: Stumm and Morgan, 1981. © John Wiley & Sons, Inc. Reprinted with permission.

been reached, fermentation reactions and reduction of SO_4^{2-} and CO_2 may occur almost simultaneously (Stumm and Morgan, 1981).

Since the reactions considered, with the possible exception of the reduction of $MnO_2(s)$, and $FeOOH(s)$, are biologically mediated, the chemical reaction sequence is paralleled by an ecological succession of microorganisms (aerobic heter-

TABLE 7.5 Type of Organic Functional Groups That Are Generally Resistant to Hydrolysis

Alkanes	Aromatic amines
Alkenes	Alcohols
Alkynes	Phenols
Benzenes/biphenyls	Glycols
Polycyclic aromatic hydrocarbons	Ethers
Heterocyclic polycyclic aromatic hydrocarbons	Aldehydes
Halogenated aromatics/PCBs	Ketones
Dieldrin/aldrin and related halogenated hydrocarbon pesticides	Carboxylic acids
Aromatic nitro compounds	Sulfonic acids

Source: Lyman et al., 1990.

otrophs, denitrifiers, fermenters, sulfate reducers, and methane bacteria; see Chapter 8).

7.9 HYDROLYSIS

Hydrolysis is a chemical transformation process in which an organic molecule, RX, reacts with water. The reaction results in the introduction of a hydroxyl group ($-OH$) into the organic compound:

$$RX + H_2O \rightarrow ROH + H^+ + X^- \tag{7.22}$$

and

$$CH_3CH_2CH_2Br + H_2O \rightarrow CH_3CH_2CH_2OH + H^+ + Br^- \tag{7.23}$$

Hydrolysis is likely to be the most important reaction of organic compounds with water in aqueous environments and is a significant environmental fate process for many organic chemicals (Lyman et al., 1990). Many organic functional groups (see Table 7.5) are relatively or completely inert with respect to hydrolysis. Table 7.6 lists other functional groups that may hydrolyze under certain environmental conditions.

Typical hydrolysis reactions are first order with respect to the concentration of the chemical reactant:

$$-\frac{d[RX]}{dt} = k_T [RX] \tag{7.24}$$

where k_T is the hydrolysis rate constant. The first-order dependence is important, because it implies that the hydrolysis half-life is independent of the RX concentration.

Reaction half-lives range from several seconds to thousands of years. The rate expression in Eq. (7.24) is an oversimplification for most organic hydrolysis

**TABLE 7.6 Types of Organic Functional Groups
That Are Potentially Susceptible to Hydrolysis**

Alkyl halides	Nitriles
Amides	Phosphonic acid esters
Amines	Phosphoric acid esters
Carbamates	Sulfinic acid esters
Carboxylic acid esters	Sulfuric acid esters
Epoxides	

Source: Lyman et al., 1990.

reactions. The rate constant k_T is a pseudo first-order rate constant, which may include contributions from acid- and base-catalyzed hydrolysis:

$$k_T = k_H[H^+] + k_0 + k_{OH}[OH^-] \tag{7.25}$$

where k_H is the rate constant for specific acid-catalyzed hydrolysis or catalysis by hydronium ion, H^+; k_o is the rate constant for neutral hydrolysis; and k_{OH} is the rate constant for specific base-catalyzed hydrolysis or catalysis by the hydroxide ion, OH^-.

7.10 VOLATILIZATION

Volatilization is a process of liquid- or solid-phase evaporation that occurs when contaminants present either as nonaqueous phase liquids or dissolved in water contact a gas phase (Domenico and Schwartz, 1990). Volatilization can occur in the saturated and unsaturated zones. Knowledge of volatilization rates is necessary to determine the amount of chemical that enters the atmosphere and the change of pollutant concentrations in ground water or in the unsaturated zone.

Volatilization is an important process in contaminant fate and transport for the following reasons:

- It accounts for losses of organic contaminants from the dissolved phase.
- It might cause some problems in sampling and analysis. If samples are exposed to the atmosphere, volatilization reduces the concentration of the contaminant exponentially.
- It might cause a vapor plume to form in the unsaturated zone. The vapor plume can potentially migrate offsite and into the atmosphere.
- It produces soil gases containing volatiles. These soil gases, when allowed to accumulate in the basement of structures, can cause fires, explosions, or health problems.
- It forms the basis for detecting and monitoring contamination in the subsurface.

The factors that control volatilization are the solubility, molecular weight, and vapor pressure of the chemical and the nature of the medium (saturated and unsaturated zone) through which it must pass. The vapor pressure of a liquid or solid is the pressure of the gas in equilibrium with respect to the liquid or solid

TABLE 7.7 Volatilization Parameters for Selected Chemicals

Chemical	M (g/mol)	Solubility[a]		Vapor pressure at 20°C	
		mg/L	mol/m^3	mm Hg	atm
					Low volatility
3-Bromo-*f*-propanol	139	1.7×10^{-5}	1233	0.1	1.3×10^{-4}
Dieldrin	381	0.25	6.6×10^{-4}	1×10^{-7}	1.3×10^{-0}
					Middle range
Lindane	290.9	7.3	2.5×10^{-2}	9.4×10^{-6}	1.2×10^{-8}
m-Bromonitrobenzene	170	104	58.8	0.07	9.2×10^{-5}
Pentachlorophenol	266	14	5.3×10^{-2}	1.4×10^{-4}	1.8×10^{-7}
4-*t*-Butylphenol	150	10^3	6.7	0.046	6.1×10^{-5}
Aldrin	365	0.2	5.5×10^{-4}	6×10^{-6}	7.9×10^{-9}
Nitrobenzene	123	2×10^3	16.3	0.27	3.5×10^{-4}
DDT	354.5	1.2×10^{-3}	3.4×10^{-6}	1×10^{-7}	1.3×10^{-10}
Phenanthrene	178	1.29	7×10^{-3}	2.1×10^{-4}	2.8×10^{-7}
Acetylene tetrabromide	344	650	1.9	0.3	3.9×10^{-4}
Aroclor 1242	254	0.24	9.5×10^{-4}	4.1×10^{-4}	5.3×10^{-7}
Ethylene dibromide	188	4.3×10^3	22.9	11.6	1.5×10^{-2}
					High volatility
Ethylene dichloride	99	8.0×10^3	80.8	67	0.09
Naphthalene	128	33	0.026	0.23	3×10^{-4}
Biphenyl	154	7.5	0.05	0.06	7.5×10^{-5}
Methylene chloride	85	1.3×10^4	155	349	0.46
Chlorobenzene	113	472	4.2	11.8	1.6×10^{-2}
Chloroform	119	8×10^3	67	246	0.32
o-Xylene	106	175	1.7	6.6	8.7×10^{-3}
Benzene	78	1780	22.8	95.2	1.25×10^{-1}
Toluene	92	515	5.6	28.4	3.7×10^{-2}
Perchloroethylene	166	400	2.4	14.3	2×10^{-2}
Ethyl benzene	106	152	1.43	9.5	1.25×10^{-2}
Trichloroethylene	131	1×10^3	7.6	60	8×10^{-2}
Mercury	201	3×10^{-2}	1.5×10^{-4}	1.3×10^{-3}	1.7×10^{-6}
Methyl bromide	95	1.3×10^4	137	1.4×10^3	1.8
Cumene (isopropyl benzene)	120	50	0.416	4.6	6.1×10^{-3}
1,1,1-Trichloroethane	133	950	7.1	100	0.13
Carbon tetrachloride	154	800	5.2	91	0.12
Methyl chloride	50.5	7.4×10^3	146	3.6×10^3	4.74
Ethyl bromide	109	900	8.3	460	6.1×10^{-1}
Vinyl chloride	62.5	90	1.44	2580	3.4
2,2,4-Trimethyl pentane	114	2.44	2.1×10^{-2}	49.3	6.5×10^{-2}
n-Octane	114	0.66	5.8×10^{-3}	14.1	1.85×10^{-2}
Fluoro-trichloromethane	137				
Ethylene	28	131	4.7		>40

[a] Source: Mackay and Leinonen, 1975; Mackay et al., 1979; Mackay and Wolkoff, 1973; Verschueren, 1977.
Source: Lyman et al., 1990.

Henry's law Const. H	H	Phase Exchange Coeff.	(cm/hr)	Mass Transfer Coefficient K_L	Half-life
Atm-m³/mol	Non-dim	Liquid k_l	Gas k_g	(cm/hr)	$\tau_{1/2}$ (hr)
($H < 3 \times 10^{-7}$)					
1.1×10^{-7}	4.6×10^{-6}	16	1600	7.4×10^{-3}	9400
2×10^{-7}	8.9×10^{-6}	12	990	8.8×10^{-3}	7840
($3 \times 10^{-7} < H < 10^{-3}$)					
4.8×10^{-7}	2.2×10^{-5}	14	1130	0.025	2760
1.6×10^{-6}	7.4×10^{-5}	16	1500	0.11	626
3.4×10^{-6}	1.5×10^{-4}	12	1150	0.17	406
9.1×10^{-6}	3.8×10^{-4}	17	1600	0.59	117
1.4×10^{-5}	6.1×10^{-4}	12	1810	1.0	68
2.2×10^{-5}	9.3×10^{-4}	20	1700	1.5	45
3.8×10^{-5}	1.7×10^{-3}	13	1025	1.5	45
3.9×10^{-5}	1.7×10^{-3}	18	1450	2.2	31
2.1×10^{-4}	8.9×10^{-3}	13	1000	5.3	13.1
5.6×10^{-4}	2.4×10^{-2}	15	1210	9.9	7
6.6×10^{-4}	2.8×10^{-2}	16	1400	11.4	6.1
($H > 10^{-3}$)					
1.1×10^{-3}	4×10^{-2}	22	1900	17.1	4
1.15×10^{-3}	4.9×10^{-2}	21	1765	16.9	4.1
1.5×10^{-3}	6.8×10^{-2}	19	1554	16.2	4.3
3×10^{-3}	0.13	25	2000	22.8	3
3.7×10^{-3}	0.165	22	1820	15.1	4.6
4.8×10^{-3}	0.2	20	1700	18.9	3.7
5.1×10^{-3}	0.22	23	1870	22	3.2
5.5×10^{-3}	0.24	27	2180	26	2.7
6.6×10^{-3}	0.28	25	2010	24	2.9
8.3×10^{-3}	0.34	17	1450	16.4	4.2
8.7×10^{-3}	0.37	23	1870	22.3	3.1
1×10^{-2}	0.42	21	1700	20.4	3.4
1.1×10^{-2}	0.48	17	1360	16.6	4.2
1.3×10^{-2}	0.56	23	2000	22.5	3.1
1.5×10^{-2}	0.62	22	1760	21.6	3.2
1.8×10^{-2}	0.77	19	1650	18.7	3.7
2.3×10^{-2}	0.97	19	1500	18.8	3.7
2.4×10^{-2}	0.36	30	2600	29	2.4
7.3×10^{-2}	3.1	22	1800	22	3.1
2.4	99	28	2400	28	2.5
3.1	129	22	1810	22	3.1
3.2	136	22	1810	22	3.1
5		11.3	1090	11.3	6.1
>8.6	~360	40	3700	40	1.7

at a given temperature. Vapor pressure represents a compound's tendency to evaporate and is essentially the solubility of an organic solvent in a gas. At equilibrium, Raoult's law describes the equilibrium partial pressure of a volatile organic in the atmosphere above an ideal solvent (like benzene):

$$P_{org} = x_{org} \cdot P_{org}^{\circ} \tag{7.26}$$

where P_{org} is the partial pressure of the vapor in the gas phase, x_{org} is the mole fraction of the organic solvent, and P_{org}° is the vapor pressure of the pure organic solvent.

Volatilization of the dissolved organic solutes from water is described by Henry's law. The Henry's law constant (K_H), expressed commonly as atmospheres-cubic meters/mole, is equal to P° in atmospheres divided by the solubility of the compound in water (moles/m^3). The values of K_H for different chemicals give some insight into the rate of mass transfer from the liquid to the gas phase. If K_H is less than about 3×10^{-7} atm-m^3/mol, the chemical could be considered essentially nonvolatile (Lyman et al., 1982). Where $K_H > 10^{-7}$ atm-m^3/mol, volatilization might be a significant mass-transfer mechanism.

Thus, the most common convention treats volatilization of pure solvents (for example, benzene) using Raoult's law and the volatilization of solutes (for example, benzene dissolved in water) using Henry's law. Complex mixtures such as gasoline are treated with modified forms of Raoult's law to account for the nonideal effects of the mixture. Table 7.7 lists vapor pressures and Henry's law constants for a selected group of organics.

7.11 ACID–BASE REACTIONS

Acid–base reactions affect the pH and the ion chemistry of ground water. An acid is a substance with a tendency to lose a proton, while a base is a substance with a tendency to gain a proton. A typical acid–base reaction has the following form:

$$\text{Acid}_1 + \text{base}_2 = \text{acid}_2 + \text{base}_1 \tag{7.27}$$

The proton lost by acid$_1$ is gained by base$_2$, and the proton lost by acid$_2$ is gained by base$_1$. The following equilibrium reaction illustrates this point:

$$HCO_3^- + H_2O = H_3O^+ + CO_3^{2-} \tag{7.28}$$

In Eq. (7.28), protons are transferred from HCO_3^- to H_2O, which in this reaction functions as a base, and from the acid H_3O^+ to the base CO_3^{2-}.

Applying the law of mass action to Eq. (7.28) yields

$$K = \frac{[H^+] [CO_3^{2-}]}{[HCO_3^-]} \tag{7.29}$$

since $[H_2O] = 1$ and $[H_3O^+] = [H^+]$ (Domenico and Schwartz, 1990).

The strength of an acid or base refers to the extent to which protons are lost

TABLE 7.8 Important Weak Acid–Base Reactions in Natural Water Systems

Reaction	Mass law equation	$-\log K(25°C)$
$H_2O = H^+ + OH^-$	$K_w = (H^*)(OH^-)$	14.0
$CO_2(g) + H_2O = H_2CO_3{}^{a}$	$K_{CO_2} = \dfrac{H_2CO_3^*}{P_{CO_2}(H_2O)}$	1.46
$H_2CO_3^* = HCO_3^- + H^+$	$K_1 = \dfrac{(HCO_3^-)(H^+)}{H_2CO_3^*}$	6.35
$HCO_3^- = CO_{2-} + H^+$	$K_2 = \dfrac{(CO_3^{2-})(H^+)}{HCO_3^-}$	10.33
$H_2SiO_3 = HSiO_3^- + H^+$	$K = \dfrac{(HSiO_3^-)(H^+)}{H_2SiO_3}$	9.86
$HSiO_3^- = SiO_{2-} + H^+$	$K = \dfrac{(SiO_3^{2-})(H^+)}{HSiO_3^-}$	13.1

Source: Domenico and Schwartz, 1990.

[a] From F. M. M. Morel, 1983. *Principles of Aquatic Chemistry*. Copyright © 1983 by John Wiley & Sons, Inc. Reprinted by permission of John Wiley and Sons, Inc.

or gained, respectively. Consider the following generalized ionization reaction for an acid HA in water (Glasstone and Lewis, 1960):

$$HA + H_2O = H_3O^+ + A^- \tag{7.30}$$

where A^- is an anion like Cl^-. The strength of the acids and bases depends on whether equilibrium in the reaction is established to the right or left side of Eq. (7.30). The few weak acid–base reactions that are important in natural ground waters are listed in Table 7.8. The reactions involving CO_2 in Table 7.8 show that CO_2 dissolved in water partitions among H_2CO_3, HCO_3^-, and CO_3^{2-}. When the pH of a solution is fixed, the mass law equations in Table 7.8 let us determine the concentrations of the species (Domenico and Schwartz, 1990).

Alkalinity

In ground water systems, the pH and carbonate speciation are interdependent, a function of not only the ionization equilibria for the carbonate species and water but also of strong bases added through the dissolution of carbonate and silicate minerals (Domenico and Schwartz, 1990). Alkalinity is defined as the net concentration of strong base in excess of strong acid with a pure CO_2–water system as the point of reference (Morel, 1983):

$$Alk = \sum (i^+)_{sb} - \sum (i^-)_{sa} = -(H^*) + (OH^-) + (HCO_3^-) + 2(CO_3^{2-}) \tag{7.31}$$

where (i^+) and (i^-) are the positively and negatively charged species from the strong acids and bases. When the alkalinity is zero, Eq. (7.31) becomes the charge balance equation for a pure CO_2–water system (Domenico and Schwartz, 1990).

As Eq. (7.31) shows, increasing alkalinity increases the net positive charge on the left side of the equation. This increase is not simply balanced by an increase in one of the negative ions on the right side, because equilibrium relationships in the solution must be maintained. The increase in alkalinity is matched by an increase in the concentration of negatively charged species that comes from the ionization of HCO_3^- to CO_3^{2-} and an increase in pH. Thus, increasing the alkalinity with a strong base ultimately leads to an increase in pH. This behavior is commonly observed as ground waters evolve by dissolving minerals along a flow path (Domenico and Schwartz, 1990).

7.12 ION EXCHANGE

Ion exchange is a specific category of adsorption that is referred to as adsorbent-motivated sorption (Knox et al., 1993); that is, the accumulation occurs due to an affinity of the solid surface for the chemical. Typically, in subsurface media, the clay fraction is most commonly involved in ion exchange. Equation (7.32) shows a general equation for the exchange of a multivalent cation (B^{n+}) for a monovalent cation (A^+) on a natural mineral surface (R^-), and Eq. (7.33) shows the exchange of calcium for sodium ions (Knox et al., 1993):

$$n R^- \ A^+ \ + \ B^{n+} \Leftrightarrow R_n^- B^{n+} \ + \ n A^+ \qquad (7.32)$$

$$2 R^- \ Na^+ \ + \ Ca^{2+} \Leftrightarrow R_2^- Ca^{2+} \ + \ 2 Na^+ \qquad (7.33)$$

The most common ions occurring naturally in soil are as follows (listed in decreasing order of occurrence for nonmarine deposits): cations, Ca^{2+}, Mg^{2+}, Na^+, and K^+; anions, SO_4^{2-}, Cl^-, PO_4^{3-}, and NO_3^- (Mitchell, 1976). The phenomenon of ion exchange is reviewed by Thomas (1977).

Valocchi (1980) and Valocchi et al. (1981b) evaluated the transport of ion-exchanging solutes during ground water recharge (Knox et al., 1993). A model was developed for analyzing these systems, and applied to a field project involving direct injection of municipal effluents subject to advanced (tertiary) treatment. The model was used to predict the transport of sodium, ammonia, potassium, magnesium, and calcium at various observation wells at the field site. The predicted results agreed very closely with the field data, helping to substantiate the fundamental modeling approach. Valocchi (1984) also discussed the use of an "effective" partitioning coefficient approach (similar to that discussed for sorption of hydrophobic chemicals on the organic content of the soil) for describing the transport of ion-exchanging contaminants.

Ceazan et al. (1989) conducted field studies with respect to the exchange and transport of ammonium and potassium in the sand and gravel aquifer at a Cape Cod, Massachusetts, site. The unconfined aquifer below the Otis Air Base has been contaminated by the discharge of secondary treated wastewater onto sand infiltration beds (see Chapter 6 and Knox et al., 1993). Caezan et al. (1989) conducted two forced gradient field tracer studies at the site.

For the forced gradient studies, chemicals were injected at a depth of 10 m below the land surface, and the appearance of the ions was monitored at an observation well 1.5 m away. For the tracer test of interest, 20 meq/L of NH_4^+ and

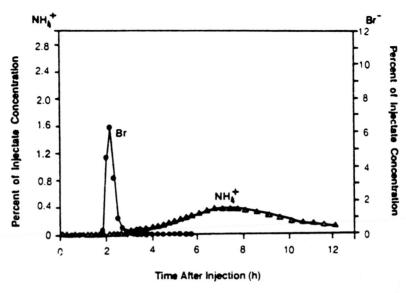

Figure 7.14 Breakthrough of bromide and ammonium in field study. Source: Knox et al., 1993.

Br$^-$ (as NH$_4$Br) was injected as a pulse over a 2-min period. Figure 7.14 shows the appearance of the NH$_4^+$ and Br$^-$ at the observation well and suggests that the NH$_4^+$ was retarded with respect to Br$^-$. As Caezan et al. (1989) observed, the duration of their study (less than 12 hr) makes it highly unlikely that bioactivity was significant. This served to reinforce the notion that adsorption (exchange) was responsible for the retardation of ammonium (Knox et al., 1993).

For ammonium to adsorb, it must exchange for cations. Monitoring of pH indicated that the ammonium was not exchanging for hydrogen ions. Monitoring for calcium and magnesium resulted in the data shown in Figure 7.15 (note that the ordinate values in Figure 7.15 have units of μeq/L). The increase in calcium and magnesium corresponds to the period of time in which the ammonium was being adsorbed, indicating that ammonium (a monovalent cation) displaced calcium and magnesium (divalent cations) in the transport process (Knox et al., 1993). Ion selectivity suggests that divalent cations would be preferentially adsorbed over monovalent cations (the opposite of observations made in this research). The observations on ion exchange are consistent with the conceptual model noted by Caezen et al. (1989) presented previously and predictive models proposed by others (Valocchi et al., 1981b; Persaud et al., 1983).

7.13 DISSOLUTION AND PRECIPITATION OF SOLIDS

The dissolution and precipitation of solids are two of the most important processes that influence solute transport in terms of their control on ground water chemistry. Extremely large quantities of mass can be transferred under some conditions be-

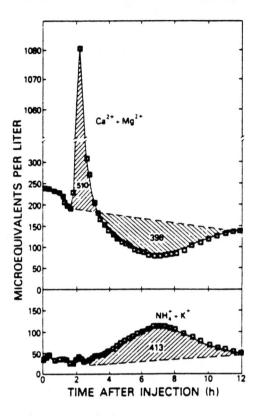

Figure 7.15 Breakthrough of calcium and magnesium in field study. Source: Knox et al., 1993.

tween the ground water and solid-mineral phases. For example, recharge derives almost its entire solute load through the dissolution of minerals along flow paths (Domenico and Schwartz, 1990). Mineral precipitation removes much of the metal present in a low-pH contaminant plume.

The solubility of a solid reflects the extent to which the reactant (solid) or products (ions and/or secondary minerals) are favored in a dissolution–precipitation reaction. In many reactions where the activity of the reacting solid is equal to 1, a comparison of the relative size of the equilibrium constant provides an indication of the solid solubility in pure water. Table 7.9 shows the chloride and sulfate salts to be the most soluble phase and the sulfide and hydroxide groups to be the least soluble. Minerals in the carbonate and silicate and aluminum silicate groups have a small solubility (Domenico and Schwartz, 1990).

When other ions are present in the solution, the solubility of a solid can be different from its value in pure water. Solubility increases due to solution nonidealities and decreases due to the common ion effect. Generally, the solubility of a solid increases with increasing ionic strength, because other ions in solution reduce the activity of the ions involved in the reaction (Domenico and Schwartz, 1990).

The common ion effect occurs when a solution contains the same ions that will be released when the solid dissolves. The presence of the common ion means that less solid is able to dissolve before the solution reaches saturation with respect

TABLE 7.9   Some Common Mineral Dissolution Reactions and Associated Equilibrium Constants

Mineral or solid	Reaction	log K(25°C)	Source
Chlorides and Sulfates			
Halite	$NaCl = Na^+ + Cl^-$	1.54	1
Sylvite	$KCl = K^+ + Cl^-$	0.98	1
Gypsum	$CaSO_4 \cdot 2H_2O = Ca^+ + SO_4^{2-} + 2H_2O$	-4.62	1
Carbonates			
Magnesite	$MgCO_3 = Mg^{2+} + CO_3^{2-}$	-7.46	1
Aragonite	$CaCO_3 = Ca^{2+} + CO_3^{2-}$	-8.22	1
Calcite	$CaCO_3 = Ca^{2+} + CO_3^{2-}$	-8.35	1
Siderite	$FeCO_3 = Fe^{2+} + CO_3^{2-}$	-10.7	1
Dolomite	$CaMg(CO_3) = Ca^{2+} + Mg^{2+} + 2CO_3^{2-}$	-16.7	2
Hydroxides			
Brucite	$Mg(OH)_2 = Mg^{2+} + 2OH^-$	-11.1	1
Ferrous hydroxide	$Fe(OH)_2 = Fe^{2+} + 2OH^-$	-15.1	1
Gibbsite	$Al(OH)_3 = Al^{3+} + 3OH^-$	-33.5	1
Sulfides			
Pyrrhotite	$FeS = Fe^{2+} + S^{2-}$	-18.1	1
Sphalerite	$ZnS = Zn^{2+} + S^{2-}$	-23.9	3
Galena	$PbS = Pb^{2+} + S^{2-}$	-27.5	1
Silicates and aluminum silicates			
Quartz	$SiO_2 + H_2O = H_2SiO_3$	-4.00	1
Na–Montmorillonite	$3Na\text{-}Mont + 11\frac{1}{2}H_2O = 3\frac{1}{2} Kaol + 4H_4SiO_4 + Na^+$	-9.1	2
Kaolinite	$Kaol + 5H_2O = 2Al(OH)_3 + 2H_4SiO_4$	-9.4	2

Source: Domenico and Schwartz, 1990.

1. Morel, 1983.
2. Stumm and Morgan, 1981.
3. Matthess, 1982.

to that ion. Thus, the solubility of a solid is less in a solution containing a common ion than it would be in water alone (Domenico and Schwartz, 1990).

7.14 COMPLEXATION REACTIONS

Complexation reactions are important in determining the saturation state of ground water. Complexation, for example, facilitates the transport of potentially toxic metals such as cadmium, chromium, copper, lead, uranium, or plutonium. Such reactions also influence some types of surface reactions. A complex is an ion that forms by combining simpler cations, anions, and sometimes molecules. The cation or central atom is typically one of the large number of metals making up the periodic table (Domenico and Schwartz, 1990). The anions, often called ligands, include many of the common inorganic species found in ground water, such as Cl^-, F^-, Br^-, SO_4^{2-}, PO_4^{3-}, and CO_3^{2-}. The ligand might also be an organic molecule such as an amino acid (Domenico and Schwartz, 1990).

The simplest complexation reaction involves the combination of a metal and ligand such as MN^{2+} and Cl^- as follows (Domenico and Schwartz, 1990):

$$Mn^{2+} + Cl^- = MnCl^+ \tag{7.34}$$

A more complicated manifestation of complexation is the reaction series that forms when complexes themselves combine with ligands. An example is the hydrolysis reaction of Cr3+ (Domenico and Schwartz, 1990):

$$Cr^{3+} + OH^- = Cr(OH)^{2+}$$

$$Cr(OH)^{2+} + OH^- = Cr(OH)_2^+ \tag{7.35}$$

$$Cr(OH)_2^+ + OH^- = Cr(OH)_3^0$$

and so on. The metal is distributed among at least three complexes. Such series involve reactions not only with $(OH)^-$, but with other ligands such as Cl^-, F^-, and Br^-.

7.15 CHEMISTRY OF METALS

Metals have fairly limited mobility in soil and ground water because of cation exchange or sorption on the surface of mineral grains. They can also form precipitates of varying solubility under specific E_H–pH conditions. Metals are mobile in ground water if soluble ions exist and the soil has a low cation-exchange capacity (Dowdy and Volk, 1983). They can also be mobile if they are chelated or if they are attached to a mobile colloid. Conditions that promote mobility include an acidic, sandy soil with low organic and clay content (Fetter, 1993). Discharge of a metal in an acidic solution would keep the metal soluble and promote mobility. Only two metals, chromium and cadmium, are discussed briefly in this text. The reader is referred to Fetter (1993) for additional information.

Chromium

Chromium in natural waters occurs in a +3 and a +6 valance state. Stable ionic forms in aqueous systems include Cr^{3+}, $CrOH^{2+}$, $Cr(OH)_2^+$, CrO_7^{2+} and CrO_4^{2-} (Fetter, 1993). Chromous hydroxide, $Cr(OH)_3$ is a possible precipitate under reducing conditions. Under some conditions, chromate might react with ferrous iron to produce a chromous hydroxide precipitate (Robertson, 1975).

$$CrO_4^{2-} + 3Fe^{2+} + 8H_2O \Leftrightarrow 3Fe(OH)_3 + Cr(OH)_3 + 4H^+$$

In general, the hexavalent chromium in ground water is soluble and mobile, and trivalent chromium will be insoluble and immobile. Industrial discharges of hexavalent chromium are common from metal-plating industries (Fetter, 1993). A hexavalent chromium spill on Long Island, New York, traveled more than 3000 ft from a waste-discharge pond to a stream (Perlmutter et al., 1963). Hexavalent chromium from a natural source has been found in ground water in Paradise Valley, Arizona (Robertson, 1975).

Cadmium

Cadmium has a very low maximum contaminant level (MCL) in drinking water (10 μg/L) due to its toxicity (Fetter, 1993). It exists in aqueous solution in the +2 valence state. Although cadmium carbonate has a very low solubility product, $10^{-13.7}$, it can be mobile in the environment. On Long Island, New York, a metal-plating waste containing cadmium and chromium traveled about 3000 ft in a shallow aquifer (Perlmutter et al., 1963).

7.16 RADIOACTIVE ISOTOPES

Certain isotopes of elements undergo spontaneous decay, resulting in the release of energy and energetic particles and consequent formation of different isotopes. Humans have created nuclear isotopes through the detonation of nuclear weapons and the construction of nuclear reactors (Fetter, 1993). Table 7.10 lists the sources of environmentally important isotopes.

The cationic radionuclides may sorb onto mineral or organic surfaces in the soil. The following transition metals have low mobilities in waters that are in the neutral range: ^{60}Co, ^{59}Ni, ^{63}Zn, ^{93}Zr, ^{107}Pd, ^{110}Ag, ^{114}Ce, ^{147}Pm, ^{151}Sm, ^{152}Eu, and ^{154}Eu. Many do not desorb significantly. The degree of sorption is strongly related to the pH of the solution. Insoluble metal hydroxides may also be formed (Fetter, 1993). The discussion on radioactive isotopes will be limited to radium, radon, and tritium. The reader is referred to Fetter (1993) and Domenico and Schwartz (1990) for information on other isotopes.

Radium

Radium occurs naturally in four isotopes: ^{223}Ra, ^{224}Ra, ^{226}Ra, and ^{228}Ra. One isotope, ^{226}Ra, has a longer half-life than the others, 1599 yr. Because of their short half-lives, the radium isotopes are strongly radioactive (Hem, 1985). The aqueous chemistry of radium is reportedly similar in chemical behavior to barium (Hem, 1985) and calcium (Kathren, 1984). Radium is more soluble than uranium or thorium and can be bioconcentrated by plants (Fetter, 1993).

TABLE 7.10   Sources of Environmentally Important Radioactive Isotopes

Source	Radionuclides
Naturally occurring	^{40}K, ^{222}Rn, ^{226}Ra, $^{230,232}Th$, $^{235,238}U$
Cosmic irradiation	^{3}H, ^{7}Be, ^{14}C, ^{22}Na
Nuclear weapons tests	^{3}H, ^{90}Sr, ^{137}Cs, $^{239,240}Pu$
Mining waste: uranium, phosphate, coal	^{222}Rn ^{226}Ra, $^{230,232}Th$, $^{235,238}U$
Industrial wastes: e.g., nuclear power plants, weapons manufacturing, research and medical waste	$^{59,63}Ni$, ^{60}Co, ^{90}Sr, $^{93,99}Zr$, ^{99}Tc, ^{107}Pd, ^{129}I, ^{137}Cs, ^{144}Ce, ^{151}Sm, $^{152,154}Eu$, ^{237}Np, $^{239,240,242}Pu$, $^{241,243}Am$

Source: Gee et al., 1983.

Radium can be strongly exchanged in the cation exchange series. According to Kathren (1984), the cation exchange sequence for soils is

$$Sr^{2+} < Ra^{2+} < Ca^{2+} < Mg^{2+} < Cs^{2+} < Rb^{2+} < K^+ < NH^{4+} < Na^+ < Li^+$$

^{228}Ra has a much shorter half-life (5.8 yr) than ^{226}Ra; however, both isotopes are found in ground water.

Wells with high radium levels in ground water have been concentrated in two areas of the United States: the Piedmont and coastal plain of the Middle Atlantic states and the upper Midwestern states of Minnesota, Iowa, Illinois, Missouri, and Wisconsin (Hess et al., 1985). Fetter (1993) states that the radium content of ground water is a function of the rock type of the aquifer. Igneous rocks, such as granites, contain the highest proportion of uranium and thorium, the parent isotopes of radium. Granitic rock aquifers and sands and sandstones derived from the weathering of granites have the potential to have high radium. Phosphate rock is also very high in uranium. Radium is not only a naturally occurring problem, but there are also localized areas of radium contamination from industrial operations. These areas are associated with uranium mill tailings, as well as facilities where radioluminescent paints were prepared and used.

Radon

There are several isotopes of radon, but ^{222}Rn is the only one that is important environmentally. The other isotopes have half-lives of less than 1 min, whereas the half-life of ^{222}Rn is 3.8 days. ^{222}Rn is produced by the decay of ^{226}Ra, so it is associated with rocks that are high in uranium. Radon can be associated with water that is low in dissolved ^{226}Ra, because it comes primarily from the decay of the radium in the rock. Radon does not undergo any chemical reactions, nor is it sorbed onto mineral matter. Radon is lost from water by diffusion into the atmosphere and by radioactive decay through a series of short-lived daughter products of ^{210}Pb, which has a half-life of 21.8 yr (Fetter, 1993).

In addition to drinking-water standards, there is also a health concern for excessive radon accumulation in homes. Radon can enter homes through emanations from the soil, and by diffusion from tap water with a high radon content. Owners of private water systems are at risk from radon in drinking water. Public water-supply systems normally have storage facilities so that the residence time for the water in these facilities allows the radon to both diffuse and decay. Private water systems rely on wells and usually only have a very small storage facility used to maintain pressure (Fetter, 1993).

Brutsaert et al. (1981) studied radon in ground water in Maine. They found radon levels in private wells of up to 122,000 pCi/L. The average ^{222}Rn content of wells obtaining water from granites was 22,100 pCi/L, from metasedimentary rocks, it was 13,600 pCi/L, and from chlorite-grade metasedimentary rocks, it was 1100 pCi/L. The high radon values in the high-grade metamorphic terrains were believed to be due to metamorphic pegmatites and associated uranium mineralization (Fetter, 1993).

Tritium

Tritium, ^3H, is produced naturally by cosmic-ray bombardment of the atmosphere, by thermonuclear detonations, and in nuclear reactors (Fetter, 1993). Because it is an isotope of hydrogen, it can form a water molecule or be incorporated into living tissue. Tritium has a half-life of 12.6 yr. Much of the radiation at low-level radioactive waste sites is due to tritium (Kathren, 1984). Low-level radioactive waste is disposed by shallow burial on land. If waste packages at these sites leak and if the landfill is not secure, tritium is likely to escape. Tritium migration via ground water flow from low-level waste disposal has been detected at sites at the Savannah River facility, Los Alamos National Laboratory (Overcamp, 1982), the Sheffield, Illinois, commercial disposal site (Foster et al., 1984), and the Hanford, Washington, site (Levi, 1992).

7.17 MODELS INCORPORATING CHEMICAL REACTIONS

The modeling of equilibrium aqueous chemistry has become well established. Special models have been developed to predict concentrations of the ions and complexes that will be formed, their thermodynamic activities, and the saturation state of the water with respect to different species and minerals. A variety of different chemical reactions are modeled, including complexation, sorption, dissolution–precipitation, redox, acid–base, mineral alteration, and gas-solution equilibria.

The joining of hydrologic transport models and aqueous chemical models is an active area of research by a number of research groups. To date, most of the efforts along this line have been restricted to the case of a constant-velocity flow field in one or two dimensions. Generally, a sophisticated coupled hydrochemical model should incorporate a fully three-dimensional flow field with the capability to simulate heterogeneities and complex time-dependent boundary conditions. It would also incorporate a large number of chemical reactions, including nonequilibrium reaction kinetics. Such a model would represent a powerful tool for numerical simulation of hydrochemical systems. However, the lack of superlarge computers, adequate data bases, and experienced code users usually limits the development and utilization of such very complex models.

Almost all current modeling efforts distinguish between two broad classes of chemical processes (Rubin, 1983): (1) those that are "sufficiently fast" and reversible, so that local chemical equilibrium may be assumed to exist, and (2) those that are "insufficiently fast" and/or irreversible reactions, where the local equilibrium formulation is inappropriate. Rubin (1983) gives two further levels of classification of chemical reactions within each of the two preceding classes (other classification schemes are also possible). The first distinguishes reactions by the number of phases involved, either homogeneous (within a single phase) or heterogeneous (more than one phase). Second, within the heterogeneous class of reactions are the surface reactions (that is, adsorption and ion-exchange reactions) and the "classical" chemical reactions (precipitation–dissolution, oxidation–re-

TABLE 7.11 Geochemical Models

	ECHEM	EQUILIB	EQ3/6	MINTEQ	PHREEQE	SOILCHEM	TRANSCHEM	WATEQ4F
Authors	Morrey (1988)	Morrey and Shannon (1981)	Wolery (1979)	Felmy et al. (1983a)	Parkhurst et al. (1980)	Sposito and Coves (1988)	Scrivner et al. (1986)	Ball et al. (1987)
Institutions[a] Type	EPRI–PNL Speciation	EPRI–PNL Speciation	LLNL EQ3: Speciation EQ6: Reaction path	EPRI–PNL Speciation	USGS Reaction path	UCB Speciation	OLI Systems Speciation	USGS Speciation
Elements	>34	26	47	31	19	47	100	32
Aqueous species	>520	200	686	373	120	1853	300	245
Organics	No	No	<10	No	Yes	889	50	12
Gases	7	7	11	3	3	3	50	2
Redox elements	>38	9	>25	8	3	11	10	7
Sorption[b]	8 Models	No	No	6 Models	IE	SC	IE	No
Precipitation–dissolution minerals	>340	186	713	328	21	250	50	321
Temperature, °C	0–100	0–300	0–300	25	0–100	25	–25–200	0–100
Pressure, bars	1	1	1 and saturation pressure above 100°C	1	1	1	1–200	1
Earlier and related models	MINTEQ	PATHI	PATHI, MINEQL, WATEQ	REDEQL	MIX	GEOCHEM, REDEQL2	ECES	WATEQ series

Source: Mangold and Tsang, 1991.

All models employ chemical equilibrium calculations, and EQ3/6 includes some kinetics as well. ECHEM is part of the FASTCHEM hydrochemical package described in Table 7.12.

[a] EPRI-PNL denotes Electric Power Research Institute, Pacific Northwest Laboratory; LLNL, Lawrence Livermore National Laboratories; USGS, U.S. Geological Survey; and UCB, University of California, Berkeley.

[b] IE denotes ion exchange; SC, surface complexation model.

TABLE 7.12 Hydrochemical Models

	FIESTA	CTMID	Unnamed Model	CHMTRNS	DYNAMIX	TRANQL	THCC	FASTCHEM
Authors	Theis et al. (1983)	Morrey and Hostetler (1985)	Ortoleva (1985)	Noorishad et al. (1987)	Narasimhan et al. (1985)	Cederberg et al. (1985)	Camahan (1987)	Hostetler and Erikson (1989)
Institution[a]	Notre Dame	PNL	GeoChem Research Associates	LBL	LBL	Stanford	LBL	EPRI-PNL
Dimensions	1	1	1	1	3	2	1	1
Aqueous speciation	Yes	Yes	No	Yes	Yes	Yes	Yes	Yes
Sorption[c]	L	IE, TL	No	IE, SC	IE	IE, SC	IE	IE, SC
Precipitation–Dissolution	No	Yes	Yes	Yes	Yes	No	Yes	Yes
Temperature	Constant	Constant	Constant	Variable	Constant	Constant	Variable	Constant

	Unnamed Model	Unnamed Model	Unnamed Model	Unnamed Model	CHEMTRN	Unnamed Model	CPT	MININR
Authors	Rubin and James (1973)	Grove and Wood (1979)	Valocchi et al. (1981a)	Jennings et al. (1982)	Walsh et al. (1982)	Miller and Benson (1983)	Schulz and Reardon (1983)	Mangold and Tsang (1983)	Felmy et al. (1983b)
Institution[a]	USGS	USGS	Stanford	Notre Dame	University of Texas	LBL	Kiel Univ., Germany; Univ. of Waterloo	LBL	PNL
Dimensions	1	1	2	1	1	1	2	3	1
Aqueous speciation	No	Yes	No	Yes	Yes	Yes	No	No	Yes
Activity coefficient[b]	No	D-H	No	No	Davies	Davies	No	No	Davies and D-H
Sorption[c]	IE	IE	IE	SC	L	IE, SC	IE	No	L
Precipitation–Dissolution	No	Yes	No	No	Yes	Yes	Yes	Yes	Yes
Temperature	Constant	Constant	Constant	Constant	Constant	Constant	Constant	Variable	Constant

Source: Mangold and Tsang, 1991.

All models employ chemical equilibrium calculations; the *Ortoleva* (1985) model includes kinetics, and the *Noorishad et al.* (1987) model also includes kinetics and ^{13}C fractionation. The models of *Noorishad et al.* (1987), *Narasimhan et al.* (1985), and *Carnahan* (1987) include oxidation–reduction reactions. The FASTCHEM package consists of separate modules for calculating the flow, geochemistry, and coupled transport. The geochemistry module is ECHEM.

[a] USGS denotes U.S. Geological Survey; LBL, Lawrence Berkeley Laboratory; PNL, Pacific Northwest Laboratory; and EPRI, Electric Power Research Institute.

[b] D-H denotes extended Debye Huckel equation.

[c] IE denotes ion exchange; L, Langmuir adsorption; SC, surface complexation model; and TL, triple-layer model.

duction or redox, and complex formation, although redox and complex formation are not always heterogeneous).

Three types of chemical reactions are considered significant for chemical transport: ion complexation in the aqueous solution (liquid phase), sorption on solid surfaces (interphase boundaries), and precipitation–dissolution of solids (solid phase). Derivation of the examples that have been used to represent these chemical reactions are beyond the scope of this chapter. The reader is referred to Mangold and Tsang (1991) for mathematical details.

Geochemical Models

Table 7.11 summarizes basic reference information, solution methods, the number of various chemical constituents allowed, method for determining activity coefficients, sorption models, number of minerals participating in precipitation–dissolution, and temperature and pressure ranges for eight geochemical models.

Hydrochemical Models

Table 7.12 gives basic information on hydrochemical models. The coupling relations and methods are listed with information on aqueous speciation, activity coefficient equations, sorption models, precipitation–dissolution capability, and the possibility for nonisothermal calculation. A few models incorporate kinetic reactions, and many of these codes are undergoing continual development. The authors of the code should be contacted directly for details.

SUMMARY

Sorption and other chemical interactions have a significant impact on the fate and transport of contaminant species in aquifers. Sorption causes contaminants to move more slowly than the flowing ground water and to accumulate on the soil matrix, thus affecting remediation efforts. Chemical interactions such as radioactive decay, hydrolysis, volatilization, precipitation, and redox reactions affect the concentrations of the contaminants within the plume, but may not necessarily have an impact on the rate of plume movement. A site characterization effort or a remediation plan has to account for these processes to be able to predict contaminant behavior at the site more accurately.

REFERENCES

BALL, J. W., NORDSTROM, D. K., and ZACHMANN, D. W., 1987, "WATEQ4F—A Personal Computer FORTRAN Translation of the Geochemical Model WATEQ2 with Revised Database," U.S. Geological Survey Open File Report 87-50.

BRIGGS, G. G., 1973, "A Simple Relationship between Soil Adsorption of Organic Chemicals and Their Octanol/Water Partition Coefficients," *Proc. 7th British Insecticide and Fungicide Conf.*, Boots Company Ltd., Nottingham, G.B., Vol. 1.

BROWN, D. S., 1979, U.S. Environmental Protection Agency, Athens, GA, personal communication.

BROWN, D. S., KARICKHOFF S. W., and FLAGG, E. W., 1981, "Empirical Prediction of Organic Pollutant Sorption in Natural Sediments," *J. Environ. Qual.* 10(3):382–386.

BRUTSAERT, W. F., NORTON, S. A., HESS, C. T., and WILLIAMS, J. S., 1981, "Geologic and Hydrologic Factors Controlling Radon-222 in Ground Water in Maine," *Ground Water* 19(4):407–417.

CARNAHAN, C. L., 1987, "Simulation of Chemically Reactive Solute Transport under Conditions of Changing Temperature," *Coupled Processes Associated with Nuclear Waste Repositories*, C. F. Tsang, ed., Academic Press, San Diego, CA, pp. 249–257.

CEAZAN, M. L., THURMAN, E. M., and SMITH, R. L., 1989, "Retardation of Ammonium and Potassium Transport through a Contaminated Sand and Gravel Aquifer: The Role of Cation Exchange," *Environ. Sci. Technol.* 23(11):1402–1408.

CEDERBERG, G. A., STREET, R. L., and LECKIE, J. O., 1985, "A Groundwater Mass Transport and Equilibrium Chemistry Model for Multicomponent Systems," *Water Resources Res.* 21:1095–1104.

CHIOU, C. T., PETERS, L. J., and FREED, V. H., 1979, "A Physical Concept of Soil-water Equilibria for Nonionic Organic Compounds," *Science* 206:831–832.

DOMENICO, P. A., and SCHWARTZ, F. W., 1990, *Physical and Chemical Hydrogeology*, Wiley, New York.

DOWDY, R. H., and VOLK, V. V., 1983, "Movement of Heavy Metals in Soil," *Chemical Mobility and Reactivity in Soil Systems*, Soil Science Society of America, Madison, WI, pp. 229–239.

FELMY, A. R., GIRVIN, D., and JENNE, E. A., 1983a, "MINTEQ: A Computer Program for Calculating Aqueous Geochemical Equilibria," U.S. Environmental Protection Agency Report, Washington, DC.

FELMY, A. R., REISENAUER, A. E., ZACHORA, J. M., and GEE, G. W., 1983b "MININR: A Geochemical Computer Program for Inclusion in Water Flow Models—An Applications Study," *Report PNL-4921,* Pacific Northwest Laboratory, Richland, WA.

FETTER, C. W., 1993, *Contaminant Hydrogeology,* Macmillan, New York.

FOSTER, J. B., ERICKSON, J. R., and HEALY, R. W., 1984, "Hydrogeology of a Low-level Radioactive-waste Disposal Site near Sheffield, Illinois," U.S. Geological Survey, *Water Resources Investigations Report 83-4125.*

GEE, G. W., RAI, D., and SERNE, R. J., 1983, "Mobility of Radionuclides in Soil," *Chemical Mobility and Reactivity in Soil Systems*, Soil Science Society of America, Madison, WI, pp. 203–227.

GLASSTONE, S., and LEWIS, D., 1960, *Elements of Physical Chemistry,* Van Nostrand Reinhold, New York.

GROVE, D. A., and WOOD, W. W., 1979, "Prediction and Field Verification of Subsurface-water Quality Changes during Artificial Recharge, Lubbock, Texas," *Ground Water* 17:250–257.

HANCE, R. J., 1969, "An Empirical Relationship between Chemical Structure and the Sorption of Some Herbicides by Soils," *J. Agric. Food Chem.* 17:667–668.

HEM, J. D., 1985, *Study and Interpretation of the Chemical Characteristics of Natural Water,* U.S. Geological Survey Water Supply Paper 2254.

HESS, C. T., MICHEL, J., HORTON, T. R., PRICHARD, H. M., and CONIGLIO, W. A., 1985, "The Occurrence of Radioactivity in Public Water Supplies in the United States," *Health Physics* 48:553–586.

HOSTETLER, C. J., and ERIKSON, R. L., 1989, "FASTCHEM Package, Vol. 5, User's Guide to the EICM Coupled Geohydrochemical Transport Code," *Report EA-5870-CCM,* Electric Power Research Institute, Palo Alto, CA.

JENNINGS, A. A., KIRKNER, D. J., and THEIS, T. L., 1982, "Multicomponent Equilibrium Chemistry in Groundwater Quality Models," *Water Resources Res.* 18:1089–1096.

KAN, A. T., and TOMSON, M. B., 1990, "Effect of pH Concentration on the Transport of Naphthalene in Saturated Aquifer Media," *J. Contaminant Hydrology* 5:235–251.

KARICKHOFF, S. W., 1981, "Estimation of Sorption of Hydrophobic Semi Empirical Pollutants on Natural Sediments and Soils," *Chemosphere* 10(8):833–846.

KARICKHOFF, S. W., 1984, "Organic Pollutant Sorption in Aquatic Systems," *J. Hydraulic Engineering,* 110(6):707–735.

KARICKHOFF, S. W., BROWN, D. S., and SCOTT, T. A., 1979, "Sorption of Hydrophobic Pollutants on Natural Sediments," *Water Res.* 13:241–248.

KATHREN, R. L., 1984, *Radioactivity in the Environment,* Harwood Academic Publishers, Chur, Switzerland.

KENAGA, E. E., and GORING, C. A. I., 1978, "Relationship between Water Solubility, Soil-Sorption, Octanol-Water Partitioning, and Bioconcentration of Chemicals in Biota," prepublication copy of paper dated October 13, 1978, given at the American Society for Testing and Materials, Third Aquatic Toxicology Symposium, October 17–18, New Orleans, LA.

KNOX, R. C., SABATINI, D. A., and CANTER, L. W., 1993, *Subsurface Transport and Fate Processes,* Lewis Publishers, Boca Raton, FL.

LEVI, B. G., 1992, "Hanford Seeks Short- and Long-term Solutions to Its Legacy of Waste," *Physics Today* 45(3):17–21.

LYMAN, W. J., REEHL, W. F., and ROSENBLATT, D. H., 1982, *Handbook of Chemical Property Estimation Methods,* McGraw-Hill, New York.

LYMAN, W. J., REEHL, W. F., and ROSENBLATT, D. H., 1990, *Handbook of Chemical Property Estimation Methods,* American Chemical Society, Washington, DC.

MACKAY, D., and LEINONEN, P. J., 1975, "Rate of Evaporation of Low Solubility Contaminants from Water Bodies to Atmosphere," *Environ. Sci. Technol.* 9:1178–1180.

MACKAY, D., and WOLKOFF, A. W. , 1973, "Rate of Evaporation of Low Solubility Contaminants from Water Bodies to Atmosphere," *Envrion. Sci. Technol* 7:611–614.

MACKAY, D., SHIU, W. Y., and SUTHERLAND, R. P., 1979, "Determination of Air–Water Henry's Law Constants for Hydrophobic Pollutants," *Environ. Sci. Technol.* 13:333–337.

MANGOLD, D. C., and TSANG, C. F., 1983, "A Study of Nonisothermal Chemical Transport in Geothermal Systems by a Three-dimensional Coupled Thermal and Hydrologic Parcel Model," *Trans. Geotherm. Resources Council* 7:455–459.

MANGOLD, D. C., and TSANG, C., 1991, "A Summary of Subsurface Hydrological and Hydrochemical Models," *Rev. Geophysics,* American Geophysical Union, 29(1):51–79.

MATTHESS, G., 1982, *The Properties of Groundwater,* Wiley, New York.

McCARTY, P. L., REINHARD, M., and RITTMAN, B. E., 1981, "Trace Organics in Groundwater," *Environ. Sci. Technol.* 15(1):40–51.

MILLER, C. W., and BENSON, L. V., 1983, "Simulation of Solute Transport in a Chemically Reactive Heterogeneous System: Model Development and Application," *Water Resources Res*. 19:381–391.

MITCHELL, J. K., 1976, *Fundamentals of Soil Behavior*, Wiley, New York.

MOREL, F. M. M., 1983, *Principles of Aquatic Chemistry*, Wiley, New York.

MORREY, J. R., 1988, "FASTCHEM Package, Volume 4, User's Guide to the ECHEM Equilibrium Geochemistry Code," *Report EA-5870-CCM*, Electric Power Research Institute, Palo Alto, CA.

MORREY, J. R., and HOSTETLER, C. J., 1985, "Coupled Geochemical and Solute Transport Code Development," *Proceedings of the Conference on the Application of Geochemical Models to High-level Nuclear Waste Repository Assessment*, Oak Ridge, TN, October 2–5, 1984, G. K. Jacobs and S. K. Whatley, eds., *Report NUREG/CP-0062* and *ORNL/ TM-9585*, pp. 90–92, U.S. Nuclear Regulatory Commission, Washington, DC.

MORREY, J. R., and SHANNON, D. W., 1981, "Operator's Manual for EQUILIB—A Computer Code for Predicting Mineral Formation in Geothermal Brines," Vols. 1 and 2, *Final Report, Project RP 653-1*, Electric Power Research Institute, Palo Alto, CA.

NARASIMHAN, T. N., WHITE, A. F., and TOKUNAGA, T., 1985, "Hydrology and Geochemistry of the Uranium Mill Tailings Pile at Riverton, Wyoming, Part II, History Matching," *Report LBL-18526*, Lawrence Berkeley Laboratory, Berkeley, CA.

NOORISHAD, J., CARNAHAN, C. L., and BENSON, L. V., 1987, "A Report on the Development of the Non-Equilibrium Reactive Transport Code CHMTRNS," *Report LBL-22361*, Lawrence Berkeley Laboratory, Berkeley, CA.

OLSEN, R. L., and DAVIS, A., 1990, "Predicting the Fate and Transport of Organic Compounds in Groundwater, Part 1," *Hazardous Materials Conference*, pp. 40–64, Hazardous Materials Resource Control, Greenbelt, MD.

ORTOLEVA, P., 1985, "Modeling Water–Rock Interactions," *Proceedings of the Conference on the Application of Geochemical Models to High-level Nuclear Waste Repository Assessment*, Oak Ridge, TN, October 2–5, 1984, G. K. Jacobs and S. K. Whatley, eds., *Report NUREG/CP-0062* and *ORNL/TM-9585*, pp.87–89, U.S. Nuclear Regulatory Commission, Washington, DC.

OVERCAMP, T. J., 1982, "Low-level Radioactive Waste Disposal by Shallow Land Burial," in *Handbook of Environmental Radiation*, A. W. Klement, Jr., ed., CRC Press, Boca Raton, FL, pp. 207–267.

PANKOW, J. F., 1991, *Aquatic Chemistry Concepts*, Lewis Publishers, Chelsea, MI.

PARKHURST, D. L., THORSTENSON, D. C., and PLUMMER, L. M., 1980, "PHREEQE—A Computer Program for Geochemical Calculations," *U.S. Geological Survey Water Resources Investigations*, pp. 80–96.

PERLMUTTER, N. M., LIEBER, M., and FRAUENTHAL, H. L., 1963, *Movement of Waterborne Cadmium and Hexavalent Chromium Wastes in South Farmingdale, Nassau County, Long Island*, U.S. Geological Survey Professional Paper 475C, pp. C170–C184.

PERSAUD, N., DAVIDSON, J. M., and RAO, P. S. C., 1983, "Miscible Displacement of Inorganic Cations in a Discrete Homoionic Exchange Medium," *Soil Sci*. 136(5):269–278.

PIWONI, M. D., and KEELEY, J. W., 1990, "Basic Concepts of Contaminant Sorption at Hazardous Waste Sites," EPA/540/4-90/053.

RAO, P. S. C., and DAVIDSON, J. M., 1980, "Estimation of Pesticide Retention and Transformation Parameters Required in Nonpoint Source Pollution Models," *Environmental Impact of Nonpoint Source Pollution*, M. R. Overcash and J. M. Davidson, eds., Ann Arbor Science Publishers, Ann Arbor, MI.

ROBERTS, P. V., GOLTZ, M. N., and MACKAY, D. M., 1986, "A Natural Gradient Experiment on Solute Transport in a Sand Aquifer. 3. Retardation Estimates and Mass Balances for Organic Solutes," *Water Resources Res.* 22(13):2047–2058.

ROBERTSON, F. N., 1975, "Hexavalent Chromium in the Ground Water in Paradise Valley, Arizona," *Ground Water* 13(6):516–527.

RUBIN, J., 1983, "Transport of Reacting Solutes in Porous Media: Relation between Mathematical Nature of Problem Formulation and Chemical Nature of Reactions," *Water Resources Res.* 19(5):1231–1252.

RUBIN, J., and JAMES, R. V., 1973, "Dispersion-affected Transport of Reacting Solutes in Saturated Porous Media: Galerkin Method Applied to Equilibrium-controlled Exchange in Unidirectional Steady Water Flow," *Water Resources Res.* 9:1332–1356.

SCHULZ, H. D., and REARDON, E. J., 1983, "A Combined Mixing Cell/Analytical Model to Describe Two-dimensional Reactive Solute Transport for Unidirectional Groundwater Flow," *Water Resources Res.* 19:493–502.

SCRIVNER, N. C., BENNETT, K. E., PEASE, R. A., KOPATSIS, A., SANDERS, S. J., CLARK, D. M., and RAFAL, M., 1986, "Chemical Fate of Injected Wastes," *Ground Water Monitoring Rev.* 6(3):53–58.

SPOSITO, G., and COVES, J., 1988, "SOILCHEM: A Computer Program for the Calculation of Chemical Speciation in Soils," Report, Kearny Foundation, University of California, Riverside and Berkeley.

STUMM, W., and MORGAN, J. J., 1981, *Aquatic Chemistry,* Wiley, New York.

THEIS, T. L., KIRKNER, D. J., and JENNINGS, A. A., 1983, "Multi-solute Subsurface Transport Modeling for Energy Solid Wastes," *Report COO-10253-3,* Department of Civil Engineering, University of Notre Dame, Notre Dame, IN.

THOMAS, G. W., 1977, "Historical Developments in Soil Chemistry: Ion Exchange," *Soil Sci. Soc. Amer. J.* 41:230–238.

VALOCCHI, A. J., 1980, "Transport of Ion-exchanging Solutes during Groundwater Recharge," Ph.D. Dissertation, Stanford University, Palo Alto, CA.

VALOCCHI, A. J., 1984, "Describing the Transport of Ion-exchanging Contaminants Using an Effective Kd Approach," *Water Resources Res.* 20(4):499–503.

VALOCCHI, A. J., ROBERTS, P. V., PARKS, G. A., and STREET, R. L., 1981a, "Simulation of the Transport of Ion-exchanging Solutes Using Laboratory-determined Chemical Parameter Values," *Ground Water* 19:600–607.

VALOCCHI, A. J., STREET, R. L., and ROBERTS, P. V., 1981b, "Transport of Ion-exchanging Solutes in Groundwater: Chromatographic Theory and Field Simulation," *Water Resources Res.* 17:1517–1527.

VERSCHUEREN, K., 1977, *Handbook of Environmental Data on Organic Chemicals,* Van Nostrand Reinhold, New York.

WALSH, M. P., LAKE, L. W., and SCHECHTER, R. S., 1982, "A Description of Chemical Precipitation Mechanisms and Their Role in Formation Damage during Stimulation of Hydrofluoric Acid," *J. Petroleum Technol.* 34:2097–2112.

WOLERY, T. J., 1979, "Calculation of Chemical Equilibrium between Aqueous Solution and Minerals: The EQ3/6 Software Package," *Report UCRL-52658,* Lawrence Livermore National Laboratory, Livermore, CA.

Chapter 8 | BIODEGRADATION REACTIONS AND KINETICS

8.1 *BIOLOGICAL TRANSFORMATIONS*

The biodegradation process is a biochemical reaction that is mediated by microorganisms. In general, an organic compound is oxidized (loses electrons) by an electron acceptor, which in itself is reduced (gains electrons). Under **aerobic** or **oxic** environmental conditions, oxygen commonly acts as the electron acceptor when present. The oxidation of organic compounds coupled to the reduction of molecular oxygen is termed **aerobic heterotrophic respiration.** When oxygen is not present (**anoxic** conditions), microorganisms can use organic chemicals or inorganic anions as alternate electron acceptors under **anaerobic** conditions. Anaerobic biodegradation can occur under **fermentative, denitrifying, iron-reducing, sulfate-reducing,** or **methanogenic** conditions (Sims et al., 1991; see Table 8.1).

Fermenting organisms utilize their substrate as both an electron donor and acceptor (Suflita and Sewell, 1991). In this process an organic compound is metabolized (metabolism refers to the degradation of organic compounds to provide energy and carbon for growth), with a portion of that compound becoming a reduced end product and another becoming an oxidized product. A common example of this process is the fermentation of starch to CO_2 (oxidized product) and ethanol (reduced product). Denitrifying bacteria utilize NO_3^- as the electron acceptor and reduce it to NO_2^-, N_2O, or N_2. Iron- or sulfate-reducing bacteria utilize ferric iron or SO_4^{2-} as electron acceptors and reduce them to ferrous iron and H_2S, respectively. Methanogens are methane-producing bacteria that utilize CO_2 as an electron acceptor.

The organisms responsible for biodegradation may be divided into two categories based on their ability to function under different environmental conditions (Howard and Banerjee, 1984). **Oligotrophs** are organisms that are active in the presence of low concentrations of organic carbon, a condition common to the subsurface. In contrast, **eutrophs** are organisms that proliferate under conditions of high organic carbon, but do not function at low organic carbon concentrations. Thus, a chemical present at high concentrations would tend to be biodegraded by eutrophs. Organisms can also be classified into different groupings on the basis of their nutrition. **Chemotrophs,** for instance, capture their energy from the oxidation of organic or inorganic materials. **Autotrophs** are capable of synthesizing their cell carbon from simple compounds such as CO_2, whereas **heterotrophs** require fixed organic sources of carbon.

TABLE 8.1 Selected Types of Aerobic and Anaerobic Respiration Involved in the Microbial Metabolism of Organic Matter

Process	Electron acceptor	Metabolic products	Relative potential energy	
Aerobic heterotrophic respiration	O_2	CO_2, H_2O	↑	High
Denitrification	NO_3^-	CO_2, N_2		
Iron reduction	Fe^{3+}	CO_2, Fe^{2+}		
Sulfate reduction	SO_4^{2-}	CO_2, H_2S		
Methanogenesis	CO_2	CO_2, CH_4		Low

Source: Suflita and Sewell, 1991.

Primary and Secondary Substrates

When the contaminant is utilized as a **primary substrate,** it is a source of energy and carbon for the microorganism. Under these conditions, in situ biotransformation may be promoted by supplying the appropriate electron acceptor and nutrients to the aquifer. The cells grown on the substrate can sometimes degrade other compounds, called **secondary substrates,** even though the secondary substrates do not provide sufficient energy to sustain the microbial population (McCarty et al., 1981). To transform contaminants as secondary substrates, both the primary substrate and the electron acceptor must be added to the treatment zone. **Cometabolism** is one form of secondary substrate transformation in which enzymes produced for primary substrate oxidation are capable of degrading the secondary substrate (Brock et al., 1984).

Generally, organic compounds represent potential electron donors to support microbial metabolism. However, halogenated compounds can act as electron acceptors and thus become reduced in the reductive dehalogenation process. Specifically, **dehalogenation** by reduction is the replacement of a halogen such as chloride, bromide, fluoride, or iodide on an organic molecule by a hydrogen atom. Ground water aquifers often become anaerobic due to the depletion of oxygen by the microbial degradation of other organic matter. The potential for anaerobic biological processes to reductively dehalogenate organic compounds becomes an important consideration.

Basic Requirements for Biodegradation

There are six basic requirements for biodegradation.

1. *Presence of the appropriate organisms:* Microorganisms in ground water aquifers are almost invariably bacteria. In general, it is preferable to have indigenous microorganisms that are capable of degrading the specific contaminants at the site. Although much work has been done on inoculating aquifers with natural or genetically engineered organisms, the success of these experiments has not been fully established.
2. *Energy source:* Organic carbon is required as an energy source and is used

by the organisms for cell maintenance and growth. The organic carbon is transformed into inorganic carbon, energy, and electrons.

3. *Carbon source:* About 50% of the dry weight of bacteria is carbon. Organic chemicals serve as both carbon and energy sources. As a carbon source, organic carbon is used in conjunction with energy to generate new cells.

4. *Electron acceptor:* Some chemicals must accept the electrons released from the energy source. Typical acceptors include O_2, NO_3^-, SO_4^{2-}, and CO_2:

$$e^- + O_2 \rightarrow H_2O \tag{8.1}$$

$$e^- + NO_3^- \rightarrow N_2 \tag{8.2}$$

$$e^- + SO_4^{2-} \rightarrow H_2S \tag{8.3}$$

$$e^- + CO_2 \rightarrow CH_4 \tag{8.4}$$

5. *Nutrients:* The required nutrients include nitrogen, phosphorus, calcium, magnesium, iron, and trace elements. Nitrogen and phosphorus are needed in largest quantities. Elemental ratios of nitrogen and phosphorus in cell dry mass are given by $C_5H_7O_2NP_{0.083}$, so nitrogen is approximately 12% and phosphorus is 2% to 3% of dry mass.

6. *Acceptable environmental conditions:* Examples are temperature, pH, salinity, hydrostatic pressure, radiation, and/or the presence of a heavy metal or other toxicant materials. Often, two or more environmental factors interact to limit microbial decomposition processes. Additionally, the concentration of the organic pollutants in an aquifer can affect the microbial population. A concentration that is too high can limit microbial metabolism due to toxicity effects at that concentration. In contrast, a threshold concentration might exist for an organic compound below which no growth of the microorganisms occurs. At concentrations below this threshold value, researchers have demonstrated that the needs for cell maintenance are satisfied, but not those for growth (Alexander, 1985).

8.2 BIOLOGICAL ACTIVITY IN SUBSURFACE MATERIALS

Microbial investigations of the subsurface have revealed that all aquifers examined thus far support a microbial population. Such studies include direct microscopic, cultivation, metabolic, and biochemical evidence for microorganisms in asceptically obtained aquifer material. Typical microbial numbers range from 1×10^5 to 1×10^7 cells per gram dry weight (cells/gdw) or cells per milliliter (cells/ml) (Ghiorse and Balkwill, 1983; Wilson et al., 1983b; Balkwill and Ghiorse, 1985; and Beeman and Suflita, 1987). Relatively high numbers of microorganisms have been detected in both contaminated and pristine aquifers of varying depth and geological composition. The picture that emerges from microbiological studies is that subsurface microorganisms tend to be small, capable of response to the influx of nutrients, and primarily attached to solid surfaces (Suflita, 1989).

The main questions that need to be answered about subsurface microorganisms in relation to biodegradation of contaminated soils and ground water can be summarized as follows:

1. Are the microorganisms metabolically active?
2. How diverse is their metabolism?
3. What factors serve to stimulate and/or limit their growth and activity?
4. Can we take advantage of their metabolism for aquifer remediation?

The following sections detail the field and laboratory procedures that can be used to address these questions.

Sampling Techniques

To make valid interpretations of the data resulting from subsurface investigations, the samples analyzed must not be contaminated with nonindigenous microorganisms. Potential sources of contamination include organisms present on drilling machinery, in surface soil layers, and in drilling muds. Most current sampling efforts rely on the recovery and subsequent removal of the outer few centimeters and the top and bottom portions of aquifer cores because of possible contamination by nonindigenous bacteria. Only the center portions of an aquifer core are used for analysis.

Ideally, the dissection process of the core occurs as soon as possible after the core is removed from the ground. In the field, a sterilized paring device is used in the dissection process (McNabb and Mallard, 1984). The paring device has an inner diameter that is smaller than the diameter of the core itself. The aquifer material is extruded out of the core barrel used for sampling. For anaerobic aquifers, this field paring procedure is performed inside plastic anaerobic glovebags; the latter are purged with nitrogen to minimize exposure of the microbes to oxygen (Beeman and Suflita, 1987). Samples acquired in this manner are termed "asceptically obtained" and are suitable for microbiological analyses.

Methods for Detection, Enumeration, and Determination of Metabolic Activity

Numerous methods can be applied to detect microorganisms and estimate their biomass and metabolic activities in subsurface samples. The applicable technologies include microscopic methods, cultural methods, biochemical indicators, radioscopic methods, and microcosms. The presence of microorganisms in subsurface samples and some of their distinguishing morphological features can be established by direct light and electron microscopy. Generally, the water or sediment sample can be stained with a fluorescent dye to improve the visibility of the cells in the light microscope.

Standard cultural procedures, which include plate counts, most probable number counts, and enrichment culture procedures, are more sensitive than microscopic methods of detection. Biochemical indicators of metabolic activity and detection of specific molecular markers in natural samples such as ATP (adenosine triphosphate) provide very sensitive estimates for biomass and activity. These methods, however, cannot show directly the presence or activity of cells, and therefore results should be interpreted cautiously. Ideally, biochemical indicator methods should be supported by direct microscopic and cultural methods applied to the same samples.

Radioisotope-labeled substrate transformation and uptake methods provide another sensitive means of measuring the metabolic activity and growth of microorganisms in subsurface samples. Ideally, the rate of labeled substrate transformation should be compared to total direct counts, plate counts, or another biomass indicator to normalize the rates to population density (Ghiorse and Wilson, 1988).

Microcosms

Pritchard and Bourquin (1984) defined a microcosm study as "an attempt to bring an intact, minimally disturbed piece of an ecosystem into the laboratory for study in its natural state. It is the establishment of a physical model or simulation of part of the ecosystem in the laboratory within definable physical and chemical boundaries under a controlled set of experimental conditions." Microcosms have been used extensively to determine the biodegradability of organic contaminants at the laboratory scale and under the influence of the site-specific physical, chemical, and hydrologic characteristics.

Microcosms range from simple batch incubation systems to large and complex flow-through devices. Figure 8.1 shows three different types of microcosms designed by Dunlap et al. (1972), Bengtsson (1981), and Wilson et al. (1981). Microcosms are used to (1) identify biodegradable pollutants and (2) determine the metabolic pathways of biotic or abiotic transformations. Finally, the decay of a particular pollutant in a microcosm may relate to the rate of biotransformation in situ without relying on indirect measures of microbial biodegradation for rate prediction, such as the enumeration of catalytic microorganisms (Suflita, 1989).

Microcosms offer several experimental advantages. For example, results from microcosm studies are replicable and allow appropriate controls to be employed. Furthermore, microcosms provide a time-efficient method for evaluating biodegradation potential at a field site. Conversely, microcosms suffer from some limitations, mainly that microcosms can disturb the normal structure of an ecosystem and generally possess high surface to volume ratios. These disturbances may influence biodegradation estimates and prevent direct extrapolation of microcosm-determined biodegradation rates to the field scale.

Biodegradation of Organics in the Subsurface

The most aerobically biodegradable compounds in the subsurface have been petroleum hydrocarbons such as gasoline, crude oil, heating oil, fuel oil, lube oil waste, and aviation gas (Lee et al., 1987). In particular, benzene, toluene, xylene, and ethylbenzene are very degradable. Other compounds such as alcohols (isopropanol, methanol, ethanol), ketones (acetone, methyl ethyl ketone), and glycols (eth-

Sampling apparatus

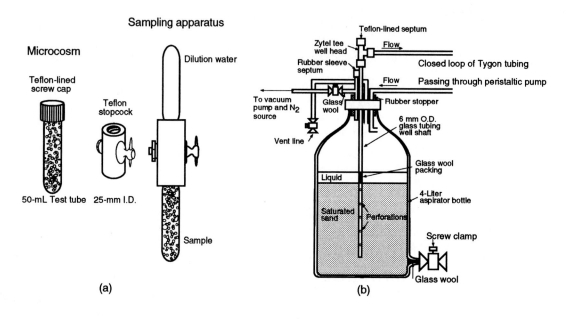

Microcosm

Teflon-lined
screw cap

Teflon
stopcock

Dilution water

50-mL Test tube   25-mm I.D.

Sample

(a)

Teflon-lined septum

Zytel tee
well head   Flow

Rubber sleeve
septum

Closed loop of Tygon tubing

Flow   Passing through peristaltic pump

To vacuum
pump and N₂
source

Glass
wool

Rubber stopper

6 mm O.D.
glass tubing
well shaft

Vent line

Glass wool
packing

Liquid

4-Liter
aspirator bottle

Saturated
sand   Perforations

Screw clamp

Glass wool

(b)

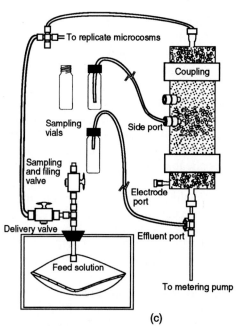

To replicate microcosms

Coupling

Sampling
vials

Side port

Sampling
and filing
valve

Electrode
port

Delivery valve

Effluent port

Feed solution

To metering pump

(c)

Figure 8.1 Examples of microcosms. Source: (a) Wilson, 1981, (b) Dunlap, 1972, (c) Bengtsson, 1981.

TABLE 8.2 Biodegradation Potential of Selected Organic Compounds

Compound	Conditions	Test	Reference
Benzene	Aerobic	In situ	Wilson et al., 1986b
	Methanogenic, microcosm	Field sample	Wilson et al., 1986c
	Aerobic, batch study	Laboratory	Tabak et al., 1990
Carbon tetrachloride	Anaerobic, batch study	Laboratory	Bouwer and McCarty, 1983a
	Methanogenic, batch study	Laboratory	Bouwer and McCarty, 1983b
	Methanogenic, continuous flow column	Laboratory	Bouwer and McCarty, 1983b
	Methanogenic, continuous flow column	Laboratory	Bouwer and Wright, 1988
	Anaerobic, continuous flow column	Laboratory	Bouwer and Wright, 1988
p-Cresol	Methanogenic, batch study	Field sample	Smolenski and Suflita, 1987
	Sulfate reducing, batch study	Field sample	Smolenski and Suflita, 1987
	Aerobic, batch study	Laboratory	Desai et al., 1990
	Iron reducing, batch study	Laboratory, pure culture	Lovley and Lonergan, 1990
1,3-Dichlorobenzene	Anaerobic, batch study	Laboratory, pure culture	de Bont et al., 1986
Ethylbenzene	Methanogenic, microcosm	Field sample	Wilson et al., 1986c
	Anaerobic, continuous flow column	Field sample	Kuhn et al., 1988
	Aerobic, batch study	Laboratory	Tabak et al., 1990
Naphthalene	Aerobic	In situ	Wilson et al., 1985
	Anaerobic, continuous flow column	Field sample	Kuhn et al., 1988
	Aerobic, batch study	Field sample	Swindoll et al., 1988
Phenol	Methanogenic, batch study	Field sample	Gibson and Suflita, 1986
	Sulfate reducing, batch study	Field sample	Gibson and Suflita, 1986
	Aerobic, batch study	Laboratory	Aelion et al., 1987
	Aerobic, microcosm	Laboratory	Smith and Novak, 1987
	Aerobic, batch study	Field sample	Swindoll et al., 1988
	Aerobic, batch study	Laboratory	Desai et al., 1990
	Iron reducing, batch study	Laboratory, pure culture	Lovley and Lonergan, 1990
Tetrachloroethylene (PCE)	Methanogenic, batch study	Laboratory	Bouwer and McCarty, 1983b
	Methanogenic, continuous flow column	Laboratory	Bouwer and McCarty, 1983b
	Aerobic, microcosm	Field sample	Wilson et al., 1983a,b
	Anaerobic, batch study	Laboratory	Gosset, 1985
	Methanogenic, continuous flow column	Laboratory	Vogel and McCarty, 1985
	Methanogenic, batch study	Laboratory	Fogel et al., 1986
	Methanogenic, continuous flow column	Laboratory	Bouwer and Wright, 1988
	Methanogenic, batch study	Laboratory	Freedman and Gosset, 1989
Toluene	Aerobic, microcosm	Field sample	Wilson et al., 1983b
	Aerobic	In situ	Wilson et al., 1986a
	Aerobic	In situ	Wilson et al., 1986b
	Methanogenic, microcosm	Field sample	Wilson et al., 1986c
	Anaerobic, continuous flow column	Field sample	Zeyer et al., 1986
	Aerobic, batch study	Field sample	Swindoll et al., 1988
	Anaerobic, continuous flow column	Field sample	Kuhn et al., 1988
	Iron reducing, batch study	Laboratory, pure culture	Lovley and Lonergan, 1990
	Aerobic, batch study	Laboratory	Tabak et al., 1990
	Methanogenic, microcosm	Field sample	Beller et al., 1991
Trichloroethylene (TCE)	Aerobic, microcosm	Field sample	Wilson et al., 1983a,b
	Anaerobic, continuous flow column	Laboratory	Wilson and Wilson, 1985
	Methanogenic, batch study	Laboratory	Fogel et al., 1986
	Anaerobic, batch study	Field sample	Fliermans et al., 1988
	Methanogenic, batch study	Laboratory, pure culture	Little et al., 1988
	Methanogenic, batch study	Laboratory	Freedman and Gosset, 1989
	Aerobic	Laboratory	Govind et al., 1992
	Aerobic	In situ	Hopkins et al., 1992
	Aerobic	Laboratory, pure culture	Shields et al., 1992
Vinyl chloride	Methanogenic, batch study	Laboratory	Fogel et al., 1986
	Anaerobic, microcosm	Field sample	Barrio-Lage et al., 1990
m-Xylene	Aerobic	In situ	Wilson et al., 1986b
	Anaerobic, continuous flow column	Field sample	Kuhn et al., 1988
	Aerobic, batch study	Laboratory	Tabak et al., 1990

ylene glycol) are also aerobically biodegradable. Recent studies have expanded the list of aerobically degraded compounds to include methylated benzenes and chlorinated benzenes (Kuhn et al., 1985), chlorinated phenols (Suflita and Miller, 1985), methylene chloride (Jhaveri and Mazzacca, 1983), and naphthalene, methyl-naphthalenes, dibenzofuran, fluorene, and phenanthrene (Wilson et al., 1985; Lee and Ward, 1985). Additionally, the general literature is replete with studies of the biodegradation potential of a variety of compounds under aerobic and anaerobic conditions. Table 8.2 lists some of these studies for a selected number of organic compounds.

8.3 MICROBIAL DYNAMICS

The growth of microorganisms in a limited environment, a ground water aquifer in this case, is depicted in Figure 8.2. A **limited environment** refers to an environment that has a limited supply of nutrients, carbon, and electron acceptors. The initial phase of the closed growth cycle, known as the **lag phase,** shows no detectable increase in population size, and so the specific growth rate is zero. The **acceleration phase** marks the beginnings of a gradual increase in the specific growth rate. The **exponential phase** may be described by the basic growth equations. The **retardation phase** is established once the specific growth rate begins to decline.

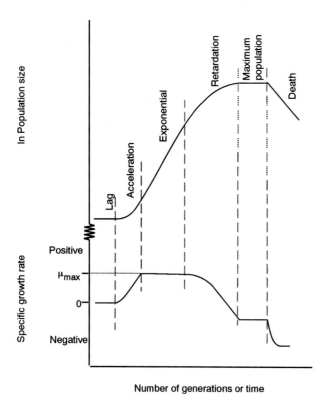

Figure 8.2 Phases of growth occurring during a closed culture growth cycle in a limited environment, showing the changes in the organism's specific growth rate. Source: Bull, 1974.

In the **maximum population** phase, the population is metabolically active, but active growth stops. During the **death phase** the numbers of microorganisms decline.

One of the most used models to describe growth in a closed environment is the **logistic** model. This model may be used from the beginning of the exponential phase through the maximum population phase (Lynch and Poole, 1979):

$$\frac{dx}{dt} = \mu_{max} \cdot x \left(1 - \frac{x}{x_f}\right) \tag{8.5}$$

where x is the population size, μ_{max} is the maximum growth rate in units of time^{-1}, and x_f is the population size in the maximum population phase. The term $1 - (x/x_f)$ models the decrease in the rate of increase of the population as the population density increases. When the population density, x, is small (at the beginning of exponential growth), x/x_f is also small, and Eq. (8.5) simplifies to the basic growth equation for exponential growth. However, with increasing growth time, x becomes larger and x/x_f approaches unity, hence reducing the value of the maximum specific growth rate. When $x = x_f$ (at the end of the retardation phase), $dx/dt = 0$, as expected for the maximum population phase.

8.4 KINETICS OF BIODEGRADATION

Kinetic expressions have been developed to describe the rate of biotransformation of organic contaminants. Most kinetic studies in ground water have been mainly confined to uptake kinetics of organic compounds in a variety of environmental systems or to activity measurements of indigenous microorganisms. Effort has not been directed toward developing rate equations to describe biodegradation of these organic compounds in aquifers. Unfortunately, these rate equations are necessary for extrapolating laboratory rate data to actual subsurface environments.

Several types of kinetic expressions may be appropriate for predicting rates of biodegradation in ground water aquifers. The most common expression is the hyperbolic saturation function presented by Monod (1942) and referred to as Monod or Michaelis–Menten kinetics:

$$\mu = \mu_{max} \frac{C}{K_c + C} \tag{8.6}$$

where μ is the growth rate (time^{-1}), μ_{max} is the maximum growth rate (time^{-1}), and C is the concentration of the growth-limiting substrate (mg/L). The term K_c is known as the half-saturation constant and is defined as the growth-limiting substrate concentration that allows the microorganism to grow at half the maximum specific growth rate (Figure 8.3). The saturation constant, which has the same units as substrate concentration, is a measure of the affinity the microorganism has for the growth-limiting substrate. The lower the value of K_c is, the greater

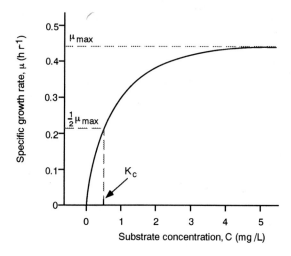

Figure 8.3 Relationship between specific growth rate and growth-limiting substrate concentration. Source: Lynch and Poole, 1979.

the capacity of the microorganism to grow rapidly in an environment with low growth-limiting substrate concentrations.

The rate equation describing μ as a function of C is fairly complex and contains first-order, mixed-order, and zero-order regions. When $C \gg K_c$, $K_c + C$ is almost equal to C, and the reaction approaches zero order, with

$$\mu = \mu_{max} \tag{8.7}$$

and μ_{max} becomes the limiting maximum reaction rate. When $C \ll K_c$, Eq. (8.6) reduces to

$$\mu = \frac{\mu_{max}}{K_c} \cdot C \tag{8.8}$$

and the reaction approaches first order with μ_{max}/K_c = first-order rate constant.

In ground water, we are concerned with relating the Monod growth function with the rate of decrease of an organic compound. This is done by utilizing a yield coefficient, Y, where Y is a measure of the organisms formed per substrate utilized. The change in substrate concentration can then be expressed as follows:

$$\frac{dC}{dt} = \frac{\mu_{max} \, MC}{Y(K_c + C)} \tag{8.9}$$

where M is the microbial mass in mg/L. Because of the relationship between substrate utilization and the growth of microbial mass, Eq. (8.9) is accompanied by an expression of the change in microbial mass as a function of time:

$$\frac{dM}{dt} = \mu_{max} \, MY \frac{C}{K_c + C} - b \cdot M \tag{8.10}$$

where b is a first-order decay coefficient that accounts for cell death.

A simple alternative to using growth functions for determining the rate of degradation of a chemical involves the use of a first-order equation of the form

$$C = C_o \cdot e^{-kt} \tag{8.11}$$

where C is the biodegraded concentration of the chemical, C_o is the starting concentration, and k is the rate of decrease of the chemical. First-order rate constants are particularly useful since they allow half-lives for chemicals to be estimated:

$$t_{1/2} = \frac{0.693}{k} \tag{8.12}$$

Although there are certain advantages to assuming first-order kinetics in laboratory systems, a more rigorous theoretical basis for extrapolating laboratory rate constants to the subsurface is necessary. The advantage of using Monod kinetics, for example, is that the constants K_c and μ_{max} uniquely define the rate equation for mineralization of a specific compound. The ratio μ_{max}/K_c also represents the first-order rate constant for degradation when $C \ll K_c$. This rate constant incorporates both the activity of the degrading population and the substrate dependency of the reaction. It therefore takes into account both population and substrate levels and provides a theoretical basis for extrapolating laboratory rate data to the environment.

Biodegradation Rates

Much research has been undertaken to determine biodegradation rates of organics in the subsurface. Field biodegradation rates generally refer to the rate of mass loss of contaminants as a function of time. Laboratory biodegradation rates similarly refer to the rate of removal of contaminants during the controlled experiment. Overall, laboratory rates are easily determined; however, their usefulness may be limited because of the limitations of microcosm studies discussed earlier. Additionally, laboratory degradation rates are very dependent on the soil and ground water used in the experiment and may vary from one location to another in the same site. A methodology has not yet been established that would allow the transfer of laboratory-determined biodegradation rates to field situations with any degree of confidence.

The main difficulty in determining field rates of biodegradation is due to the complicating transport processes, such as advection, dispersion, and sorption. Additionally, biodegradation under many field situations is limited by the transport of the required nutrients into the plume. Outside of conducting controlled field experiments for which the total initial mass of contaminants is known and extensive monitoring allows calculating contaminant mass at any time during the experiment, it would be very difficult to determine the field rates of biodegradation. Often, as will be seen in the next section, researchers and practitioners indirectly verify the occurrence of biodegradation at the field scale and calculate an "apparent" rate of biodegradation based on the changes in total mass in the plume as a function of time. Such a rate may incorporate the effects of the other physical and chemical processes occurring at the site.

8.5 FIELD STUDIES OF NATURAL ATTENUATION

A number of field studies of natural attenuation have been presented in the general literature over the past few years. A few of these studies are reviewed to illustrate the type of field measurements that can be made to support their laboratory counterparts.

United Creosoting Company (UCC) Site in Conroe, Texas

An injection-production test was conducted at the United Creosoting Company (UCC) Superfund site in Conroe, Texas (Borden and Bedient, 1987). The aim of the field test was to obtain an accurate evaluation of the effects of adsorption and biotransformation on the fate and transport of the organic contaminants present at the site. The United Creosoting Company site was operated as a wood-preserving facility from 1946 to 1972. In 1972, the facility was closed and the land sold off for use by a small commercial operation and later subdivided for a housing development. Monitoring of the site has shown elevated levels of organic contaminants in the soil and ground water. Wastes generated by the wood creosoting operation were disposed of in two unlined ponds (see Figure 8.4). These wastes were composed of predominantly polycyclic aromatic hydrocarbons and pentachlorophenol

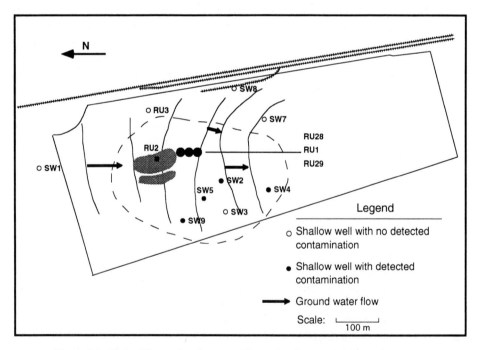

Figure 8.4 United Creosoting Company site map. Source: Borden and Bedient, 1987. © AWRA.

(PCP). In addition to the organic wastes, elevated levels of chloride are present in the ground water.

The geology at the UCC site is heterogeneous, with numerous changes between relatively clean well-sorted sand and sandy clay. Two separate water-bearing zones have been identified: (1) a lower-permeability unconfined zone, and (2) a slightly higher permeability semiconfined zone. These two zones are separated by a thin semiconfining clay layer. The injection-production test was performed in the semiconfined zone approximately 75 m south of the former ponds (Figure 8.4). This zone is composed of medium-fine clayey sand with a permeability of approximately 1 m/day. The piezometric gradient in the area is low, resulting in ground water velocities in the range of 5 m/yr.

Ground water was produced from the two outside wells (RU28 and RU29) and injected into the center well (RU1) for 7 days. A nonreactive tracer, chloride, and naphthalene and paradichlorobenzene (pDCB), were added to the injection stream for the first 24 hr. The two organics have approximately equal affinities for adsorption (log octanol water partition coefficient, K_{ow} = 3.37 for naphthalene and 3.39 for pDCB), but show considerably different affinities for biodegradation.

Breakthrough curves for chloride, pDCB, and naphthalene in wells RU28 and RU29 were compared by Borden and Bedient (1987) to obtain estimates of retardation and biotransformation. Normalized concentration distributions for chloride, naphthalene, and pDCB in RU28 are shown in Figure 8.5. The breakthrough curves for naphthalene and pDCB shown in Figure 8.5 have the same general appearance as the chloride curve, but the area under the organic curves is significantly less than that under chloride. The observed retardation factors and fractional recoveries for naphthalene and pDCB are shown in Table 8.3.

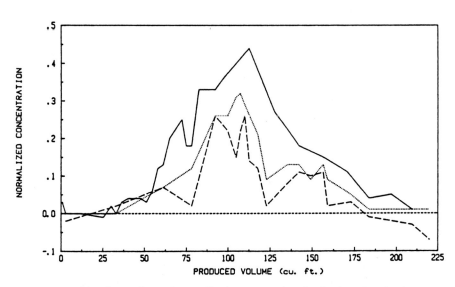

Figure 8.5 Comparison of normalized concentration distributions for chloride (——), pDCB (····), and naphthalene (––––) in RU28.   Source: Borden and Bedient, 1987. © AWRA.

TABLE 8.3   Summary of Normalized Recovery Data for Well RU29

Moments	Chloride	Conductivity	Naphthalene	pDCB
R	1.00	1.02	1.03	0.97
Fraction recovery	1.00	1.15	0.35	0.52

Source: Borden and Bedient, 1987.

The results in Table 8.3 indicated that there was essentially very little retardation of the organic compounds (the aquifer has very low organic carbon content, which is below the analytical detection limit of 0.01% by weight), yet there was a significant loss of the organic compounds from solution. The observed loss of mass (65% for naphthalene and 48% for pDCB) was attributed by Borden and Bedient (1987) to biotransformation in the aquifer. The extent of biodegradation was expected to be greater for naphthalene than for pDCB, since the microbial population had been adapted to naphthalene but not to pDCB.

Borden Landfill in Canada

Barker et al. (1987) introduced 7.6 mg/L total BTX below the water table (in addition to chloride as a tracer) into the sandy aquifer at the Borden landfill. Average concentrations of benzene, toluene, p-xylene, m-xylene, and o-xylene (BTX) were 2360, 1750, 1080, 1090, and 1290 µg/L, respectively. The migration of the contaminants was monitored using a dense sampling network. The horizontal aspect of the plume, produced by using vertically averaged concentrations of each constituent at each multilevel piezometer, is shown in Figure 8.6. Chloride behaves conservatively and illustrates the effects of advection and dispersion. Although the organics follow the same path, their centers of mass are delayed. This is mainly due to sorption at the site.

The apparent decline in total mass is shown in Figure 8.7. All the injected mass of BTX was lost within 434 days due to biodegradation. The rates of mass loss appear to be similar for all the monoaromatics; benzene was the only component to persist beyond 270 days. Laboratory microcosm studies confirmed the conclusions regarding biodegradation drawn from the field experiment.

Gas Plant Facility in Michigan

Soluble hydrocarbon and dissolved oxygen (DO) were characterized in a shallow aquifer beneath a gas plant facility in Michigan by Chiang et al. (1989). The distributions of benzene, toluene, and xylene (BTX) in the aquifer had been monitored in 42 wells for a period of 3 yr. A general site plan including the locations of the wells is shown in Figure 8.8. The site geology is characterized as a medium to coarse sand with interbeds of small gravel and cobbles. The general direction of ground water flow is northwesterly. The depth to water table ranges from 10 to 25 ft below land surface, and the slope of the water table was estimated as 0.006. Based on ground water and soil-sampling data, Chiang et al. (1989) concluded that

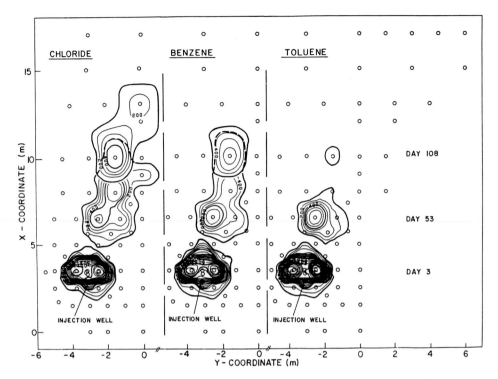

Figure 8.6 Contour plots of vertically averaged concentrations. Contour units are milligrams per milliliters for chloride and micrograms per milliliter for the organics. Source: Barker et al., 1987.

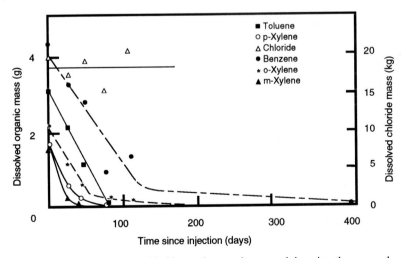

Figure 8.7 Masses of chloride and organics remaining in the ground water. Source: Barker et al., 1987.

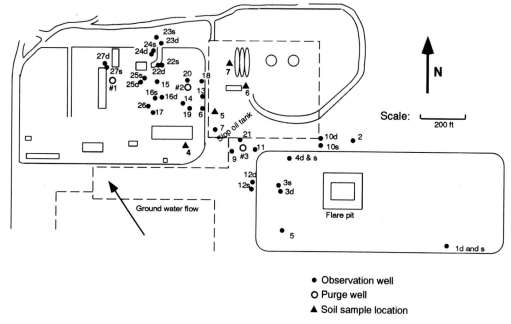

Figure 8.8 General plan of gas plant facility. Source: Chiang et al., 1989.

the flare pit was the major source of the hydrocarbons found in the aquifer, while the slope oil tank was a secondary source.

Results from the 3-yr sampling period showed a significant reduction in total benzene mass with time in ground water (Figure 8.9). The plume sampled in 1984 contained an approximate total mass of 9.83 kg, while the plume sampled in 1985 and 1986 contained 5.66 and 2.27 kg, respectively. Chiang et al. (1989) determined the attenuation rates of the soluble benzene and determined the effects of DO on

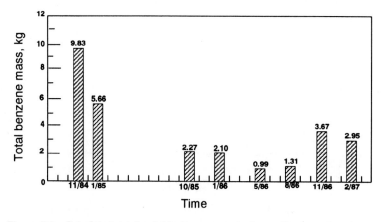

Figure 8.9 Calculated total soluble benzene mass in aquifer as a function of time. Source: Chiang et al., 1989.

the biodegradation of BTX through a combination of material balance, statistical analyses, soluble transport modeling, and laboratory microcosm experiments.

Dissolved oxygen concentrations in the ground water were measured by the Winkler titration method and with a field DO probe on two occasions (February and July 1987). Figure 8.10 shows the total BTX and DO concentration distribu-

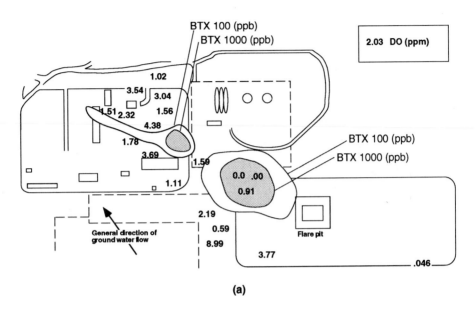

(a)

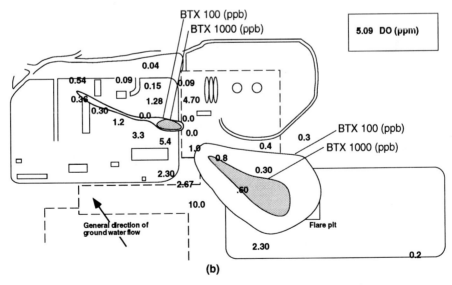

(b)

Figure 8.10 Total BTX and dissolved oxygen concentration distributions in ground water. (a) Measurements taken in February 1987. (b) Measurements taken in July 1987. Source: Chiang et al., 1989.

tions of the February and July analyses, respectively. It can be seen from Figure 8.10 that the DO concentrations are high at low BTX concentrations, and vice versa. The DO concentrations increase from their low values of < 1.0 ppm inside the 1000-ppb BTX contour to higher concentrations of >1.0 ppm outside the 100-ppb BTX contour line.

Chiang et al. (1989) also studied the decay rates of BTX in soil microcosms at an initial DO concentration of 0 to 6 ppm and concluded that a minimum DO (threshold) level of about 1 to 2 ppm may exist in ground water, which could sustain the natural biodegradation of BTX by microorganisms. Chiang et al. (1989) also conducted a detailed modeling analysis of their field data using the BIOPLUME II model, as will be seen in Section 8.6.

Cliffs–Dow Superfund Site

Ground water at the Cliffs–Dow site is contaminated with low levels of phenolic and polycyclic compounds. The aquifer sediments at the site consist of mostly coarse sands and gravels. The hydraulic conductivity ranges between 3.5×10^{-3} and 4.6×10^{-2} cm/sec. The principal contaminants found at the site near the source area include phenol, several methyl-substituted phenols, and naphthalene at concentrations ranging from 220 to 860 µg/L. Results of the characterization of the ground water are shown in Table 8.4.

Based on the analysis of samples obtained from monitoring wells, Klecka et al. (1990) found that the levels of organic contaminants are reduced to near or below the detection limit within a distance of 100 m downgradient from the source.

TABLE 8.4   Characteristics of Ground Water from the Cliffs–Dow Site

Parameter	Well			
	85-1	B-3A	85-3	B-6
Temperature (°C)[a]	—	9.2	8.0	8.5
Dissolved oxygen (ppm)[a]	1.2	<0.1	0.1[b]	1.6
Total organic C (ppm)	3.2	131.7	2.6	2.8
Total inorganic C (ppm)	39.2	381.0	48.0	17.4
Total phenolics (ppb)	—	47[c]	<4	—
Total polycyclics (ppb)	—	<40	<4	—
pH	6.1	6.4	5.5	6.1
Conductivity (µS)	220	1180	335	65
Phosphate (ppm)	<1	<1	<1	<1
Nitrate (ppm)	1.7	<0.1	0.1	<0.1
Sulfate (ppm)	48	7	124	14
Chloride (ppm)	28	14	70	3

Source: Klecka et al., 1990.

Well 85-1 is an upgradient well, wells B-3A and 85-3 are near the source area, and well B-6 is downgradient.

[a] Temperature and dissolved oxygen analyzed during sampling.

[b] Degassing of sample noted during sampling.

[c] Phenolic contaminant identified as 2,4-dimethylphenol.

Further analyses of the ground water chemistry were used to verify that biodegradation was occurring at the site and causing the disappearance of the contaminants.

Klecka et al. (1990) attributed the reduction of dissolved oxygen from 1.2 mg/L in an upgradient well to less than 0.1 mg/L at downgradient wells to biodegradation. This conclusion, Klecka et al. (1990) argue, is supported by the increase in total inorganic carbon at well B-3A. Microcosm studies were used by Klecka et al. (1990) to simulate the aerobic biodegradation of phenols at the site. Klecka et al. (1990) used the BIO1D to simulate the aerobic biodegradation of contaminants at the site, as will be seen in Section 8.6.

8.6 MODELING BIODEGRADATION

The problem of quantifying biodegradation in the subsurface can be addressed by using models that combine physical, chemical, and biological processes. Developing such models is not simple, however, due to the complex nature of microbial kinetics, the limitations of computer resources, the lack of field data on biodegradation, and the need for robust numerical schemes that can simulate the physical, chemical, and biological processes accurately.

The main expressions that have been utilized for modeling biodegradation kinetics include the following:

1. Monod kinetics
2. First-order decay kinetics
3. Instantaneous reaction assumption

Monod kinetics have been discussed in Section 8.4. The reduction of contaminant concentrations using Monod kinetics can be expressed as

$$\Delta C = M_t \mu_{max} \frac{C}{K_c + C} \Delta t \tag{8.13}$$

where C = contaminant concentration, M_t is the total microbial concentration, μ_{max} = maximum contaminant utilization rate per unit mass microorganisms, K_c = contaminant half-saturation constant, and Δt is the time interval being considered.

Incorporating Eq. (8.13) into the one-dimensional transport equation, for example, results in

$$\frac{\partial C}{\partial t} = D_x \frac{\partial^2 C}{\partial x^2} - v \frac{\partial C}{\partial x} - M_t \mu_{max} \frac{C}{K_c + C} \tag{8.14}$$

where v is the seepage velocity and D_x is the dispersion coefficient.

For aerobic biodegradation, and assuming that oxygen and the contaminant are the only substrates required for growth, the change in contaminant and oxygen concentrations due to biodegradation is given by

$$\Delta C = M_t \mu_{\max} \frac{C}{K_c + C} \frac{O}{K_o + O} \Delta t \tag{8.15}$$

$$\Delta O = M_t \mu_{\max} F \frac{C}{K_c + C} \frac{O}{K_o + O} \Delta t \tag{8.16}$$

where O = oxygen concentration, K_o = oxygen half-saturation constant, and F = ratio of oxygen to contaminant consumed.

Incorporating Eqs. (8.15) and (8.16) into the general form of the transport equation results in a system of partial differential equations as follows (Borden and Bedient, 1986):

$$\frac{\partial C}{\partial t} = \frac{1}{R_c} \nabla \cdot (D \nabla C - vC) - M_t \frac{\mu_{\max}}{R_c} \frac{C}{K_c + C} \frac{O}{K_o + O} \tag{8.17}$$

$$\frac{\partial O}{\partial t} = \nabla \cdot (D \nabla O - vO) - M_t \mu_{\max} F \frac{C}{K_c + C} \frac{O}{K_o + O} \tag{8.18}$$

$$\frac{\partial M_s}{\partial t} = \frac{1}{R_m} \nabla \cdot (D \nabla M_s - vM_s)$$

$$+ M_s \mu_{\max} Y \frac{C}{K_c + C} \frac{O}{K_o + O} + \frac{k_c Y(OC)}{R_m} - b M_s \tag{8.19}$$

where C is the contaminant concentration, O is the oxygen concentration, D is a dispersion tensor, v is the ground water velocity, R_c is the retardation coefficient for the contaminant, M_s and M_t are the concentrations of microbes in solution and the total microbial concentration, respectively ($M_t = R_m \cdot M_s$, where R_m is the microbial retardation factor), $\mu_{\max}$ is the maximum contaminant utilization rate per unit mass of microorganisms, Y is the microbial yield coefficient, K_c is the half-saturation constant for the contaminant, K_o is the half-saturation constant for oxygen, OC is the natural organic carbon concentration, F is the ratio of oxygen to hydrocarbon consumed, and b is the microbial decay rate.

First-order kinetics represent an exponential decay and are a simplification of Monod kinetics; see Eq. (8.11). The instantaneous reaction model, a model first proposed by Borden and Bedient (1986), was used in the BIOPLUME II model, as will be seen later. The main assumption in the model is that microbial biodegradation kinetics are fast in comparison with the transport of oxygen and that the growth of microorganisms and utilization of oxygen and organics in the subsurface can be simulated as an instantaneous reaction between the organic contaminant and oxygen.

From a practical standpoint, the instantaneous reaction model assumes that the rate of utilization of the contaminant and oxygen by the microorganisms is very high and that the time required to mineralize the contaminant is very small, or almost instantaneous. Biodegradation is calculated using the expression

$$\Delta C_R = -\frac{O}{F} \tag{8.20}$$

where ΔC_R is the change in contaminant concentration due to biodegradation, O

is the concentration of oxygen, and F is the ratio of oxygen to contaminant consumed. The three expressions discussed here have been used in a variety of ground water models, as we will see in the following section.

Example 8.1 BIODEGRADATION EXPRESSIONS

The purpose of this example is to illustrate the differences between the three expressions that can be used to simulate biodegradation: first-order decay, Monod kinetics, and an instantaneous reaction. Assume that the dissolved benzene concentration at a downgradient location in a given aquifer is 12.0 mg/L. Also assume that aerobic biodegradation is occurring in the aquifer and that 8.0 mg/L of oxygen is available for utilization by the microorganisms over a period of 10 days. A simple calculation can be made using each of the three biodegradation expressions to estimate the anticipated reduction in benzene concentrations due to the presence of the 8.0 mg/L of oxygen.

Solution

Instantaneous reaction expression
Assuming that 3.0 mg/L of oxygen is required to biodegrade 1.0 mg/L of contaminant:

$$\text{Benzene reduction} = 8.0/3.0 = 2.67 \text{ mg/L}$$

$$\text{Resulting benzene concentration} = 12.0 - 2.67 = 9.33 \text{ mg/L}$$

Monod kinetic expression
Assuming an oxygen half-saturation constant of 0.1 mg/L (Borden et al., 1986), a benzene half-saturation constant of 22.16 mg/L, a maximum utilization rate of 9.3 days^{-1} (Tabak et al., 1990), and a microorganism population of 0.05 mg/L;

$$\text{Benzene reduction} = 9.3 \times \frac{12}{12 + 22.16} \times \frac{8}{8 + 0.1} \times 10 \times 0.05 = 1.59 \text{ mg/L}$$

$$\text{Resulting benzene concentration} = 12 - 1.59 \text{ mg/L} = 10.4 \text{ mg/L}$$

First-order decay expression
Assuming a half-life of benzene of 5 days (Howard et al., 1991),

$$\text{First-order decay rate from Eq. (8.12)} = \frac{0.693}{t_{1/2}} = 0.1386 \text{ day}^{-1}$$

$$\text{Resulting benzene concentration} = 12 \times e^{-0.1386 \times 10} = 3.0 \text{ mg/L}$$

These calculations show that the Monod kinetic model is the most conservative model in predicting the amount of biodegradation that occurs. Only 1.59 mg/L of benzene concentration reduction is attributed to biodegradation. The instantaneous reaction model assumes total utilization of the oxygen available during the 10-day period, thus resulting in a predicted reduction in benzene concentration of 2.67 mg/L. Finally, the first-order decay expression predicts a resulting concentration of benzene of 3.0 mg/L after 10 days, which is unrealistic because there is not enough oxygen in the aquifer to reduce the benzene concentration to the predicted level. Therefore, it is important to recognize that the first-order expression does not incorporate the electron acceptor limitation, and thus care should be taken when using this expression.

TABLE 8.5 Biodegradation Models

Name	Dimension	Description	Author(s)
—	1	aerobic, microcolony, Monod	Molz et al. (1986)
BIOPLUME	1	aerobic, Monod	Borden et al. (1986)
—	1	analytical first-order	Domenico (1987)
BIO1D	1	aerobic and anaerobic, Monod	Srinivasan & Mercer (1988)
—	1	cometabolic, Monod	Semprini & McCarty (1991)
—	1	aerobic, anaerobic, nutrient limitations, microcolony, Monod	Widdowson et al. (1988)
—	1	aerobic, cometabolic, multiple substrates, fermentative, Monod	Celia et al. (1989)
BIOPLUME II	2	aerobic, instantaneous	Rifai et al. (1988)
—	2	Monod	MacQuarrie et al. (1990)
BIOPLUS	2	aerobic, Monod	Wheeler et al. (1987)
ULTRA	2	first-order	Tucker et al. (1986)
—	2	denitrification	Kinzelbach et al. (1991)
—	2	Monod, biofilm	Odencrantz et al. (1990)

Developed Biodegradation Models

Many biodegradation models have been developed in recent years, most of which utilize some form of the three expressions presented earlier (see Table 8.5). This section will focus on a few of the models listed in Table 8.5 and on their application to field biodegradation case studies.

Biofilm Model

McCarty et al. (1984) believe that the nature of the ground water environment (low substrate concentration and high specific surface area) dictates that the predominant type of bacterial activity will be bacteria attached to solid surfaces in the form of biofilm. The attached bacteria remain generally fixed in one place and obtain energy and nutrients from the ground water that flows by.

Figure 8.11 is an illustration of an idealized biofilm having a uniform cell density of X_f and a locally uniform thickness of L_f. An idealized biofilm is a homogeneous matrix of bacteria and their extracellular polymers that bind them together and to the inert surface (McCarty et al., 1981). Ground water flows past the biofilm in the x direction, while substrates are transported from the water to the biofilm in the z direction. The distance L represents the thickness of a mass-transport diffusion layer through which substrate must pass in order to go from the bulk liquid into the biofilm, where utilization occurs.

Within the biofilm, two processes occur simultaneously: (1) utilization of the substrate by the bacteria, assumed to follow a Monod-type relation, and (2) diffusion of the substrate through the biofilm according to Fick's law. If steady-state substrate utilization is set equal to molecular diffusion, the following differential equation can be written:

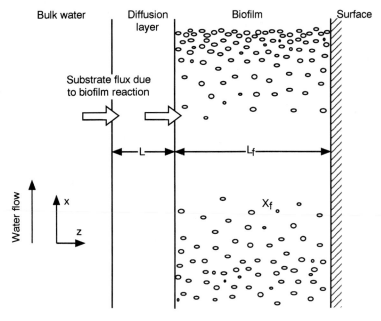

Figure 8.11 Idealized biofilm illustrating uniform cell density (X_f), thickness (L_f), water flow, and substrate flux into biofilm. Source: McCarty et al., 1984. Reprinted by permission of John Wiley & Sons, Inc.

$$D_f \frac{d^2 S_f}{dz^2} = \frac{k X_f S_f}{K_s + S_f} \tag{8.21}$$

where S_f = concentration of rate-limiting substrate at a point within the biofilm ($M_S L^{-3}$), D_f = molecular diffusivity of the substrate within the biofilm ($L^2 T^{-1}$), k = maximum specific rate of substrate utilization by the bacteria ($M_S M_x^{-1} T^{-1}$), K_s = Monod half-maximum rate concentration ($M_S L^{-3}$), and z = distance normal to the surface of the biofilm (L). For the substrate to reach the biofilm surface, mass transport across the diffusion layer occurs according to the following equation:

$$J = -D \frac{dS}{dz} = D \frac{S - S_s}{L} \tag{8.22}$$

where J = flux of rate-limiting substrate across the diffusion layer and into the biofilm ($M_S L^{-2} T^{-1}$), D = molecular diffusivity of the rate-limiting substrate in water, and S and S_s = substrate concentrations in the bulk liquid and at the biofilm surface, respectively.

Figure 8.12 shows the interaction of the three processes: substrate utilization, molecular diffusion within the biofilm, and mass transport across the diffusion layer. For a thick biofilm (case A), the substrate concentration approaches zero and the biofilm is called deep. If the biofilm is very thin (case C), almost no substrate utilization occurs and the biofilm is essentially fully penetrated at the surface concentration S_s. The remaining cases are termed shallow (case B). The

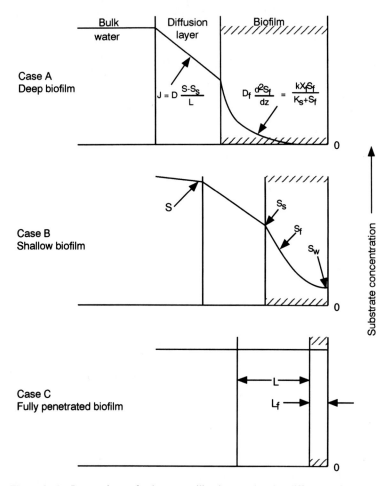

Figure 8.12 Interactions of substrate utilization, molecular diffusion within biofilm, and mass transport across diffusion layer. Source: McCarty et al., 1984. Reprinted with permission of John Wiley & Sons, Inc.

biofilm concept was not directly incorporated into the ground water transport equations and has not been specifically applied to a field site.

Microcolony Models

Molz et al. (1986) and Widdowson et al. (1987) developed one- and two-dimensional models for aerobic biodegradation of organic contaminants in ground water coupled with advective and dispersive transport. A microcolony approach was utilized in the modeling; microcolonies of bacteria were represented as disks of uniform radius and thickness attached to aquifer sediments. A boundary layer of a given thickness was associated with each colony across which substrate and oxygen are transported by diffusion to the colonies.

Their results indicate that biodegradation would be expected to have a major effect on contaminant transport when proper conditions for growth exist. Simulations of two-dimensional transport suggested that under aerobic conditions microbial degradation reduces the substrate concentration profile along longitudinal sections of the plume and retards the lateral spread of the plume. Anaerobic conditions developed in the plume center due to microbial consumption and limited oxygen diffusion into the plume interior.

Widdowson et al. (1988) extended their 1986 and 1987 studies to simulate oxygen- and/or nitrate-based respiration. Basic assumptions incorporated into the model included a simulated particle-bound microbial population comprised of heterotrophic, facultative bacteria in which metabolism is controlled by lack of either an organic carbon-electron donor source (substrate), electron acceptor (O_2 and/or NO_3), or mineral nutrient (NH_4^+), or all three simultaneously. Transport of substrate and oxygen in the porous medium is assumed to be governed by advection–dispersion equations with surface adsorption. Microbial degradation enters the two basic transport equations as sink terms.

Based on aerobic assumptions, five coupled, nonlinear equations govern microbial growth dynamics in porous media. Three of these are partial differential equations and two are nonlinear algebraic equations:

$$\frac{\partial S}{\partial t}\left(1 + \frac{\alpha_s A_s}{n}\right) = -v_x \frac{\partial S}{\partial x} + D_s \frac{\partial^2 S}{\partial x^2} - \frac{D_{sb}}{n}\frac{S - s}{\delta} N_c \pi r_c^2 \tag{8.23}$$

$$\frac{\partial O}{\partial t}\left(1 + \frac{\alpha_o A_s}{n}\right) = -v_x \frac{\partial O}{\partial x} + D_o \frac{\partial^2 O}{\partial x^2} - \frac{D_{ob}}{n}\frac{O - o}{\delta} N_c \pi r_c^2 \tag{8.24}$$

$$D_{sb}\frac{S - s}{\delta}\pi r_c^2 = \frac{\mu_m m_c}{Y}\frac{s}{K_s + s}\frac{o}{K_o + o} \tag{8.25}$$

$$D_{ob}\frac{O - o}{\delta}\pi r_c^2 = \gamma \mu_m m_c \frac{s}{K_s + s}\frac{o}{K_o + o} + \alpha K_d m_c \frac{o}{K_o' + o} \tag{8.26}$$

$$\frac{1}{N_c}\frac{\partial N_c}{\partial t} = \mu_m \frac{s}{K_s + s}\frac{o}{K_o + o} - K_d \tag{8.27}$$

where S is the substrate concentration in the pore fluid, s is the substrate concentration within the colony, D_s is the substrate dispersion coefficient, n is the porosity, t is the time, x is length, O is the oxygen concentration in the pore fluid, o is the oxygen concentration within the colony, D_o is the oxygen dispersion coefficient, A_s is the effective specific surface of the aquifer matrix, N_c is the number of colonies per unit volume of aquifer, μ_m is the maximum specific growth rate of heterotrophic microorganisms, δ is the thickness of the boundary layer, γ is the oxygen use coefficient for synthesis of heterotrophic biomass, α_s, α_o are the ratios of the adsorbed concentration to that in solution for the substrate and oxygen, M_c is the cell mass per colony, K_d is the distribution coefficient, K_s is the substrate half-saturation constant, and K_o is the oxygen half-saturation constant. The microcolony approach has not been applied as of yet to a field site.

Monod Kinetic Models: BIO1D Model

Srinivasan and Mercer (1988) presented a one-dimensional, finite difference model for simulating biodegradation and sorption processes in saturated porous media. The model formulation allowed for accommodating a variety of boundary conditions and process theories. Aerobic biodegradation was modeled using a modified Monod function; anaerobic biodegradation was modeled using Michaelis–Menten kinetics. In addition, first-order degradation can be simulated for both substances. Sorption can be incorporated using linear, Freundlich, or Langmuir equilibrium isotherms for either substance.

The Srinivasan and Mercer (1988) model is an extension of that presented by Borden and Bedient (1986). The governing equations are

$$f = D \frac{\partial^2 S}{\partial x^2} - V \frac{\partial S}{\partial x} - B(S, O) - [1 + A(S)] \frac{\partial S}{\partial t} = 0 \qquad (8.28)$$

$$g = D \frac{\partial^2 O}{\partial x^2} - V \frac{\partial O}{\partial x} - F \cdot B(S, O) - [1 + A(O)] \frac{\partial O}{\partial t} = 0 \qquad (8.29)$$

For aerobic conditions,

$$B(S, O) = Mk \frac{S}{k_s + S} \frac{O}{k_o + O} \frac{S - S_{min}}{S} \qquad (8.30)$$

for $S \geq S_{min}$ and $O \geq O_{min}$. Otherwise,

$$B(S, O) = 0$$

For anaerobic conditions, $B(S, O)$ reduces to $B(S)$ and only one equation is solved for S.

$$B(S) = M_n k_n \frac{S}{k_{sn} + S} \qquad (8.31)$$

where S is the substrate concentration in the pore fluid (ML^{-3}), O is the oxygen concentration in the pore fluid (ML^{-3}), D is the longitudinal hydrodynamic dispersion coefficient (L^2T^{-1}), x is the distance, V is the interstitial fluid velocity (LT^{-1}), $B(S, O)$ is a biodegradation term expressed as a function of the dependent variables S and O ($ML^{-3}T^{-1}$), $A(S)$ is the adsorption term expressed as a function of S (the term $[1 + A(S)]$ is the retardation factor), t is the time, M is the microbial mass, k is the maximum substrate utilization rate per unit mass of microorganisms, k_s is the substrate half-saturation constant, k_o is the oxygen half-saturation constant, S_{min} is the minimum substrate concentration that permits growth and decay, O_{min} is the minimum oxygen concentration that permits growth and decay, and F is the ratio of oxygen to substrate consumed. Note that M_n, k_n, and k_{sn} are counterparts of M, k, and k_s under anaerobic conditions. The BIO1D model was used by Klecka et al. (1990) at the Cliffs–Dow Superfund site, as will be seen later.

Instantaneous Reaction Models: The BIOPLUME II Model

The BIOPLUME II model was developed by modifying a two-dimensional transport model developed by the U.S. Geological Survey and known as the method of characteristics (MOC) model (Konikow and Bredehoeft, 1978; see Chapter 10). The basic concept applied to modify the USGS MOC model and to develop the BIOPLUME II model includes the use of a dual-particle mover procedure to simulate the transport of oxygen and contaminants in the subsurface. The transport equation is solved twice at every time step to calculate the oxygen and contaminant distributions (Rifai et al., 1988):

$$\frac{\partial (Cb)}{\partial t} = \frac{1}{R_c} \left[\frac{\partial}{\partial x_i} \left(bD_{ij} \frac{\partial C}{\partial x_j} \right) - \frac{\partial}{\partial x_i} (bCV_i) \right] - \frac{C'W}{n} \tag{8.32}$$

$$\frac{\partial (Ob)}{\partial t} = \left[\frac{\partial}{\partial x_i} \left(bD_{ij} \frac{\partial O}{\partial x_j} \right) - \frac{\partial}{\partial x_i} (bOV_i) \right] - \frac{O'W}{n} \tag{8.33}$$

where C and O = concentration of contaminant and oxygen, respectively, C' and O' = concentration of contaminant and oxygen in a source or sink fluid, n = effective porosity, b = saturated thickness, t = time, x_i and x_j = cartesian coordinates, W = volume flux per unit area, V_i = seepage velocity in the direction of x_i, R_c = retardation factor for contaminant, and D_{ij} = coefficient of hydrodynamic dispersion.

The two plumes are combined using the principle of superposition to simulate the instantaneous reaction between oxygen and the contaminants, and the decrease in contaminant and oxygen concentrations is calculated from

$$\Delta C_{RC} = \frac{O}{F}; \qquad O = 0 \text{ where } C > \frac{O}{F} \tag{8.34}$$

$$\Delta C_{RO} = C \cdot F; \qquad C = 0 \text{ where } O > C \cdot F \tag{8.35}$$

where ΔC_{RC} and ΔC_{RO} are the calculated changes in concentrations of contaminant and oxygen, respectively, due to biodegradation.

Figure 8.13 is a conceptual schematic of the BIOPLUME II model. At the left of the figure, a plan view of the contaminant and oxygen plumes with and without biodegradation is shown. After the two plumes are superimposed, the contaminant plume is reduced in size and concentrations. The dissolved oxygen is depleted in zones of high contaminant concentrations and reduced in zones of relatively moderate contaminant concentrations. The schematics on the right in Figure 8.13 present transects down the plume center line and help to illustrate the distributions of contaminant and oxygen concentration with and without biodegradation. Field data have verified the correlation between oxygen and contaminant concentrations at sites.

Two methods can be used to simulate biodegradation in the BIOPLUME II model: first-order decay and instantaneous reaction. For the first-order decay model, the reaction rate, k_2, is required as input. The model input parameters

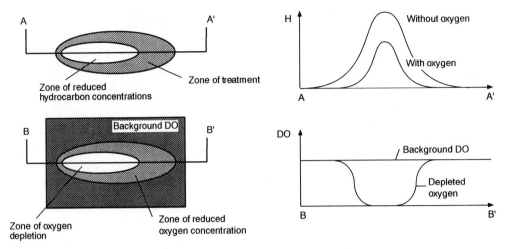

Figure 8.13 Principle of superposition for organics and oxygen in BIOPLUME II model. Source: Rifai et al., 1988. Reproduced by permission of ASCE.

required for the instantaneous reaction include the amount of dissolved oxygen in the aquifer prior to contamination and the oxygen demand of the contaminant determined from a stoichiometric relationship. Modeling the biodegradation of several components such as BTX at a site requires that an average stoichiometric coefficient for the three components be calculated.

Two additional sources of oxygen can be specified in BIOPLUME II. Injection of oxygen in a bioremediation project can be simulated by using injection wells or infiltration galleries. Reaeration from the unsaturated zone can be simulated in an indirect way by specifying a first-order decay rate for the contaminants at the site.

The output from the MOC/BIOPLUME II model consists generally of a head map and a chemical concentration map for each node in the grid. Immediately following the head and concentration maps is a listing of the hydraulic and transport errors. If observation wells have been specified, a concentration history for those wells is included in the output. The sensitivity of aerobic biodegradation to some of the model parameters has been analyzed in detail by Rifai et al. (1988). Their analyses indicate that biodegradation is mostly sensitive to the hydraulic conductivity (Figure 8.14). This result verifies some field observations about the applicability of bioremediation for systems with relatively large hydraulic conductivities. Biodegradation was not sensitive to the retardation factor or dispersion. The BIOPLUME II model has been applied to several sites (see next section).

Example 8.2 MODELING BIOREMEDIATION USING BIOPLUME II

A modeling analysis using BIOPLUME II is presented to demonstrate how the model might be used for designing bioremediation systems. Bioremediation involves the injection of oxygen and other limiting nutrients to enhance biodegradation and accelerate the remediation of contaminated sites (see Chapter 13).

A hypothetical aquifer with the parameters shown in Table 8.6 is being bioremediated for 2 yr using three injection wells and three pumping wells; each well

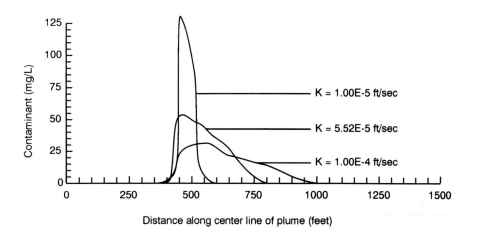

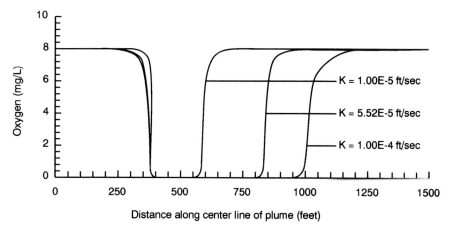

Figure 8.14 Variation of contaminant and oxygen concentrations with hydraulic conductivity predicted with BIOPLUME II model. Source: Rifai et al., 1988. Reproduced by permission of ASCE.

TABLE 8.6   Model Parameters Used in Example 8.2

Grid size	22 × 22
Cell size	50 ft × 50 ft
Transmissivity	0.002 ft²/sec
Aquifer thickness	10 ft
Hydraulic gradient	0.001
Longitudinal dispersivity	10 ft
Transverse dispersivity	3 ft
Effective porosity	30%

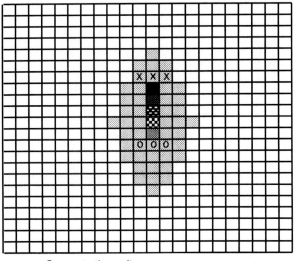

Concentration units

■	1.00e + 3	x	Injection well
■	8.89e + 2	0	Pumping well
▦	7.78e + 2		
▩	6.67e + 2		
▨	2.22e + 2		
▢	1.11e + 2		
□	0.00e + 0		

Values represent upper limits for corresponding pattern

Figure 8.15 Initial contaminant plume for Example 8.2.

is pumping or injecting at rate of 1 gpm. The initial plume and well locations are shown in Figure 8.15. Three different scenarios were modeled: (1) No oxygen was injected and biodegradation was due only to a background oxygen concentration of 3 mg/L. (2) 20 mg/L of oxygen was injected throughout the pumping period. (3) In a different simulation, 40 mg/L was injected into the wells. The latter two concentrations were selected because these values can be obtained by the injection of liquid oxygen.

Figure 8.16 shows the extent of the contaminant plume when pumping has occurred without enhanced biodegradation. The highest contaminant concentration after 2 yr of pumping is 20 mg/L, down from the original maximum of 1000 mg/L. Figure 8.17 is the output for the plume after 20 mg/L of oxygen was injected. The maximum concentration in this case is 15 mg/L. Fewer cells have concentrations greater than 5 mg/L, and the resulting plume is smaller in size.

When 40 mg/L of oxygen is injected, the resulting plume shown in Figure 8.18 is only slightly different from that in Figure 8.17. The maximum concentration, however, is now 9 mg/L instead of 15 mg/L. This indicates limited benefits from doubling the oxygen concentration. The main reason for this observation is that oxygen is not getting transported to the contamination areas, where it can be most useful. One way around this is to move the injection wells closer to the con-

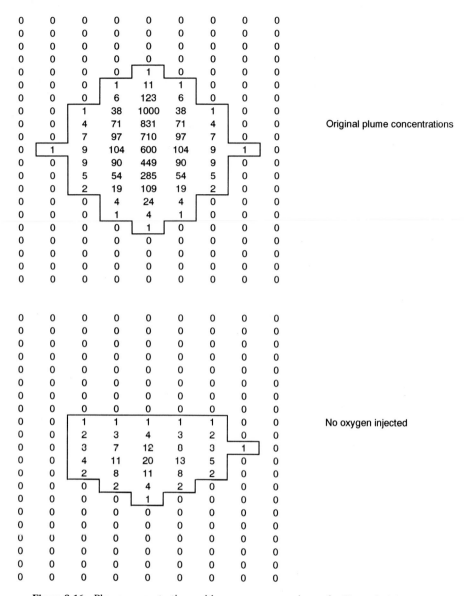

0	0	0	0	0	0	0	0	0	
0	0	0	0	0	0	0	0	0	
0	0	0	0	0	0	0	0	0	
0	0	0	0	0	0	0	0	0	
0	0	0	0	1	0	0	0	0	
0	0	0	1	11	1	0	0	0	
0	0	0	6	123	6	0	0	0	
0	0	1	38	1000	38	1	0	0	
0	0	4	71	831	71	4	0	0	Original plume concentrations
0	0	7	97	710	97	7	0	0	
0	1	9	104	600	104	9	1	0	
0	0	9	90	449	90	9	0	0	
0	0	5	54	285	54	5	0	0	
0	0	2	19	109	19	2	0	0	
0	0	0	4	24	4	0	0	0	
0	0	0	1	4	1	0	0	0	
0	0	0	0	1	0	0	0	0	
0	0	0	0	0	0	0	0	0	
0	0	0	0	0	0	0	0	0	
0	0	0	0	0	0	0	0	0	
0	0	0	0	0	0	0	0	0	

0	0	0	0	0	0	0	0	0	
0	0	0	0	0	0	0	0	0	
0	0	0	0	0	0	0	0	0	
0	0	0	0	0	0	0	0	0	
0	0	0	0	0	0	0	0	0	
0	0	0	0	0	0	0	0	0	
0	0	0	0	0	0	0	0	0	
0	0	0	0	0	0	0	0	0	
0	0	1	1	1	1	1	0	0	No oxygen injected
0	0	2	3	4	3	2	0	0	
0	0	3	7	12	8	3	1	0	
0	0	4	11	20	13	5	0	0	
0	0	2	8	11	8	2	0	0	
0	0	0	2	4	2	0	0	0	
0	0	0	0	1	0	0	0	0	
0	0	0	0	0	0	0	0	0	
0	0	0	0	0	0	0	0	0	
0	0	0	0	0	0	0	0	0	
0	0	0	0	0	0	0	0	0	
0	0	0	0	0	0	0	0	0	

Figure 8.16 Plume concentrations with no oxygen amendment for Example 8.2.

taminated zones. The other alternative is to inject oxygen concentrations in gradual increments.

In conclusion, there is an optimal level of oxygen that will be of benefit in cleanup operations. The overall extent of biodegradation for this hypothetical scenario can be summed up in Figure 8.19, which shows contaminant concentrations across the center line (from top to bottom) of the resultant plumes. It is obvious that biodegradation can be of immense benefit when conditions for its use prevail.

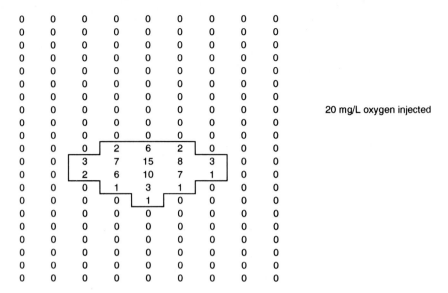

```
0   0   0   0   0   0   0   0   0
0   0   0   0   0   0   0   0   0
0   0   0   0   0   0   0   0   0
0   0   0   0   0   0   0   0   0
0   0   0   0   0   0   0   0   0
0   0   0   0   0   0   0   0   0
0   0   0   0   0   0   0   0   0       20 mg/L oxygen injected
0   0   0   0   0   0   0   0   0
0   0   0   0   0   0   0   0   0
0   0   0   2   6   2   0   0   0
0   0   3   7   15  8   3   0   0
0   0   2   6   10  7   1   0   0
0   0   0   1   3   1   0   0   0
0   0   0   0   1   0   0   0   0
0   0   0   0   0   0   0   0   0
0   0   0   0   0   0   0   0   0
0   0   0   0   0   0   0   0   0
0   0   0   0   0   0   0   0   0
0   0   0   0   0   0   0   0   0
0   0   0   0   0   0   0   0   0
```

Figure 8.17   Plume concentrations with 20 mg/L oxygen injected in Example 8.2.

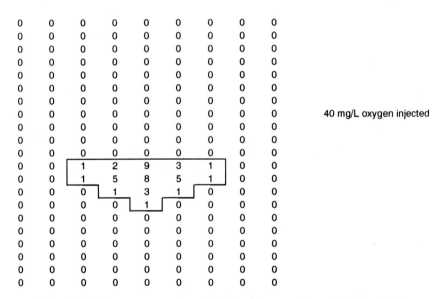

```
0   0   0   0   0   0   0   0   0
0   0   0   0   0   0   0   0   0
0   0   0   0   0   0   0   0   0
0   0   0   0   0   0   0   0   0
0   0   0   0   0   0   0   0   0
0   0   0   0   0   0   0   0   0
0   0   0   0   0   0   0   0   0       40 mg/L oxygen injected
0   0   0   0   0   0   0   0   0
0   0   0   0   0   0   0   0   0
0   0   1   2   9   3   1   0   0
0   0   1   5   8   5   1   0   0
0   0   0   1   3   1   0   0   0
0   0   0   0   1   0   0   0   0
0   0   0   0   0   0   0   0   0
0   0   0   0   0   0   0   0   0
0   0   0   0   0   0   0   0   0
0   0   0   0   0   0   0   0   0
0   0   0   0   0   0   0   0   0
0   0   0   0   0   0   0   0   0
```

Figure 8.18   Plume concentrations with 40 mg/L oxygen injected in Example 8.2.

Example 8.3   COMPARISON OF BIO1D AND BIOPLUME II MODELS

The two contaminant transport models BIO1D and BIOPLUME II can be used to simulate the transport and biodegradation of a contaminant in a specified porous medium. A comparison between the two models is made using a simple one-dimensional example in order to illustrate what difference might be expected between an instantaneous reaction assumption and a dual-Monod kinetics biodegradation expression. Initially, the concentration distribution resulting from the injection of a conser-

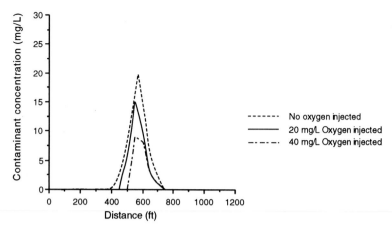

Figure 8.19 Contaminant concentrations along plume center line for Example 8.2.

vative tracer into a one-dimensional column is simulated with the two models (see Table 8.7 for input data), and the resulting graphs are shown in Figure 8.20. The models yield similar results, except BIOPLUME II shows a slightly higher concentration around the center of mass of the plume.

Figure 8.21 compares the concentration distribution of a conservative tracer and benzene as calculated by BIOPLUME II. The benzene plume shows a slightly lower (2 to 3 mg/L lower) concentration over the entire length of the plume. This seems consistent with the assumption of the model, because oxygen reacts with the organic contaminant in a fixed stoichiometric relation. The initial background concentration of 8 mg/L therefore reacted with the contaminant, reducing it by a relatively fixed amount across the full length of the plume.

Figure 8.22 compares the concentration distribution of a conservative tracer and benzene as calculated by BIO1D. Oddly, the center of mass of the benzene plume has a slightly higher concentration than the conservative plume. While this is physically very unlikely, it appears to be the mathematical result of benzene accumulation in the absence of oxygen (note that this appears in the center of the anoxic

TABLE 8.7   Input Parameters for BIOPLUME II and BIO1D Models in Example 8.3

BIOPLUME II model input	
Transmissivity	0.0001 ft^2/sec
Aquifer thickness	10 ft
Hydraulic gradient	0.001
Effective porosity	0.3
Oxygen concentration	8 mg/L
Contaminant initial concentration	100 mg/L
Additional data for BIO1D	
O$_2$ saturation constant	0.1 mg/L
Benzene saturation constant	0.1 mg/L
μ_{max}	3390/yr

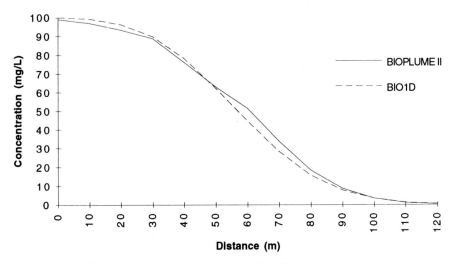

Figure 8.20 Conservative tracer plume in Example 8.3.

cross section of the plume, as demonstrated by the lack of oxygen shown in Figure 8.23).

Note also the difference between the concentration of benzene at the source of the plume between the two models. Because BIOPLUME II assumes an instantaneous reaction, the concentration of benzene at this point is immediately reduced to 96 mg/L. BIO1D, however, does not make this assumption. The concentration at the source remains 100 mg/L, but it reduces rapidly over the first few meters.

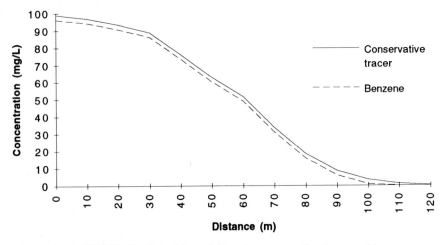

Figure 8.21 BIOPLUME II model predictions for conservative tracer and benzene in Example 8.3.

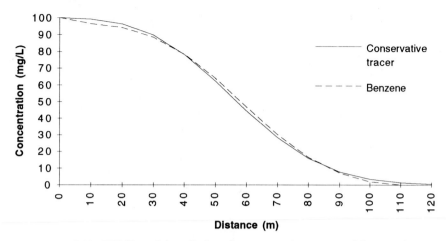

Figure 8.22 BIO1D model predictions for conservative tracer and benzene in Example 8.3.

Modeling of Biodegradation at Field Sites

Conroe Superfund Site, Texas

Borden et al. (1986) applied the first version of the BIOPLUME model to simulate biodegradation at the Conroe Superfund site in Texas. Oxygen exchange with the unsaturated zone was simulated as a first-order decay in hydrocarbon concentration. The loss of hydrocarbon due to horizontal mixing with oxygenated ground water and the resulting biodegradation were simulated by generating oxygen and hydrocarbon distributions independently and then combining by superpo-

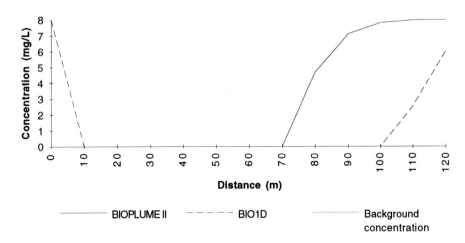

Figure 8.23 BIOPLUME II and BIO1D predictions for oxygen concentrations in Example 8.3.

sition. Simulated oxygen and hydrocarbon concentrations closely matched the observed values.

Traverse City Site, Michigan

The Traverse City field site is a U.S. Coast Guard Air Station located in Grand Traverse County in the northwestern portion of the lower peninsula of Michigan. A modeling effort of natural attenuation at the site was completed with the BIOPLUME II model (see Chapter 13 for details).

Gas Plant Facility in Michigan

Chiang et al. (1989) evaluated a first-order decay biodegradation approach and the BIOPLUME II model for simulating biodegradation at the gas plant facility. Using the model and assuming first-order decay, several simulations were made to match the observed benzene concentration distribution of January 22, 1985, by setting the observed concentration distribution of November 1, 1984, as the initial condition. The variables involved included the distribution of the leakage and spill rates between the flare pit and the slop oil tanks and macrodispersivities of the aquifer.

The BIOPLUME II model was used to simulate the July 1987 data by setting the observed concentration distribution of February 1987 as an initial condition. Figure 8.24 shows correlations between the measured and the simulated soluble BTX concentrations of July 1987. As can be seen from Figure 8.24, the correlations for BTX were reasonable. The correlations, however, for oxygen were not as

```
0   0   0   0  [3 ]  0   0   0   0   0   0   0   0   0   0   0   0
                [0 ]
0   0   0   0  [2 ]  0   0   0   0   0   0   0   0   0   0       0
                [0 ]
0   0   0 [158][4 ][5  ][0 ] 0   0   0   0   0   0   0   0   0   0
          [63][0 ][532][0 ]
0   0   0 [6 ][82 ][923 ][5317] 0   0   0   0   0   0   0   0   0   0
          [0 ][0  ][1503][3421]
0   0   0   0   0   0 [10690] 0   0   0   0   0   0   0   0   0
                      [10544]
0   0   0   0   0   0 [0 ][38 ][12]  0   0   0   0   0   0   0
                      [0 ][0  ][0 ]      (Slop Oil tank)
0   0   0   0   0   0   0  [4 ][1387][5876]  0   0   0   0   0   0
                           [0 ][1573][5024]
0   0   0   0   0   0   0   0   0 [1301]  0 [0    0 ] 0   0   0
                                  [1398]    [ Flare pit]
0   0   0   0   0   0   0   0   0   0   0   0 [0    0 ] 0   0   0
0   0   0   0   0   0   0   0   0   0 [0 ] 0   0   0   0   0
                                      [0 ]
```

Figure 8.24 Predicted (bottom number) and observed (top number) BTX by BIOPLUME II (ppb). Source: Chiang et al., 1989.

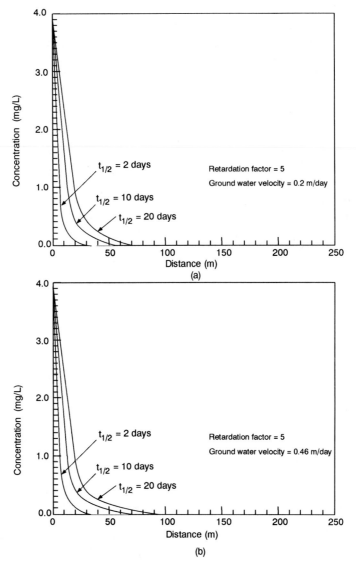

Figure 8.25 BIO1D simulations for the Cliff–Dow Superfund site. Source: Klecka et al., 1990.

similar. The authors attributed the differences to the fact that the BIOPLUME II model assumes a requirement of 3 ppm of oxygen for 1 ppm of benzene, whereas the actual requirement is in the range from 1 to 3 ppm.

Cliffs-Dow Superfund Site

The migration of organic constituents in the aquifer was simulated using the BIO1D model and assuming a first-order decay expression. Half-lives for the contaminants at the site were estimated from the results of soil microcosm experiments based on the time required for 50% disappearances of the parent compound. The velocity was varied over a range from 0.2 to 0.46 m/day, which is representative of the range of ground water flow rates at the site.

Figure 8.25 illustrates the impact of biodegradation on contaminant concentrations at the site. Model simulations performed using a half-life of 2 days indicated that the levels of the phenolic components were reduced by greater than 99% within a distance of 30 m downgradient of the source. When the half-life was increased by a factor of 10, the concentrations were reduced to a similar extent within 75 m. Because of the dominance of biodegradation, increases in ground water velocity from 0.2 to 0.46 m/day had minor effects on the level of attenuation predicted with the model.

SUMMARY

Biodegradation processes attenuate contaminant concentrations in ground water. A variety of laboratory and field procedures is necessary to verify and quantitate the contribution of these processes to contaminant transport in ground water aquifers. Biodegradation models can then be used to predict contaminant behavior for different conditions and scenarios.

REFERENCES

AELION, C. M., SWINDOLL, C. M., and PFAENDER, F. K., 1987, "Adaptation to and Biodegradation of Xenobiotic Compounds by Microbial Communities from a Pristine Aquifer," *Applied Environ. Microbiology* 53(9):2212–2217.

ALEXANDER, M., 1985, "Biodegradation of Organic Chemicals," *Environ. Sci. Technology* 2(18):106–111.

BALKWILL, D. L., and GHIORSE, W. C., 1985, "Characterization of Subsurface Bacteria Associated with Two Shallow Aquifers in Oklahoma," *Applied Environ. Microbiology* 3(50):580–588.

BARKER, J. F., PATRICK, G. C., and MAJOR, D., 1987, "Natural Attenuation of Aromatic Hydrocarbons in a Shallow Sand Aquifer," *Ground Water Monitoring Rev.* 7:64–71.

BARRIO-LAGE, G. A., PARSONS, F. Z., NARBAITZ, R. M., and LORENZO, P. A., 1990,

"Enhanced Anaerobic Biodegradation of Vinyl Chloride in Ground Water," *Environ. Toxicology Chem.* 9:403–415.

BEEMAN, R. E., and SUFLITA, J. M., 1987, "Microbial Ecology of a Shallow Unconfined Ground Water Aquifer Polluted by Municipal Landfill Leachate," *Microbial Ecology* 14:39–54.

BELLER, H. R., EDWARDS, E. A., GRBIC-GALIC, D., HUTCHINS, S. R., and REINHARD, M., 1991, "Microbial Degradation of Alkylbenzenes under Sulfate-reducing and Methanogenic Conditions," EPA/600/S2-91/027, Robert S. Kerr Environmental Research Laboratory, Ada, OK.

BENGTSSON, S. G. L., 1981, "Microcosm for Ground Water Research." Presentation at the First International Conference on Ground Water Quality Research, October 7–10, Rice University, Houston, TX.

BITTON, G., and GERBA, C. P., eds., 1984, *Groundwater Pollution Microbiology,* Wiley, New York.

BORDEN, R. C., and BEDIENT, P. B., 1986, "Transport of Dissolved Hydrocarbons Influenced by Oxygen-limited Biodegradation: 1. Theoretical Development," *Water Resources Res.* 13:1973–1982.

BORDEN, R. C., and BEDIENT, P. B., 1987, "In Situ Measurement of Adsorption and Biotransformation at a Hazardous Waste Site," *Water Resources Bulletin,* American Water Resources Association, 4:629–636.

BORDEN, R. C., BEDIENT, P. B., LEE, M. D., WARD, C. H., and WILSON, J. T., 1986, "Transport of Dissolved Hydrocarbons Influenced by Oxygen-limited Biodegradation: 2. Field Application," *Water Resources Res.* 13:1983–1990.

BOUWER, E. J., and McCARTY, P. L., 1983a. "Transformations of Halogenated Organic Compounds under Denitrification Conditions," *Applied Environ. Microbiology* 45(4): 1295–1299.

BOUWER, E. J., and McCARTY, P. L., 1983b, "Transformations of 1- and 2-carbon Halogenated Aliphatic Organic Compounds under Methanogenic Conditions," *Applied Environ. Microbiology* 45(4):1286–1294.

BOUWER, E. J., and WRIGHT, J. P., 1988, "Transformations of Trace Aliphatics in Anoxic Biofilm Columns," *J. Contaminant Hydrology* 2:155–169.

BROCK, T. D., SMITH, D. W., and MADIGAN, M. T., 1984, *Biology of Microorganisms,* 4th ed., Prentice Hall, Englewood Cliffs, NJ.

BULL, A. T., 1974, "Microbial Growth," in *Companion to Biochemistry,* p. 415 (ed. by A. T. Bull, J. R. Lagnado, J. O. Thomas, and K. F. Tipton), Longmans, London.

CELIA, M. A., KINDRED, J. S., and HERRERA, I., 1989, "Contaminant Transport and Biodegradation: 1. A Numerical Model for Reactive Transport in Porous Media," *Water Resources Res.* 25(6):1141–1148.

CHIANG, C. Y., SALANITRO, J. P., CHAI, E. Y., COLTHART, J. D., KLEIN, C. L., 1989, "Aerobic Biodegradation of Benzene, Toluene, and Xylene in a Sandy Aquifer—Data Analysis and Computer Modeling," *Ground Water* 6:823–834.

DE BONT, J. A. M., VORAGE, J. A. W., HARTMANS, S., and VAN DEN TWEEL, W. J. J., 1986, "Microbial Degradation of 1,3-Dichlorobenzene," *Applied Environ. Microbiology* 52(4):677–680.

DESAI, S., GOVIND, R., and TABAK, H., 1990, "Determination of Monod Kinetics of Toxic Compounds by Respirometry for Structure-Biodegradability Relationships," EPA/600/D-490/134, National Technical Information Service, Springfield, VA.

DOMENICO, P. A., 1987, "An Analytical Model for Multidimensional Transport of a Decaying Contaminant Species," *J. Hydrology* 91:49–58.

DUNLAP, W. J., COSBY, R. L., MCNABB, J. F., BLEDSOE, B. E., and SCALF, M. R., 1972, "Probable Impact of NTA on Ground Water," *Ground Water* 10(1):107–117.

FLIERMANS, B., PHELPS, T. J., RINGELBERG, D., MIKELL, A. T., and WHITE, D. C., 1988, "Mineralization of Trichloroethylene by Heterotrophic Enrichment Cultures," *Applied Environ. Microbiology* 54(7):1709–1714.

FOGEL, M. M., TADDEO, A. R., and FOGEL, S., 1986, "Biodegradation of Chlorinated Ethenes by a Methane-utilizing Mixed Culture," *Applied Environ. Microbiology* 51(4): 720–724.

FREEDMAN, D. L., and GOSSETT, J. M., 1989, "Biological Reductive Dechlorination of Tetrachloroethylene and Trichloroethylene to Ethylene under Methanogenic Conditions," *Applied Environ. Microbiology* 55(9):2144–2151.

GHIORSE, W. C., and BALKWILL, D. L., 1983, "Enumeration and Morphological Characterization of Bacteria Indigenous to Subsurface Environments," *Develop. Industrial Microbiology* 24:213–225.

GHIORSE, W. C., and WILSON, J. T., 1988, "Microbial Ecology of the Terrestrial Subsurface," *Advances Applied Microbiology* 33:107–172.

GIBSON, S. A., and SUFLITA, J. M., 1986, "Extrapolation of Biodegradation Results to Groundwater Aquifers: Reductive Dehalogenation of Aromatic Compounds," *Applied Environ. Microbiology* 52(4):681–688.

GOSSET, J. M., 1985, "Anaerobic Degradation of C1 and C2 Chlorinated Hydrocarbons," U.S. Air Force Report ESL-TR-85-38, National Technical Information Service, Springfield, VA.

GOVIND, R., UTGIKAR, V., SHAN, Y., and ZHAO, W., 1992, "Fundamental Studies on the Treatment of VOCs in a Biofilter," EPA/600/R-92/126, p. 75, U.S. EPA, Office of Research and Development, Washington, DC.

HOPKINS, G. D., SEMPRINI, L., and MCCARTY, P. L., 1992, "Evaluation of Enhanced in Situ Aerobic Biodegradation of cis- and trans-1-Trichloroethylene and cis- and trans-1,2-Dichloroethylene by Phenol-utilizing Bacteria," EPA/600/R-92/126, pp. 71–73, U.S. EPA, Office of Research and Development, Washington, DC.

HOWARD, P. H., and BANERJEE, S., 1984, "Interpreting Results from Biodegradability Tests of Chemicals in Water and Soil," *Environ. Toxicology Chem.* 3:551–562.

HOWARD, P. H., BOETHLING, R. S., JARVIS, W. F., MEYLAN, W. M., and MICHALENKO, E. M., 1991, *Handbook of Environmental Degradation Rates,* Lewis Publishers, Chelsea, MI.

JHAVERI, V., and MAZZACCA, A. J., 1983, "Bioreclamation of Ground and Groundwater. A Case History," *Proceedings of 4th National Conference on Management of Uncontrolled Hazardous Waste Sites,* Washington, DC.

KINZELBACH, W., SCHÄFER, W., and HERZER, J., 1991, "Numerical Modeling of Natural and Enhanced Denitrification Processes in Aquifers," *Water Resources Res.* 27(6): 1123–1135.

KLECKA, G. M., DAVIS, J. W., GRAY, D. R., and MADSEN, S. S., 1990, "Natural Bioremediation of Organic Contaminants in Ground Water: Cliffs–Dow Superfund Site," *Ground Water* 4:534–543.

KONIKOW, L. F., and BREDEHOEFT, J. D., 1978, "Computer Model of Two-dimensional Solute Transport and Dispersion in Ground Water, Automated Data Processing and Computations," *Techniques of Water Resources Investigations of the U.S.G.S.,* U.S. Geological Survey, Washington, DC.

KUHN, E. P., and SUFLITA, J., 1989, "Anaerobic Biodegradation of Nitrogen-substituted and Sulfonated Benzene Aquifer Contaminants," *Hazardous Waste Hazardous Materials* 6(2):121–133.

KUHN, E. P., COLBERG, P. J., SCHNOOR, J. L., WANNER, O., ZEHNDER, A. J. B., and SCHWARTZENBACH, R. P., 1985, "Microbial Transformations of Substituted Benzenes during Infiltration of River Water to Groundwater: Laboratory Column Studies," *Environmental Sci. Technology* 19(10):961–967.

KUHN, E. P., ZEYER, J., EICHER, P., and SCHWARZENBACH, R. P., 1988, "Anaerobic Degradation of Alkylated Benzenes in Denitrifying Laboratory Aquifer Columns," *Applied Environ. Microbiology* 54(2):490–496.

LEE, M. D., and WARD, C. H., 1985, "Microbial Ecology of a Hazardous Waste Disposal Site: Enhancement of Biodegradation," in *Proceedings of Second International Conference on Ground Water Quality,* OSU University Printing Services, Stillwater, OK, pp. 25–27.

LEE, M. D., JAMISON, V. W., and RAYMOND, R. L., 1987, "Applicability of in Situ Bioreclamation as a Remedial Action Alternative," in *Proceedings of the Petroleum Hydrocarbons and Organic Chemicals in Ground Water: Prevention, Detection and Restoration,* National Water Well Association, Houston, TX, pp. 167–185.

LITTLE, C. D., PALUMBO, A. V., HERBES, S. E., LIDSTROM, M. E., TYNDALL, R. L., and GILMER, P. J., 1988, "Trichloroethylene Biodegradation by a Methane-utilizing Bacterium," *Applied Environ. Microbiology* 54(4):951–956.

LOVLEY, D. R., and LONERGAN, D. J., 1990, "Anaerobic Oxidation of Toluene, Phenol, and *p*-Cresol by the Dissimilatory Iron Reducing Organism, GS-15," *Applied Environ. Microbiology* 56(6):1858–1864.

LYNCH, J. M., and POOLE, N. J., eds., 1979, *Microbial Ecology: A Conceptual Approach,* Wiley, New York.

MACQUARRIE, K. T. B., SUDICKY, E. A., and FRIND, E. O., 1990, "Simulation of Biodegradable Organic Contaminants in Groundwater: 1. Numerical Formulation in Principal Directions," *Water Resources Res.* 26(2):207–222.

MCCARTY, P. L., REINHARD, M., and RITTMAN, B. E., 1981, "Trace Organics in Groundwater," *Environ. Sci. Technology* 15(1):40–51.

MCCARTY, P. L., RITTMAN, B. E., and BOUWER, E. J., 1984, "Microbiological Processes Affecting Chemical Transformations in Groundwater," in *Groundwater Pollution Microbiology,* G. Bitton and C. P. Gerba, eds., Wiley, New York, pp. 89–115.

MCNABB, J. F., and MALLARD, G. E., 1984, "Microbiological Sampling in the Assessment of Groundwater Pollution," in G. Bitton and C. P. Gerba, eds., *Groundwater Pollution Microbiology,* Wiley, New York.

MOLZ, F. J., WIDDOWSON, M. A., and BENEFIELD, L. D., 1986, "Simulation of Microbial Growth Dynamics Coupled to Nutrient and Oxygen Transport in Porous Media," *Water Resources Res.* 22(8):1207–1216.

MONOD, J., 1942, "Recherches sur la Croissance des Cultures Bacteriènnes," Herman & Cie, Paris.

ODENCRANTZ, J. E., VALOCCHI, A. J., and RITTMAN, B. E., 1990, "Modeling Two-dimensional Solute Transport with Different Biodegradation Kinetics," *Proceedings of Petroleum Hydrocarbons and Organic Chemicals in Ground Water: Prevention, Detection and Restoration,* October 31–November, 1990, National Water Well Association, Houston, TX.

PRITCHARD, P. H., and BOURQUIN, A. W., 1984, "The Use of Microcosms for Evaluation of Interactions between Pollutants and Microorganisms," *Advances Microbial Ecology,* K. C. Marshall, ed., Plenum Publishing, New York, 7:133–215.

RIFAI, H. S., BEDIENT, P. B., WILSON, J. T., MILLER, K. M., and ARMSTRONG, J. M., 1988, "Biodegradation Modeling at Aviation Fuel Spill Site," *J. Environ. Engineering* 114(5):1007–1029.

SEMPRINI, L., and McCARTY, P. L., 1991, "Comparison between Model Simulations and Field Results for in Situ Biorestoration of Chlorinated Aliphatics: Part 1. Biostimulation of Methanotrophic Bacteria," *Ground Water* 29(3):365–374.

SHIELDS, M., SCHAUBHUT, R., and REAGIN, M., 1992, "Bacterial Degradation of Trichloroethylene in a Gas-phase Bioreactor," EPA/600/R-92/126, pp. 83–85, U.S. EPA Office of Research and Development, Washington, DC.

SIMS, J. L., SUFLITA, J. M., and RUSSELL, H. H., 1991, "Reductive Dehalogenation of Organic Contaminants in Soils and Ground Water," EPA/540/4-90/054, Robert S. Kerr Environmental Research Laboratory, Ada, OK.

SMITH, J. A., and NOVAK, J. T., 1987, "Biodegradation of Chlorinated Phenols in Subsurface Soils," *Water, Air, Soil Pollution* 33:29–42.

SMOLENSKI, W., and SUFLITA, J. M, 1987, "The Microbial Metabolism of Cresols in Anoxic Aquifers," *Applied Environ. Microbiology* 53:710–716.

SRINIVASAN, P., and MERCER, J. W., 1988, "Simulation of Biodegradation and Sorption Processes in Ground Water," *Ground Water* 26(4):475–487.

SUFLITA, J. M., 1989, "Microbiological Principles Influencing the Biorestoration of Aquifers," in *Transport and Fate of Contaminants in the Subsurface,* EPA/625/4-89/019, Robert S. Kerr Environmental Research Laboratory, U.S. Environmental Protection Agency, Ada, OK, pp. 85–99.

SUFLITA, J. M., and MILLER, G. D., 1985, "Microbial Metabolism of Chlorophenolic Compounds in Groundwater Aquifers," *Environ. Toxicology Chem.* 4:751–758.

SUFLITA, J. M., and SEWELL, G. W., 1991, "Anaerobic Biotransformation of Contaminants in the Subsurface," EPA/600/90/024, Robert S. Kerr Environmental Research Laboratory, Ada, OK.

SWINDOLL, C. M., AELION, C. M., and PFAENDER, F. K., 1988, "Influence of Inorganic and Organic Nutrients on Aerobic Biodegradation and on the Adaptation Response of Subsurface Microbial Communities," *Applied Environ. Microbiology* 54(1):212–217.

TABAK, H. H., DESAI, S., and GOVIND, R., 1990, "Determination of Biodegradability Kinetics of RCRA Compounds Using Respirometry for Structure–Activity Relationships," EPA/600/D-90/136. National Technical Information Service, Springfield, VA.

TUCKER, W. A., HUANG, C. T., BRAL, J. M., and DICKINSON, R. E., 1986, "Development and Validation of the Underground Leak Transport Assessment Model (ULTRA)," *Proceedings of Petroleum Hydrocarbons and Organic Chemicals in Ground Water: Prevention, Detection and Restoration,* National Water Well Association, Houston, TX, pp. 53–75.

VOGEL, T. M., and McCARTY, P. L., 1985, "Biotransformation of Tetrachloroethylene to Trichloroethylene, Dichloroethylene, Vinyl Chloride and Carbon Dioxide under Methanogenic Conditions," *Applied Environ. Microbiology* 49(5):1080–1083.

WHEELER, M. F., DAWSON, C. N., BEDIENT, P. B., CHIANG, C. Y., BORDEN, R. C., and

RIFAI, H. S., 1987, "Numerical Simulation of Microbial Biodegradation of Hydrocarbons in Ground Water," in *Proceedings of the Solving Ground Water Problems with Models Conference,* February 10–12, Denver, CO, National Water Well Association, Dublin, OH.

WIDDOWSON, M. A., MOLZ, F. J., and BENFIELD, L. D., 1987, "Development and Application of a Model for Simulating Microbial Growth Dynamics Coupled to Nutrient and Oxygen Transport in Porous Media," in *Proceedings of the Solving Ground Water Problems with Models Conference,* February 10–12, Denver, CO, National Water Well Association, Dublin, OH, pp. 28–51.

WIDDOWSON, M. A., MOLZ, F. J., and BENEFIELD, L. D., 1988, "A Numerical Transport Model for Oxygen- and Nitrate-based Respiration Linked to Substrate and Nutrient Availability in Porous Media," *Water Resources Res.* 24(9):1553–1565.

WILSON, B. H., BLEDSOE, B. E., CAMPBELL, D. H., WILSON, J. T., ARMSTRONG, J. M., and SAMMONS, J. H., 1986a, "Biological Fate of Hydrocarbons at an Aviation Gasoline Spill Site," in *Proceedings of the Petroleum Hydrocarbons and Organic Chemicals in Ground Water: Prevention, Detection, and Restoration,* National Water Well Association, Houston, TX, pp. 78–89.

WILSON, B. H., SMITH, G. B., and REES, J. F., 1986c, "Biotransformation of Selected Alkylbenzenes and Halogenated Aliphatic Hydrocarbons in Methanogenic Aquifer Material: A Microcosm Study," *Environmental Sci. Technology* 20(10):997–1002.

WILSON, J. T., McNABB, J. F., BALKWILL, D. L., and GHIORSE, W. C., 1983b, "Enumeration and Characterization of Bacteria Indigenous to a Shallow Water-table Aquifer," *Ground Water* 2:134–142.

WILSON, J. T., McNABB, J. F., COCHRAN, J. W., WANG, T. H., TOMSON, M. B., and BEDIENT, P. B., 1985, "Influence of Microbial Adaption on the Fate of Organic Pollutants in Ground Water," *Environ. Toxicology Chem.* 4:721–726.

WILSON, J. T., McNABB, J. F., WILSON, B. H., and NOONAN, M. J., 1983a, "Biotransformation of Selected Organic Pollutants in Ground Water," *Develop. Industrial Microbiology* 24:225–233.

WILSON, J. T., MILLER, G. D., GHIORSE, W. C., and LEACH, F. R., 1986b, "Relationship between the ATP Content of Subsurface Material and the Rate of Biodegradation of Alkylbenzenes and Chlorobenzene," *J. Contaminant Hydrology* 1:163–170.

WILSON, J. T., NOONAN, M. J., and McNABB, J. F., 1981, Presentation at the First International Conference on Ground Water Quality Research, October 7–10, Rice University, Houston, TX.

WILSON, J. T., and WILSON, B. H., 1985, Biotransformation of Trichloroethylene in Soil, Applied and Environmental Microbiology, 49(1):242–243.

ZEYER, J., KUHN, E. P., and SCHWARZENBACH, R. P., 1986, "Rapid Microbal Mineralization of Toluene and 1,3-Dimethylbenzene in the Absence of Molecular Oxygen," *Applied Environ. Microbiology* 52(4):944–947.

FLOW AND TRANSPORT IN THE UNSATURATED ZONE

9.1 CAPILLARITY

Flow and transport mechanisms in the unsaturated zone are much more complex than in the saturated zone. This is due to the effect of capillary forces and nonlinear soil characteristics. In an unsaturated porous medium, part of the pore space is filled with water and part is filled with air, and the total porosity is defined as the sum of the two moisture contents,

$$n = \theta_w + \theta_a \tag{9.1}$$

The moisture content θ_w is defined as the ratio of the volume of water to the total volume of porous media sample. The unsaturated or vadose zone constitutes that part of the soil profile where water content is less than the soil porosity and where the soil water pressure is negative. The water-saturated capillary fringe just above the water table is also considered part of the unsaturated zone.

Figure 9.1 shows the variation of moisture content and pressure head ψ with depth. Near the water table, a capillary fringe can occur where ψ is a small negative pressure corresponding to the air entry pressure. In this zone, the pores are saturated, but the pressure is slightly less than atmospheric. This capillary zone is small for sandy soils, but can be up to 2 m for fine-grained soils. By definition, pressure head is negative (under tension) at all points above the water table and is positive for points below the water table. The value of ψ is greater than zero in the saturated zone below the water table and equals zero at the water table. Soil physicists refer to $\psi < 0$ as the tension head or capillary suction head, which can be measured by an instrument called a tensiometer (Section 9.5).

The water in the unsaturated zone is adsorbed as a film on the surface of the grains and held strongly by suction pressure. As more water is added to the porous structure, water movement is restricted due to the strong sorption and capillary forces. Eventually, a continuous wetting phase will form, and the air can become trapped in the larger pores, thus restricting the air phase. Full water saturation is generally not achieved in the unsaturated zone because of the trapped residual air.

We can consider actual porous media to be similar to a bundle of thin glass tubes of radius r. A curved surface will tend to develop at the interface between the water and air, and the difference in pressure across the interface is called

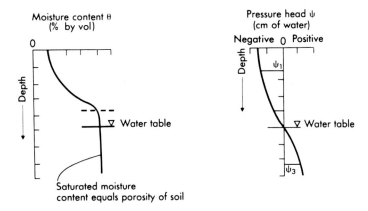

Figure 9.1 Typical θ and ψ relationships with depth in the unsaturated zone. Source: Bedient & Huber, 1992. © Addison Wesley Publishing Co., Inc. Reprinted by permission.

capillary pressure; it is directly proportional to the interfacial tension and inversely proportional to the radius of curvature. The capillary pressure P in a thin tube can be determined from a balance between the weight of the water column in the tube and capillary forces pulling upward, or

$$P = \frac{2\sigma \cos \gamma}{r} \tag{9.2}$$

where σ is the interfacial tension, r is the radius of the tube, and γ is the interface angle between the two liquids. Capillary forces and wettability of a fluid, the tendency of the fluid to preferentially spread over a solid surface, are described in detail in Section 11.3.

9.2 SOIL-WATER CHARACTERISTIC CURVES

Hydraulic conductivity $K(\theta)$ relates velocity and hydraulic gradient in Darcy's law. Moisture content θ is defined as the ratio of the volume of water to the total volume of a unit of porous media. To complicate the analysis of unsaturated flow, both θ and K are functions of the capillary suction ψ. Also, it has been observed experimentally that the θ–ψ relationships differ significantly for different types of soil. Figure 9.2a shows the characteristic drying and wetting curves that occur in soils that are draining water or receiving infiltration of water. The nonlinear nature of these curves for several selected soils is shown in Figure 9.2b. The curves reflect the fact that the hydraulic conductivity and moisture content of an unsaturated soil increase with decreasing capillary suction. The sandier soils show a very different response compared to the tighter loam and clay soils.

At atmospheric pressure, a soil is saturated when water content is equal to θ_s. The soil will initially remain saturated as the capillary pressure (matric potential) is gradually decreased. Eventually, water will begin to drain from the soil as the

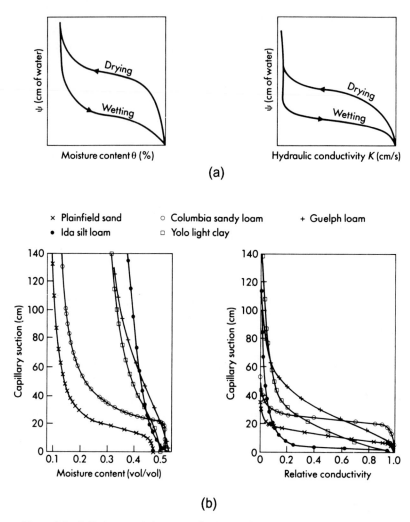

Figure 9.2 Soil characteristic curves for (a) wetting and drying and (b) different soil types.   Source: Bedient and Huber, 1992. © Addison Wesley Publishing Co., Inc. Reprinted by permission.

pressure is lowered, and the moisture content will continue to decline until it reaches some minimum water content θ_r. Further reductions in the capillary pressure will not result in any additional moisture, as shown by the ψ-θ curve in Figure 9.2a. Thus, the sandier soils are drained (or lose moisture) more quickly than the tighter soils, because the sands have larger pores.

Some simple empirical expressions used to relate water content of a soil to the capillary pressure head include the one from Brooks and Corey (1964):

$$\theta = \theta_r + (\theta_s - \theta_r)\left(\frac{\psi}{\psi_b}\right)^{-\lambda} \tag{9.3}$$

where θ = volumetric water content

$\quad \theta_s$ = volumetric water content at saturation

$\quad \theta_r$ = irreducible minimum water content on ψ-θ curve

$\quad \psi$ = matric potential or capillary suction

$\quad \psi_b$ = bubbling pressure (where $\theta = \theta_s$ on ψ-θ curve)

$\quad \lambda$ = experimentally derived parameter

Brooks and Corey (1964) also defined an effective saturation, S_e, as

$$S_e = \frac{S_w - \theta_r}{1 - \theta_r} \qquad (9.4)$$

where $S_w = \theta/\theta_s$, the saturation ratio.

Van Genuchten (1980) also derived an empirical relationship between capillary pressure head and volumetric water content, defined by

$$\theta = \theta_r + \frac{\theta_s - \theta_r}{[1 + (\alpha\psi)^n]^m} \qquad (9.5)$$

where α, m, and n are constants.

Generally, these equations work well for medium- and coarse-textured soils; predictions for fine-textured materials are usually less accurate. These equations are often used in computer models to represent soil characteristics for flow in the unsaturated zone. For large capillary heads, the Brooks and Corey and Van Genuchten models become identical if $\lambda = mn$ and $\psi_b = 1/\alpha$.

A drying curve occurs when we allow an initially saturated sample to desorb water by applying suction (or decreasing capillary pressure). If the sample is then resaturated with water, thus decreasing the suction, it will follow a wetting curve. The fact that the drying curve and wetting curve will generally not be the same produces the hysteretic behavior of the soil water retention curve, as illustrated in Figure 9.3. The soil-water content is not a unique function of capillary pressure,

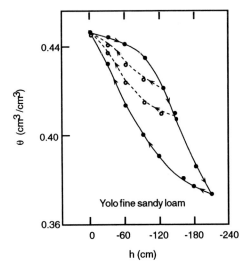

Figure 9.3 Water-retention curves for a sample of Yolo fine, sandy loam. The solid curves are eye-fitted through measured data along the two main boundary curves. Dashed curves represent primary wetting scanning curves. Arrows indicate the direction at which the pressure head changes are imposed. Source: Nielson et al., 1986. © American Geophysical Union.

but depends on the previous history of the soil. The hysteretic nature of θ is due to the presence of different contact angles during wetting and drying cycles and to geometric restrictions of single pores (Nielson et al., 1986). For example, the contact angle between the water and the soil surface is greater during the advance of a water front than during its retreat. Air trapped in pores during a wetting cycle will reduce the water content of a soil during wetting, but eventually that trapped air will dissolve. Hysteresis effects are usually augmented by the presence of trapped air and soil shrinking and swelling (Davidson et al., 1966). Hysteresis creates a significant problem for the prediction of flow rates and transport phenomena in the unsaturated zone.

9.3 UNSATURATED HYDRAULIC CONDUCTIVITY

Evidence suggests that Darcy's law is still valid for unsaturated flow, except that hydraulic conductivity is now a function of moisture content. Darcy's law is then used with the unsaturated value for K and can be written

$$v = -K(\theta)\,\frac{\partial h}{\partial z} \tag{9.6}$$

where v = Darcy velocity
z = depth below surface
h = potential or head = $z + \psi$
ψ = tension or suction
$K(\theta)$ = unsaturated hydraulic conductivity
θ = volumetric moisture content

Unsaturated hydraulic conductivity can be determined by both field methods and laboratory techniques, both of which are time consuming and tedious. Estimates arc often made from soil parameters obtained from soil-water retention relationships, such as those from Brooks and Corey (1964) and Van Genuchten (1980). Figure 9.4 shows observed values and calculated curves for relative hydraulic conductivity as a function of capillary pressure for two soil types.

Flow and mass transport in the vadose zone

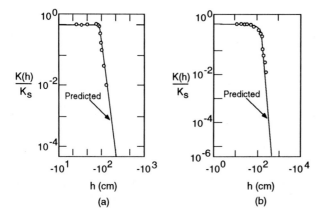

Figure 9.4 Observed values (open circles) and calculated curves (solid lines) for relative hydraulic conductivity of (a) hygiene sandstone and (b) touchet silt loam G.E.3. Source: Van Genuchten, 1980.

9.4 GOVERNING EQUATION FOR UNSATURATED FLOW

The **water table** defines the boundary between the unsaturated and saturated zones and is defined by the surface at which the fluid pressure P is exactly atmospheric, or $P = 0$. Hence, the total hydraulic head $h = \psi + z$, where $\psi = P/\rho g$, the capillary pressure head.

Considerations of unsaturated flow include the solution of the governing equation of continuity and Darcy's law in an unsaturated porous media. The resulting governing equation, originally derived by Richards in 1931, is based on substituting Darcy's law, Eq. (9.6), into the unsaturated continuity equation (Freeze and Cherry, 1979):

$$-\left[\frac{\partial(\rho v_x)}{\partial x} + \frac{\partial(\rho v_y)}{\partial y} + \frac{\partial(\rho v_z)}{\partial z}\right] = \frac{\partial}{\partial t}\rho\theta \qquad (9.7)$$

The resulting equation is

$$\frac{\partial\theta}{\partial t} = -\frac{\partial}{\partial z}\left[K(\theta)\frac{\partial\psi(\theta)}{\partial z}\right] - \frac{\partial K(\theta)}{\partial z} \qquad (9.8)$$

where θ = volumetric moisture content
  z = distance below the surface (L)
  θ = capillary suction (pressure) $(L$ of water),
  $K(\theta)$ = unsaturated hydraulic conductivity (L/T)

Equation (9.8) is called Richards (1931) equation and is a nonlinear partial differential equation that is very difficult to solve. The Richards equation assumes that the presence of air can be ignored, that water is incompressible, and that the soil matrix is nondeformable. Both numerical and analytical solutions exist for certain special cases. The most difficult part of the procedure is determining the characteristic curves for a soil (Figure 9.2). The characteristic curves reduce to the fundamental hydraulic parameters K and n in the saturated zone and remain as functional relationships in the unsaturated zone.

A number of analytical and numerical solution techniques have been developed for Eq. 9.8. Analytical approaches require simplified conditions for boundaries and are generally restricted to one-dimensional vertical systems. Early work by Philip and de Vries (1957) provided much of the physical and mathematical groundwork for subsequent analyses. The analytical approaches are useful for deriving physically based expressions for infiltration rates in unsaturated soils. A more recent approach based on the method of characteristics has been applied to gravity-dominated flow in the unsaturated zone (Smith, 1983; Charbeneau, 1984).

Charbeneau (1984) considers the kinematic theory of soil moisture and solute transport in the vertical direction for unsaturated ground water recharge. It is assumed that the soil starts at and eventually drains to field capacity. Analytical expressions are developed for water content and moisture flux as a function of depth and time, and the approach is extended for an arbitrary sequence of surface boundary conditions. In this approach, dissipative terms are neglected in the governing equations, thus allowing the powerful method of characteristics to be used in the solution. The approach is described in more detail in Section 9.8.

Numerical methods are more suited to handling actual laboratory and field situations. Finite difference methods were proposed by Rubin and Steinhardt (1963) and Freeze (1969, 1971). Finite difference numerical techniques are presented in detail in Chapter 10 for flow and transport in the saturated zone. Freeze's work was significant, since for the first time a single equation was used to describe transient, unsaturated–saturated flow, thus accounting for transient changes in both zones.

Finite element solution techniques became available in the late 1960s and have been applied to the unsaturated zone by Neuman (1983) and Frind et al. (1977). The advantage of the finite element method over the finite difference method is the ability to more accurately describe irregular system boundaries and heterogeneous properties. Conflicting evidence exists for the relative accuracy of both methods and, for one-dimensional problems, finite difference methods are quite satisfactory (Nielson, et al., 1986).

Two extreme cases should be considered. For large values of moisture content, $\partial\psi/\partial\theta$ is approximately zero, and the continuity equation becomes

$$\frac{\partial\theta}{\partial t} = \frac{-\partial K(\theta)}{\partial z} \qquad (9.9)$$

Equation (9.9) leads to the kinematic theory of modeling the unsaturated flow, where capillary pressure gradients are neglected. The theory is also applicable if ψ = constant within the profile. Thus, Darcy's law predicts that flow is downward under a unit gradient. The second extreme case occurs when capillary forces completely dominate gravitational forces, and a nonlinear diffusion equation results. This latter form is useful for modeling evaporation processes.

To summarize the properties of the unsaturated zone as compared to the saturated zone, Freeze and Cherry (1979) indicate that:

For the unsaturated zone (vadose zone):

1. It occurs above the water table and above the capillary fringe.
2. The soil pores are only partially filled with water; the moisture content θ is less than the porosity n.
3. The fluid pressure p is less than atmospheric; the pressure head ψ is less than zero.
4. The hydraulic head h must be measured with a tensiometer.
5. The hydraulic conductivity K and the moisture content θ are both functions of the pressure head ψ.

For the saturated zone:

1. It occurs below the water table.
2. The soil pores are filled with water, and the moisture content θ equals the porosity n.

3. The fluid pressure p is greater than atmospheric, so the pressure head ψ (measured as gauge pressure) is greater than zero.

4. The hydraulic head h must be measured with a piezometer.

5. The hydraulic conductivity K is not a function of the pressure head ψ.

Finally, more details on the unsaturated zone can be found in Freeze and Cherry (1979), Fetter (1993), Rawls et al. (1993), and Charbeneau and Daniel (1993).

9.5 MEASUREMENT OF SOIL PROPERTIES

Moisture characteristic curves for a particular soil can be determined by any of three approaches (Charbeneau and Daniel, 1993). The first technique is to estimate the curve (water content versus capillary pressure head) from published data for similar soils. For example, Rawls et al. (1983) collected and analyzed data from over 500 soils. Gupta and Larson (1979) describe the use of grain-size distribution data, organic content, and bulk density to estimate moisture characteristic curves. The second technique is to assume an analytic function such as that of Brooks and Corey (1964) or Van Genuchten (1980). These equations were presented in Section 9.2. The empirical coefficients in these functions are usually estimated based on correlations of various soil characteristics.

The third approach is to actually measure the soil moisture characteristic curve directly. Incremental equilibrium methods allow the soil to come to equilibrium at some moisture content θ, and then capillary pressure ψ is measured. Next, θ is changed and time is allowed for equilibrium to reestablish, and ψ is remeasured. The process is repeated until a sufficient number of ψ–θ points have been measured to define the entire curve (Figure 9.2). We can also allow ψ to come to equilibrium and measure θ for changes in ψ.

The pressure plate is the most commonly used method of measurement and involves placing a chamber around soil samples that have been soaked with water. Positive air pressure is used to force water out of the soil samples, and the outflow is monitored to confirm equilibrium. After equilibrium is established, the chamber is disassembled and the soil samples are oven dried to determine θ.

Hydraulic conductivity in unsaturated soils is determined from K_s, the value at saturation and the relative permeability k_r, which is a function of capillary pressure head, volumetric water content, or degree of water saturation. The relative permeability k_r can be estimated from other soil properties or can be directly measured in the laboratory. Klute and Dirksen (1986) summarize several methods of unsaturated hydraulic conductivity in the laboratory. They report on steady-state methods for a steady flow through a soil column with controlled pressure at both ends of the column. This technique is not practical for soils with low K because the flow rate cannot be measured accurately. Olson and Daniel (1981) report on transient tests in which soil is placed in a pressure plate device, a step increase in air pressure is imposed, the rate of water flow out of the soil is measured versus time, and K is computed from the resulting data.

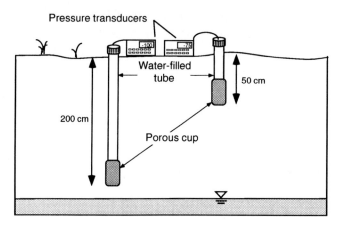

Figure 9.5 Two tensiometers used to determine the gradient of the soil-water potential.

Hydraulic conductivity at saturation K_s is measured in the laboratory using rigid-wall and flexible-wall permeameters. Hydraulic conductivity tests may be performed with a constant head, falling head, or constant flux (Olson and Daniel, 1981; Klute and Dirksen, 1986). Differences have been noted between laboratory-determined values and those estimated from field tests. Olson and Daniel compared laboratory- and field-measured K_s values for 72 data sets involving clayey soils and found significant differences, since hydraulic conductivity tends to increase with increasing scale of measurement relating to structural features of the soil.

Capillary pressure head can be measured in an undisturbed soil sample in the laboratory, or the measurement can be obtained in the field. In both cases, a tensiometer consisting of a porous element inserted into the soil and a pressure-sensing device at the surface is often used. The tensiometer is initially saturated with a liquid, and when brought into contact with the soil, and the soil will pull water out of the tensiometer, creating a negative pressure that can be measured (Charbeneau and Daniel, 1993). A tensiometer generally cannot read ψ more negative than 0.9 bar due to problems of water cavitation. Figure 9.5 depicts a tensiometer for field use.

9.6 INFILTRATION MODELS

Infiltration is the process of vertical movement of water into a soil from rainfall, snowmelt, or irrigation. Infiltration of water plays a key role in surface runoff, ground water recharge, evapotranspiration, and transport of chemicals into the subsurface. Models to characterize infiltration for field applications usually employ simplified concepts that predict the infiltration rate, assuming surface ponding begins when the surface application rate exceeds the soil infiltration rate. Empirical, physically based, and physical models have all been developed for the infiltration process. A more detailed review of infiltration can be found in Rawls et al. (1993).

The Richards Eq. (9.8) is the physically based infiltration equation used for describing water flows in soils. Philip and de Vries (1957) solved the equation analytically for the condition of excess water at the surface and given characteristic curves. Water flow coefficients can be predicted in advance from soil properties and do not have to be fitted to field data. However, the more difficult case when the rainfall rate is less than the infiltration capacity cannot be handled by Philip's equation. Another limitation is that it does not hold valid for extended time periods. Swartzendruber (1987) presented a solution to Richards equation that holds for both small, intermediate, and large times.

A number of operational infiltration models have been developed over the years and are covered in detail in hydrology textbooks such as Chow et al. (1988), Bedient and Huber (1992), and Rawls et al. (1993). These include the SCS runoff curve number method, the Horton method, and the Holtan method (Soil Conservation Service, 1986; Horton, 1940; Holtan, 1961).

One of the most interesting and useful approaches to solving the Richards equation for the problem of infiltration was originally advanced by Green and Ampt (1911). In this method, water is assumed to move into dry soil as a sharp wetting front that separates the wetted and unwetted zones. At the location of the front, the average capillary suction head $\psi = \psi_f$ is used to represent the characteristic curve. The moisture content profile at the moment of surface saturation is shown in Figure 9.6a. The area above the moisture profile is the amount of infiltration up to surface saturation F and is represented by the shaded area of depth L in Figure 9.6a. Thus, $F = (\theta_s - \theta_i)L = M_d L$, where θ_i is the initial moisture content, θ_s is the saturated moisture content, and $M_d = \theta_s - \theta_i$ the initial moisture deficit.

Darcy's law is then applied as an approximation to the saturated conditions between the soil surface and the wetting front, as indicated in Figure 9.6b (Bedient and Huber, 1992).

The volume of infiltration down to the depth L is given by

$$F = L(\theta_s - \theta_i) = L M_d \tag{9.10}$$

Neglecting the depth of ponding at the surface, the original form of the Green–Ampt equation is

$$f = K_s \frac{1 - (\theta_s - \theta_i)\psi_f}{F} \tag{9.11}$$

$$= K_s \frac{1 - M_d \psi_f}{F}$$

Remembering that ψ_f is negative, Eq. (9.11) indicates that the infiltration rate is a value greater than the saturated hydraulic conductivity, as long as there is sufficient water at the surface for infiltration, as sketched in curves C and D of Figure 9.6c. Functionally, the infiltration rate decreases as the cumulative infiltration increases.

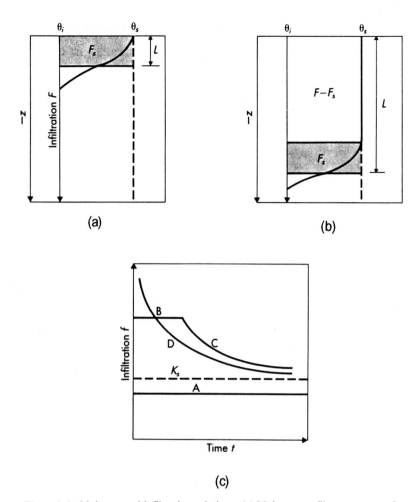

Figure 9.6 Moisture and infiltration relations. (a) Moisture profile at moment of surface saturation. (b) Moisture profile at later time. (c) Infiltration behavior under rainfall.   Source: Mein and Larson, 1973. © American Geophysical Union.

The rainfall intensity, i, is often less than the potential infiltration rate given by Eq. (9.11), in which case $f = i$. Let the corresponding volume of infiltration be F_s. With $f = i$, Eq. (9.11) can then be solved for F_s, the volume of infiltration at the time of surface saturation [t_s, the time at which Eq. (9.11) becomes valid]:

$$F_s = \frac{(\theta_s - \theta_i)\psi_f}{1 - i/K_s} = \frac{M_d \psi_f}{1 - i/K_s} \tag{9.12}$$

We require $i > K_s$ in Eq. (9.12) and remember that ψ_f is negative. The Green–Ampt infiltration prediction is thus the following:

1. If $i \leq K_s$, then $f = i$ (curve A in Figure 9.6c).
2. If $i > K_s$, then $f = i$ until $F = it_s = F_s$ [Eq. (9.12)].

3. Following surface saturation, $f = K_s$, $[1 - M_d\psi_f/F]$ [Equation (9.11) for $i > K_s$] and $f = i$ for $i \leq K_s$.

The combined process is sketched in curves B and C of Figure 9.6c. As long as the rainfall intensity is greater than the saturated hydraulic conductivity, the infiltration rate asymptotically approaches K_s, as a limiting lower value. Mein and Larson (1973) found excellent agreement between this Green–Ampt method, numerical solutions of the Richards equation, and experimental soils data. If the rainfall rate starts above, drops below, and then again rises above K_s during the infiltration computation, the use of Green–Ampt becomes more complicated, making it necessary to redistribute the moisture in the soil column, rather than maintaining the assumption of saturation from the surface down to the wetting front shown in Figure 9.6b. The use of the Green–Ampt procedures for unsteady rainfall sequences is illustrated by Skaggs and Khaleel (1982).

Equation (9.11) predicts infiltration rate f as a function of cumulative infiltration F, not time. Because $f = \partial F/\partial t$, the equation can be converted into a differential equation, the solution of which can be solved iteratively for $F(t)$ (Chow et al., 1988). Then Eq. (9.11) can be used to determine $f(t)$.

A major advantage of the Green–Ampt model is that, in principle, the necessary parameters K_s, ψ_f, and $M_d = \theta_s - \theta_i$ can be determined from physical measurements in the soil, rather than empirically as for the Horton parameters. For example, saturated hydraulic conductivity (often loosely called permeability) is tabulated by the U.S. Soil Conservation Service (SCS) for a large number of soils as part of that agency's Soil Properties and Interpretation sheets (available from local SCS offices). An increasing quantity of tension versus moisture content data (of the type shown in Figure 9.2) is also available, from which a value of ψ_f can be obtained by integration over the moisture content of interest. For example, several volumes of such information have been assembled for Florida soils (see Carlisle et al., 1981). In practice, the Green–Ampt parameters are often calibrated, especially when used in continuous simulation models.

A useful source of information on Green–Ampt parameters is provided by Rawls et al. (1983) who present data for a large selection of soils from across the United States. These data are shown in Table 9.1. Two porosity (θ_s) values are given: total and effective. Effective porosity accounts for trapped air and is the more reasonable value to use in computations. It can be seen in Table 9.1 that, as the soil particles get finer from sands to clays, the saturated hydraulic conductivity, K_s, decreases, the average wetting front suction, ψ_f, increases (negatively), and porosity, θ_s, is variable. Table 9.1 provides valuable estimates for Green–Ampt parameters, but local data (for example, Carlisle et al., 1981) are preferable if available. Missing is the initial moisture content, θ_i, since it depends on antecedent rainfall and moisture conditions. Typical values for $M_d = \theta_s - \theta_i$ are given in SCS Soil Properties and Interpretation sheets and are usually termed "available water (or moisture) capacity, in./in." Values usually range from 0.03 to 0.30. The value to use for a particular soil in question must be determined from a soil test. Otherwise, a conservative (low) M_d value could be used for design purposes (for example, 0.10).

TABLE 9.1 Green–Ampt Infiltration Parameters for Various Soil Texture Classes

Soil class	Porosity η	Effective porosity θ_e	Wetting front soil suction head ψ (cm)	Hydraulic conductivity K (cm/hr)	Sample size
Sand	0.437 (0.374–0.500)	0.417 (0.354–0.480)	4.95 (0.97–25.36)	11.78	762
Loamy sand	0.437 (0.363–0.506)	0.401 (0.329–0.473)	6.13 (1.35–27.94)	2.99	338
Sandy loam	0.453 (0.351–0.555)	0.412 (0.283–0.541)	11.01 (2.67–45.47)	1.09	666
Loam	0.463 (0.375–0.551)	0.434 (0.334–0.534)	8.89 (1.33–59.38)	0.34	383
Silt loam	0.501 (0.420–0.582)	0.486 (0.394–0.578)	16.68 (2.92–95.39)	0.65	1206
Sandy clay loam	0.398 (0.332–0.464)	0.330 (0.235–0.425)	21.85 (4.42–108.0)	0.15	498
Clay loam	0.464 (0.409–0.519)	0.309 (0.279–0.501)	20.88 (4.79–91.10)	0.10	366
Silty clay loam	0.471 (0.418–0.524)	0.432 (0.347–0.517)	27.30 (5.67–131.50)	0.10	689
Sandy clay	0.430 (0.370–0.490)	0.321 (0.207–0.435)	23.90 (4.08–140.2)	0.06	45
Silty clay	0.479 (0.425–0.533)	0.423 (0.334–0.512)	29.22 (6.13–139.4)	0.05	127
Clay	0.475 (0.427–0.523)	0.385 (0.269–0.501)	31.63 (6.39–156.5)	0.03	291

Source: Rawls et al., 1983.

The numbers in parentheses below each parameter are one standard deviation around the parameter value given.

Example 9.1 GREEN–AMPT TIME TO SURFACE SATURATION

Guelph loam has the following soil properties (Mein and Larson, 1973) for use in the Green–Ampt equation:

$$K_s = 3.67 \times 10^{-4} \text{ cm/sec}$$

$$\theta_s = 0.523$$

$$\psi_f = -31.4 \text{ cm water}$$

For an initial moisture content of $\theta_i = 0.3$, compute the time to surface saturation for the following storm rainfall:

$$i = 6K_s \text{ for 10 min}$$

$$i = 3K_s \text{ thereafter}$$

Solution. The initial moisture deficit $M_d = 0.523 - 0.300 = 0.223$. For the first rainfall segment, we compute the volume of infiltration required to produce saturation from Eq. (9.12):

$$F_s = \frac{\psi_f M_d}{1 - i/K_s} = \frac{(-31.4 \text{ cm})(0.223)}{1 - 6K_s/K_s} = 1.40 \text{ cm}$$

The rainfall volume during the first 10 min is

$$10i = (10 \text{ min})(6 \times 3.67 \times 10^{-4} \text{ cm/sec})(60 \text{ sec/min}) = 1.31 \text{ cm}$$

Since $1.31 < 1.40$, all rainfall infiltrates, surface saturation is not reached, and $F(10 \text{ min}) = 1.31$ cm.

The volume required for surface saturation during the lower rainfall rate of $i = 3K_s$ is

$$F_s = \frac{(-31.4 \text{ cm})(0.223)}{1 - 3K_s/K_s} = 3.50 \text{ cm}$$

Thus, an incremental volume of $\Delta F = F_s - F(10 \text{ min}) = 3.50 - 1.31 = 2.19$ cm must be supplied before surface saturation occurs. This requires an incremental time of

$$\Delta t = \Delta F/i = \frac{2.19 \text{ cm}}{3 \times 3.67 \times 10^{-4} \text{ cm/sec}} = 1989 \text{ sec}$$

$$= 33.15 \text{ min}$$

Thus, the total time to surface saturation is $10 + 33.15 = 43.15$ min.

9.7 TRANSPORT PROCESSES IN THE UNSATURATED ZONE

Interest in vadose zone transport processes has recently focused on source characterization for ground water contamination problems. The parameters of interest include the travel time from the ground surface to the water table, the amount of contaminant that actually reaches the water table, and the rate at which it enters the ground water aquifer. Soluble contaminants such as gasolines and fuels are released into the subsurface as a pool or floating lens on the surface of the water table and slowly solubilize into the ground water. Solvents that are denser than water will pool at the bottom of aquifers and also slowly solubilize into the ground water. Finally, residuals of these chemicals left behind in the soil as the free phase mobilize from the unsaturated zone to the ground water and provide a possible long-term source of contamination (see Chapter 11). The residuals act as an intermittent source that leaches during rainfall events and seasonal changes in the water table. The following sections will focus on different models for predicting contaminant transport in the vadose zone, assuming a variety of hydraulic and soil conditions.

Simple Source Model

This model applies to cases where a contaminant has been released in the unsaturated zone such that it occupies a certain volume of soil, for example, a zone of hydrocarbon residual or a land-treatment zone. The conceptual model for predicting transport of contaminants with a simple source is shown in Figure 9.7. Basically, water at a constant velocity, V_a, is flowing through the soil and carrying a dissolved component at a specific concentration, C_a. This component can be sorbed by the soil or biodegraded by microorganisms (Chapters 7 and 8). The

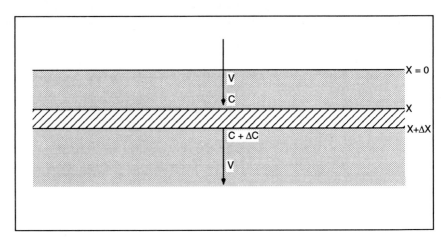

Figure 9.7 Conceptual model for land treatment. Source: Short, 1986.

concentration of the component will be a function of both depth and time. The advection–dispersion equation for this case (Short, 1986) is

$$\frac{\partial C_a}{\partial t} = D_a \frac{\partial^2 C_a}{\partial X^2} - V_a \frac{\partial C_a}{\partial X} - \frac{\rho}{\theta} \frac{\partial C_s}{\partial t} - \mu_a C_a - \frac{\rho}{\theta} \mu_s C_s \qquad (9.13)$$

where t is time, D_a is the dispersion coefficient, X is the depth, ρ is the bulk density of the soil, θ is the volumetric water content of the soil, C_s is the concentration of the contaminant in the soil phase (weight of component per unit weight of soil), μ_a is the first-order degradation constant in the aqueous phase, and μ_s is the first-order decay constant in the soil phase. Equation (9.13) assumes that sorption is linear such that

$$C_s = K_D C_a \qquad (9.14)$$

where K_D is the sorption constant.

Equation (9.13) can be rearranged into a familiar form and rewritten as follows:

$$\frac{\partial C_a}{\partial t} = \frac{D_a}{R} \frac{\partial^2 C_a}{\partial X^2} - \frac{V_a}{R} \frac{\partial C_a}{\partial X} - \frac{\mu C_a}{R} \qquad (9.15)$$

where R and μ are the retardation factor and a new degradation coefficient, respectively, given by

$$R = 1 + \rho \frac{K_D}{\theta} \qquad (9.16)$$

$$\mu = \mu_a + \mu_s \rho \frac{K_D}{\theta} \qquad (9.17)$$

Equation (9.15) can be solved for a variety of different cases, and two will be presented: (1) plug flow with no degradation and (2) plug flow with degradation.

Plug flow means that the flow is advection dominated and that dispersion can be neglected.

Plug Flow with No Degradation

If dispersion and degradation are neglected, Eq. (9.15) reduces to

$$R \frac{\partial C_a}{\partial t} = -V_a \frac{\partial C_a}{\partial X} \qquad (9.18)$$

The solution to Eq. (9.18) for a **step concentration input** is given by

$$\frac{C_a}{C_i} = S, \quad \text{where} \quad S = 1 \text{ for } X \leq \frac{V_a t}{R}$$

$$(9.19)$$

$$\text{and} \quad S = 0 \text{ for } X > \frac{V_a t}{R}$$

where C_i is the initial concentration and Eq. 9.18 is subject to the following boundary conditions:

$$C_a = 0, \quad \text{for all } X > 0 \text{ at } t = 0 \qquad (9.20)$$

$$C_a = C_i, \quad \text{for all } t > 0 \text{ at } X = 0 \qquad (9.21)$$

$$C_a \rightarrow 0, \quad \text{as } X \rightarrow \infty \qquad (9.22)$$

Figure 9.8a is a graphical representation of this solution. The relative concentration at any given point is either 0 or 1 depending on whether or not breakthrough has occurred. The velocity with which the contaminant moves is equal to the water flow velocity divided by the retardation coefficient.

The solution to Eq. (9.18) for a **slug concentration input** (Figure 9.8b) is given by

$$\frac{C_a}{C_i} = P, \quad \text{where} \quad P = 0, X \leq \frac{V_a t}{R} - L \qquad (9.23)$$

$$\text{and} \quad P = 1, \frac{V_a t}{R} - L \leq X \leq \frac{V_a t}{R} \qquad (9.24)$$

$$\text{and} \quad P = 0, X > \frac{V_a t}{R} \qquad (9.25)$$

where L is the length of pulse. This solution is subject to the following boundary conditions:

$$C_a = 0, \quad \text{for all } X > 0 \text{ at } t = 0 \qquad (9.26)$$

$$C_a = C_i, \quad \text{for all } t \geq 0 \text{ and } t \leq \frac{L}{V_p} \text{ at } X = 0 \qquad (9.27)$$

$$C_a = 0, \quad \text{for } t > \frac{L}{V_p} \text{ at } X = 0 \qquad (9.28)$$

where V_p is the contaminant velocity.

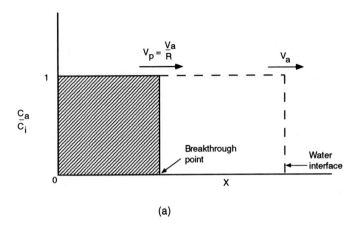

(a)

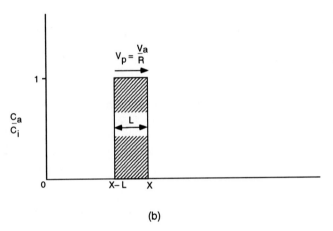

(b)

Figure 9.8 Concentration profile for (a) step input and (b) pulse input.   Source: Short, 1986.

Plug Flow with Degradation

For plug flow problems with first-order biodegradation and without dispersion, Eq. (9.15) reduces to

$$R \frac{\partial C_a}{\partial t} = -V_a \frac{\partial C_a}{\partial X} - \mu C_a \qquad (9.29)$$

This equation can be solved for the case of step and pulse inputs, respectively (see Figure 9.9):

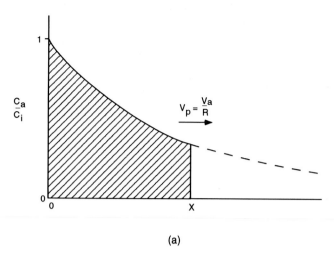

(a)

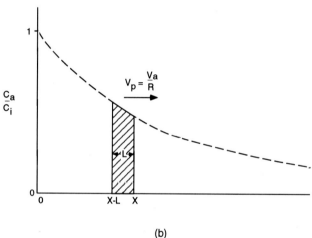

(b)

Figure 9.9 Concentration profile for (a) step input and (b) pulse input with biodegradation. Source: Short, 1986.

$$\frac{C_a}{C_i} = Se^{-\mu X/V_a} \qquad \text{where} \quad S = 1 \text{ for } X \le \frac{V_a t}{R}$$

$$\text{and} \qquad S = 0 \text{ for } X > \frac{V_a t}{R} \tag{9.30}$$

$$\frac{C_a}{C_i} = Pe^{-\mu X/V^a}, \qquad \text{where} \quad P = 0, X \le \frac{V_a t}{R} - L \tag{9.31}$$

$$\text{and} \qquad P = 1, \frac{V_a t}{R} - L \le X \le \frac{V_a t}{R} \tag{9.32}$$

$$\text{and} \qquad P = 0, X > \frac{V_a t}{R} \tag{9.33}$$

As can be seen from Figure 9.9, the concentration profiles for the step and slug input are similar to the nondegrading case except that the concentrations are attenuated as the contaminant moves through the soil. These simplified cases allow for the computation of unsaturated zone transport in relatively simplistic terms. If dispersion coefficients are included in the analysis, the resulting solutions include error functions (erfc) and appear similar to those in Chapter 6.

Example 9.2 CONCENTRATION OF LEACHATE FROM A HYDROCARBON SPILL

Consider the release of a contaminant that initially occupies an area with plan view A and depth z^*. For example, this could come from a hydrocarbon spill of volume V_{spill} over an area A_{spill}. The depth of the spilled material L_o is then found from

$$L_o = \frac{V_{spill}}{\theta_{or} A_{spill}}$$

where θ_{or} is the residual oil content. The problem also corresponds to a land-treatment system with a zone of incorporation depth L_o. There is a finite amount of mass available near the ground surface, and we want to estimate the rate at which it is leached from the contaminated zone.

Solution. The bulk concentration of contaminant is

$$m = B_w c_w$$

where $B_w = (\theta_w + \theta_a K_H + \theta_o K_o + \rho_b K_d)$
$\theta_{w,a,o}$ = volumetric phase content for water, air, and oil
K_H = Henry's law constant
K_o = oil–water partition coefficient
K_d = soil–water partition coefficient
ρ_b = bulk density of soil
c_w = concentration in water phase

and the total mass present is

$$M = A L_o B_w c_w$$

The mass advective flux from the contaminated region is given by

$$JA = U c_w A$$

where U = Darcy's velocity in the vertical direction. The continuity equation then implies that

$$\frac{dM}{dt} = A L_o B_w \frac{dc_w}{dt} = -U A c_w$$

which may be integrated to find

$$c_w = c_{leachate} = c_{wo} \exp\left(-\frac{Ut}{L_o B_w}\right)$$

where c_{wo} is the initial aqueous phase concentration. Thus, the leachate concentration decreases exponentially, as does the total mass within the region (example adapted from Charbeneau, 1990).

Example 9.3 TRAVEL TIME IN VADOSE ZONE

To estimate the travel time for contaminants within the vadose zone, three pieces of information are required. First, we must know the net ground water recharge through the unsaturated zone. This requires us to perform a water balance at the ground surface including rainfall, infiltration, runoff, detention storage, evapotranspiration, and soil-water redistribution. We can assume that the mean ground water recharge rate is known and given by

$$U_{mean} = W$$

Next, we need to determine the average water content. Near the ground surface the water content is quite variable over time, and both gravitational and capillary pressure forces are important. However, below the upper meter or so (for all but a very coarse soil), the water content does not vary significantly with either depth or time (assuming a homogeneous soil). A nearly uniform water content dictates a nearly constant capillary pressure, so capillary pressure gradients may be expected to be small below the upper zone. The Darcy velocity in the vertical direction is

$$U = K \left(\frac{\partial \psi}{\partial z} + 1 \right)$$

and since the capillary pressure gradients are assumed negligible [based on the kinematic theory proposed by Charbeneau (1984)], the relation

$$U = K(\theta)$$

may be used to estimate the mean water content. If we assume the Brooks and Corey (1964) power law model discussed earlier, we have

$$U = W = K_s \left(\frac{\theta - \theta_r}{n - \theta_r} \right)^\epsilon$$

where θ = volumetric water content
θ_r = irreducible minimum water content
ϵ = coefficient to be determined
K_s = saturated hydraulic conductivity
n = volumetric water content at saturation

So

$$\theta = \theta_r + (n - \theta_r) \left(\frac{W}{K_s} \right)^{1/\epsilon}$$

In this case, the seepage velocity is

$$v = \frac{U}{\theta} = \frac{W}{\theta_r + (n - \theta_r)(W/K_s)^{1/\epsilon}}$$

Finally, if L_{vz} is the thickness of the vadose zone from the base of the contaminated region to the water table and if R is the retardation factor for the given water content $[R = 1 + (\rho_b/\theta)K_d]$, then the travel time t_{vz} is given by

$$t_{vz} = \frac{RL_{vz}}{v} = \frac{[\theta_r + (n - \theta_r)(W/K_s)^{1/\epsilon} + \rho_b K_d]L_{vz}}{W}$$

If, during transport in the vadose zone, the contaminant is being degraded at a first-order rate constant λ, then it will have decayed by a fraction $e^{-\lambda t_{vz}}$. Thus, the concentration reaching the water table can be expressed in terms of the leachate concentration from the region of contamination as

$$c_{wt}(t) = c_{\text{leach}}(t - t_{vz})e^{-\lambda t_{vz}}$$

where $c_{\text{leach}}(t - t_{vz})$ represents the leachate concentration at time $t - t_{vz}$. With the results from the previous example, this gives

$$c_{wt}(t) = c_{wo} \exp\left\{ -\left[\frac{U(t - t_{vz})}{L_o B_w} + \lambda t_{vz} \right] \right\}$$

A number of assumptions contained in this formula should be noted. The formula does not account for diffusion or volatilization from soils (example adapted from Charbeneau, 1990).

Multiphase Transport

In reality, transport in the unsaturated zone is complex mainly because there are numerous phases of interest: soil, water, air, and contaminant. Investigations of the fate and transport of chemicals in the unsaturated zone must inherently deal with multiphase issues. Most contaminants are soluble and some contaminants volatilize. If the transport of contaminants through aqueous, soil, vapor, and immiscible oil phases is considered, a differential material balance for each phase can be written assuming local equilibrium and first-order kinetics. Local equilibrium implies that the net interphase transport is zero; that is, the rate of mass transfer of contaminant from the aqueous phase to the solid phase is exactly equal to the rate of mass transfer from the solid phase to the aqueous phase. Also, the assumption is made that the relative amounts of each phase remain constant through the soil profile. The pore space is filled by the sum of the fluids present, so porosity equals the sum of the volumetric fluid contents. The concentrations of a constituent in the aqueous, vapor, and oil phases are designated C_a, C_v, and C_o, all on a mass per unit volume basis. The chemical soil phase concentration is designated as mass of chemical sorbed per unit mass of soil, C_s. More details on multiphase flow are presented in Chapter 11 where nonaqueous phase liquids (NAPLs) are described. The governing transport equations are given below.

$$\textbf{Aqueous phase:} \quad \theta \frac{\partial C_a}{\partial t} = \theta D_a \frac{\partial^2 C_a}{\partial X^2} - \theta V_a \frac{\partial C_a}{\partial X} - \mu_a \theta C_a \tag{9.34}$$

where θ is the volumetric water content.

$$\textbf{Soil phase:} \quad \rho \frac{\partial C_s}{\partial t} = -\mu_s \rho C_s \tag{9.35}$$

where ρ is the density of the soil phase, C_s is the concentration of the contaminant in the soil phase, and μ_s is a first-order rate constant for degradation in the soil phase.

Immiscible oil phase: $\quad \phi \dfrac{\partial C_o}{\partial t} = \phi D_o \dfrac{\partial^2 C_o}{\partial X^2} - \phi V_o \dfrac{\partial C_o}{\partial X} - \mu_a \phi C_o$ (9.36)

where ϕ is the volumetric oil content, C_o is the concentration of the contaminant in the oily phase, D_o is the dispersion coefficient in the oil phase, V_o is the velocity of the oily phase, and μ_o is the first-order constant for degradation in the oil phase.

Vapor phase: $\quad \eta \dfrac{\partial C_v}{\partial t} = \eta D_v \dfrac{\partial^2 C_v}{\partial X^2} - \eta V_v \dfrac{\partial C_v}{\partial X} - \mu_v \eta C_v$ (9.37)

where η is the volumetric vapor content.

Equations (9.34) through (9.37) can be combined to yield

$$
\theta \dfrac{\partial C_a}{\partial t} + \phi \dfrac{\partial C_o}{\partial t} + \eta \dfrac{\partial C_v}{\partial t} + \rho \dfrac{\partial C_s}{\partial t}
$$

$$
= \theta D_a \dfrac{\partial^2 C_a}{\partial X^2} + \phi D_o \dfrac{\partial^2 C_o}{\partial X^2} + \eta D_v \dfrac{\partial^2 C_v}{\partial X^2}
$$

$$
- \theta V_a \dfrac{\partial C_a}{\partial X} - \phi V_o \dfrac{\partial C_o}{\partial X} - \eta V_v \dfrac{\partial C_v}{\partial X}
$$

$$
- \mu_o \phi C_o - \mu_a \theta C_a - \mu_v \eta C_v - \mu_s \rho C_s
$$

(9.38)

The total concentration of contaminant is equal to the sum of its concentration in the four phases:

$$ C_T = \theta C_a + \phi C_o + \eta C_v + \rho C_s \tag{9.39} $$

The partitioning between the soil and aqueous phase is assumed to be linear, between the aqueous and vapor phases to follow Henry's law, and between the oil and aqueous phase to be linear so that

$$ C_s = K_D C_a \tag{9.40} $$
$$ C_v = K_H C_a \tag{9.41} $$
$$ C_o = K_L C_a \tag{9.42} $$

where K_D is the soil–water partitioning coefficient, K_H is the Henry's law constant, and K_L is the oil–water partitioning coefficient.

The total concentration can be rewritten in terms of the individual phase concentration using Eq. (9.40) through (9.42):

$$ C_T = B_a C_a = B_o C_o = B_v C_v = B_s C_s \tag{9.43} $$

where

$$ B_a = \theta + \phi K_L + \eta K_H + \rho K_D \tag{9.44} $$

$$ B_o = \dfrac{\theta}{K_L} + \phi + \eta \dfrac{K_H}{K_L} + \rho \dfrac{K_D}{K_L} \tag{9.45} $$

$$ B_v = \dfrac{\theta}{K_H} + \phi \dfrac{K_L}{K_H} + \eta + \rho \dfrac{K_D}{K_H} \tag{9.46} $$

$$ B_s = \dfrac{\theta}{K_D} + \phi \dfrac{K_L}{K_D} + \eta \dfrac{K_H}{K_D} + \rho \tag{9.47} $$

The transport equation (9.38) can be rewritten in terms of the total concentration in the system:

$$\frac{\partial C_T}{\partial t} = D_E \frac{\partial^2 C_T}{\partial X^2} - V_E \frac{\partial C_T}{\partial X} - \mu_E C_T \qquad (9.48)$$

where D_E, V_E, and μ_E are the effective dispersion coefficient, effective velocity, and degradation rate constant, respectively, defined by

$$D_E = \frac{D_a \theta}{B_a} + \frac{D_o \phi}{B_o} + \frac{D_v \eta}{B_v} \qquad (9.49)$$

$$V_E = \frac{V_a \theta}{B_a} + \frac{V_o \phi}{B_o} + \frac{V_v \eta}{B_v} \qquad (9.50)$$

$$\mu_E = \frac{\mu_a \theta}{B_a} + \frac{\mu_o \phi}{B_o} + \frac{\mu_o \eta}{B_v} + \frac{\mu_s \rho}{B_s} \qquad (9.51)$$

Equation (9.48) is a simple one-dimensional advective transport equation that can be solved semianalytically or numerically for multiphase transport in the unsaturated zone (Short, 1988).

Volatile Emissions from Soil

The modeling of emissions to the atmosphere from soil processes has been dealt with by Jury et al. (1983). Their approach is similar to the equations presented previously except that an immiscible phase is not considered. Jury et al. (1983) assume that (1) the chemical is incorporated uniformly to a specified depth L; (2) volatilization of the chemical takes place through a stagnant air layer just above the soil surface; (3) there is a steady flux of water through the soil; and (4) there is an infinite depth of uniform soil below the zone of incorporation. The boundary conditions considered by Jury et al. (1983) are as follows:

$$C_T = C_{T_i}, \qquad 0 < X < L \text{ for } t = 0 \qquad (9.52)$$

$$C_T = 0, \qquad X > L \text{ for } t = 0 \qquad (9.53)$$

The transport of the chemical through the stagnant boundary is equal to the flux at the soil surface such that

$$-D_E \frac{\partial C_T}{\partial X} + V_E C_T = -H_E C_T, \qquad \text{for } X = 0 \text{ at all } t \qquad (9.54)$$

where H_E is the effective mass transfer coefficient defined by

$$H_E = \frac{h}{B_v} \qquad (9.55)$$

where h is the mass transfer coefficient across the stagnant air film.

The vapor flux at the surface of the soil is estimated using

$$J(0, t) = -H_E C_T \qquad (9.56)$$

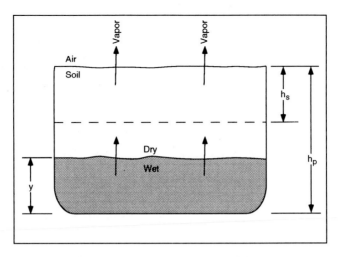

Figure 9.10 Thibodeaux-Hwang land treatment volatilization model. Source: Short, 1986.

Jury et al. (1983) solved the appropriate equations analytically and obtained an expression for the vapor flux of organics from soils.

Thibodeaux and Hwang (1982) have also presented solutions for estimating volatile emissions from land treatment. Their approach (illustrated in Figure 9.10) assumes that there is no movement of liquids in the soil and that no adsorption or biodegradation takes place. It is assumed that the soil is contaminated between depths h_s and h_p. As the contaminant evaporates and the vapor diffuses upward to the surface, a "dried-out" zone develops. The flux through the surface (ignoring mass transfer through the stagnant boundary layer above the soil) is given by

$$J(0, t) = -\frac{D_V C_{WZ}}{h_p - y} \tag{9.57}$$

where D_v is the effective diffusivity in the air-filled pores and C_{WZ} is the vapor phase concentration of the contaminant in the wet zone. A mass balance on the system using the appropriate boundary conditions yields an expression for the flux through the wet–dry interface:

$$J(0, t) = D_V C_{WZ} \sqrt{h_s^2 + 2(h_p - h_s) D_V C_{WZ} A t M_A} \tag{9.58}$$

where M_A is the mass of contaminant applied, and A is the soil surface area over which the contaminant is applied.

9.8 VADOSE ZONE FLOW AND TRANSPORT MODELS

In recent years, there has been increased interest in developing models for simulating ground water and soil contamination by immiscible organics in the vadose zone. Kuppusamy et al. (1987), Corapcioglu and Baehr (1987), Panday and Corap-

TABLE 9.2   Selected Models for Unsaturated Zone Analysis

Model name	Model description	Model processes	Author(s)
Flow Models			
FEMWATER/ FECWATER (1987)	Two-dimensional finite-element model to simulate transient, cross-sectional flow in saturated–unsaturated anisotropic, heterogeneous porous media	Capillarity, infiltration, ponding	G. T. Yeh D. S. Ward
SWATRE, SWATR-CROPR (1981)	Finite-difference model for simulation of the water balance of agricultural soil	Precipitation, capillarity, evapotranspiration	R. A. Feddes
UNSAT1D (1981)	Finite-difference model for one-dimensional simulation of unsteady vertical unsaturated flow	Capillarity, evapotranspiration, plant uptake	S. K. Gupta C. S. Simmons
UNSAT2 (1979)	Two-dimensional finite-element model for horizontal, vertical, or axisymmetric simulation of transient flow in variably saturated, nonporous medium	Capillarity, evapotranspiration, plant uptake	S. P. Neuman
VS2D (1987)	Two-dimensional finite-difference code for the analysis of flow in variably saturated porous media; model considers recharge, evaporation, and plant root uptake	Evaporation, recharge, plant uptake	E. G. Lappala R. W. Healy E. P. Weeks
WATERFLO (1985)	One-dimensional finite-difference solution for the Richards equation to simulate water movement through soils		D. L. Nofziger
Transport Models			
CHEMRANK (1988)	A package that utilizes four schemes for screening of organic chemicals relative to their potential to leach into ground water	Leaching, degradation, mobility	D. L. Nofziger R. S. C. Rao A. G. Hornsby
FEMTRAN (1984)	Two-dimensional finite-element model to simulate cross-sectional radionuclide transport in saturated–unsaturated porous media	Capillarity, advection, decay	M. J. Martinez
FLAMINCO (1985)	Three-dimensional finite-element code for analyzing water flow and solute transport in saturated–unsaturated porous media	Evapotranspiration, decay, dispersion, diffusion, adsorption, advection	P. S. Huyakorn

TABLE 9.2 *(continued)*

Model name	Model description	Model processes	Author(s)
MOTIF (1986)	Finite-element model for one-, two-, and three-dimensional saturated–unsaturated ground water flow, heat transport, and solute transport in fractured porous media; facilitates single-species radionuclide transport and solute diffusion from fracture to rock matrix	Convection, dispersion, diffusion, adsorption, decay, advection	V. Guvanasen
NITROSIM (1981)	Finite-difference solution for simulation of transport and plant uptake of nitrogen and transformations nitrogen and carbon in the root zone	Capillarity, precipitation, ion exchange, decay, evapotranspiration, convection, dispersion, diffusion, adsorption	P. S. C. Rao
PRZM (1984)	The Pesticide Root Zone Model simulates the vertical movement of pesticides in the unsaturated zone	Runoff, erosion, plant uptake, leaching, decay, foliar washoff, volatilization	R. F. Carsel C. N. Smith L. A. Mulkey J. D. Dean P. Jowise
RITZ (1988)	The Regulatory and Investigative Treatment Zone Model is an interactive model for simulation of the movement and fate of hazardous chemicals during land treatment of oily wastes	Volatilization, degradation, leaching	D. L. Nofziger J. R. Williams T. E. Short
SESOIL (1984)	Describes flow, sediment transport, pollutant transport and transformation, pollutant migration to ground water, and soil quality		M. Bonazountas
SUTRA (1984)	Finite-element simulation model for two-dimensional, transient or steady-state, saturated–unsaturated, density-dependent ground water flow with transport of energy or chemically reactive single-species solutes	Capillarity, convection, dispersion, diffusion, adsorption reactions	C. I. Voss
VAM2D (1988)	Two-dimensional finite-element model to simulate flow and contaminant transport in variably saturated porous media	Recharge, infiltration, evapotranspiration, advection, dispersion, adsorption, degradation	P. S. Huyakorn
VAM3D (1988)	Three-dimensional finite-element model that simulates flow and contaminant transport in variably saturated porous	Advection, dispersion, adsorption, degradation	P. S. Huyakorn

cioglu (1989), Faust et al. (1989), Kaluarachchi and Parker (1989, 1990), Mayer and Miller (1989), and many others have presented a variety of multiphase flow and transport models (see Table 9.2 for additional listings). Many of these models require the solution of complex partial differential equations for which the values of parameters may not be well known in the subsurface.

The purpose of a screening model is to evaluate the behavior of a chemical and to determine processes that control fate and transport. Noteworthy are several public-domain vadose models that utilize simplified assumptions and allow relatively straightforward analyses to be completed. Three such models are discussed in detail: RITZ (Nofziger et al., 1988), CHEMFLO (Nofziger et al., 1989), and HSSM (Weaver and Charbeneau, 1992).

RITZ, or the Regulatory and Investigative Treatment Zone model, was developed to help decision makers estimate the movement and fate of hazardous chemicals during land treatment of oily wastes. The model considers the downward movement of the contaminant through the soil while solubilizing, volatilizing, and degrading, as shown in Figure 9.11. The model incorporates the influence of oil on the transport and fate of the contaminant. The main assumptions in the RITZ model include an immobile oil phase at residual saturation in the waste material, uniform soil properties, uniform water flux throughout the treatment site, and insignificant dispersion. Constituents from the oily phase may be lost due to volatilization and biodegradation. It is also assumed that the water content of the soil is related to the hydraulic conductivity as described by Clapp and Hornberger (1978).

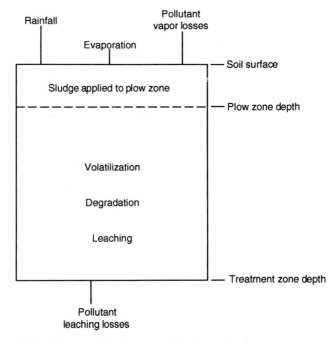

Figure 9.11 Diagram of land-treatment site.

As time proceeds, a dry zone develops near the ground surface, which is influenced by volatilization. The base of the dry zone proceeds downward at a rate controlled by the volatile flux and by the leaching rate of the pollutant. The concentration of the leachate from the wet zone changes with time due to first-order degradation losses. The pollutant is leached from the wet zone and migrates downward below the oily zone (Charbeneau and Daniel, 1993). The travel time from the region of initial concentration to the water table is computed in the model. The time for the pollutant to be completely removed from the soil profile is the time at which the dry zone interface reaches the water table.

Table 9.3 lists the input parameters and typical values required to run the RITZ model. The input parameters consist of data on soil properties and data on the contaminant and oil properties. Additional input data include information about the physical and hydrologic configuration of the treatment unit. The output from RITZ is extensive and includes mass balance for the contaminant, indicating the portions leached, volatilized, and degraded, and charts and tables of the

TABLE 9.3   Input Parameters Describing Land-Treatment Site

Input data: Soil properties	
Fraction organic carbon	0.0050
Bulk density (kg/m³)	1500.0
Saturated water content (m³/m³)	0.4100
Saturated hydraulic conductivity (m/day)	0.50000
Clapp and Hornberger constant	4.9000
Input data: Oil and pollutant properties	
Concentration of pollutant in the sludge (g/kg)	1.0000
Organic carbon partition constant (m³/kg)	2.2000×10^{-2}
Oil–water partition coefficient	5.0000×10^{1}
Henry's law constant	5.5000×10^{-5}
Diffusion coefficient of pollutant in air (m²/day)	4.3000×10^{-1}
Half-life of the pollutant (day)	3.0000×10^{1}
Concentration of oil in the sludge (g/kg)	2.5000×10^{2}
Density of the oil (kg/m³)	1.0000×10^{3}
Half-life of the oil (day)	4.5000×10^{1}
Diffusion coefficient of water vapor in air (m²/day)	2.0000
Input data: Operational and environmental factors	
Sludge application rate (kg/ha)	1.5000×10^{5}
Plow zone depth (m)	0.150
Treatment zone (m)	1.500
Recharge rate (m/day)	0.0060
Evaporation rate (m/day)	0.0025
Air temperature (°C)	25.0
Relative humidity	0.500
Identification code: Site #1	
Soil name: Tipton Sandy Loam	
Compound name: Pollutant #1	
CAS number: 123-4567	

Source: Nofziger et al., 1988.

TABLE 9.4 Typical Output from RITZ Model

| Time, days | Total pollutant, g/m³ | Concentration profile Pollutant in: | | | | Oil content, m³/m³ |
		Water, g/m³	Soil, g/kg	Vapor, g/m³	Oil, g/m³	
Depth = 0.250 m						
0.00	0.00E + 00	0.00E + 00	0.00E + 00	0.00E + 00	0.00E + 00	0.00E + 00
10.00	2.20E + 01	4.80E + 01	5.30E − 03	2.60E − 03	0.00E + 00	0.00E + 00
20.00	1.90E + 01	4.20E + 01	4.70E − 03	2.30E − 03	0.00E + 00	0.00E + 00
30.00	1.70E + 01	3.70E + 01	4.10E − 03	2.10E − 03	0.00E + 00	0.00E + 00
40.00	1.50E + 01	3.30E + 01	3.60E − 03	1.80E − 03	0.00E + 00	0.00E + 00
50.00	0.00E + 00	0.00E + 00	0.00E + 00	0.00E + 00	0.00E + 00	0.00E + 00
75.00	0.00E + 00	0.00E + 00	0.00E + 00	0.00E + 00	0.00E + 00	0.00E + 00
100.00	0.00E + 00	0.00E + 00	0.00E + 00	0.00E + 00	0.00E + 00	0.00E + 00
Depth = 0.500 m						
0.00	0.00E + 00	0.00E + 00	0.00E + 00	0.00E + 00	0.00E + 00	0.00E + 00
10.00	0.00E + 00	0.00E + 00	0.00E + 00	0.00E + 00	0.00E + 00	0.00E + 00
20.00	0.00E + 00	0.00E + 00	0.00E + 00	0.00E + 00	0.00E + 00	0.00E + 00
30.00	1.40E + 01	3.00E + 01	3.40E − 03	1.70E − 03	0.00E + 00	0.00E + 00
40.00	1.20E + 01	2.70E + 01	3.00E − 03	1.50E − 03	0.00E + 00	0.00E + 00
50.00	1.10E + 01	2.40E + 01	2.60E − 03	1.30E − 03	0.00E + 00	0.00E + 00
75.00	0.00E + 00	0.00E + 00	0.00E + 00	0.00E + 00	0.00E + 00	0.00E + 00
100.00	0.00E + 00	0.00E + 00	0.00E + 00	0.00E + 00	0.00E + 00	0.00E + 00

Identification code: Site #1
Soil name: Tipton Sandy Loam
Compound name: Pollutant #1
CAS number: 123-4567

Source: Nofziger et al., 1988.

leaching, volatilization, and concentration of the contaminant as a function of time (and depth for concentration). Examples of output corresponding to the input data shown in Table 9.4 are presented in Figure 9.12.

CHEMFLO is similar in concept to RITZ except that it deals with the movement of water and chemicals into and through soils. Water movement is modeled using Richards equation [Eq. (9.8)]. Chemical transport is modeled using the advection–dispersion equation. The equations are solved numerically in one-dimension using finite differences. The chemicals are applied to the soil in two different ways: (1) an inflowing solution that has a constant concentration of a given chemical and (2) a constant concentration in surface soils. Results of the water model can be displayed in the form of graphics of water content and flux density of water versus distance or time. Results from the transport equation are displayed as graphs of concentration and flux density of chemical as functions of distance or time. Typical input and output from CHEMFLO are shown in Table 9.5 and Figure 9.13, respectively.

HSSM, or the Hydrocarbon Simulation and Spill Exposure Assessment Modeling program, is aimed at predicting the fate of LNAPLs in aquifers (Weaver and Charbeneau, 1992). The HSSM model consists of three components: KOPT, OILENS, and TSGPLUME. The first two components simulate flow of LNAPL through the unsaturated zone and NAPL spreading along the water table. The acronym KOPT stands for kinematic oily pollutant transport. An emphasis of the OILENS model is the determination of the size of the spreading oil lens on the

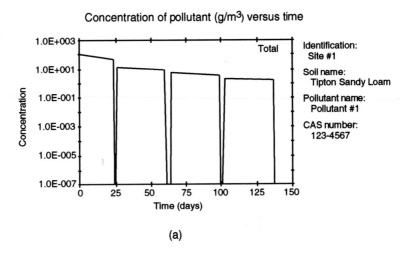

(a)

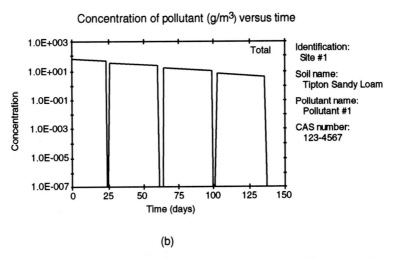

(b)

Figure 9.12 Typical output from the RITZ model. (a) Concentration of total pollutant as a function of time for depths of 0.1, 0.5, 1.0, and 1.5 m. (b) Concentration of pollutant in water as a function of time for depths of 0.1, 0.5, 1.0, and 1.5 m. Source: Nofziger et al., 1988.

TABLE 9.5 Input Data for the CHEMFLO Model

Definition of Soil System	(SCR: 1/5)
Name of the soil	YOLO CLAY
Orientation of flow system	0.0
Finite or semi-infinite soil system (F or S)	F
Length of soil system (cm)	30.00

F1 - Help; ESC - Abort; F10 - End editing; PgUp - Prev screen; PgDn - Next screen; F8 - Save input parameters in file; F9 - Read input parameters from file;

Definition of Soil System	(SCR: 2/5)
Boundary condition for water at the UPPER surface (P, F, R, M)	P
Desired matric potential (cm) at the UPPER soil surface	1.00
That matric potential corresponds to water content (cc/cc)	0.495
Boundary condition for water at the LOWER surface (P, F, M)	P
Desired matric potential (cm) at the LOWER surface	-50.00
That matric potential corresponds to water content (cc/cc)	0.406
Uniform initial matric potential throughout the soil (Y/N)	Y
Matric potential (cm) throughout the soil before simulation	-50.00
That matric potential corresponds to water content (cc/cc)	0.406

F1 - Help; ESC - Abort; F10 - End editing; PgUp - Prev screen; PgDn - Next screen; F8 - Save input parameters in file; F9 - Read input parameters from file;

Soil and Chemical Parameters	(SCR: 4/5)
Soil bulk density (g/cc)	1.4000
Water: Soil partition coefficient (cc/g soil)	0.0000
Diffusion coefficient of chemical in water (cm^2/hr)	$3.00E - 002$
Dispersivity (cm)	$2.00E + 000$
First-order degradation constant, liquid (1/hr)	$0.00E + 000$
First-order degradation constant, solid (1/hr)	$0.00E + 000$
Zero-order rate constant, liquid (μg/cc/hr)	$0.00E + 000$

F1 - Help; ESC - Abort; F10 - End editing; PgUp - Prev screen; PgDn - Next screen; F8 - Save input parameters in file; F9 - Read input parameters from file;

surface of the water table and the calculation of the mass flux of contaminants into the aquifer. The last module, TSGPLUME, uses as input the output files from KOPT/OILENS and calculates the spread of contaminants in the saturated zone resulting from the LNAPL pollution.

The flow system in the KOPT model is idealized as consisting of a circular source region overlying an aquifer at specified depth. The KOPT model treats transport through the unsaturated zone as one dimensional. Lateral spreading by capillary forces is neglected, as is spreading due to heterogeneity, since the soil is assumed to be heterogeneous. The main multiphase flow variables tracked by

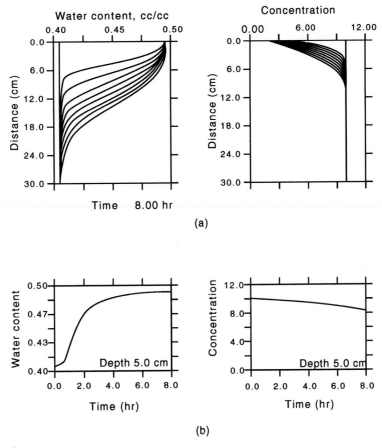

Figure 9.13 Water content and chemical concentration from CHEMFLO (a) as functions of depth at 1-hr intervals and (b) as functions of time at the 5-cm position. Source: Nofziger et al., 1989.

the model are the phase saturations (defined as the percent of the pore space filled by each fluid) and pressure heads.

The spill or release of the LNAPL may be simulated in three ways (Figure 9.14). The first method simulates the release of a known LNAPL flux for a specified duration; the second simulates a constant depth of ponded LNAPL for a known duration; finally, a known volume of LNAPL may be placed over a specified depth of soil. In all three scenarios, if a large enough volume of LNAPL is supplied, the LNAPL reaches the water table. If sufficient head is available, the water table is displaced downward, lateral spreading begins (Figure 9.15), and the OILENS portion of the model is triggered. Table 9.6 contains the parameters for an example problem with HSSM and Figures 9.16 through 9.18 present typical output from HSSM.

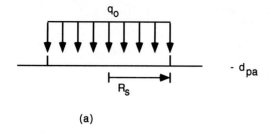

(a)

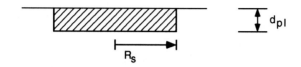

(b)

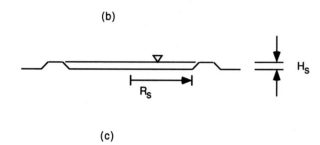

(c)

Figure 9.14 Source representation in HSSM. (a) Flux source representation. (b) Volume source representation. (c) Constant head source representation. Source: Weaver and Charbeneau, 1992.

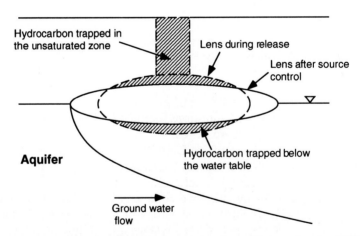

Figure 9.15 Spreading and dissolution from an oil lens in the HSSM model. Source: Weaver and Charbeneau, 1992.

TABLE 9.6 Parameters for Example Simulation for the HSSM Model

Soil and aquifer characteristics	
Saturated horizontal hydraulic conductivity (water)	5.0 m/day
Saturated vertical hydraulic conductivity (water)	1.0 m/day
Pore size distribution (Brooks and Corey)	1.0
Residual water saturation	0.10
Thickness of capillary fringe	0.10 m
Porosity	0.40
Bulk density	1.65 g/cm^3
Longitudinal dispersivity	10 m
Transverse dispersivity	2 m
Vertical dispersivity	0.10 m
Ground water seepage velocity	0.25 m/day
Average water infiltration rate	0.001 m/day
Aquifer thickness	5 m
Penetration depth	2.8 m
Hydrocarbon characteristics	
Density	0.70 g/cm^3
Dynamic viscosity	0.30 cp
Residual oil saturation above lens	0.05
Residual oil saturation below lens	0.15
Hydrocarbon solubility	200 mg/L
Surface tension	30 dyn/cm^2
Contaminant characteristics	
Initial concentration in gasoline	7000 mg/L
Hydrocarbon–water partition coefficient	100
Soil–water partition coefficient	2.0 L/kg
Contaminant solubility in water	175 mg/L
Contaminant theoretical oxygen demand	0.032

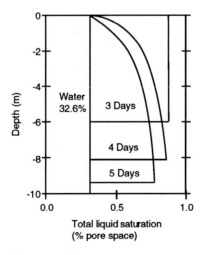

Figure 9.16 Total liquid saturation profiles in the unsaturated zone using HSSM model. Source: Weaver and Charbeneau, 1992.

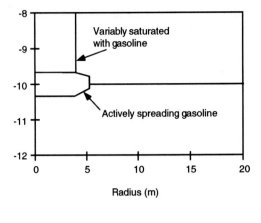

Figure 9.17 Gasoline lens profile at 7.5 days using HSSM model. Source: Weaver and Charbeneau, 1992.

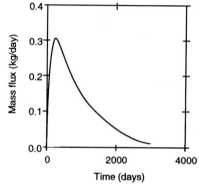

Figure 9.18 Xylene mass flux into aquifer using HSSM model. Source: Weaver and Charbeneau, 1992.

SUMMARY

In conclusion, flow and transport mechanisms in the unsaturated zone often control the amount of contamination in the saturated zone. However, the unsaturated zone is less well understood and much more complex. The governing equations are highly nonlinear and dependent on soil wetting and drying conditions. Measurement techniques in the unsaturated zone are less well developed compared to the saturated zone. Finally, a number of computer models have been recently developed that help to simulate processes in the unsaturated zone and the interrelationships between the vadose and saturated zones.

REFERENCES

BEDIENT, P. B., and HUBER, W. C., 1992, *Hydrology and Floodplain Analysis,* Addison-Wesley, Reading, MA.

BROOKS, R. H., and COREY, A. T., 1964, "Hydraulic Properties of Porous Media," *Hydrol. Pap. 3,* Colorado State University, Fort Collins, CO, p. 27.

CARLISLE, V. W., HALLMARK, C. T., SODEK, F., III, CALDWELL, R. E., HAMMOND, L. C., and BERKHEISER, V. E., 1981, "Characterization Data for Selected Florida Soils," Soil Characterization Laboratory, Soil Science Department, University of Florida, Gainesville.

CHARBENEAU, R. J., 1984, "Kinematic Models for Moisture and Solute Transport," *Water Resources Res.* 20(6):699, 706.

CHARBENEAU, R. J., 1990, "Groundwater Hydraulics and Pollutant Transport," Course Notes, University of Texas, Austin, Texas.

CHARBENEAU, R. J., and DANIEL, D. E., 1993, "Contaminant Transport in Unsaturated Flow," *Handbook of Hydrology,* Chapter 15, D. R. Maidment, ed., McGraw-Hill, New York.

CHOW, V. T., MAIDMENT, D. R., and MAYS, L. W., 1988, *Applied Hydrology,* McGraw-Hill, New York.

CLAPP, R. B., and HORNBERGER, G. M., 1978, "Empirical Equations for Some Soil Hydraulic Properties," *Water Resources Res.* 14:601–604.

CORAPCIOGLU, M. Y., and BAEHR, A. L., 1987, "A Compositional Multiphase Model for Groundwater Contamination by Petroleum Products. 1. Theoretical Considerations," *Water Resources Res.* 23:191–200.

DAVIDSON, J. M., NIELSON, D. R., and BIGGAR, J., 1966, "The Dependency of Soil Water Uptake and Release upon the Applied Press Increment," *Soil Sience Society Amer. J. Proc.* 30(3):298–304.

FAUST, C. R., GUSWOR, J. H., and MERCER, J. W., 1989, "Simulation of Three-dimensional Flow of Immiscible Fluids within and below the Unsaturated Zone," *Water Resources Res.* 25:2449–2464.

FETTER, C. W., 1993, *Contaminant Hydrogeology,* Macmillan, New York.

FREEZE, R. A., 1969, "The Mechanism of Natural Groundwater Recharge and Discharge: 1. One-dimensional, Vertical, Unsteady, Unsaturated Flow above a Recharging or Discharging Groundwater Flow System," *Water Resources Res.* 5:153–171.

FREEZE, R. A., 1971, "Three-dimensional, Transient, Saturated–Unsaturated Flow in a Groundwater Basin," *Water Resources Res.* 7:347–366.

FREEZE, R. A., and CHERRY, J. A., 1979, *Groundwater,* Prentice Hall, Englewood Cliffs, NJ.

FRIND, E. O., GILLHAM, R. W., and PICKENS, J. F., 1977, "Application of Unsaturated Flow Properties in the Design of Geologic Environments for Radioactive Waste Storage Facilities," *Finite Elements in Water Resources,* W. G. Gray, G. F. Pinder, and C. A. Brebbia eds., Pentech, London, pp. 3.133–3.163.

GREEN, W. H., and AMPT, G. A., 1911, "Studies of Soil Physics, 1: The Flow of Air and Water through Soils," *J. Agric. Sci.* 4(1):1–24.

GUPTA, S. C., and LARSON, W. E., 1979, "Estimating Soil Water Retention Characteristics from Particle Size Distribution, Organic Matter Percent, and Bulk Density," *Water Resources Res.* 15:1633–1635.

HOLTAN, H. N., 1961, "A Concept for Infiltration Estimates in Watershed Engineering," *U.S. Department of Agriculture Bulletin,* pp. 41–51.

HORTON, R. E., 1940, "An Approach toward a Physical Interpretation of Infiltration Capacity," *Soil Science Society Amer. J.* 5:399–417.

JURY, W. A., SPENCER, W. F., and FARMER, W. J., 1983, "Chapter 4: Chemical Transport

Modeling: Current Approaches and Unresolved Problems," *Chemical Mobility and Reactivity in Soil Systems,* American Society of Agronomy, Madison, WI, pp. 49–64.

KALUARACHCHI, J. J., and PARKER, J. C., 1989, "An Efficient Finite Element Method for Modeling Multiphase Flow," *Water Resources Res.* 25:43–54.

KALUARACHCHI, J. J., and PARKER, J. C., 1990, "Modeling Multicomponent Organic Chemical Transport in Three-fluid-phase Porous Media," *J. Contaminant Hydrology* 5: 349–374.

KLUTE, A., and DIRKSEN, C., 1986, "Hydraulic Conductivity and Diffusivity—Laboratory Methods," in A. Klute, ed., *Methods of Soil Analysis, Part I—Physical and Mineralogical Methods,* American Society of Agronomy Monograph 9, 2d ed., pp. 687–734.

KUPPUSAMY, T., SHENG, J., POULEER, J. C., and LENHARD, R. J., 1987, "Finite-element Analysis of Multiphase Immiscible Flow through Soils," *Water Resources Res.* 23: 625–631.

MAYER, A. S., and MILLER, C. T., 1989, "Equilibrium and Mass Transfer Limited Approaches to Modeling Multiphase Ground Water Systems," C. R. Omelia, ed., *Environ. Eng. Proc.* 1990 Specialty Conf., New York, pp. 314–321.

MEIN, R. G., and LARSON, C. L., 1973, "Modeling Infiltration during a Steady Rain," *Water Resources Res.* 9(2):384–394.

NEUMAN, S. P., 1983, "Saturated–Unsaturated Seepage by Finite Elements," *J. Hydraulic Engineering* 99(HY12):2233–2251.

NIELSON, D. R., VAN GENUCHTEN, M. TH., and BIGGAR, J. W., 1986, "Water Flow and Solute Transport Processes in the Unsaturated Zone," *Water Resources Res.* 22(9): 89S–108S.

NOFZIGER, D. L., RAJENDER, K., NAYUDU, S. K., and SU, P., 1989, "Chemflo" One-dimensional Water and Chemical Movement in Unsaturated Soils," EPA/600/8-89/076. U.S. EPA Robert S. Kerr Environmental Research Laboratory, Ada, OK.

NOFZIGER, D. L., WILLIAMS, J. R., and SHORT, T. E., 1988, "Interactive Simulation of the Fate of Hazardous Chemicals during Land Treatment of Oily Wastes," EPA/600/8-88/001. U.S. EPA Robert S. Kerr Environmental Research Laboratory, Ada, OK.

OLSON, R. S., and DANIEL, D. E., 1981, "Measurement of Hydraulic Conductivity of Fine-grained Soils," in T. F. Zimmie and C. O. Riggs, eds., *Permeability and Groundwater Contaminant Transport,* STP 746, American Society of Testing and Materials, Philadelphia, pp. 18–64.

PANDAY, S., and CORAPCIOGLU, M. T., 1989, "Reservoir Transport Equations by Compositional Approach," *Trans. Porous Media* 4:369–393.

PHILIP, J. R., and D. A, DE VRIES, 1957, "Moisture Movement in Porous Media under Temperature Gradients," *Eos Trans.,* American Geophysical Union 38(2):222–232.

RAWLS, W. J., BRAKENSIEK, D. L., and MILLER, N., 1983, "Green—Ampt Infiltration Parameters from Soils Data," *J. Hydraulic Engineering,* American Society of Civil Engineers 109(1):62–70.

RAWLS, W. J., AHUJA, L. R., BRAKENSIEK, A., and SHIRMOHAMMADI, A., 1993, "Infiltration and Soil Water Movement," *Handbook of Hydrology,* Chapter 5, D. R. Maidment, ed., McGraw-Hill, New York.

RICHARDS, L. A., 1931, "Capillary Conduction of Liquids in Porous Mediums," *Physics* 1:318–333.

RUBIN, J., and STEINHARDT, R., 1963, "Soil Water Relations during Rain Infiltration, I, Theory," *Soil Scientific Society Amer. Proc.,* 27(3):246–251.

SHORT, T. E., 1986, "Modeling of Processes in the Unsaturated Zone," in R. C. Loehr and J. F. Malina, Jr., eds., *Land Treatment: A Hazardous Waste Management Alternative,* Water Resources Symposium No. 13, U.S. Environmental Protection Agency and University of Texas at Austin, pp. 211–240

SHORT, T. E., 1988, "Movement of Contaminants from Oily Wastes during Land Treatment," in E. J. Calabrese and P. J. Kostecki, eds., *Soils Contaminated by Petroleum: Environment and Public Health Effects,* Wiley, New York, pp. 317–330.

SKAGGS, R. W., and KHALEEL, R., 1982, "Chapter 4, Infiltration," *Hydrologic Modeling of Small Watersheds,* C. T. Haan, J. P. Johnson, and D. L. Brakensiek, eds., Monograph No. 5, American Society of Agricultural Engineers, St. Joseph, MI.

SMITH R. E., 1983, "Approximate Soil Water Movement by Kinetic Characteristics," *Soil Scientific Society Amer. J.* 47(1):3–8.

SOIL CONSERVATION SERVICE, 1986, *Urban Hydrology for Small Watershelds,* Technical Release 55, pp. 2.5–2.8.

SWARTZENDRUBER, D., 1987, "A Quasi Solution of Richards' Equation for Downward Infiltration of Water into Soil," *Water Resources Res.* 5:809–817.

THIBODEAUX, L. J., and HWANG, S. T., 1982, "Landfarming of Petroleum Wastes—Modeling of the Air Emissions Problem," *Environ. Progress* 1(1):42–46.

VAN GENUCHTEN, M. TH., 1980, "A Closed-form Equation for Predicting the Hydraulic Conductivity of Unsaturated Soils," *Soil Scientific Society Amer. J.* 44:892–898.

WEAVER, T. W., and CHARBENEAU, R. J., 1992, "Hydrocarbon Spill Exposure Assessment Modeling," *User's Manual,* Robert S. Kerr Environmental Research Laboratory, Ada, OK.

Chapter 10

NUMERICAL MODELING OF CONTAMINANT TRANSPORT

A ground water model is a tool designed to represent a simplified version of a real field site. It is an attempt to take our understanding of the physical, chemical, and biological processes and translate them into mathematical terms. The resulting model is only as good as the conceptual understanding of the processes. The goal of modeling is to predict the value of an unknown variable, such as head in an aquifer system or the concentration distribution of a given chemical in the aquifer in time and space. Models can be used as follows:

1. *Predictive tools:* These are site-specific applications of models with the objective of determining future conditions or the impact of a proposed action on existing conditions in the subsurface.
2. *Interpretive or research tools:* These models are usually used for studying system dynamics and understanding processes.
3. *Generic or screening tools:* These models generally incorporate uncertainty in aquifer parameters and are used in a regulatory mode for the purpose of developing management standards and guidelines.

In developing a ground water flow or solute transport model, the first step is to develop a *conceptual model* consisting of a description of the physical, chemical, and biological processes that are thought to be governing the behavior of the system being analyzed (Istok, 1989). The next step is to translate the conceptual model into mathematical terms, or a *mathematical model,* that is, a set of partial differential equations and an associated set of auxiliary boundary conditions. Finally, solutions of the equations subject to the auxiliary conditions can be obtained using analytical or numerical methods.

If numerical methods are used, the collection of partial differential equations, auxiliary conditions, and the numerical algorithms are referred to as a *numerical model.* A computer program that implements the numerical model is referred to as a *computer code* or *computer model.* Computer models are essential for analyzing subsurface flow and contamination problems because they are designed to incorporate spatial variability within the aquifer, as well as spatial and temporal trends in hydrologic parameters.

Prior to initiating the modeling process, the following questions need to be addressed:

1. What is the problem that is being solved or answered through modeling?
2. Is modeling the most appropriate method for establishing the answer to the problem at hand?
3. What level of sophistication in modeling would be required to answer the question?
4. What level of confidence can be associated with the available field data and the anticipated results from the modeling effort?

The responses to these questions will allow the prospective modeler to determine the magnitude of the modeling effort, that is, whether the model is one, two, or three dimensional and analytical or numerical and whether a steady-state or transient analysis is necessary. Answering question 4 allows the modeler to anticipate the benefits against the costs that would be incurred for the modeling study. Finally, keep in mind that modeling is only one component in a hydrogeologic assessment and not an end in inself. The proposed modeling effort should be integrated within the framework for action at a given field site.

10.1 INTRODUCTION TO NUMERICAL MODELING

The field of ground water flow and transport modeling has grown tremendously over the past 15 years. This is mostly due to the need for quantitative estimates of flow and mass transport in the subsurface. Many articles and books have been written about the science of modeling in ground water. The reader is referred to Mercer and Faust (1981), Wang and Anderson (1982), Huyakorn and Pinder (1983), Hunt (1983), Javandel et al. (1984), Bear and Verruijt (1987), Istok (1989), van der Heijde et al. (1988), National Research Council (1990), and Anderson and Woessner (1992) as a starting point.

Anderson and Woessner (1992) propose a modeling protocol that can be summarized as follows:

1. Establish the purpose of the model.
2. Develop a conceptual model of the system.
3. Select the governing equation and a computer code. Both the governing equation and code should be verified. Verification of the governing equation demonstrates that it describes the physical, chemical, and biological processes occurring. Code verification can be accomplished by comparing the model results to an analytical solution of a known problem.
4. Design the model. This step includes selection of a grid design, time parameters, initial and boundary conditions, and developing estimates of model parameters.
5. Calibrate the designed model. Calibration refers to the process of determining a set of model input parameters that approximates field measured heads,

flows, and/or concentrations. The purpose of calibration is to establish that the model can reproduce field-measured values of the unknown variable.

6. Determine the effects of uncertainty on model results. This is sometimes referred to as a sensitivity analysis. The model parameters are varied individually within a range of possible values, and the effect on model results is evaluated.
7. Verify the designed and calibrated model. This step involves testing the model's ability to reproduce another set of field measurements using the model parameters that were developed in the calibration process.
8. Predict results based on the calibrated model.
9. Determine the effects of uncertainty on model predictions.
10. Present modeling design and results.
11. Postaudit and redesign the model as necessary. As more data are collected beyond model development, it is possible to compare the model predictions against the new field data. This may lead to further modifications and refinements of the site model.

Steps 1 through 4 of Anderson and Woessner's protocol are discussed in this section. The remaining steps are discussed in Section 10.8 within the context of applying models to field sites.

Purpose of Model

It is essential to identify clearly the purpose of the modeling effort at the outset in order to maximize the benefits from the analysis. Stating the purpose of modeling also helps focus the study, determine expectations, and limit the unnecessary expenditure of resources. Typical objectives from modeling studies include the following:

1. Testing a hypothesis, or improving knowledge of a given aquifer system.
2. Understanding physical, chemical, or biological processes.
3. Designing remediation systems.
4. Predicting future conditions or the impact of a proposed stress on a ground water system.
5. Resource management.

Conceptual Models

A key step in the modeling process is to formulate a conceptual model of the system being modeled. A conceptual model is a pictorial representation of the ground water flow and transport system, frequently in the form of a block diagram or a cross section (Anderson and Woessner, 1992). The nature of the conceptual model will determine the dimensions of the numerical model and the design of the grid.

The purpose of building a conceptual model is to simplify the field problem and make it more amenable to modeling. For example, the geologic framework shown in Figure 10.1 is very complex from a modeling standpoint. A conceptual model of the system, however, can be constructed by identifying the pertinent hydrologic features of the geologic framework as shown in Figure 10.1.

Formulating a conceptual model for flow and/or transport includes (1) defining hydrogeologic features of interest, that is, the aquifers to be modeled; (2) defining the flow system and sources and sinks of water in the system such as recharge from infiltration, baseflow to streams, evapotranspiration, and pumping; and (3) defining the transport system and sources and sinks of chemicals in the system.

Discretization

In numerical models, the physical layout of the area in question is replaced with a discretized model domain referred to as a grid and consisting of cells, blocks, or elements depending on whether finite difference or finite element methods are used. Generally, the grid should be drawn on an overlay of a map of the area to be modeled. It is preferable to align the horizontal plane of the grid such that the x and y axes are colinear with K_x and K_y, respectively (where K_x and K_y are the hydraulic conductivities in the x and y directions, respectively). The vertical axis of the model, when present, should be aligned with K_z (Anderson and Woessner, 1992). Selecting the size of the cells and elements to be used is a critical step in grid design; it depends on many factors, such as spatial variability in model parameters, physical boundaries of system, type of model being used (finite difference or finite element), computer model limitations, data-handling limitations, runtime, and associated computer costs.

Discretization decisions also need to be made for the time parameter. The majority of numerical models calculate results at time T by subdividing the total time into time steps, Δt. Generally, smaller time steps are preferable, but the computational time and cost involved in the modeling process may increase as the time step is decreased. Time steps may be influenced by the requirements of the model. Some models suffer from *numerical instabilities* that cause unrealistic oscillating solutions if a sufficiently small time step is not used. It is a good modeling practice to test the sensitivity of the model results to the size of the time step.

Dimensionality

Another issue that is closely linked to discretization is that of the dimensionality of the problem: Is a one-dimensional model sufficient to achieve the purposes of modeling or is it necessary to develop a two- or three-dimensional model? Would an analytical solution provide the required answer or is it necessary to utilize a numerical model? Is a steady-state assumption adequate or does the problem necessitate a transient analysis?

A good rule of thumb when resolving dimensionality issues is to avoid complexity if at all possible. So, for example, whereas modeling air pollution might

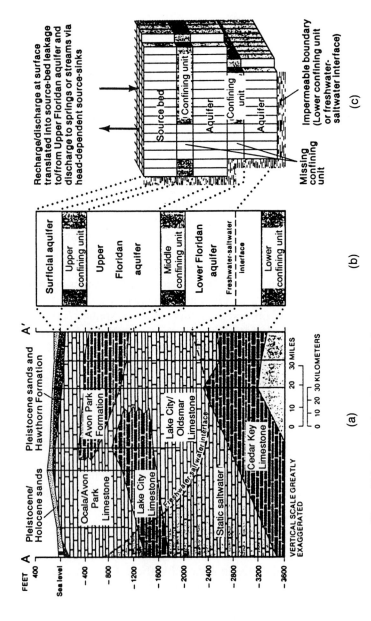

Figure 10.1 Developing a hydrogeologic conceptual model. (a) Geologic units in hydrogeologic framework. (b) Hydrogeologic units in conceptual model. (c) Equivalent units in digital ground water flow model. Source: Anderson and Woessner, 1992.

292

require a three-dimensional analysis of pollutant dispersion and diffusion, the problem of ground water contamination at a field site where the data have been collected using conventional monitoring wells can only be simulated as a two-dimensional problem, because there is no three-dimensional definition of the plume of contamination.

Boundary and Initial Conditions

The governing equation alone is not sufficient to describe a specific physical system. This is because a general solution of an nth-order differential equation will involve n independent arbitrary constants or functions. To define uniquely a given physical problem, the values of the constants or forms of the functions must be specified. Initial and boundary conditions can be used to provide this required additional information. Generally, boundary conditions specify the value of the dependent variable, or the value of the first derivative of the dependent variable, along the boundaries of the system being modeled.

Boundary conditions are typically derived from physical and/or hydraulic boundaries of ground water flow systems, for example, the presence of an impermeable body of rock or a river in connection with the ground water aquifer. Hydrogeologic boundaries are represented by three types of mathematical formulations: specified head, specified flux, and head-dependent flux boundaries.

Specified Head (Dirichlet Conditions)

A specified head boundary is simulated by setting the head at the relevant locations equal to known values:

$$H(x, y, z) = Ho \qquad (10.1)$$

where $H(x, y, z)$ is the head at a point with coordinates (x, y, z) and Ho is a specified head value. It is important to recognize that a specified head boundary represents an inexhaustible supply of water.

Specified Flux Boundaries (Neumann Conditions)

Flux boundaries are defined by specifying the derivative of the head across the boundary:

$$q_x = \frac{\partial H}{\partial x} = C \qquad (10.2)$$

where C is a constant. This type of boundary is used to describe fluxes to surface water bodies, spring flow, and seepage to and from bedrock underlying the system. A special type of specified flux boundaries is a no-flow boundary, which is set by specifying flux to be zero. A no-flow boundary may represent impermeable bedrock, an impermeable fault zone, a ground water divide, or a streamline.

Head-dependent Flux Boundaries (Cauchy or Mixed Conditions)

For this type of boundary, the flux across the boundary is calculated given a boundary head value:

$$\frac{\partial H}{\partial x} + aH = C \qquad (10.3)$$

where a and C are constants. Leakage to or from a river, for instance, can be simulated using this type of boundary condition.

In some instances, it may not be possible to use physical boundaries and regional ground water divides. Other hydraulic boundaries can be defined from information on the configuration of the flow system. However, care must be taken when establishing such boundaries to ensure that the model boundaries will not cause the solution to differ significantly from the response that would occur in the field. For example, hydraulic boundaries may be defined from a water-table map of the area to be modeled. The model grid is superimposed on the water-table contour map, and specified head boundary conditions can be interpolated. It is important to verify, however, that these boundary conditions will not be affected by stresses imposed on the model, such as pumping from a location that is near to the boundary.

The above-mentioned boundary conditions used to represent flow may also be used as sources of chemicals into a ground water aquifer. For example, a specified head boundary may be used to represent a contaminant source releasing chemicals into the aquifer at a specified concentration. Similarly, a flux boundary can be used to simulate the flux of contaminants across the boundary. More typically, however, injection wells are used to represent sources of contamination, as discussed in the following paragraphs.

Sources and Sinks

Water as well as chemicals may enter the grid in one of two ways: (1) through the boundaries, as determined by the boundary conditions, or (2) through sources and sinks within the interior of the grid. Even though the same model options may be used to represent boundary sources and sinks as are used to represent internal sources and sinks, the reader should remember that internal sources and sinks are not boundary conditions. For example, specified head cells are used to represent specified head boundary conditions, but specified head nodes may be placed within the grid to represent lakes and rivers or some other type of source.

An injection or pumping well is a point source or sink and is represented in a ground water model by specifying an injection or pumping rate at a designated node or cell. In a two-dimensional model, an assumption of a fully penetrating well over the aquifer thickness is made. The prospective modeler is cautioned when modeling wells with models that only allow a uniform grid (that is, all cells have the same size) and when the cell size in the model greatly exceeds the actual diameter of the well. The head calculated by the model is not an accurate approximation of the head measured in the well, but rather the head value predicted by the model is closer to an average of the heads measured as we move outward from the well toward the edge of the cell.

Source and Types of Errors: Accuracy of Numerical Models

A key component in computer modeling of ground water flow and contaminant transport that is often neglected is an assessment of the error introduced due to the selected modeling methodology. Two types of errors resulting from a modeling study need to be clearly distinguished:

1. *Computational errors:* These errors occur because of the numerical approximation procedures that are used to solve the governing equations subject to the given boundary and initial conditions. Computational errors are usually estimated by applying the continuity equation or the principle of conservation of mass (input $-$ output $=$ accumulation).

2. *Calibration errors:* These errors occur due to model assumptions and limitations in parameter estimation. Calibration errors can be quantified by comparing model predicted values to observed values of the unknown variable (see Section 10.8).

Limitations of Models

Mathematical models have several limitations, which can be conceptual or application related. Conceptual limitations relate to representation of the actual process or system with a mathematical model. For example, analytical models are limited by the simplifying assumptions that are required to develop the solution. The analytical models that are available are limited to certain idealized conditions and may not be applicable to a field problem with complex boundary conditions. Another limitation of analytical models is that spatial or temporal variation of system properties such as permeability and dispersivity cannot be handled (Javandel et al., 1984).

Application-related limitations have to do with the solution procedure that is utilized in the model development or with the amount of effort required to implement the solution. For example, the approximation of a differential equation by a numerical solution introduces two types of computational errors: numerical and residual. The numerical errors are due to the solution method used in solving the differential equations. The residual errors are a result of approximating the differential equation with a series of mathematical expressions.

Numerical models are also limited by their complexity, so the user must have a certain level of knowledge to be able to apply these models. Achievement of the required familiarity level is time consuming and could be prohibitive when funding is limited or when dealing with a time constraint. Preparation of input data for numerical models often takes a long time. Computer efficiency (a relatively fast computer) is necessary when using a numerical model.

10.2 NUMERICAL METHODS

Numerical or computerized solutions of the flow and/or transport equation in two dimensions are the most plentiful and commonly used techniques. These solutions are generally more flexible than analytical solutions because the user can approximate complex geometries and combinations of recharge and withdrawal wells by judicious arrangement of grid cells. The general method of solution is to break up the flow field into small cells, approximate the governing partial differential equations by differences among the values of parameters over the network of time t, and then predict new values for time $t + \Delta t$. This continues forward in time in small increments Δt.

The most common mathematical formulations for approximating the partial differential equations of flow and solute transport include the following:

- Finite difference methods
- Finite element methods
- Collocation methods
- Method of characteristics
- Boundary element methods

Before defining these methods, it is necessary to define the different types of partial differential equations (PDEs). All PDEs of the form $L(u) = f$ can be classified as *elliptic, parabolic,* or *hyperbolic.* The PDE can be written as

$$a \frac{\partial^2 u}{\partial x^2} + 2b \frac{\partial^2 u}{\partial x\, \partial y} + c \frac{\partial^2 u}{\partial y^2} = F\left(x,\, y,\, u,\, \frac{\partial u}{\partial x},\, \frac{\partial u}{\partial y}\right) \qquad (10.4)$$

where a, b, and c are functions of x and y only and the equation is linear if F is linear. The PDE is referred to as

Hyperbolic	if	$b^2 - ac > 0$
Parabolic	if	$b^2 - ac = 0$
Elliptic	if	$b^2 - ac < 0$

The finite difference method is the most popular method for simulating problems of ground water flow and transport. Finite difference methods are conceptually straightforward and easily understood. The primary disadvantage of these methods is that the truncation error in approximating the partial differential equations can be significant (Anderson, 1979). More details about the finite difference method are included in Section 10.3, and the reader is referred to Peaceman and Rachford (1955, 1962), Forsythe and Wasow (1960), Aziz and Settari (1979), Crichlow (1977), Peaceman (1977), Remson et al. (1965), Freeze and Witherspoon (1966), Bittinger et al. (1967), Pinder and Bredehoeft (1968), Cooley (1971), Freeze (1971), Brutsaert (1973), Green et al. (1970), Shamir and Harleman (1967), Oster et al. (1970), Tanji et al. (1967), and Wierenga (1977) for additional information.

The finite element method also operates by breaking the flow field into elements, but in this case the elements may vary in size and shape. A disadvantage of the finite element method is the need for formal mathematical training to understand the procedures properly. Finite element methods generally have higher computing costs (Pinder and Gray, 1977; Pinder, 1973; Wang and Anderson, 1982). More details on the finite element method are presented in Section 10.4, and the reader is additionally referred to Bathe and Wilson (1976), Chung (1979), Clough (1960), Cook (1974), Finlayson (1972), Huebner (1975), Hinton and Owen (1979), Norrie and DeVries (1978), Segerlind (1976), Zienkiewicz (1977), Gallagher (1975), and Weinstock (1952).

The method of characteristics (MOC) is a variant of the finite difference method and is particularly suitable for solving hyperbolic equations. The method of characteristics was developed to simulate advection-dominated transport by Garder et al. (1964). In ground water hydrology, Pinder and Cooper (1970) and Reddell and Sunada (1970) used the method to solve the density-dependent transport equations. Later, MOC was used widely to simulate the movement of contaminants in the subsurface (Bredehoeft and Pinder, 1973). The MOC is most useful where solute transport is dominated by advective transport. This procedure is computationally efficient and minimizes numerical dispersion problems (Konikow and Bredehoeft, 1978). The method of characteristics is discussed in more detail in Section 10.5.

The collocation method and the boundary element method will not be discussed in detail in this book. For details on the collocation method, the reader is referred to Finlayson (1972), Chang and Finlayson (1977), Houstis et al. (1979), Douglas and Dupont (1974), Sincovec (1977), Allen and Pinder (1983), Pinder et al. (1978), Frind and Pinder (1979), Lapidus and Pinder (1982), and Celia and Pinder (1980). For details on the boundary element method, the reader is referred to Jawson and Ponter (1963), Liggett (1977), Brebbia and Walker (1980), Lapidus and Pinder (1982), Liggett and Liu (1983), Banerjee et al. (1981), Rizzo and Shippy (1970), Dubois and Buysee (1980), Jawson and Symm (1977), Lennon et al. (1979), and Liggett and Liu (1979).

FUNDAMENTAL CONCEPTS

Iteration means that a process is repeated until an answer is achieved. Iterative techniques can be used to find roots of equations, solutions of linear and nonlinear systems of equations, and solutions of differential equations. Iteration methods require that an initial guess be made and that a rule or function for computing successive terms exist.

Convergence is characterized by the question, "Does the numerical solution approach the true solution as a chosen numerical procedure is applied?" The convergence problem may arise from using an iterative technique or a numerical technique that involves truncation of an infinite series.

In a **stable** numerical procedure, the solution marches forward in time, the errors are not amplified such that the solution becomes invalid.

Curve-fitting techniques attempt to fit a given set of data with a mathematical expression or function. For example, one of the most common curve-fitting methods is a **least squares fit,** where the function $f(x)$ is approximated using

$$f(x) \approx \sum_{k=0}^{n} a_k \phi_k(x) \equiv y(x) \qquad (10.5)$$

where $\phi_0(x), \ldots, \phi_n(x)$ are $n + 1$ appropriately chosen functions. The exact values of $f(x)$ are known over a certain domain, which consists of a discrete set of points $x_0, x_1, \ldots, x_n$ or of a continuous interval (a, b). The least squares approximation is defined to be that for which the a_k's are determined so that the sum of $w(x)r^2(x)$ over the domain is as small as possible, where $w(x)$ is a nonnegative weighting function and $r(x) = f(x) - y(x)$.

Gaussian elimination and backward substitution refers to the simplest method of solving a set of equations of the form:

$$a_{11}x_1 + a_{12}x_2 + a_{13}x_3 + \cdots + a_{1n}x_n = b_1$$

$$a_{21}x_1 + a_{22}x_2 + a_{23}x_3 + \cdots + a_{2n}x_n = b_2$$

$$a_{31}x_1 + a_{32}x_2 + a_{33}x_3 + \cdots + a_{3n}x_n = b_3 \qquad (10.6)$$

$$\vdots$$

$$a_{n1}x_1 + a_{n2}x_2 + a_{n3}x_3 + \cdots + a_{nn}x_n = b_n$$

Gaussian elimination consists of dividing the first equation by a_{11} and using the result to eliminate x_1 from all succeeding equations. Next, the modified second equation is used to eliminate x_2 from the succeeding equations, and so on. After this elimination has been effected n times, the resultant set of equations is solved by backward substitution.

Numerical Solution of Partial Differential Equations

The fundamental idea on which the numerical solution of partial differential equations is based is this: Each partial derivative that appears in the equation is replaced by a finite difference approximation. When these differences are evaluated at each mesh point, the result is a set of simultaneous equations that can be solved either directly or by various iterative procedures.

Specifically, in a plane grid, if the coordinates of the mesh points (named neutrally for the moment) are $p_i = p_o + ih$ and $q_j = q_o + jk$, then from the usual difference quotient approximation to the first derivative we have

$$\left.\frac{\partial u}{\partial p}\right|_{p_i, q_j} = \frac{u_{p_{i+1}, q_j} - u_{p_i, q_j}}{h} = \frac{u_{i+1, j} - u_{i, j}}{h} \qquad (10.7)$$

and, similarly,

$$\left.\frac{\partial u}{\partial q}\right|_{p_i, q_j} = \frac{u_{i, j+1} - u_{i, j}}{k} \qquad (10.8)$$

Furthermore, for the case of a second derivative, we obtain

$$\left.\frac{\partial^2 u}{\partial p^2}\right|_{p_i, q_j} = \frac{u_{p_{i+1}, q_j} - 2u_{p_i, q_j} + u_{p_{i-1}, q_j}}{h^2} = \frac{u_{i+1, j} - 2u_{i, j} + u_{i-1, j}}{h^2} \qquad (10.9)$$

and, similarly,

$$\left.\frac{\partial^2 u}{\partial q^2}\right|_{p_i, q_j} = \frac{u_{i,j+1} - 2u_{i,j} + u_{i,j-1}}{k^2} \tag{10.10}$$

Elliptic Equations (Laplace's Equation in Two Dimensions)

Using Eqs. (10.9) and (10.10) to approximate each partial derivative in the two-dimensional form of Laplace's equation, that is,

$$\frac{\partial^2 u}{\partial x^2} + \frac{\partial^2 u}{\partial y^2} = 0 \tag{10.11}$$

we obtain, as a difference equation approximating the actual equation,

$$\frac{u_{i+1,j} - 2u_{i,j} + u_{i-1,j}}{h^2} + \frac{u_{i,j+1} - 2u_{i,j} + u_{i,j-1}}{k^2} = 0 \tag{10.12}$$

or, making the natural and convenient assumption that $h = k$ and solving for $u_{i,j}$,

$$u_{i,j} = \frac{u_{i+1,j} + u_{i,j+1} + u_{i-1,j} + u_{i,j-1}}{4} \tag{10.13}$$

This asserts that *the value of* u *at any mesh point is equal to the average of the values of* u *at the four adjacent mesh points.*

If Eq. (10.13) is evaluated at each mesh point that is not a boundary point, where the value of the solution u is initially given; the result is a system of simultaneous linear equations in the unknown functional values u_{ij}. The number of equations is equal to the number of mesh points at which the value of u is to be calculated, and (at least for rectangular regions) it can be shown that this system of equations always has a unique nontrivial solution.

Example 10.1 STEADY-STATE HEAD IN A SQUARE REGION

To illustrate the formulation of a numerical system, let us attempt to approximate the steady-state head distribution $u(x, y)$ in the square region shown in Figure 10.2, using the grid obtained by dividing each edge into four equal parts. The unknowns in this problem are the heads at the nine points of the grid that are not boundary points.

Solution. At the outset, we note from symmetry that $u_{11} = u_{31}$, $u_{12} = u_{32}$, and $u_{13} = u_{33}$, so our problem actually involves only six equations in the six unknowns u_{11}, u_{12}, u_{13}, u_{21}, u_{22}, and u_{23}. Applying Eq. (10.13) at each of the six mesh points P_{11}, P_{12}, P_{13}, P_{21}, P_{22}, and P_{23} and taking into account the symmetries we have just noted and the known values of u on the boundary, we have at P_{11}

$$4u_{11} - u_{01} - u_{10} - u_{21} - u_{12} = 0$$

or, noting that by hypothesis $u_{01} = u_{10} = 0$,

$$4u_{11} - u_{21} - u_{12} = 0 \tag{10.14}$$

Similarly, at P_{12}, P_{13}, P_{21}, P_{22}, P_{23} we have, respectively,

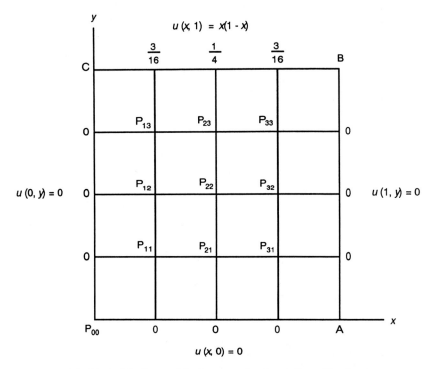

Figure 10.2 Typical lattice used in the approximate solution of Laplace's equation. Source: Wylie and Barrett, 1982. Reproduced with permission of McGraw-Hill.

$$4u_{12} - u_{11} - u_{22} - u_{13} = 0 \qquad (10.15)$$

$$4u_{13} - u_{12} - u_{23} = \frac{3}{16} \qquad (10.16)$$

$$4u_{21} - 2u_{11} - u_{22} = 0 \qquad (10.17)$$

$$4u_{22} - u_{21} - 2u_{12} - u_{23} = 0 \qquad (10.18)$$

$$4u_{23} - u_{22} - 2u_{13} = \frac{1}{4} \qquad (10.19)$$

Using Eqs. (10.14), (10.15), and (10.16) to eliminate u_{21}, u_{22}, and u_{23} from Eqs. (10.17), (10.18), and (10.19), we obtain the following system, which is easily solved by elimination:

$$15u_{11} - 8u_{12} + u_{13} = 0$$

$$-8u_{11} + 16u_{12} - 8u_{13} = -\frac{3}{16}$$

$$15u_{13} - 8u_{12} + u_{11} = 1$$

yielding $u_{11} = 0.0154$, $u_{12} = 0.0396$, and $u_{13} = 0.0872$.

There is another way in which the finite difference approximation to the Lapla-

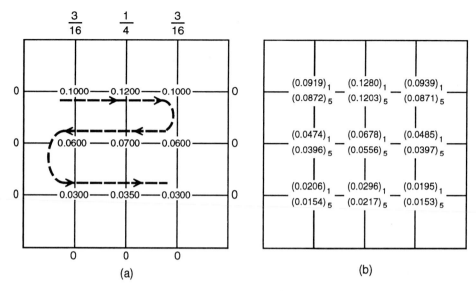

Figure 10.3 Data from an iterative solution of Laplace's equation. Source: Wylie and Barrett, 1982. Reproduced with permission of McGraw-Hill.

cian can be used to determine the value of the solution at the points of the grid. It is a simple iterative method that proceeds as follows: We first recall that the finite difference approximation to the Laplacian [Eq. (10.13)] expresses the value of the solution at any mesh point as the average of the values at the four adjacent points. Thus, after an initial estimate for the value of the solution at each mesh point has been made, it can be corrected and improved by systematically moving through the grid and replacing each value according to Eq. (10.13). In doing this, each value as soon as it is corrected should be used in all subsequent calculations.

As an illustration of this method, let us reconsider the problem we have just worked. Beginning with the estimates shown in Figure 10.3a, we have for the first refinement of u_{13} the value

$$\frac{0.1875 + 0.0000 + 0.1200 + 0.0600}{4} = 0.0919$$

Continuing through the grid as indicated, using the corrected values as soon as they become available (but taking no advantage of the symmetry of the problem), we obtain the values shown in Figure 10.3b. Values bearing the subscript 1 were obtained by a single iteration; values bearing the subscript 5 were obtained after five iterations.

Parabolic Equations (One-dimensional Transport Equation)

For the one-dimensional transport equation

$$\frac{\partial^2 c}{\partial x^2} = a^2 \frac{\partial c}{\partial t} \tag{10.20}$$

the region of the xt plane over which a solution is sought is always infinite because of the infinite increase of time. As a finite-difference approximation to Eq. (10.20), we have, using Eqs. (10.8) and (10.9),

$$\frac{1}{h^2}(c_{i+1,j} - 2c_{i,j} + c_{i-1,j}) = \frac{a^2}{k}(c_{i,j+1} - c_{i,j}) \qquad (10.21)$$

or, setting $m = k/a^2h^2$ and solving for $c_{i,j+1}$,

$$c_{i,j+1} = mc_{i+1,j} + (1 - 2m)c_{i,j} + mc_{i-1,j} \qquad (10.22)$$

Clearly, it would be convenient to choose h and k so that the value of m is 0.5. The values of c on the boundary are provided by the data of the problem. Thus, the given initial condition $c(x, 0)$ provides the values of $c_{00}, c_{10}, c_{20}, \ldots$. Similarly, end conditions of the form

$$c(0, t) = g_1(t), \qquad c(l, t) = g_2(t) \qquad (10.23)$$

where g_1 and g_2, usually but not necessarily constant, furnish the values of c_{01}, $c_{02}, c_{03}, \ldots$ and $c_{l1}, c_{l2}, c_{l3}, \ldots$. No-flow boundary conditions can be handled, as we outlined previously in our discussion of Laplace's equation.

The determination of the solution over the rest of the grid proceeds in a straightforward way, using the extrapolation pattern provided by Eq. (10.21). First, the values of $c_{11}, c_{21}, \ldots, c_{l-1,1}$ are calculated from the known values of $c_{00}, c_{10}, c_{20}, \ldots, c_{l0}$. Then, using these new values and the boundary values c_{01} and c_{l1}, the solution is "marched" forward by calculating the values of c at the grid points in the third row, and so on for the remainder of the grid.

10.3 FINITE DIFFERENCE METHODS

In general, a finite difference model is developed by superimposing a system of nodal points over the problem domain. In the finite difference method, nodes may be located inside cells (block centered, Figure 10.4) or at the intersection of grid lines (mesh centered, Figure 10.5). Aquifer properties and head values are assumed to be constant within each cell in a block-centered finite difference model. An equation is written in terms of each nodal point in finite difference models because the area surrounding a node is not directly involved in the development of the finite difference equations (Wang and Anderson, 1982).

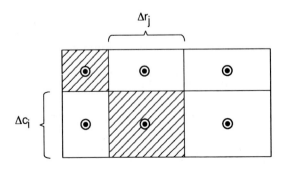

Figure 10.4 Block-centered grid system.

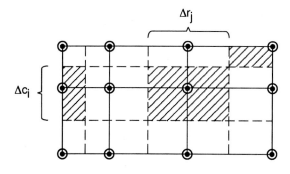

Figure 10.5 Point or mesh-centered grid system.

The principles behind finite difference approximations will be illustrated using Laplace's equation in two dimensions for steady ground water flow:

$$\frac{\partial^2 h}{\partial x^2} + \frac{\partial^2 h}{\partial y^2} = 0 \qquad (10.24)$$

Consider a finite set of points on a regularly spaced grid (Figure 10.6). In the finite difference approximation, derivatives are replaced by differences taken between nodal points. A central approximation to $\partial^2 h/\partial x^2$ at (x_0, y_0) is obtained by approximating the first derivative at $(x_0 + \Delta x/2, y_0)$ and at $(x_0 - \Delta x/2, y_0)$ and then obtaining the second derivative by taking a difference between the first derivatives of those points:

$$\frac{\partial^2 h}{\partial x^2} \approx \frac{\dfrac{h_{i+1,j} - h_{i,j}}{\Delta x} - \dfrac{h_{i,j} - h_{i-1,j}}{\Delta x}}{\Delta x} \qquad (10.25)$$

which simplifies to

$$\frac{\partial^2 h}{\partial x^2} \approx \frac{h_{i-1,j} - 2h_{i,j} + h_{i+1,j}}{\Delta x^2} \qquad (10.26)$$

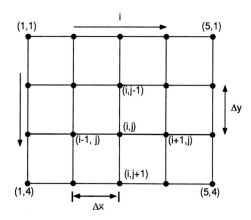

Figure 10.6 Finite difference grid system. Source: Bedient and Huber, 1992.

Similarly,

$$\frac{\partial^2 h}{\partial y^2} \approx \frac{h_{i,j-1} - 2h_{i,j} + h_{i,j+1}}{\Delta y^2} \tag{10.27}$$

By adding the expressions in Eqs. (10.26) and (10.27) and assuming that $\Delta x = \Delta y$, we obtain the finite difference approximation to Laplace's equation

$$h_{i-1,j} + h_{i+1,j} + h_{i,j-1} + h_{i,j+1} - 4h_{i,j} = 0 \tag{10.28}$$

The generalized form of Eq. (10.28) is the most widely used equation in finite difference solutions of steady-state problems. There will be one equation of the form of Eq. (10.28) for each interior point (i, j) of the problem.

Explicit Finite Difference Approximation

For transient conditions, the head in an aquifer is a function of time; therefore, in addition to the spatial finite difference approximation for head, a finite difference approximation for $\partial h / \partial t$ is also needed. A *forward* or *explicit* difference approximation is given by

$$\frac{\partial h}{\partial t} \approx \frac{h_{i,j}^{n+1} - h_{i,j}^{n}}{\Delta t} \tag{10.29}$$

where n and $n + 1$ represent two conservative time levels. Similarly, a backward difference approximation is given by

$$\frac{\partial h}{\partial t} \approx \frac{h_{i,j}^{n} - h_{i,j}^{n-1}}{\Delta t} \tag{10.30}$$

Finally, a central difference approximation in time is given by

$$\frac{\partial h}{\partial t} \approx \frac{h_{i,j}^{n+1} - h_{i,j}^{n-1}}{2\Delta t} \tag{10.31}$$

The central difference approximation was found to be unconditionally stable by Remson et al. (1971) and therefore should be avoided. For the case of transient flow given by

$$\frac{\partial^2 h}{\partial x^2} + \frac{\partial^2 h}{\partial y^2} = \frac{S}{T} \frac{\partial h}{\partial t} \tag{10.32}$$

the explicit approximation yields a stable solution if the value of the ratio $(T\Delta t)/[S(\Delta x)^2]$ is kept sufficiently small. In the one-dimensional case, where flow occurs only in the x direction, the parameter $(T\Delta t)/[S(\Delta x)^2]$ must be ≤ 0.5 (Remson et al., 1971) to ensure numerical stability. For the two-dimensional case, where $\Delta x = \Delta y$, $(T\Delta t)/[S(\Delta x)^2]$ must be ≤ 0.25 (Rushton and Redshaw, 1979).

Implicit Finite Difference Approximation

An *implicit* (or *backward*) difference formulation is one for which the heads h are evaluated at a time between n and $n + 1$. This is done by using a weighted average of the approximations at n and $n + 1$. The weighting parameter is represented by

α, and it lies between 0 and 1. If the time step $n + 1$ is weighted by α and time step n is weighted by $(1 - \alpha)$, then

$$\frac{\partial^2 h}{\partial x^2} \approx \alpha \frac{h_{i+1,j}^{n+1} - 2h_{i,j}^{n+1} + h_{i-1,j}^{n+1}}{(\Delta x)^2} + (1 - \alpha) \frac{h_{i+1,j}^n - 2h_{i,j}^n + h_{i-1,j}^n}{(\Delta x)^2} \tag{10.33}$$

A similar expression is written for $\partial^2 h / \partial y^2$. The parameter α is selected by the modeler. For $\alpha = 1$, the space derivatives are approximated at $n + 1$, and the finite difference scheme is said to be fully implicit. If a value of 0.5 is selected for α, then the scheme is referred to as the *Crank–Nicolson method*.

Alternating Direction Implicit (ADI)

The derivation and solution of the finite difference equation and the use of the iterative ADI have been discussed extensively by Pinder and Bredehoeft (1968), Prickett and Lonnquist (1971), and Trescott et al. (1976). In general, the basis of the ADI method is to obtain a solution to the flow equation by alternately writing the finite difference equation, first implicitly along columns and explicitly along rows, and then vice versa. To reduce the errors that may result from the ADI method, an iterative procedure is added so that, within a single time step, the solution will converge within a specified error tolerance. The ADI method will be illustrated by approximating the two-dimensional transient equation for a confined aquifer:

$$\frac{\partial^2 h}{\partial x^2} + \frac{\partial^2 h}{\partial y^2} = \frac{S}{T} \frac{\partial h}{\partial t} \tag{10.34}$$

where S is the storage coefficient and T is the transmissivity. Assuming $\Delta x = \Delta y = a$, the fully implicit finite difference approximation is

$$h_{i+1,j}^{n+1} + h_{i-1,j}^{n+1} + h_{i,j+1}^{n+1} + h_{i,j-1}^{n+1} - 4h_{i,j}^{n+1} = \frac{Sa^2}{T} \frac{h_{i,j}^{n+1} - h_{i,j}^n}{\Delta t} \tag{10.35}$$

In the first step of ADI, Eq. (10.35) is rewritten such that heads along columns are on one side of the equation and heads along rows are on the other side (also referred to as rewriting the equation, first implicitly along columns and explicitly along rows), which results in

$$h_{i,j-1}^{n+1} + \left(-4 - \frac{Sa^2}{T\Delta t}\right) h_{i,j}^{n+1} + h_{i,j+1}^{n+1} = -\frac{Sa^2}{T\Delta t} h_{i,j}^n - h_{i+1,j}^n - h_{i-1,j}^n \tag{10.36}$$

Equation (10.36) will yield a tridiagonal coefficient matrix (one that has non-zero entries only along the three center diagonals) along any column (see Example 10.2). The second step of ADI involves rewriting Eq. (10.35) implicitly along rows and explicitly along columns:

$$h_{i-1,j}^{n+1} + \left(-4 - \frac{Sa^2}{T\Delta t}\right) h_{i,j}^{n+1} + h_{i+1,j}^{n+1} = -\frac{Sa^2}{T\Delta t} h_{i,j}^n - h_{i,j+1}^n - h_{i,j-1}^n \tag{10.37}$$

The explicit approximation along columns uncouples one row from another. Therefore, Eq. (10.37) will also generate a set of matrix equations (one for each interior row) with tridiagonal coefficient matrices. Alternating the explicit approximation between column and rows is an attempt to compensate for errors generated in either direction.

Iterative Methods

A set of simultaneous finite difference equations could be solved directly; however, in problems having a large number of nodes, simultaneous solutions are impractical. Instead, iterative procedures can be used where an initial guess of the solution is made. Further improvements on the initial guess are then calculated. There are three commonly used iterative techniques: the *Jacobi iteration*, *Gauss–Seidel iteration*, and *successive over relaxation* (SOR). Of the three, Jacobi iteration is the least efficient and successive over relaxation is the most efficient.

Jacobi Iteration

If Eq. (10.28) were solved for $h_{i,j}$, then

$$h_{i,j} = \frac{h_{i-1,j} + h_{i+1,j} + h_{i,j-1} + h_{i,j+1}}{4} \tag{10.38}$$

The value of $h_{i,j}$ at any point is the average value of head computed from its four nearest neighbors.

The Jacobi iteration associates an *iteration index* (m) with the finite difference equation for head:

$$h_{i,j}^{m+1} = \frac{h_{i-1,j}^m + h_{i+1,j}^m + h_{i,j-1}^m + h_{i,j+1}^m}{4} \tag{10.39}$$

For $m = 1$, an initial guess of $h_{2,2}^1$, $h_{3,2}^1$, $h_{2,3}^1$, and $h_{3,3}^1$ is made. Equation (10.39) is used to calculate head values for $m = 2, 3, \ldots, n$. Iteration continues until the solution converges to the preset error tolerance level; that is, the difference between the answers for $m = n$ and $m = n + 1$ is less than the convergence criterion.

Gauss–Seidel Iteration

The Gauss–Seidel iteration formula is

$$h_{i,j}^{m+1} = \frac{h_{i-1,j}^{m+1} + h_{i,j-1}^{m+1} + h_{i+1,j}^m + h_{i,j+1}^m}{4} \tag{10.40}$$

Successive over Relaxation (SOR)

The SOR iteration is tied to the residual (c) or change between two successive Gauss–Seidel iterations:

$$c = h_{i,j}^{m+1} - h_{i,j}^m \tag{10.41}$$

The Gauss–Seidel iteration eliminates or *relaxes* the residual (c) by replacing $h_{i,j}^m$ with $h_{i,j}^{m+1}$ after each calculation. In the SOR method, on the other hand, the residual is multiplied by a relaxation factor ω, where $\omega \geq 1$. The value of $h_{i,j}^{m+1}$ is given by

$$h_{i,j}^{m+1} = h_{i,j}^m + \omega c \tag{10.42}$$

A value of ω between 1 and 2 has been recommended (Wang and Anderson, 1982).

Example 10.2 TRIDIAGONAL MATRICES

Consider the one-dimensional transient flow equation

$$\frac{\partial^2 h}{\partial x^2} = \frac{S}{T}\frac{\partial h}{\partial t} \tag{10.43}$$

The implicit or backward finite difference approximation, where the space derivative is evaluated at the $(n + 1)$ time level, is (Wang and Anderson, 1982)

$$\frac{h_{i-1}^{n+1} - 2h_i^{n+1} + h_{i+1}^{n+1}}{(\Delta x)^2} = \frac{S}{T}\frac{h_i^{n+1} - h_i^n}{\Delta t} \tag{10.44}$$

Assume that we have a problem domain with six nodes where the first and last nodes are boundary nodes of known head. We wish to write the set of algebraic equations that would be generated by applying Eq. (10.44) to these nodes, and we wish to write it in matrix form.

Solution. First, we rearrange Eq. (10.44) and put unknowns, that is, heads at the $(n + 1)$ time level, on the left side and put knowns on the right side.

$$h_{i-1}^{n+1} + \left(-2 - \frac{S(\Delta x)^2}{T\,\Delta t}\right)h_i^{n+1} + h_{i+1}^{n+1} = -\frac{S(\Delta x)^2}{T\,\Delta t}\,h_i^n \tag{10.45}$$

If the head values h_1 and h_6, which are known from the boundary conditions, are transferred to the right-hand side, then the matrix form of the set of algebraic equations for the six-node problem is

$$
\begin{bmatrix}
-2 - \dfrac{S(\Delta x)^2}{T\,\Delta t} & 1 & 0 & 0 \\[2ex]
1 & -2 - \dfrac{S(\Delta x)^2}{T\,\Delta t} & 1 & 0 \\[2ex]
0 & 1 & -2 - \dfrac{S(\Delta x)^2}{T\,\Delta t} & 1 \\[2ex]
0 & 0 & 1 & -2 - \dfrac{S(\Delta x)^2}{T\,\Delta t}
\end{bmatrix}
\left\{
\begin{array}{c}
h_2^{n+1} \\[2ex]
h_3^{n+1} \\[2ex]
h_4^{n+1} \\[2ex]
h_5^{n+1}
\end{array}
\right\}
=
\left\{
\begin{array}{c}
-\dfrac{S(\Delta x)^2}{T\,\Delta t}\,h_2^n - h_1 \\[2ex]
-\dfrac{S(\Delta x)^2}{T\,\Delta t}\,h_3^n \\[2ex]
-\dfrac{S(\Delta x)^2}{T\,\Delta t}\,h_4^n \\[2ex]
-\dfrac{S(\Delta x)^2}{T\,\Delta t}\,h_5^n - h_6
\end{array}
\right\}
$$

The coefficient matrix has nonzero entries only along the three center diagonals. This type of matrix is known as a tridiagonal matrix, which is easily solved.

Example 10.3 FINITE DIFFERENCE SOLUTION OF THE DIFFUSION EQUATION

A finite difference approximation to the equation that describes steady-state diffusion of a dissolved substance into an aquifer in which a first-order reaction occurs has been derived by Celia and Gray (1992) as follows:

$$D\frac{d^2u}{dx^2} - Ku = 0, \qquad 0 < x < 1 \text{ cm}$$

$$u(0) = 0 \qquad\qquad\qquad (10.46)$$

$$u(1) = C_1$$

where $u(x)$ (mg/L) is the concentration of the dissolved substance, D (ft^2/sec) is the diffusion coefficient, K (1/sec) is the reaction rate, and C_1 (mg/L) is the concentration at the right boundary. The coefficients D, K, and C_1 are constants that will be assigned the following values in this example: $D = 0.01$ cm^2/sec, $K = 0.1$ sec^{-1}, and $C_1 = 1.0$ g/cm^3.

The first step in deriving an approximate solution to Eq. (10.46) is the discretization step. Initially, three nodes are chosen, one at each boundary point and a third at $x = 0.5$. Thus, three discrete values, U_1, U_2, U_3, will be computed to approximate the true solution values: $[u(0), u(0.5), u(1)] = (u_1, u_2, u_3)$. Since three unknowns are to be determined, three algebraic equations are needed. Two equations can be obtained directly from the boundary conditions:

$$U_1 = 0 \qquad\qquad\qquad (10.47)$$

$$U_3 = C_1 \qquad\qquad\qquad (10.48)$$

The finite difference approximation to the exact solution is required to exactly satisfy the first-type boundary (or Dirichlet) conditions. For all other nodes (in this case, only node 2), finite difference equations are written. Because the governing equation must hold at all points in the region $0 < x < 1$, it must be the case that the equation holds at $x = x_2 = 0.5$. Thus,

$$D\frac{d^2u}{dx^2}\Big|_{x_2} - Ku\Big|_{x_2} = 0$$

Taylor series expansion for the second derivative leads to the equality

$$D\left[\frac{u_1 - 2u_2 + u_3}{(\Delta x)^2} + \text{T.E.}\right] - Ku_2 = 0 \qquad\qquad (10.49)$$

where T.E. is the truncation error, $u|_{x_2}$ is represented without error as u_2, and the constant spacing is denoted by Δx, with $\Delta x = 0.5$. Equation (10.49) involves exact nodal values u_j ($j = 1, 2, 3$). To write the appropriate finite difference approximation, the truncation error terms are neglected, resulting in the finite difference equation

$$D\frac{U_1 - 2U_2 + U_3}{(\Delta x)^2} - KU_2 = 0 \qquad\qquad (10.50)$$

(Note that for this problem $U_1 = u_1$ and $U_3 = u_3$ while $U_2 \approx u_2$.) Equation (10.50) is the algebraic equation used to solve for the nodal (finite difference) approximations U_j ($j = 1, 2, 3$). Combination of Eqs. (10.47), (10.48), and (10.49) leads to

$$\begin{bmatrix} 1 & 0 & & 0 \\ \dfrac{D}{(\Delta x)^2} & -K - \dfrac{2D}{(\Delta x)^2} & \dfrac{D}{(\Delta x)^2} \\ 0 & 0 & & 1 \end{bmatrix} \begin{bmatrix} U_1 \\ U_2 \\ U_3 \end{bmatrix} = \begin{bmatrix} 0 \\ 0 \\ C_1 \end{bmatrix} \qquad (10.51)$$

The solution of this set of equations is $U_1 = 0$, $U_2 = C_1 D/[2D + K(\Delta x)^2]$, and $U_3 = C_1$. Using values of $D = 0.01$, $K = 0.1$, $C_1 = 1$, and $\Delta x = 0.5$, the solution is $U_2 = 0.222$. This compares to the exact solution $u(0.5) = 0.171$.

If four nodes are chosen instead such that $\Delta x = 1/3$, then the boundary conditions produce

$$U_1 = 0 \qquad (10.52)$$

$$U_4 = 1 \qquad (10.53)$$

and the two interior finite difference equations must be written, one corresponding to each interior node. The finite difference approximations analogous to Eq. (10.50) are

$$D\frac{U_1 - 2U_2 + U_3}{(\Delta x)^2} - KU_2 = 0 \qquad (10.54)$$

$$D\frac{U_2 - 2U_3 + U_4}{(\Delta x)^2} - KU_3 = 0 \qquad (10.55)$$

Given $\Delta x = 1/3$ and the previous values of D, K, and C_1, the approximation step produces the following set of linear algebraic equations:

$$\begin{bmatrix} 1 & 0 & 0 & 0 \\ 0.09 & -0.28 & 0.09 & 0 \\ 0 & 0.09 & -0.28 & 0.09 \\ 0 & 0 & 0 & 1 \end{bmatrix} \begin{bmatrix} U_1 \\ U_2 \\ U_3 \\ U_4 \end{bmatrix} = \begin{bmatrix} 0 \\ 0 \\ 0 \\ 1 \end{bmatrix} \qquad (10.56)$$

The solution step then produces the approximate solution $U = (0, 0.094, 0.291, 1)$; this compares to the analytical solution $(0, 0.083, 0.320, 1)$. It can be seen that using four nodes improved the accuracy of the approximate solution when compared to using three nodes.

Example 10.4   GAUSS–SEIDEL ITERATIVE SOLUTION OF LAPLACE'S EQUATION

The following FORTRAN program has been written to solve Laplace's equation [Eq. (10.24)] using the Gauss–Seidel iterative method. Given the head values along the boundaries shown next, the program is used to solve for the values of head at the interior points of the grid as shown in Table 10.1.

20.00	20.00	20.00	20.00	20.00	20.00	20.00	20.00	20.00	20.00
19.55									20.00
19.07									20.00
18.55									20.00
17.95									20.00
17.22									20.00
16.29									20.00
14.95									20.00
12.68									20.00
7.68	12.68	14.95	16.29	17.22	17.92	18.55	19.07	19.55	20.00

TABLE 10.1 Computed Head Values from Example 10.4

20.00	20.00	20.00	20.00	20.00	20.00	20.00	20.00	20.00	20.00
19.55	19.55	19.58	19.62	19.68	19.74	19.80	19.87	19.93	20.00
19.07	19.09	19.14	19.23	19.34	19.47	19.60	19.73	19.87	20.00
18.55	18.58	18.67	18.82	19.00	19.20	19.40	19.60	19.80	20.00
17.95	18.00	18.16	18.38	18.65	18.92	19.20	19.47	19.74	20.00
17.22	17.32	17.57	17.91	18.28	18.65	19.00	19.35	19.68	20.00
16.29	16.48	16.90	17.41	17.91	18.38	18.82	19.23	19.62	20.00
14.95	15.41	16.15	16.90	17.57	18.16	18.68	19.14	19.58	20.00
12.68	14.04	15.41	16.48	17.32	18.00	18.58	19.09	19.55	20.00
7.68	12.68	14.95	16.29	17.22	17.95	18.55	19.07	19.55	20.00

Source: Bedient and Huber, 1992.

Program Gauss

```
C
C      This program solves Laplace's equation Equation 10.24 using
C      a Gauss-Seidel iterative procedure. The program will
C      solve for the unknown interior values of head in a grid,
C      with I used to denote grid columns and J used to denote
C      grid rows.
C
       REAL H(25,25),MAX,OLDH,ERROR,TOL
       INTEGER N,UNIT,NI,NJ
       CHARACTER*1 ANS
       CHARACTER*4 UNITS
C
       WRITE(*,*)'  GAUSS-SEIDEL SOLUTION OF EQUATION. 10.24
       WRITE(*,*)
       WRITE(*,*)
       WRITE(*,*)'(E)nglish or (M)etric units?'
       READ(*,'(A)')ANS
       UNITS = '(ft)'
       IF ((ANS .EQ. 'M') .OR. (ANS .EQ. '(m)')) UNITS = '(m)'
C
C      Determine size of grid
       WRITE (*,*) 'Enter the number of columns'
       READ (*,*) NI
       WRITE (*,*) 'Enter the number of rows'
       READ (*,*) NJ
C
C      Enter the boundary conditions
       DO 20 J = 1,NJ
       DO 10 I = 1, NI
         IF ((J .EQ. 1) .OR. (J .EQ. NJ)) THEN
           WRITE (*,1000) UNITS,I,J
           READ (*,*) H(I,J)
         ELSEIF ((I .EQ. 1) .OR. (I .EQ. NI)) THEN
           WRITE (*,1000) UNITS,I,J
           READ (*,*) H(I,J)
         ENDIF
10    CONTINUE
```

```
      20 CONTINUE
C
C      Enter initial guess for interior nodes
          WRITE (*,*) 'Enter the initial guess of the head values'.
     & 'at the interior nodes.'
          DO 40 J = 2, NJ-1
            DO 30 I = 2,NI-1
               WRITE (*,1000) UNITS,I,J
               READ (*,*) H(I,J)
      30  CONTINUE
      40 CONTINUE
C
C      Enter error tolerance
          WRITE (*,*) 'Enter the allowable error', UNITS
          READ (*,*) TOL
C
C      Sweep interior points with a 5-point operator, keeping track of
C      the number of iterations needed.
        N = 0
      50 MAX = 0
        N = N + 1
        DO 70 J = 2,NJ-1
          DO 60 I = 2,NI-1
             OLDH = H(I,J)
             H(I,J) = (H(I-1,J) + H(I+1,J)+H(I,J-1) + H(I,J+1))/4.
             ERROR = ABS(H(I,J) − OLDH)
             IF (ERROR .GT. MAX) = ERROR
      60    CONTINUE
      70 CONTINUE
        IF (MAX .GT. TOL) GOTO 50
C
C      Print results on screen
          WRITE(*,*)
          WRITE(*,*)  'The number of iterations needed was ', N
          WRITE(*,*)
        WRITE(*,*)  'The array of head values is:'
        WRITE(*,*)
        DO 80 J = 1,NJ
          WRITE (*,1010) (H(I,J), I = 1,NI)
        80 CONTINUE
1000 FORMAT (1X, 'Enter head' ,A4,'at node',I3,',',I3)
1010 FORMAT (1X, 10F7.2)
        END
```

Example 10.5   DRAWDOWN TO A WELL IN A CONFINED AQUIFER, STEADY STATE

Bedient and Huber (1992) present a program that solves Poisson's equation for drawdown to a well in a confined aquifer under steady-state conditions. Poisson's equation is written as

$$\frac{\partial^2 h}{\partial x^2} + \frac{\partial^2 h}{\partial y^2} = -\frac{R(x, y)}{T} \tag{10.57}$$

where $R(x, y)$ is the rate of pumping. A well is located in the lower-left corner of a 10 by 10 grid. The well pumps water at a rate of 10,000 ft^3/day. The transmissivity

TABLE 10.2 Computed Drawdowns and Heads for Example 10.5

The number of iterations made was 65.

The drawdown (ft, m) array is:

0.00	0.00	0.00	0.00	0.00	0.00	0.00	0.00	0.00	0.00
0.45	0.44	0.41	0.35	0.28	0.18	0.00	0.00	0.00	0.00
0.93	0.90	0.83	0.73	0.59	0.42	0.22	0.00	0.00	0.00
1.45	1.41	1.30	1.13	0.93	0.70	0.46	0.22	0.00	0.00
2.05	1.99	1.82	1.58	1.30	1.00	0.70	0.42	0.18	0.00
2.78	2.67	2.41	2.06	1.68	1.30	0.93	0.59	0.28	0.00
3.71	3.51	3.08	2.57	2.06	1.58	1.13	0.73	0.35	0.00
5.05	4.59	3.84	3.08	2.41	1.82	1.30	0.83	0.41	0.00
7.32	5.96	4.59	3.51	2.67	1.99	1.41	0.90	0.44	0.00
12.32	7.32	5.05	3.71	2.78	2.05	1.45	0.93	0.45	0.00

An option of the program allows us to compute head instead of drawdown. The following values are obtained:

The head (ft, m) array is:

20.00	20.00	20.00	20.00	20.00	20.00	20.00	20.00	20.00	20.00
19.55	19.56	19.59	19.65	19.72	19.82	20.00	20.00	20.00	20.00
19.07	19.10	19.17	19.27	19.41	19.58	19.78	20.00	20.00	20.00
18.55	18.59	18.70	18.87	19.07	19.30	19.54	19.78	20.00	20.00
17.95	18.01	18.18	18.42	18.70	19.00	19.30	19.58	19.82	20.00
17.22	17.33	17.59	17.94	18.32	18.70	19.07	19.41	19.72	20.00
16.29	16.49	16.92	17.43	17.94	18.42	18.87	19.27	19.65	20.00
14.95	15.41	16.16	16.92	17.59	18.18	18.70	19.17	19.59	20.00
12.68	14.04	15.41	16.49	17.33	18.01	18.59	19.10	19.56	20.00
7.68	12.68	14.95	16.29	17.22	17.95	18.55	19.07	19.55	20.00

Source: Bedient and Huber, 1992.

of the aquifer is 500 ft²/day and the initial head is 20 ft. The zone of influence of the pumping extends 1000 ft radially beyond the well. Using a value of 1.8 for ω, an error tolerance of 0.001, and 200 iterations, drawdown near the well can be determined with the model. The computed drawdowns and heads are shown in Table 10.2.

Example 10.6 DRAWDOWN TO A WELL, TRANSIENT CONDITIONS

A FORTRAN program for solving the governing equation for drawdown to a well in a confined aquifer under transient conditions is given by Bedient and Huber (1992). First, the transient flow equation given by

$$\frac{\partial^2 h}{\partial x^2} + \frac{\partial^2 h}{\partial y^2} = \frac{S}{T}\frac{\partial h}{\partial t} - \frac{R(x, y, t)}{T}$$

can be written in finite difference form as (Wang and Anderson, 1982)

$$\alpha \left(\overline{h}_{i,j}^{n+1} - h_{i,j}^{n+1} \right) + (1 - \alpha) \left(\overline{h}_{i,j}^{n} - h_{i,j}^{n} \right)$$
$$= \frac{a^2 S}{4T}\frac{h_{i,j}^{n+1} - h_{i,j}^{n}}{\Delta t} - \frac{a^2 R_{i,j}^{n}}{4T} \tag{10.58}$$

where $\overline{h}_{i,j}^{n+1} = 1/4[h_{i-1,j}^{n} + h_{i+1,j}^{n} + h_{i,j-1}^{n} + h_{i,j+1}^{n}]$. The iterative form of Eq. (10.58) is

TABLE 10.3   Computed Drawdown for Example 10.6

Time = 6.96
The drawdown array is:

0.88	0.88	0.87	0.85	0.82	0.80	0.78	0.76	0.75	0.74
0.90	0.89	0.88	0.86	0.83	0.81	0.78	0.76	0.75	0.75
0.93	0.93	0.91	0.89	0.86	0.83	0.80	0.78	0.76	0.76
1.00	0.99	0.97	0.93	0.90	0.86	0.82	0.80	0.78	0.78
1.10	1.09	1.05	1.00	0.95	0.90	0.86	0.83	0.81	0.80
1.24	1.22	1.16	1.09	1.02	0.95	0.90	0.86	0.83	0.82
1.45	1.40	1.30	1.19	1.09	1.00	0.93	0.89	0.86	0.85
1.76	1.65	1.47	1.30	1.16	1.05	0.97	0.91	0.88	0.87
2.32	1.98	1.65	1.40	1.22	1.09	0.99	0.93	0.89	0.88
3.56	2.32	1.76	1.45	1.25	1.10	1.00	0.94	0.90	0.89

Source: Bedient and Huber, 1992.

$$
h_{i,j}^{n+1} = \frac{[\alpha \bar{h}_{i,j}^{n+1}] + \dfrac{a^2 S}{4T \, \Delta t} h_{i,j}^n + (1 - \alpha)\left[\bar{h}_{i,j}^n - h_{i,j}^n + \dfrac{a^2 R_{i,j}^n}{4T}\right]}{\dfrac{a^2 S}{4T \, \Delta t} + \alpha} \tag{10.59}
$$

Assume that flow to the well in the aquifer described in Example 10.5 is transient. The well is pumping at a rate of 2500 ft³/day. If the storativity of the aquifer is 0.005, the drawdown near the well after 1 week can be determined. A starting time period of 0.05 day, a factor of 1.4 to increase the time increment (12 time iterations will equal about 7 days), the Crank–Nicolson scheme with $\alpha = 0.5$, a tolerance of 0.001, and no more than 200 iterations in the drawdown computation will be used for each time step. The computed drawdowns from the model are shown in Table 10.3. The solution can be compared against the analytical Theis method at various points in time. It can be shown that the selection of Δx, Δy, Δt, and α will strongly influence the accuracy of the numerical results.

For the explicit case, if Δt is too large, the scheme becomes unstable and yields useless answers. Boundary conditions and choice of error tolerance in the iterative method also contribute to numerical errors. Results in Table 10.4 show the comparison of a numerical example with $Q = 1000$ m³/day, $T = 4.50$ m²/day, and $S = 0.0005$ for various values of Δt and α. Values of $\Delta x = \Delta y = 100$ m were chosen.

10.4 FINITE ELEMENT METHODS

A difficult problem in flow through porous media involves *sharp fronts*. A sharp front refers to a large change in a dependent variable over a small distance. Sharp-front problems are encountered in both miscible (advective–dispersive flow) and immiscible (multifluid and multiphase flow) problems. The most common complaint about low-order finite difference methods applied to sharp-front problems is that the computed front is "smeared out" (Mercer and Faust, 1977). The process by which the front becomes smeared is referred to as *numerical dispersion*.

In general, for linear problems, the finite element method can track sharp fronts more accurately, which reduces considerably the numerical diffusion prob-

TABLE 10.4 Comparison of Drawdown at 100 m from a Well Using Numerical Solutions (Example 10.6) with Selected α and Theis Analytical Solutions

Time (days)	Δt	$\alpha = 0$	$\alpha = 0.5$	$\alpha = 1$	Theis
		\multicolumn{4}{c}{Drawdown in feet for:}			
0.05	0.05	0.00	0.19 (4)	0.33 (4)	0.011
0.12	0.07	0.68	1.03 (4)	1.28 (5)	0.56
0.24	0.11	2.66	2.98 (5)	3.20 (5)	3.00
0.41	0.17	6.57	6.41 (5)	6.37 (6)	6.90
0.66	0.25	12.35	11.40 (6)	10.85 (8)	11.82
1.04	0.38	18.95	17.54 (7)	16.43 (10)	17.56
1.61	0.57	25.90	24.26 (9)	22.76 (12)	23.79
2.46	0.85	32.60	31.21 (11)	29.52 (16)	30.30
3.74	1.28	42.29	38.25 (14)	36.48 (21)	37.06

Source: Bedient and Huber, 1992.

The numbers in parentheses indicate the number of iterations used.

lem. The finite element method, however, has several numerical problems, which include numerical oscillation, instability, and large computation time requirements, as will be seen later. The finite element analysis of a physical problem can be described as follows (Huyakorn and Pinder, 1983):

1. The physical system is subdivided into a series of finite elements that are connected at a number of nodal points. Each element is identified by its element number and the lines connecting the nodal points situated on the boundaries of the element.
2. A matrix expression, known as the *element matrix,* is developed to relate the nodal variables of each element. The element matrix may be obtained via a mathematical formulation that makes use of either a variational or weighted residual method.
3. The element matrices are combined or assembled to form a set of algebraic equations that describe the entire system. The coefficient matrix of this final set of equations is called the global matrix.
4. Boundary conditions are incorporated into the global matrix equation.
5. The resulting set of simultaneous equations is solved using a variety of techniques, such as Gauss elimination.

Discretizing the Problem Domain

Discretizing the physical system using finite elements is achieved by replacing the domain with a collection of *nodes* (or *nodal points*) and *elements* referred to as the *finite element mesh* (Figure 10.7). There are different element types for one-, two-. or three-dimensional problems (Figure 10.8). Elements may be of any size, the size and shape of each element in the mesh can be different, and several different types of elements can be used in a single mesh.

When preparing the finite element mesh, it is important to remember that

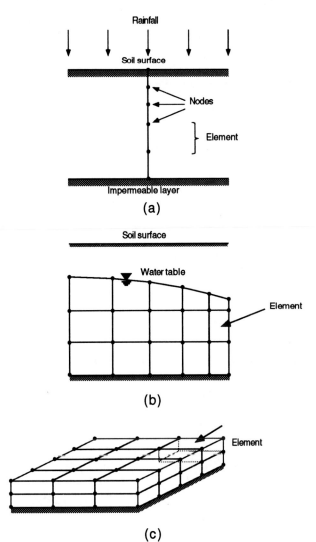

Rainfall

Soil surface

Nodes

Element

Impermeable layer

(a)

Soil surface

Water table

Element

(b)

Element

(c)

Figure 10.7 Discretization of (a) one-, (b) two-, and (c) three-dimensional problem domains. Source: Istok, 1989. © American Geophysical Union.

the precision of the solution obtained and the level of computational effort required to obtain a solution will be determined to a great extent by the number of nodes in the mesh. A *coarse mesh* has a smaller number of nodes and will give a lower precision than a *fine mesh*. However, the larger the number of nodes in the mesh is, the greater will be the required computational effort and cost. It is usually recommended to start with a coarse mesh and refine it in later stages of the modeling in those parts of the mesh where it might be necessary to do so.

Developing the Element Matrix

As mentioned in step 2, formulation of the element matrix equations can be achieved using either *variational* methods or *weighted residual* techniques. The variational method will not be discussed in this book. More emphasis will be placed

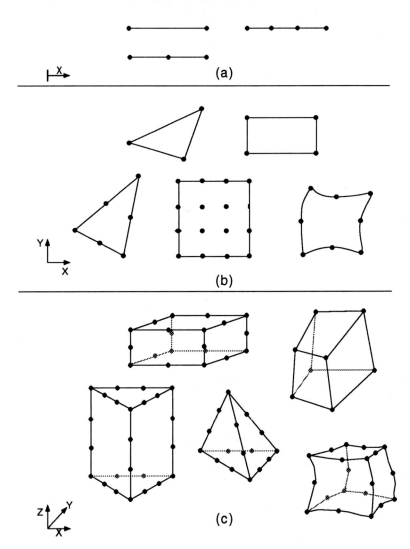

Figure 10.8 Some types of finite elements. (a) One-dimensional elements. (b) Two-dimensional elements. (c) Three-dimensional elements.

on weighted residual techniques, specifically the Galerkin approach, because it is the most commonly used technique for simulating problems of ground water flow and contaminant transport.

Galerkin's method is a special case of the method of weighted residuals. The method of weighted residuals can be described as follows. Consider a problem governed by the differential equation

$$L(u) - f = 0 \qquad (10.60)$$

in the region R enclosed by the boundary B. The unknown function u can be approximated by a trial function $\hat{u}$ of the form

$$\hat{u} = \sum_{I=1}^{n} N_I C_I \qquad (10.61)$$

where N_I are linearly independent *basis functions* defined over the entire solution domain and C_I are the unknown parameters to be determined. Generally, the n basis functions are selected in such a way that all essential boundary conditions are satisfied.

The trial function $\hat{u}$ is only an approximation; hence, substituting Eq. (10.60) in (10.61) results in an error or residual:

$$\epsilon = L(\hat{u}) - f \qquad (10.62)$$

The method of residuals attempts to determine the parameters C_I in such a way that the error is minimized in some specified manner. This is accomplished by forming a weighted integral of ϵ over the entire solution domain and then setting this integral or weighted residual to zero:

$$\int_R W_I \epsilon\, dR = \int_R W_I[L(\hat{u}) - f]\, dR = 0, \qquad \text{for } I = 1, 2, \ldots, n \qquad (10.63)$$

Once the functions (W_I) are specified, a simultaneous set of equations in the n unknowns (C_I) results from combining Eqs. (10.63) and (10.61). The final step in this procedure is to solve the equations for C_I and obtain an approximate representation of the unknown function u.

In the Galerkin method, the weighting functions are chosen to be identical to the basis functions; that is, $W_I = N_I$. The weighted residual equation then becomes

$$\int_R N_I \epsilon\, dR = 0 \qquad (10.64)$$

or

$$\int_R N_I[L(\hat{u}) - f]\, dR = 0, \qquad I = 1, 2, \ldots, n \qquad (10.65)$$

Numerical Integration

In forming the element matrix equation, it is necessary to evaluate integrals over each element subdomain. Several types of numerical integration procedures are available. The most common procedure and the one that will be discussed in this section is the *Gaussian quadrature* for lines, squares, and cubes and other quadrature formulas for triangles and tetrahedra.

Using the Gaussian quadrature, a definite integral of the function f is approximated by the weighted sum of values of f at selected points. The Gaussian quadrature for the definite integral $\int_{-1}^{1} f(\epsilon)d(\epsilon)$ is given by

$$\int_{-1}^{1} f(\epsilon)d(\epsilon) = \sum_{i=1}^{m} W_i f(\epsilon_i) \qquad (10.66)$$

where $f(\epsilon_i)$ is the value of the function f at the Gaussian point ϵ_i, m is the number of Gaussian points, and W_i $(i = 1, 2, \ldots, m)$ are the weighting coefficients.

10.5 METHOD OF CHARACTERISTICS

The method of characteristics (MOC) was developed by Garder et al. in 1964 mainly to overcome the numerical dispersion problem resulting from solving the advection–dispersion equation with conventional finite difference techniques. The MOC has been widely used for simulating the transport of miscible compounds in ground water (Reddell and Sunada, 1970; Bredehoeft and Pinder, 1973; Konikow and Bredehoeft, 1974 and 1978).

The method of characteristics will be illustrated using the one-dimensional form of the transport equation for a conservative tracer:

$$D_x \frac{\partial^2 C}{\partial x^2} - V \frac{\partial C}{\partial x} = \frac{\partial C}{\partial t} \tag{10.67}$$

where C is the tracer concentration, V is the velocity, D_x is the coefficient of hydrodynamic dispersion, and t is the time.

Equation (10.67) approaches a hyperbolic equation as the dispersion term becomes small with respect to the advection term. According to MOC, a simplified set of equations in terms of an arbitrary curve parameter is associated with the hyperbolic equation (10.67). Therefore, Eq. (10.67) can be simplified into the following system of ordinary differential equations (ODEs):

$$\frac{dx}{dt} = V \tag{10.68}$$

$$\frac{dC}{dt} = D_x \frac{\partial^2 C}{\partial x^2} \tag{10.69}$$

The solutions of the simplified set of equations are called the characteristic curves of the differential equation. The numerical procedure proposed by Garder et al. (1964) involves both a stationary grid and a set of moving points. The stationary grid is obtained by subdividing the axis into intervals such that

$$x_i = i \, \Delta x, \quad \text{for} \quad i = 0, 1, \ldots, m - 1 \tag{10.70}$$

A set of moving points or representative fluid particles with density of P points per grid interval is introduced into the numerical solution. The rate of change in concentration in the ground water is observed in the aquifer when moving with the fluid particle.

The location of each moving point is specified by its coordinate in the finite difference grid. Initially, the moving points are uniformly distributed throughout the grid system. The initial concentration assigned to each point is the initial concentration associated with the interval containing the point. At each time interval, the moving points in the system are relocated in the flow field in proportion to the flow velocity at their respective location:

$$x_{p,n+1} = x_{p,n} + \Delta t_{n+1/2} \cdot V \qquad (10.71)$$

where $x_{p,n+1}$ and $x_{p,n}$ are the locations of particle p at time $n + 1$ and n, respectively. After moving each point, the coordinates for the points are examined to determine which interval the point lies in. Each interval is assigned a concentration $\overset{*}{C}_{i,n}$ equal to the concentration of all the points that lie in the interval, after having been moved.

Next, the change in concentration due to dispersion is calculated for each interval:

$$\Delta C_{i,n+1/2} = \Delta t_{n+1/2} \cdot D_x \cdot \Delta_x^2 \cdot \overset{*}{C}_{i,n} \qquad (10.72)$$

Each moving point is then assigned a new concentration:

$$C_{p,n+1} = C_{p,n} + \Delta C_{i,n+1/2} \qquad (10.73)$$

All points falling within an interval at a given time undergo the same change in concentration due to dispersion. Finally, the concentrations at the stationary grid points are computed for the new time step:

$$C_{i,n+1} = \overset{*}{C}_{i,n} + \Delta C_{i,n+1/2} \qquad (10.74)$$

This completes the step from t_n to t_{n+1}, and the procedure is repeated for each subsequent time step.

Example 10.7   NUMERICAL TESTING OF MOC

Khaleel and Reddell (1986) tested MOC for four cases: (1) longitudinal dispersion in one-dimensional flow, (2) longitudinal dispersion in two-dimensional flow, (3) longitudinal and lateral dispersion in one-dimensional flow, and (4) longitudinal and lateral dispersion in two-dimensional flow.

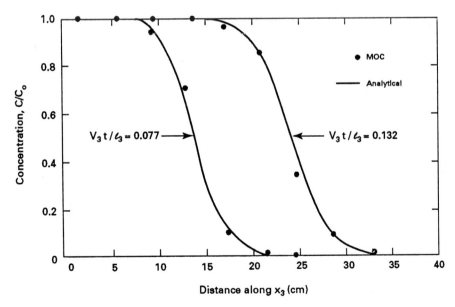

Figure 10.9   Comparison of analytical and numerical solutions in one dimension for MOC.   Source: Khaleel and Reddell, 1986.

For example, results from the one-dimensional MOC solution were compared to the analytical solution for the same problem provided by Ogata and Banks (1961) (see Chapter 6). Khaleel and Reddell's (1986) results are shown in Figure 10.9. As can be seen, there was good agreement between the analytical solution and the MOC numerical solution.

Similar results were obtained for the three other test cases. The reader is referred to the paper for more details. It should be mentioned, however, that Khaleel and Reddell found it necessary to use a coordinate transformation to simulate two-dimensional flow fields more accurately. The coordinate axes were rotated so that an angle of 45° existed between the velocity vector and the transformed coordinate axes.

10.6 NUMERICAL FLOW MODELS

Computer models to simulate saturated ground water flow are typically two or three dimensional. Two-dimensional models may be used to simulate flow in the *xy* plane or flow in a vertical cross section of the subsurface and may simulate unsaturated or saturated water flow. Table 10.5 is a list of some of the available flow models. It is not possible to review every flow model that has been developed in this chapter; however, the detailed presentation of the MODFLOW model that follows illustrates the main concepts involved in flow modeling.

MODFLOW

MODFLOW is a modular three-dimensional finite difference flow code developed by the U.S. Geological Survey (McDonald and Harbaugh, 1988). The model, which simulates saturated flow in three-dimensions, was designed such that the user can select a series of packages or modules (from a total of 10) to be used during a given simulation. Each module or package deals with a specific feature of the hydrologic system that is to be simulated.

MODFLOW can be used to simulate fully three-dimensional systems and quasi-three-dimensional systems in which the flow in aquifers is horizontal and flow through confining beds is vertical. The model can also be used in a two-dimensional mode for simulating flow in a single layer or two-dimensional flow in a vertical cross section. An aquifer can be confined, unconfined, or mixed confined–unconfined. Flow from external stresses, such as flow to wells, areal recharge, evapotranspiration, flow to drains, and flow through riverbeds, can also be simulated.

Aquifer thickness is not specified directly, but is incorporated in the transmissivity distribution specified for each layer (Figure 10.10). For unconfined aquifers, the elevations of the top and bottom of the aquifer are input and used in the model to calculate the saturated thickness based on the location of the water table within the aquifer. Low-conductivity units or clay units are typically not included in the vertical discretization of the system, but rather are included through the use of a conductance term between the upper and lower units separated by the

TABLE 10.5 Saturated Flow Models: Summary Listings

Model Name	Model Description	Model Processes	Author(s)
PLASM (1971)	Finite difference two-dimensional or quasi-three-dimensional, transient, saturated flow model for single- or multilayered confined, leaky confined, or water-table aquifer systems with optional evapotranspiration and recharge from streams	Evapotranspiration, leakage	T. A. Prickett C. G. Lonnquist
AQUIFEM (1979)	Finite element model to simulate transient, areal ground water flow in an isotropic, heterogeneous, confined, leaky-confined or water-table aquifer	Leakage, infiltration	G. F. Pinder C. I. Voss
USGS-3D-FLOW (1982)	Finite difference model to simulate transient, three-dimensional and quasi-three-dimensional, saturated flow in anisotrophic flow in anisotropic, heterogeneous ground water systems	Evapotranspiration, leakage	P. C. Trescott S. P. Larson
USGS-2D-FLOW (1982)	Finite difference model to simulate transient, two-dimensional horizontal or vertical flow in an anisotropic and heterogeneous, confined, leaky-confined, or water-table aquifer	Leakage, evapotranspiration	P. C. Trescott G. F. Pinder S. P. Larson
SWIFT (1982)	Cross-sectional finite element model for transient horizontal flow of salt and fresh water and analysis of upconing of an interface in a homogeneous aquifer	Buoyancy, leakage	A. Verruijt J. B. S. Gan
FE3DGW (1985)	Transient or steady-state, finite element three-dimensional simulation of flow in a large multilayered ground water basin	Leakage, delayed yield, compaction, infiltration	S. K. Gupta C. R. Cole F. W. Bond
AQUIFEM-1 (1979)	Two-dimensional, finite element model for transient, horizontal ground water flow	Leakage	L. R. Townley J. L. Wilson A. S. Costa
MODFLOW (1988)	Modular three-dimensional finite difference ground water model to simulate transient flow in anisotropic, heterogeneous, layered aquifer systems	Evapotranspiration, drainage	M. G. McDonald A. W. Harbaugh
HELP (1987)	Water budget model for the hydrologic evaluation of landfill performance	Surface storage, runoff, infiltration, percolation, evapotranspiration, soil moisture, storage, lateral drainage	P. R. Schroeder J. M. Morgan T. M. Walski A. C. Gibson

(continued)

TABLE 10.5   (continued)

Model Name	Model Description	Model Processes	Author(s)
UNSAT2 (1979)	Two-dimensional finite element model for horizontal, vertical, or axisymmetric simulation of transient flow in a variably saturated, nonuniform, anisotropic porous media	Capillarity, evapotranspiration, plant uptake	S. P. Neuman
FEMWATER/ FECWATER (1987)	Two-dimensional finite element model to simulate transient, cross-sectional flow in saturated–unsaturated anisotropic, heterogeneous porous media	Capillarity, infiltration, ponding	G. T. Yeh D. S. Ward
VS2D (1987)	Two-dimensional finite difference code for the analysis of flow in variably saturated porous media; model considers recharge, evaporation, and plant root update	Evaporation, recharge, plant uptake	E. G. Lappaia R. W. Healy E. P. Weeks

clay. Boundary conditions handled by the model include constant head and no flow and flux boundaries. Simulation time is divided into stress periods, which are in turn divided into time steps.

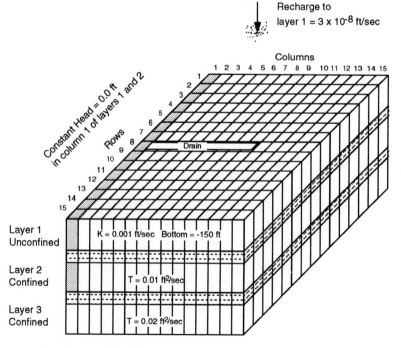

Figure 10.10   Schematic of an aquifer system in MODFLOW.   Source: McDonald and Harbaugh, 1988.

The primary output from the model is head distribution. In addition, a volumetric water budget is provided as a check on the numerical accuracy of the simulation. The model was designed to provide a cell-by-cell flow distribution if required by the user.

Example 10.8 THEIS SOLUTION USING MODFLOW

One of the most widely used analytical techniques by hydrologists is the Theis solution (see Chapter 3). The MODFLOW model will be used in this example to simulate radial flow to a well. The results from MODFLOW will then be compared to the Theis analytical solution (Andersen, 1993).

The MODFLOW model domain for this example is assumed to be uniform, homogeneous, and isotropic. A single layer is used to model a confined aquifer. The potentiometric surface of the aquifer is monitored with time at an observation well 55 m from the pumping well. Model parameters are as follows:

Initial head	0.0 m
Transmissivity	0.0023 m^2/sec
Storage coefficient	0.00075
Pumping rate	4×10^{-3} m^3/sec
Final time	86,400 sec
Number of time steps	20
Time step expansion factor	1.3
Closure criterion	0.0001
Maximum number of iterations	50

Because of the symmetrical nature of the problem, it is not necessary to model the whole aquifer domain. Rather, a quadrant can be simulated with the well located in the lower-left corner of the domain (Figure 10.11). The following grid spacing can be used for discretizing the quadrant:

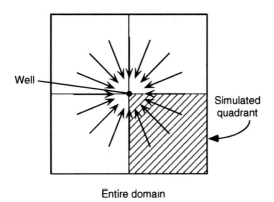

Well

Simulated quadrant

Entire domain

Figure 10.11 Configuration of MODFLOW for simulating radial flow. Arrows denote ground water flow direction.

Row number, i (= column number, j)	Grid spacing in meters (symmetrical along rows and columns)
1	1
2	1.413
3	2
4	2.83
5	4
6	5.65
7	8
8	11.3
9	12
10	14.62
11	20
12	28.3
13	40
14	56.5
15	80
16	110
17	150
18	200
19	252.89

The resulting drawdowns versus time from MODFLOW are tabulated next and plotted in Figure 10.12.

Time step	Time (sec)	Drawdown (m)	
		Analytic	MODFLOW
1	137.1	0.009	0.014
2	315.3	0.044	0.043
3	547.1	0.086	0.079
4	848.6	0.129	0.120
5	1,239.9	0.170	0.160
6	1,748.9	0.210	0.201
7	2,410.7	0.249	0.241
8	3,271.1	0.288	0.280
9	4,389.5	0.326	0.320
10	5,843.4	0.364	0.358
11	7,733.6	0.401	0.397
12	10,190.7	0.438	0.434
13	13,385.1	0.475	0.471
14	17,537.7	0.512	0.508
15	22,936.1	0.549	0.545
16	29,954.0	0.586	0.582
17	39,077.4	0.622	0.619
18	50,937.7	0.659	0.656
19	66,356.1	0.695	0.697
20	86,400.0	0.731	0.738

As can be seen from Figure 10.12, the MODFLOW results compare well with the analytic solution. The numerical results are generally within 0.005 m of the analytic.

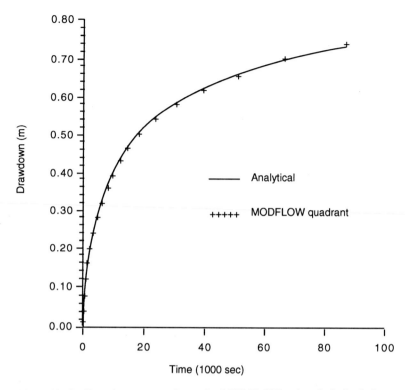

Figure 10.12 Drawdown versus time using MODFLOW and analytical solution.

10.7 NUMERICAL TRANSPORT MODELS

Many numerical transport models have been developed over the past 20 years. These models utilize a variety of numerical techniques and solve various forms of the governing transport equation subject to a certain set of boundary conditions. In addition, these models simulate transport in one, two, or three dimensions; in the saturated or unsaturated zone; and for miscible or immiscible transport. Table 10.6 provides examples of some of these models. The prospective modeler should consult the EPA document *Groundwater Modeling: An Overview and Status Report* (van der Heijde et al., 1988) for additional listings of available models.

Reviewing each transport model listed in Table 10.6 is beyond the scope of this chapter. One class of transport models, however, that is currently evolving and needs to be briefly mentioned is three-dimensional particle tracking models. Two such models, MODPATH (Pollock, 1988, 1989) and PATH3D (Zheng, 1989; see Table 10.6), have been recently developed. Heads obtained from the MODFLOW model are required as input for both. PATH3D has transient modeling abilities, whereas MODPATH is only designed for steady-state simulations. The remainder of the section will focus on reviewing the U.S. Geological Survey two-dimensional transport model in an effort to demonstrate the utility of ground water

TABLE 10.6   Saturated Solute Transport Models: Summary Listings

Model Name	Model Description	Model Processes	Author(s)
SEFTRAN (1985)	Two-dimensional finite element model for simulation of transient flow and transport of heat or solutes in anisotropic, heterogeneous porous media	Advection, dispersion, diffusion, adsorption, decay	P. Huyakom
USGS-2D-TRANSPORT/ MOC (1988)	Simulate transient, two-dimensional, horizontal ground water flow and solute transport in confined–semiconfined aquifers using finite differences and method of characteristics	Advection, conduction, dispersion, diffusion, adsorption	L. F. Konikow J. D. Bredehoeft
MOTIF (1986)	Finite element model for one-, two-, and three-dimensional saturated–unsaturated ground water flow, heat transport, and solute transport in fractured porous media; facilitates single-species radionuclide transport and solute diffusion from fracture to rock matrix	Convection, dispersion, diffusion, adsorption, decay, advection	V. Guvanasen
CFEST (1987)	Three-dimensional finite element model to simulate coupled transient flow, solute- and heat-transport in saturated porous media	Advection, dispersion, diffusion, adsoption, decay	S. K. Gupta C. T. Kinkaid P. R. Meyer C. A. Newbill C. R. Cole
Random Walk (1981)	Simulates one- or two-dimensional, steady–nonsteady flow and solute transport problems in a heterogeneous aquifer with water table and/or confined or semiconfined conditions, using a random-walk technique	Advection, dispersion, diffusion, adsorption, decay, chemical reaction	T. A. Prickett T. G. Naymik C. G. Lonnquist
FEMWASTE/ FECWASTE (1981)	Two-dimensional finite element model for transient simulation of areal or cross-sectional transport of dissolved constitutents for a given velocity field in an anisotropic, heterogeneous porous medium	Capillarity, convection, dispersion, diffusion, adsorption, decay	G. T. Yeh D. S. Ward
SUTRA (1984)	Finite element simulation model for two-dimensional, transient or steady-state, saturated–unsaturated, density-dependent ground water flow with transport of energy or chemically reactive single-species solutes	Capillarity, convection, dispersion, diffusion, adsorption, reactions	C. I. Voss

(continued)

TABLE 10.6   (continued)

Model Name	Model Description	Model Processes	Author(s)
SWIFT (1981)	Three-dimensional finite difference model for simulation of coupled, transient, density-dependent flow and transport of heat, brine, tracers or radionuclides in anisotropic, heterogeneous saturated porous media	Advection, dispersion, diffusion, adsorption, ion exchange, decay, chemical reaction	R. T. Dillon R. M. Cranwell R. B. Lantz S. B. Pahwa M. Reeves
HST3D (1988)	Density-dependent, three-dimensional, finite difference model for simulation of heat and solute transport	Advection, dispersion, diffusion, retardation, decay	K. L. Kipp

contaminant transport models. This model is relatively simple to use and requires a reasonable amount of input data.

U.S. Geological Survey Two-dimensional Solute Transport Model: MOC

The method of characteristics (MOC) is used in the USGS model to solve the solute transport equation (Figure 10.13). To apply the model to a field site, it is necessary to superimpose a block-centered finite difference grid over the site by specifying the number of cells in the x and y directions (note that the y axis should be oriented along the main direction of flow at the site). The values of the various parameters in the model can be uniform over the whole domain or can vary over each cell in the domain.

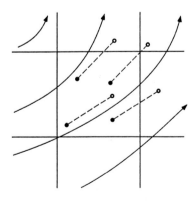

• Initial location of particle
∘ New location of particle
▸ Flow line and direction of flow
--- Computed path of particle

Figure 10.13   Particle tracking in MOC.   Source: Konikow and Bredehoeft, 1978.

The finite difference flow equation is solved numerically in MOC using an iterative alternating-direction implicit (ADI) procedure. The aquifer-specific parameters in MOC include porosity, longitudinal and transverse dispersivity, thickness of the aquifer, transmissivity, and recharge. Several model parameters relate to the numerical methods used in MOC. The reader is referred to Konikow and Bredehoeft (1978) for details on these parameters.

The length of time for which modeling is required is specified in MOC using three parameters: the number of pumping periods in the simulation time, the actual time in years for each pumping period, and the number of time steps in a pumping period. The number of pumping periods is usually more than one in cases where the hydraulic conditions at a site have changed through additional pumping or injection of water into the aquifer. The number of time steps is usually a function of how often the output is desired from the model. The number of time steps has some implications on runtime and numerical errors and can be manipulated for those purposes in some cases.

Source parameters include injection wells, constant-concentration cells, and recharge cells. Injection wells and recharge cells basically define a source that leaks into an aquifer, that is, a source that has a flow rate and a concentration associated with it ($Q = Q_0$ and $C = C_0$). A constant-concentration cell ($C = C_0$ boundary condition) simulates a source that adds contaminant mass at natural gradients into the aquifer. The MOC model allows the specification of up to five observation wells or monitoring wells at a given site. The history of chemical concentrations in these wells is included in the output from the model.

Boundary conditions in MOC are specified by the user. Types of boundary conditions that can be used include constant-head cells or constant-flux cells. A constant-flux boundary can be used to represent aquifer underflow, well withdrawals, or well injection. A constant-head boundary can represent parts of the aquifer where the head will not change in time. Constant-head boundaries are simulated by using a high leakage term (1.0 sec^{-1}). The resulting rate of leakage into or out of the constant-head cell would equal the flux required to maintain the head in the aquifer at the specified altitude. If a constant-flux or constant-head boundary represents a source, the chemical concentration must be specified.

The numerical procedure in MOC requires that a no-flow boundary surround the modeled site. No-flow boundaries simply preclude the flow of water or contaminants across the boundaries of the cell. Initial conditions in the aquifer also have to be specified: the initial water table and the initial contaminant concentrations.

The output from the MOC model consists generally of a head map and a chemical concentration map for each node in the grid. Immediately following the head and concentration maps is a listing of the hydraulic and transport errors. If observation wells had been specified, a concentration history for these wells would be included in the output. The user is encouraged to check the echo of the model input printed in the output and the results from the model run. Model verification for the MOC model has been completed by comparing the model results with two analytical solutions (Konikow and Bredehoeft, 1978).

Example 10.9 USING MOC FOR SIMULATING PUMPING

El-Kadi (1988) examined the applicability of the MOC model to remedial actions by recovery wells. The model was tested in situations dominated by radially convergent and divergent flow around wells. In the first test case, El-Kadi compared the MOC results for a recharge-pumping scenario to the analytical solution of the same problem given by Gelhar and Collins (1971). By neglecting the effects of well radius and molecular diffusion, the concentration at the withdrawal well can be calculated using

$$\frac{C}{C_0} = \frac{1}{2} \operatorname{erfc} \left\{ \frac{V}{\left[\frac{16}{3} \frac{\alpha}{R_1} (2 + |V|^{0.5} V) \right]^{0.5}} \right\}$$ (10.75)

V is given by

$$V = 1 - \frac{V_2}{V_1}$$ (10.76)

with $i = 1$ and 2 representing the indices for recharge and discharge period, respectively, where $V_i = Q_i t_i$ is the recharge or discharge volume of water, Q_i is the recharge or discharge rate, and t_i is the time. In Eq. (10.75), α is the radial dispersivity, and R_1 is given by

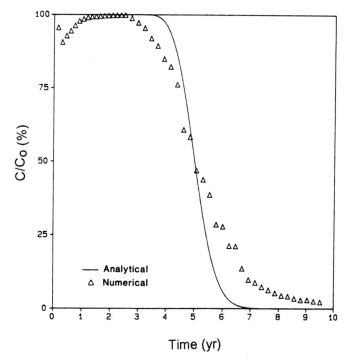

Figure 10.14 Comparison of MOC to an analytical solution: case 1.A. Source: El-Kadi, 1988.

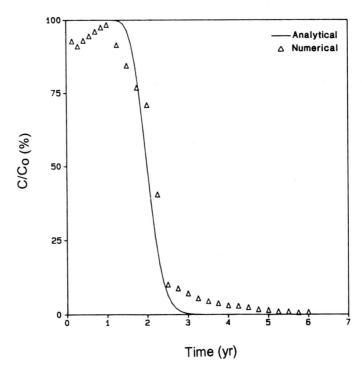

Figure 10.15 Comparison of MOC to an analytical solution: case 1.B.   Source: El-Kadi, 1988.

$$R_1 = \left(\frac{Q_1 t_1}{\pi n B}\right)^{1/2} \tag{10.77}$$

where B is the aquifer thickness and n is porosity.

A number of hypothetical experiments were simulated by El-Kadi (1988), and the results were compared to the analytical solution given by Eq. (10.75). Figures 10.14 through 10.17 illustrate results of the analysis for experiments 1.A through 1.D,

TABLE 10.7   Parameters for Cases 1.A, 1.B, 1.C, and 1.D in Example 10.9

Parameter	Symbol	Value	Units
Saturated conductivity	K	0.005	ft/sec
Aquifer thickness	B	20	ft
Porosity	n	0.30	—
Ratio of longitudinal to transverse dispersivity	α_t/α_l	1	—
Mesh increments in x direction	Δx	900	ft
Mesh increments in y direction	Δy	900	ft
Number of increments in x direction	N_x	9	—
Number of increments in y direction	N_y	11	—
Initial concentration	C_i	0	%
Concentration of injected water	C_o	100	%

Source: El-Kadi, 1988.

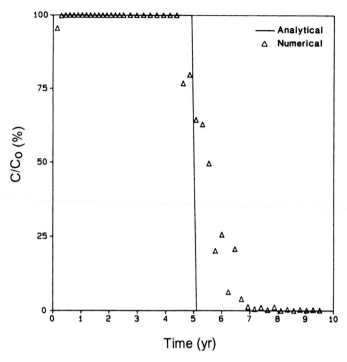

Figure 10.16 Comparison of MOC to an analytical solution: case 1.C.   Source: El-Kadi, 1988.

and the data for the runs are listed in Tables 10.7 and 10.8. As can be seen from Figures 10.14 through 10.17 and despite the severity of the test, the results are reasonable. El-Kadi (1988) attributed the errors to poor representation of radial flow near the well, that is, by inaccuracies in predicting the flow field, rather than inaccuracies in predicting the advection term (or numerical dispersion). The maximum value of error resulting from the four scenarios was about -16%, -22%, -23%, and -13% for cases 1.A, 1.B, 1.C, and 1.D, respectively.

In a second test case, El-Kadi compared MOC results for a recharge–recovery doublet to a semianalytical solution to the purely advective transport case developed by Javandel et al. (1984). Figure 10.18 illustrates the aquifer model for the test problem

TABLE 10.8   Values of α_1, t_1, and $Q_1 = Q_2$ for Cases 1.A, 1.B, 1.C, and 1.D in Example 10.9

Case	α_l (ft)	t_1 (year)	$Q_1 = Q_2$ (ft^3/sec)
1.A	100.	2.5	1.0
1.B	100.	1.0	1.0
1.C	0.001	2.5	1.0
1.D	100.	2.5	0.5

Source: El-Kadi, 1988.

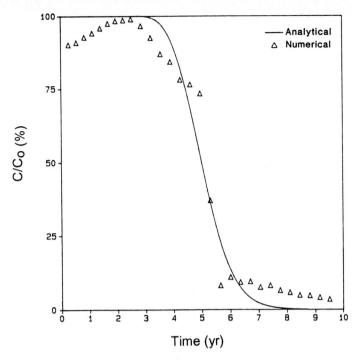

Figure 10.17 Comparison of MOC to an analytical solution: case 1.D. Source; El-Kadi, 1988.

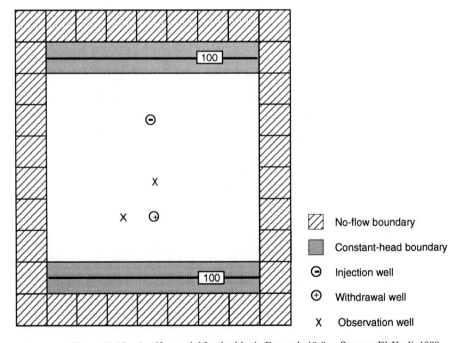

Figure 10.18 Aquifer model for doublet in Example 10.9. Source: El-Kadi, 1988.

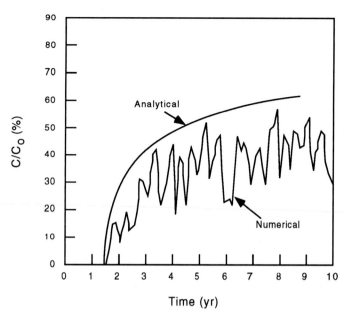

Figure 10.19 Results from model shown in Figure 10.18 and Example 10.9. Source: El-Kadi, 1988.

and the input parameters are listed in Table 10.7. Both longitudinal and transverse dispersion were set to zero. The rate of withdrawal or recharge was set to 1.0 ft^3/sec.

Figure 10.19 compares the concentration as a function of time in the withdrawal well estimated using MOC with that estimated using the analytical model. The data in Figure 10.19 show a reasonable match in the initial years (<2 yr); however, for a time greater than 2 yr, the numerical solution is not accurate and shows large fluctuations. El-Kadi attributes these errors to model weaknesses discussed by Konikow and Bredehoeft (1978). The decline in accuracy is a direct effect of the manner in which concentrations are computed at sink nodes and the method of estimating the mass of contaminant removed from the aquifer at sink nodes during each time increment. The conclusion from this example is that, whereas MOC is a very useful tool for simulating remediation scenarios, the user needs to compare the model results to analytical solutions if possible. Alternatively, the model results are to be used as a guide until field data can be obtained to refine the model.

10.8 APPLYING NUMERICAL MODELS TO FIELD SITES

Good field data are essential when using a model for simulating existing flow and/or contaminant conditions at a site or when using a model for predictive purposes. However, an attempt to model a system with insufficient data may be useful because it may serve as a method for identifying those areas where detailed field information needs to be collected. As mentioned earlier, a good modeling methodology will increase confidence in modeling results. In this section, emphasis is placed on outlining the procedure of designing and applying a selected ground water model to a field site, as described in the modeling protocol established by Anderson and Woessner (1992), which was discussed in Section 10.1.

Model Setup

Once a conceptual model has been developed for the site and a computer model selected, it becomes necessary to interpret the conceptual model and translate it into an input file that can then be used by the model. The interpretation process begins by analyzing the hydrogeologic and water-quality data collected at the site, with the objective of predicting trends in the data and estimating the parameters required to run the model.

Data Collection and Analysis

Most site characterization efforts include identification of the subsurface geology, history of contamination, and water quality at the site. The stratigraphy is determined using soil borings, well logs, and geophysical tools. The subsurface geologic data usually have to be interpreted into values of the hydraulic conductivity and/or transmissivity, thickness of unit, and porosity, which can then be used as input to the model. The elevation of the water table and/or the potentiometric surface measured at discrete monitoring wells can be interpreted by constructing water-level contours to determine the general direction of ground water flow.

Water-quality data collected at specified time intervals from monitoring wells are generally analyzed to determine the trends in the spatial and temporal distributions of chemicals at the site. In many cases, the collected water-quality data for a specific chemical are contoured to determine the extent of the plume of contamination.

An emerging tool in spatial data analysis at field sites is **geostatistics.** The impetus for using geostatistics is the recognition by ground water hydrologists of the extensive level of heterogeneity that the porous formations exhibit and the sparseness of the data collected at waste sites. The goal of geostatistical description is to represent aquifer heterogeneity by a small number of statistical parameters; usually the **mean,** the **variance** and the **correlation length.** Geostatistics can thus be viewed as a set of statistical procedures for describing the correlation of spatially distributed random variables and for performing interpolation and aerial estimation for these variables (Cooper and Istok, 1988).

The most frequently cited aquifer property when dealing with statistical analyses is the hydraulic conductivity. This is because of the paramount importance of the hydraulic conductivity as the primary driving force in flow and transport in the subsurface. The hydraulic conductivity has been described using a lognormal distribution (Freeze and Cherry, 1979). The heterogeneity of the aquifer is described by its **correlation length.** This is a measure of the average distance over which the aquifer properties are correlated (Dagan, 1986). Thus the smaller the correlation length the more heterogeneous the aquifer is.

A common need in modeling is to estimate the hydraulic conductivity at all locations within the model domain given a few field measurements of hydraulic conductivity. **Kriging** is one of the most widely used methods for this purpose. Kriging is a procedure that allows the estimation of the hydraulic conductivity (or other variables) using a linear combination of the measurements. The U.S. EPA has recently developed two geostatistical software programs, GEOPACK (Yates and Yates, 1990) and GEO-EAS (Englund and Sparks, 1991) that can be used for kriging and other statistical data analyses.

For transport models, additional parameters that describe the physical, chemical, and biological properties of the chemicals of concern are necessary. A history of the chemical release (when it happened, how much was released) is an input requirement when we are attempting to simulate an existing contaminant plume. Defining whether a chemical is subject to biotic (or biological) and/or abiotic (chemical) reactions is essential for developing more representative models.

Parameter Estimation

Obtaining the information necessary for modeling is not an easy task. Some data may be obtained from existing reports and studies, but more often it is necessary to rely on field characterization efforts. Transmissivity and the storage coefficient, for example, are typically obtained from pumping test results. Hydrologic stresses include pumping, recharge, and evapotranspiration. Of the three, pumping rates are the easiest to estimate. It is common for modelers to simulate recharge as a spatially uniform rate across the water table equal to some percentage of average annual precipitation. It is important to note, however, that this approach is very simplistic and does not take into account the spatial and temporal variations in recharge rates.

Calibration

Calibrating a model is the process of demonstrating that the model is capable of producing field-measured values of the unknown model variable. For the case of ground water flow, for example, calibration is accomplished by finding a set of parameters, boundary and initial conditions, and stresses that produces simulated values of heads and/or fluxes that match measured values within a specified range of error.

There are two ways to achieve model calibration: (1) manual trial-and-error selection of parameters and (2) automated parameter estimation. In trial-and-error calibration, parameter values are initially assigned to the grid. The initial parameter values are adjusted in sequential model runs to match simulated data to the calibration targets. Trial-and-error calibration may produce nonunique solutions because different combinations of parameters can yield essentially the same results. Also, the trial-and-error process is very subjective and is influenced by the modeler's expertise. Automated parameter estimation is sometimes referred to as solving the *inverse problem*. Very few models actually perform automated calibration. It may take a few more years before the use of these models becomes common practice.

The results of the calibration should be evaluated relative to the measured values both qualitatively and quantitatively. To date, there is no standard protocol for evaluating model calibrations. A qualitative evaluation of the calibration involves comparing trends in the simulated results to those observed from the measured data. For example, a visual comparison could be made between contour maps of measured and simulated heads or concentrations to determine the similarities and differences between them. However, contour maps of field data themselves may include some error introduced by contouring and therefore should not be used as the only evaluation measure of calibration.

A quantitative evaluation of the calibration involves listing the measured and simulated values and determining some average of the algebraic differences between them. Two methods are commonly used to express this difference:

$$\text{Mean error} = \frac{1}{n} \sum_{i=1}^{n} (x_m - x_s)_i \qquad (10.78)$$

$$\text{Root mean squared error} = \frac{1}{n} \sum_{i=1}^{n} (x_m - x_s)_i^2 \qquad (10.79)$$

where x_m and x_s are the measured and simulated values, respectively.

Sensitivity Analysis

The purpose of a sensitivity analysis is to quantify the effects of uncertainty in the estimates of model parameters on model results. During a sensitivity analysis, calibrated values for hydraulic conductivity, recharge, boundary conditions, and the like, are systematically changed within a preestablished range of applicable values. The magnitude of change in heads and/or concentrations from the calibrated solution is a measure of the sensitivity of the solution to that particular parameter. The results of the sensitivity analysis are expressed as the effects of the parameter change on the average measurer of error (mean error or root mean square error) and on the spatial distribution of heads and/or concentrations.

Model Verification

Because of uncertainties in parameter estimates for a given site, the calibrated model parameters may not accurately represent the system under a different set of boundary conditions or hydrologic stresses. In a typical verification exercise, values of parameters and hydrologic stresses determined during calibration are used to simulate a transient response for which a set of field data exists (Anderson and Woessner, 1992). Examples of transient data sets include pumping test data and changes in water levels in response to pumping. In the absence of a transient data set, however, the model can be tested using a second set of steady-state data.

Unfortunately, sometimes it is not possible to verify a model because only one data set exists and that is usually used in the calibration process. A calibrated but unverified model can still be used to make predictions as long as sensitivity analyses of both the calibrated and predictive model are performed and evaluated.

Prediction

Prediction is one of the more common motivations for modeling. In a predictive simulation, the parameters determined during calibration are used to predict future conditions or the response of the system to future events. The length of time for which prediction may be required is an important consideration in model selection and design. For example, permits for deep well injection of hazardous wastes require contaminant transport modeling horizontally and vertically for 10,000 yr. This implies that if a numerical model is used, care should be taken to ensure a large enough grid to allow for 10,000 yr of transport.

The prediction process should be associated with a sensitivity analysis similar to that completed after calibration. Even though the calibrated model has been verified and subjected to a sensitivity analysis, the model may not give accurate results when stressed in some new way.

Example 10.10   MODELING THE PUMP-AND-TREAT SYSTEM AT U.S. AIR FORCE PLANT NUMBER 44, TUCSON, ARIZONA

Trichloroethene (TCE), 1,1-dichloroethylene (DCE), and chromium (hexavalent) have been detected in the ground water beneath the U.S. Air Force Plant No. 44 site in Tucson, Arizona. Activities at this facility include development, manufacturing, testing, and maintenance of missile systems from 1952 until the present. The contaminated ground water is in an extensive alluvial aquifer that Tucson uses as its principal aquifer. The Tucson area is one of the largest metropolitan areas in the country that is totally dependent on ground water for drinking-water supplies, and the TCE contamination is viewed as a threat to the integrity of the water-supply system. The site is listed on the National Priority List and is known as the Tucson Airport Area Superfund Sites.

In 1986, prior to implementing the pump-and-treat remediation system at the site, the area of TCE contamination was approximately 5 miles long and 1.6 miles wide. Contours of the dissolved TCE plume from measurements taken in 1986 are shown in Figure 10.20. The maximum measured TCE concentration in 1986 was 2.7 mg/L, although concentrations as high as 15.9 mg/L have been observed. The cleanup standard for TCE in this case is the drinking-water standard of 5 µg/L.

TCE is a halogenated aliphatic organic compound that is typically used as an industrial cleaning solvent. It is not carcinogenic; however, it is believed to become a hazard to human health after it is processed in the liver. In addition, reductive dehalogenation through natural biodegradation may result in production of vinyl chloride, which is a known carcinogen. TCE is denser than water, has low viscosity, is sparingly soluble in water, and is fairly volatile. It has a relatively low octanol–water partition coefficient, which means that its movement in the aqueous phase is not retarded to a great degree by the organic materials in the aquifer.

Site Characterization. Results of a subsurface investigation indicate that the subsurface can be divided into four hydrogeologic units:

1. Unsaturated zone that extends from the ground surface to the water table, which is at a depth of 110 to 130 ft
2. Upper zone extending to a depth of 180 to 220 ft
3. Aquitard consisting of 100 to 150 ft of low-permeability clay
4. Lower zone of unknown thickness.

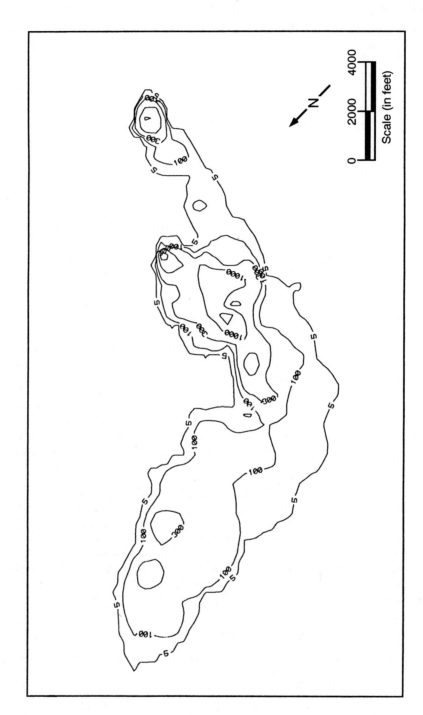

Figure 10.20 TCE plume prior to remediation at the Plant 44 site, Dec. 1986 (ppb). Source: Burgess, 1993.

338

The lower portion of the unsaturated zone consists of a laterally extensive layer of relatively low permeability perching clay, which is 20 to 40 ft thick and which partially confines the upper aquifer. The unsaturated, upper and lower zones consist of layers of relatively low conductivity clays and sandy clays, alternating with highly permeable sand and gravel units. A conceptual model of the complex subsurface stratigraphy is presented in Figure 10.21.

Hydraulic conductivity measurements were obtained in 12 locations in the immediate vicinity of current surface impoundments from pump test data. These data indicate that hydraulic conductivity ranges from 1.5×10^{-4} to 3.1×10^{-3} ft/sec for the upper-zone aquifer. For clean sands and gravels, porosity measurements of 26% to 34% have been obtained for this formation. The background hydraulic gradient, calculated from water-table elevation data, is estimated to be 0.006 in the north-northwesterly direction. Ground water velocity at the site ranges from 250 to 800 ft/yr, based on an average porosity value of 25%.

Potential sources of subsurface contamination at the site include pits, ponds, trenches, and drainage ditches in which on-site disposal of solvents, waste-waters, and sludges is believed to have occurred from approximately 1952 through 1977. Wastes generated during this time period include TCE, paint sludges, acids, cyanides, and alcohols. For the purpose of demonstrating pump- and treat-remediation in this case study, TCE is considered the contaminant of interest.

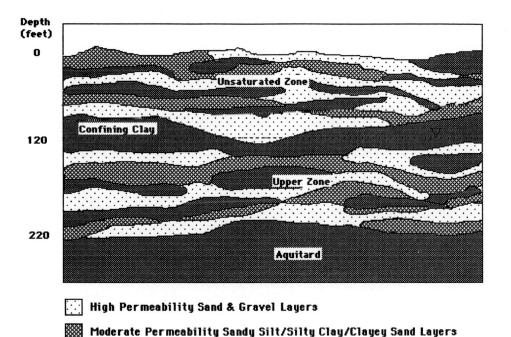

High Permeability Sand & Gravel Layers

Moderate Permeability Sandy Silt/Silty Clay/Clayey Sand Layers

Low Permeability Clay Layers

Figure 10.21 Conceptual model of subsurfac stratigraphy at the Plant 44 site. Source: Burgess, 1993.

Remediation by Pump and Treat. A ground water pump-and-treat remediation system began operating at the site in April 1987. The pump-and-treat system consists of 17 extraction wells, 13 recharge wells, and the aboveground treatment plant, which utilizes ion exchange followed by air stripping, vapor phase activated carbon and pressure sand filtration, and was designed to treat 5000 gpm of extracted ground water. Water-level elevation and contamination concentration data for the approximately 40 monitoring wells are collected monthly. The locations of the extraction and recharge wells are shown in Figure 10.22.

Monthly concentration measurements taken at the extraction and monitoring wells are used to evaluate the remediation process with respect to TCE removal. Measured TCE plumes were contoured yearly, and, as can be seen from Figure 10.23, the plume area does not appear to shrink a great deal through the 5-yr pumping period. However, concentrations do decrease through this period. Dissolved mass of the contaminant plumes may be calculated by integrating concentration contours across the area of the plume, and adjusting for porosity (Figure 10.24). The total TCE mass removed via pumping is approximately 6000 kg over 5 yr. The concentration of dissolved TCE in the extracted ground water decreases during remediation, particularly in those wells with initially high TCE concentrations. In the majority of cases, the TCE concentration appears to level out between 3 and 5 yr.

Modeling of the Pump-and-Treat System. The pump-and-treat system as it applies to the Tucson site was evaluated further through modeling with the BIOPLUME II (Rifai et al., 1988) model for transport in the aqueous phase (Burgess, 1993). All advection and dispersion were assumed to take place in the high-permeability sand and gravel layers, which were assumed continuous in the upper zone. Based on the site characteristics and modeling results, hydraulic conductivities were found to range from 2.39×10^{-4} to 5.55×10^{-3} ft/sec.

The initial distribution of potentiometric ground water elevation was determined by contouring water-table elevation measurements from December 1986. The model was calibrated by varying the hydraulic conductivity as well as the model boundaries in order to obtain the water table as it existed prior to extraction. Zones of conductivity were chosen based on the shape of the water table and matching the initial field measurements. The root mean square (rms) error of the model versus the actual potentiometric head distribution was used to determine the deviation of the simulated run from the actual data. For the flow calibration, the calculated rms error was 4.25% (a value of less than 10% to 15% is desirable). The model was validated by modeling the potentiometric elevation after approximately 5 yr of pumping and injection. The rms error for the simulation versus actual December 1991 measurements was 7.94% based on comparison of values at 68 locations.

The TCE plume as it appeared in December 1986 is shown in Figure 10.20. The objective of the transport calibration was to match the TCE plume as it appeared in 1986 prior to remediation start-up using source locations, flow rates, and TCE concentrations. Source infiltration rates used varied from 0.001 to 0.003 ft³/sec, and TCE concentrations of the infiltrating water varied from 400 to 10,000 µg/L. The duration of source infiltration was assumed to be 20 yr. Calibration is determined based on comparison of TCE concentrations at monitoring locations as well as plume area and dissolved mass. The simulated plume compares well visually with the 1986 plume. The chemical mass balance (CMB) error for the simulation, which is calculated based on the difference between the change in dissolved mass inflow minus outflow, is 2.26% after taking into account the mass that moves beyond the model boundary during simulation. (A CMB error of 5% is desirable.)

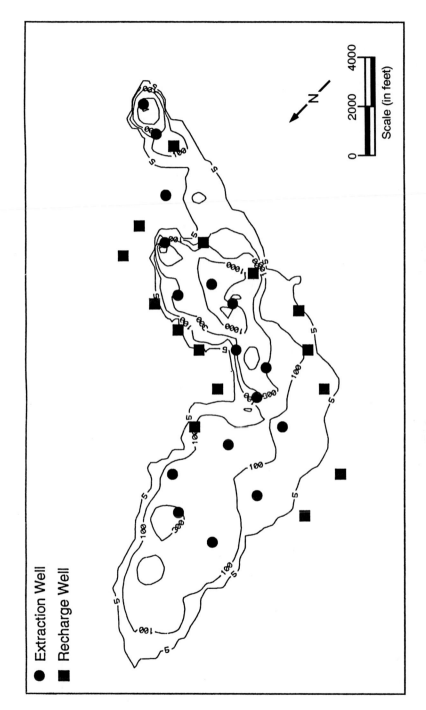

Figure 10.22 Locations of recharge and extraction wells at the Plant 44 site. Source: Burgess, 1993.

Extraction Well

Recharge Well

N

0   2000   4000
Scale (in feet)

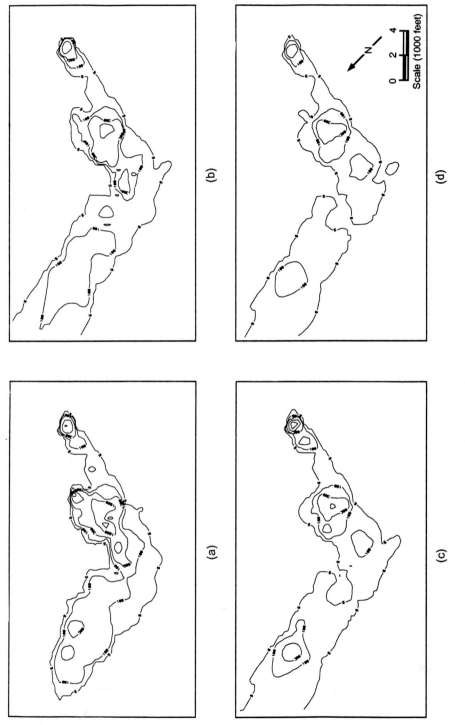

Figure 10.23 TCE plumes during remediation at the Plant 44 site, December 1986 to December 1991. Measured TCE concentrations for (a) December 1986 (ppb), (b) December 1988 (ppb), (c) December 1990 (ppb), and (d) December 1991 (ppb).   Source: Burgess, 1993.

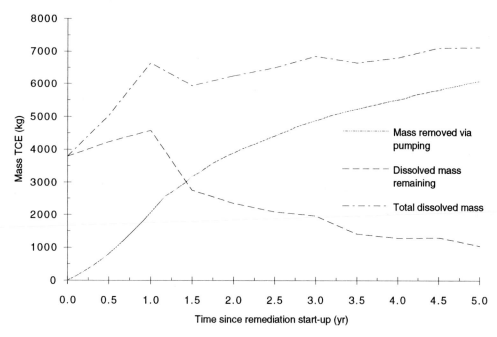

Figure 10.24 Overall mass balance for actual pump and treat at the Plant 44 site. Source: Burgess, 1993.

Results of the Tucson Model. To validate the transport calibration, the December 1986 plume was modeled using the pumping and injection pattern as in the actual remediation operation. All the other aquifer parameters remained the same as in the calibration to pumped water table with the exception of the addition of the initial TCE concentration array. The simulation was run for 5 yr in an attempt to predict the TCE plume in December 1991. The simulated plume compared well to the actual remediation. However, the simulated plume appeared to decrease more in area than the actual, a fact that may be attributed to the contouring method used. CMB errors for the remediation runs range from 0.52% to 2.36%. Overall, the model predicts that plume remediation is achieved in about 50 yr, assuming the NAPLs are not a major factor at this site, which remains for further investigation.

SUMMARY

In conclusion, computer models are very useful tools for simulating ground water flow and contaminant transport problems. Their use, however, requires a certain level of knowledge and expertise that can only be gained through experience. Many advances are being made in the way that computer models are being developed and applied. More emphasis is being placed on using models in the regulatory process. A computer model, for example, was used to establish regulatory criteria for the disposal of industrial wastes in municipal landfills. Additionally, computer models are the cornerstone of risk assessment studies.

REFERENCES

ALLEN, M. B., and PINDER, G. F., 1983, "Collocation Simulation of Multiphase Porous-medium Flow," *Soc. Pet. Eng. J.* 23, 135–142.

ANDERSEN, P. F., 1993, A Manual of Instructional Problems for the U.S.G.S. MODFLOW Model, EPA/600/R-93/010, U.S. EPA Robert S. Kerr Environmental Research Laboratory, Ada, OK.

ANDERSON, M. P., 1979, "Using Models to Simulate the Movement of Contaminants through Groundwater Flow Systems," *Critical Rev. Environ. Control* 9(2):97–156.

ANDERSON, M. P., and WOESSNER, W. W., 1992, *Applied Groundwater Modeling,* Academic Press, San Diego, CA.

AZIZ, K., and SETTARI, A., 1979, *Petroleum Reservoir Simulation,* Applied Science Publications, Essex, England.

BANERJEE, P. K., BUTTERFIELD, R., and TOMLIN, G. R., 1981, "Boundary Element Methods for Two-dimensional Problems of Transient Groundwater Flow," *Int. J. Num. Anal. Method. Geomech.* 5:15–31.

BATHE, K. J., and WILSON, E. L., 1976, *Numerical Methods in Finite Element Analysis,* Prentice Hall, Englewood Cliffs, NJ.

BEAR, J., and VERRUIJT, A., 1987, *Modeling Groundwater Flow and Pollution,* D. Reidel, Dordrecht, Holland.

BEDIENT, P. B., and HUBER, W. C., 1992, *Hydrology and Flood Analysis,* 2nd ed., Addison-Wesley, Reading, MA.

BITTINGER, M. W., DUKE, H. R., and LONGENBAUGH, R. A., 1967, "Mathematical Simulations for Better Aquifer Management," Publ. no. 72, International Association of Scientific Hydrology, pp. 509–519.

BREBBIA, C. A., and WALKER, S., 1980, *Boundary Element Techniques in Engineering,* Butterworths, London.

BREDEHOEFT, J. D., and PINDER, G. F., 1973, "Mass Transport in Flowing Groundwater," *Water Resources Res.* 9:192–210.

BRUTSAERT, W., 1973, "Numerical Solutions of Multiphase Well Flow," *ASCE J. Hydraul. Div.* 99(HY1):1981–2001.

BURGESS, K. S., 1993, *Flow and Transport Modeling of Heterogeneous Field Site Contaminated with Dense Chlorinated Solvent Waste,* thesis submitted in partial fulfillment of the requirements for the degree Master of Science, Rice University, Houston, TX.

CELIA, M. A., and GRAY, W. G., 1992, *Numerical Methods for Differential Equations,* Prentice Hall, Englewood Cliffs, NJ.

CELIA, M. A., and PINDER, G. F., 1980, "Alternating Direction Collocation Solution to the Transport Equation," *Proc of the 3rd International Conference on Finite Elements in Water Resources,* S. Y. Wang, C. V. Alsonso, C. A. Brebbia, W. G. Gray, and G. F. Pinder, eds., University of Mississippi, Oxford, MS., pp. 3.36–3.48.

CHANG, P. W., and FINLAYSON, B. A., 1977, "Orthogonal Collocation on Finite Elements for Elliptic Equations," *Advances in Computer Methods for Partial Differential Equations III,* R. Vichnevetsky, ed., International Association for Mathematics and Computers in Simulation, New Brunswick, NJ, pp. 79–86.

CHUNG, T. J., 1979, *Finite Element Analysis in Fluid Dynamics,* McGraw-Hill, New York.

CLOUGH, R. W., 1960, "The Finite Element Method in Plane Stress Analysis," *ASCE J. Struc. Div. Proc. 2nd Conf. Electronic Computation,* pp. 345–378.

COOK, R. D., 1974, *Concepts and Applications of Finite Element Analysis,* Wiley, New York.

COOLEY, R. L., 1971, "A Finite Difference Method for Unsteady Flow in Variably Saturated Porous Media: Application to a Single Pumping Well," *Water Resources Res.* 7(6): 1607–1625.

COOPER, R. M., and ISTOK, J. D., 1988, "Geostatistics Applied to Groundwater Contamination. I: Methodology," *Journal of Environmental Engineering,* 114(2):270–286.

CRICHLOW, H. B., 1977, *Modern Reservoir Engineering—A Simulation Approach,* Prentice Hall, Englewood Cliffs, NJ.

DAGAN, G., 1986, "Statistical Theory of Groundwater Flow and Transport: Pore to Laboratory, Laboratory to Formation, and Formation to Regional Scale," *Water Resources Research,* 22:120S–135S.

DOUGLAS, J., JR., and DUPONT, T., 1974, *Collocation Methods for Parabolic Equations in a Single Space Variable,* Springer-Verlag, New York.

DUBOIS, M., and BUYSEE, M., 1980, "Transient Heat Transfer Analysis by the Boundary Integral Method," *Proc. Second Int. Seminar on Recent Advances in Boundary Element Methods,* C. Brebbia, ed., CML Publications, University of Southampton, pp. 137–154.

EL-KADI, A. I., 1988, "Applying the USGS Mass-transport Model (MOC) to Remedial Actions by Recovery Wells," *Ground Water,* 26(3):281–288.

ENGLUND, E., and SPARKS, A., 1991, "Geostatistical Environmental Assessment Software," EPA 600/8-91/008, U.S. EPA Environmental Monitoring Systems Laboratory, Las Vegas, NV, 1991.

FINLAYSON, B. A., 1972, *The Method of Weighted Residuals and Variational Principles with Application in Fluid Mechanics, Heat and Mass Transfer,* Academic Press, New York.

FORSYTHE, G. E., and WASOW, W. R., 1960, *Finite-Difference Methods for Partial Differential Equations,* Wiley, New York.

FREEZE, R. A., 1971, "Three-dimensional, Transient, Saturated–Unsaturated Flow in a Groundwater Basin," *Water Resources Res.* 7(2):347–366.

FREEZE, R. A., and CHERRY, J. A., 1979, *Groundwater,* Prentice-Hall, Englewood Cliffs, NJ.

FREEZE, R. A., and WITHERSPOON, P. A., 1966, "Theoretical Analysis of Regional Groundwater Flow; 1. Analytical and Numerical Solutions to the Mathematical Model," *Water Resources Res.* 2:641–656.

FRIND, E. O., and PINDER, G. F., 1979, "A Collocation Finite Element Method for Potential Problems in Irregular Domains," *Int. J. Num. Methods Eng.* 14:681–701.

GALLAGHER, R. H., 1975, *Finite Element Analysis Fundamentals,* Prentice Hall, Englewood Cliffs, NJ.

GARDER, A. O., PEACEMAN, D. W., and POZZI, A. L., JR., 1964, "Numerical Calculation of Multidimensional Miscible Displacement by the Method of Characteristics," *Soc. Pet. Eng. J.* 4(1):26–36.

GELHAR, L. W., and COLLINS, M. A., 1971, "General Analysis of Longitudinal Dispersion in Nonuniform Flow," *Water Resources Res.* 7(6):1511–1521.

GREEN, D. W., DABIRI, H., and WEINAUG, C. F., 1970, "Numerical Modeling of Unsaturated Groundwater Flow and Comparison to a Field Experiment," *Water Resources Res.* 6:862–874.

HINTON, E., and OWEN, D. R. J., 1979, *An Introduction to the Finite Element Computations,* Chapters 1–4, Pineridge Press, Swansea, United Kingdom.

HOUSTIS, E. N., MITCHELL, W. F., and PAPATHEODOROU, T. S., 1979, "A C¹-collocation Method for Mildly Nonlinear Elliptic Equations on General 2-D Domains," *Advances Computer Methods for Partial Differential Equations III,* R. Vichnevetsky and R. S. Stepleman, eds., International Association for Mathematics and Computers in Simulations, New Brunswick, NJ.

HUEBNER, K. H., 1975, *Finite Element Method for Engineers,* Wiley, New York.

HUNT, B., 1983, *Mathematical Analysis of Groundwater Resources,* Butterworths, London.

HUYAKORN, P. S., and PINDER, G. F., 1983, *Computational Methods in Subsurface Flow,* Academic Press, San Diego, CA.

ISTOK, J., 1989, *Groundwater Modeling by the Finite Element Method,* Water Resources Monograph 13, American Geophysical Union, Washington, DC.

JAVANDEL, I., DOUGHTY, C., and TSANG, C. F., 1984, *Groundwater Transport: Handbook of Mathematical Models, Water Resources Monograph 10,* American Geophysical Union, Washington, DC.

JAWSON, M. A., and PONTER, A. R., 1963, "An Integral Equation Solution of the Torsion Problem," *Proc. R. Soc. London Ser. A* 273:237–246.

JAWSON, M. A., and SYMM, G. T., 1977, *Integral Equation Methods in Potential Theory and Elastostatics,* Academic Press, New York.

KHALEEL, R., and REDDELL, D. L., 1986, "MOC Solutions of Convective–Dispersion Problems," *Ground Water* 24(6):798–807.

KONIKOW, L. F., and BREDEHOEFT, J. D., 1974, "Modeling Flow and Chemical Quality Changes in an Irrigated Stream–Aquifer System," *Water Resources Res.* 10(3):546–562.

KONIKOW, L. F., and BREDEHOEFT, J. D., 1978, "Computer Model of Two Dimensional Solute Transport and Dispersion in Ground Water," *Techniques of Water Resources Investigation,* Book 7, Chapter C2, U.S. Geological Survey, Reston, VA.

LAPIDUS, L., and PINDER, G. F., 1982, *Numerical Solution of Partial Differential Equations in Science and Engineering,* Wiley, New York.

LENNON, G. P., LIU, P. L.-F., and LIGGETT, J. A., 1979, "Boundary Integral Equation Solution to Axisymmetric Potential Flows: 1. Basic Formulation," *Water Resources Res.* 15(5):1102–1106.

LIGGETT, J. A., 1977, "Location of Free Surface in Porous Media," *ASCE Hydraul. Div.* 103(HY4):353–365.

LIGGETT, J. A., and LIU, P. L.-F., 1979, "Unsteady Flow in Confined Aquifers—A Comparison of Two Boundary Integral Methods," *Water Resources Res.* 15(4):861–866.

LIGGETT, J. A., and LIU, P. L.-F., 1983, *The Boundary Integral Equation Method for Porous Media Flow,* George Allen and Unwin, London.

McDONALD, M. G., and HARBAUGH, A. W., 1988, *A Modular Three-dimensional Finite-Difference Ground-water Flow Model,* Book 6 Modeling Techniques, Scientific Software Group, Washington, DC.

MERCER, J. W., and FAUST, C. R., 1977, "The Application of Finite-element Techniques to Immiscible Flow in Porous Media," *Finite Elements in Water Resources,* W. G. Gray, G. F. Pinder, and C. A. Brebbia, eds., Pentech Press, Plymouth, England, pp. 1.21–1.57.

MERCER, J. W., and FAUST, C. R., 1981, *Ground-water Modeling,* 2nd ed., National Water Well Association, Columbus, OH, 1986.

NATIONAL RESEARCH COUNCIL, 1990, *Ground Water Models: Scientific and Regulatory Applications,* F. W. Schwartz et al., eds., National Academy Press, Washington, DC.

NORRIE, D. H., and DEVRIES, G., 1978, *An Introduction to Finite Element Methods,* Academic Press, New York.

OGATA, A., and BANKS, R. B., 1961, "A Solution of the Differential Equation of Longitudinal Dispersion in Porous Media," U.S. Geol. Survey Prof. Paper 411-A, U.S. Government Printing Office, Washington, DC.

OSTER, C. A., SONNICHSEN, J. C., and JASKE, P. T., 1970, "Numerical Solution of the Convective Diffusion Equation," *Water Resources Res.* 6:1746–1752.

PEACEMAN, D. W., 1977, *Fundamentals of Numerical Reservoir Simulation,* Elsevier, Amsterdam.

PEACEMAN, D. W., and RACHFORD, H. H., JR., 1955, "The Numerical Solution of Parabolic and Elliptic Differential Equations," *J. Soc. Ind. Appl. Math.* 3:28–41.

PEACEMAN, D. W., and RACHFORD, H. H., JR., 1962, "Numerical Calculation of Multidimensional Miscible Displacement," *Soc. Pet. Eng. J.* 2:327–339.

PINDER, G. F., 1973, "A Galerkin–Finite Element Simulation of Groundwater Contamination on Long Island, New York," *Water Resources Res.* 9:1657–1669.

PINDER, G. F., and BREDEHOEFT, J. D., 1968, "Application of the Digital Computer for Aquifer Evaluation," *Water Resources Res.* 4:1069–1093.

PINDER, G. F., and COOPER, H. H., JR., 1970, "A Numerical Technique for Calculating the Transient Position of the Salt Water Front," *Water Resources Res.* 6(3):875–882.

PINDER, G. F., and GRAY, W. G., 1977, *Finite Element Simulation in Surface and Subsurface Hydrology,* Academic Press, New York.

PINDER, G. F., FRIND, E. O., and CELIA, M. A., 1978, "Groundwater Simulation Using Collocation Finite Elements," *Proc. 2nd International Conference on Finite Elements in Water Resources,* C. A. Brebbia, W. G. Gray, and G. F. Pinder, eds., Pentech Press, Plymouth, England, pp. 1.171–1.185.

POLLOCK, D. W., 1988, "Semianalytical Computation of Path Lines of Finite-difference Models," *Ground Water* 26(6):743–750.

POLLOCK, D. W., 1989, " Documentation of Computer Programs to Complete and Display Pathlines Using Results from the U.S. Geological Survey Modular Three-dimensional Finite-Difference Ground-water Model," U.S. Geological Survey, Open File Report 89–381, U.S. Geological Survey, Reston, VA.

PRICKETT, T. A., and LONNQUIST, C. G., 1971, "Selected Digital Computer Techniques for Groundwater Resource Evaluation," *Illinois Water Survey Bulletin* 55, Illinois State Water Survey, Urbana, IL.

REDDELL, D. L., and SUNADA, D. K., 1970, "Numercial Simulation of Dispersion in Groundwater Aquifers," *Hydrology* Paper 41, Colorado State University, Fort Collins, CO.

REMSON, I., APPEL, C. A., and WEBSTER, R. A., 1965, "Groundwater Models Solved by Digital Computer," *ASCE J. Hydraul. Div.* 91(HY3):133–147.

REMSON, I., HORNBERGER, G. M., and MOLZ, F. J., 1971, *Numerical Methods in Subsurface Hydrology,* Wiley-Interscience, New York.

RIFAI, H. S., BEDIENT, P. B., WILSON, J. T., MILLER, K. M., and ARMSTRONG, J. M.,

1988, "Biodegradation Modeling at Aviation Fuel Spill Site," *Journal of Environmental Engineering,* 114(5):1007–1029.

RIZZO, F. J., and SHIPPY, D. J., 1970, "A Method of Solution for Certain Problems of Transient Heat Conduction," *AIAA J.* 89(11):2004–2009.

RUSHTON, K. R., and REDSHAW, S. C., 1979, *Seepage and Groundwater Flow,* Wiley, New York.

SEGERLIND, L. J., 1976, *Applied Finite Element Analysis,* Wiley, New York.

SHAMIR, U. Y., and HARLEMAN, D. F., 1967, "Dispersion in Layered Porous Media," *ASCE J. Hydraul. Div.* 93:237–260.

SINCOVEC, R. F., 1977, "Generalized Collocation Methods for Time-dependent, Nonlinear Boundary-value Problems," *Soc. Pet. Eng. J.* 345–352.

TANJI, K. K., DUTT, G. R., PAUL, J. L., and DONEN, L. D., 1967, "II. A Computer Method for Predicting Salt Concentrations in Soils at Variable Moisture Contents," *Hilgardia* 38:307–318.

TRESCOTT, P. C., PINDER, G. F., and LARSON, S. P., 1976, "Finite-difference Model for Aquifer Simulation in Two Dimensions with Results of Numerical Experiments," *U. S. Geological Survey Techniques of Water Resources Investigations,* Book 7, Chapter C1, Reston, VA.

VAN DER HEIJDE, P. K. M., EL-KADI, A. I., and WILLIAMS, S. A., 1988, *Groundwater Modeling: An Overview and Status Report,* U.S. Environmental Protection Agency, Ada, OK.

WANG, H. F., and ANDERSON, M. P., 1982, *Introduction to Ground Water Modeling Finite Difference and Finite Element Methods,* W. H. Freeman, San Francisco.

WEINSTOCK, R., 1952, *Calculus of Variations with Applications to Physics and Engineering,* McGraw-Hill, New York.

WIERENGA, P. J., 1977, "Solute Distribution Profiles Computed with Steady State and Transient Water Movement Models," *Soil. Sci. Soc. Am. J.* 41:1050–1055.

WYLIE, C. R., and BARRETT, L. C., 1982, *Advanced Engineering Mathematics,* 5th ed., McGraw-Hill, New York.

YATES, S. R., and YATES, M. V., 1990, "Geostatistics for Waste Management," EPA/600/8-90/004, U.S. EPA Robert S. Kerr Environmental Reseach Laboratory, Ada, OK.

ZHENG, C., 1989, PATH3D, S.S. Papadopulos & Assoc., Rockville, MD.

ZIENKIEWICZ, O. C., 1977, *The Finite Element Method,* 3rd ed., McGraw-Hill, New York.

Chapter 11 | NONAQUEOUS-PHASE LIQUIDS

11.1 INTRODUCTION

Nonaqueous-phase liquids (NAPLs) are hydrocarbons that do not dissolve in water. These immiscible fluids exhibit different behavior and properties in the subsurface than dissolved contaminant plumes. While components of dissolved plumes are invisible to the naked eye and travel with the flow of ground water, NAPLs in the subsurface form a visible, separate oily phase whose migration is governed by gravity, buoyancy, and capillary forces. A variety of chemical compositions results in a wide range of NAPL behaviors: after moving downward through the vadose zone, some light NAPLs (LNAPLs) can float and move on top of the water table, while more dense NAPLs (DNAPLs) can move downward past the water table and penetrate hundreds of feet into the saturated zone.

When released at the surface, **free-phase** or **mobile** NAPL is forced into the pores of the soil–aquifer matrix by the hydrostatic pressure on the continuous body of NAPL. NAPL under pressure can enter very small pores and fractures in the subsurface as long as the original NAPL source, such as a waste pond or leaking underground storage tank, provides continuing pressure. When the supply of NAPL is exhausted, pressure on the free-phase NAPL is removed, and small blobs (or ganglia) of NAPL snap off or bypass the once continuous NAPL body. These blobs then become trapped in individual pores or small groups of pores by capillary forces (see Figures 11.1 and 11.2).

Residual saturation, an important parameter in ground water remediation problems, is defined as the fraction of total pore volume occupied by residual NAPL under ambient ground water flow conditions and indicates the amount of NAPL that is trapped in the subsurface. The actual residual saturation value at a particular site is determined by the type of chemicals that comprise the NAPL, whether the NAPL is in the saturated zone or vadose zone, and, most importantly, the structure and hydrogeologic characteristics of the soil–aquifer matrix.

NAPLs have a tremendous impact on the remediation of contaminated aquifers, because removal of all residual NAPL blobs trapped in individual pores is very difficult or impossible. Although many NAPL removal technologies are currently being tested, to date there have been no field demonstrations where sufficient NAPL was removed to restore an aquifer to drinking-water quality. NAPL remaining trapped in the soil–aquifer matrix acts as a continuing source

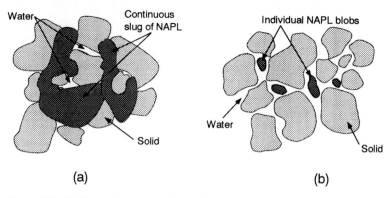

Figure 11.1   (a) Free-phase versus (b) residual NAPL.

of dissolved contaminants to ground water, greatly complicating the restoration of NAPL-affected aquifers for many years.

Key Concepts

Five key concepts are introduced next to illustrate how NAPLs act so differently from dissolved constituents in ground water.

Single- versus Multicomponent NAPLs

NAPLs are either comprised of a single chemical or a complex mixture of several or hundreds of chemicals. For example, a spill of pure trichloroethylene (TCE) can produce a single-component NAPL comprised of TCE. A gasoline spill, on the other hand, will yield a NAPL comprised of a number of aromatic and aliphatic hydrocarbons. Understanding and predicting the behavior of a mixed NAPL is much more complex than for a single-component NAPL.

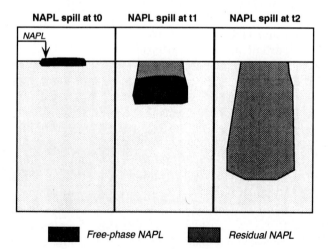

Figure 11.2   Downward percolation of free-phase NAPL from one-time spill. As NAPL moves downward, more free-phase NAPL is trapped as residual NAPL.

Saturation and Residual Saturation

Saturation (S_i) is the fraction of total pore space containing NAPL or, more specifically, the relative fraction of a particular fluid i in a representative volume of a porous media:

$$S_i = \frac{V_i}{V_{\text{pore space}}} \tag{11.1}$$

where

V_i = volume of fluid i

$V_{\text{pore space}}$ = volume of open pore space

Saturation values for free-phase, continuous masses of NAPLs are generally greater than 50% and can approach 100% in some systems.

The saturation at which a continuous NAPL becomes discontinuous and is immobilized by capillary forces is known as the **residual saturation** (S_{ir}). In unsaturated soils, residual saturation of NAPL fluids, defined as the fraction of the total pore volume occupied by NAPL under ambient conditions, typically ranges from 5% to 20%. Residual saturation values are typically higher in the saturated zone, because NAPL serves as a nonwetting fluid when in contact with water (see Section 11.3). Residual saturation values range from 15% to 50% of the total pore volume in the saturated zone (Mercer and Cohen, 1990; Schwille, 1988). Residual saturation appears to be relatively insensitive to the type of chemicals that comprise a NAPL, but is very sensitive to soil properties and heterogeneities (U.S. EPA, 1990). Figure 11.3 illustrates the difference between saturation and residual saturation.

Relative Mass of Hydrocarbons in Dissolved Phase versus NAPL Phase

At many sites the total contaminant mass in the NAPL phase is much larger than the total mass of even a large dissolved contaminant plume. To illustrate this point, consider that concentrations of dissolved hydrocarbon plumes are typically reported in units of parts per million or parts per billion (1 g of hydrocarbon per each million or billion grams of water). The amount of NAPL in an aquifer, on the other hand, is reported as a percentage of open pore space occupied by NAPL (the saturation value). For a comparison of the relative masses of hydrocarbon in the dissolved versus NAPL phases at a hypothetical hazardous-waste site, see Example 11.1.

Because many sites are contaminated by hundreds, thousands, or even millions of gallons of NAPLs in the subsurface, the relative importance of the NAPL phase cannot be overlooked. The importance of a dissolved phase plume is based on the potential for quick off-site migration. However, large quantities of NAPL in the subsurface can hinder ground water remediation efforts or cause them to fail completely.

Example 11.1 COMPARING DISSOLVED MASS VERSUS NAPL MASS

A leaking underground storage tank released 1000 gal of gasoline (density approximately 0.9 g/mL) to the subsurface. After 1 yr, the resulting dissolved-phase plume is 1000 ft long, 100 ft wide, and 10 ft deep. The average concentration in the plume

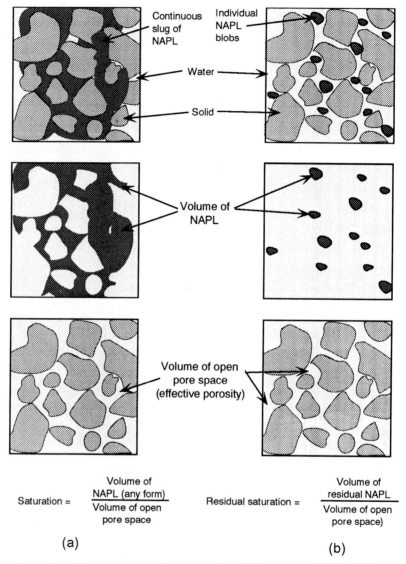

Figure 11.3   Saturation and residual saturation. (a) Free-phase DNAPL. (b) Residual NAPL.

is 1 mg/L, and the porosity of the aquifer is 0.30. If no hydrocarbon is lost due to volatilization or biodegradation, how much of the original release is in the dissolved phase and how much is in the NAPL phase?

Solution

1. Mass of gasoline released:

$$\text{Total mass} = (1000 \text{ gal})(3.78 \text{ l/gal})(1000 \text{ mL/L})(0.9 \text{ g/mL})(\text{kg}/1000 \text{ mg})$$

$$= 3402 \text{ kg} \quad (7484 \text{ lb})$$

2. Volume of contaminated ground water in plume:

$$\text{Volume} = (1000 \text{ ft})(100 \text{ ft})(10 \text{ ft})(0.30)(28.3 \text{ L/ft}^3)$$

$$= 8.49 \times 10^6 \text{ L}$$

3. Mass of dissolved hydrocarbons:

$$\text{Dissolved mass} = (8.49 \times 10^6 \text{ L})(1 \text{ mg/L})(\text{kg}/10^6 \text{ mg})$$

$$= 8.5 \text{ kg}$$

4. Comparison:

$$\text{Total mass} = 3402 \text{ kg}$$

$$\text{Dissolved mass} = 8.5 \text{ kg} \quad (0.2\% \text{ of total})$$

$$\text{NAPL mass} = 3393.5 \text{ kg} \quad (99.8\% \text{ of total})$$

Analysis. As this example shows, most of the organic mass at a NAPL site is found in the nonaqueous phase. Thus, removing even a few gallons of free-phase NAPL may prevent the contamination of millions of gallons of ground water.

Difficulty in Confirming Presence or Absence of NAPLs

Knowing whether a site is contaminated by NAPL is important in order to properly develop site characterization studies and to design appropriate ground water remediation systems. At some sites the presence of NAPLs can be confirmed visually by observing if free-phase hydrocarbons have accumulated in wells or are present in soil cores. At many sites, however, the presence of NAPLs is difficult to confirm because residual NAPL trapped in the soil pores does not flow into a monitoring well and is difficult to observe directly in soil cores. The availability of indirect methods for estimating the potential of NAPL occurrence, such as measuring the concentrations of hydrocarbons on soil, has not eliminated the difficulty in confirming the presence or absence of NAPLs at many sites.

Difficulty in Removing Trapped Residual NAPLs

Capillary forces make removal of all NAPLs that have been released to the subsurface very difficult or impossible. For example, conventional pumping of oil reservoirs typically leaves over two-thirds of the oil in the ground. Enhanced oil-recovery techniques, such as water flooding or the application of surfactants, can bring 50% to 80% of the original in-place oil (NAPL) to the surface under optimum conditions. These recovery rates are acceptable to the oil industry, where economic factors determine the ultimate amount of oil recovered from a reservoir. At hazardous-waste sites, however, removal of better than 99% of NAPL is probably required in order to restore a contaminated aquifer to drinking-water standards (that is, low part-per-billion range for most dissolved organics). At most sites, this level of recovery is impractical when using pumping and injecting alone (Wilson and Conrad, 1984) and is very difficult or impossible to achieve even when using such developing remediation technologies as in situ biodegradation. The overall level of aquifer cleanup that can be achieved is very site specific and dependent on the type and amount of NAPL present, the soil type, and the amount of resources that are devoted to the project.

11.2 TYPES OF NAPLs

Although NAPLs are associated with a diverse group of industries and have a wide range of physical properties, they are generally classified by specific gravity (density relative to water) as LNAPLs (light NAPLs) or DNAPLs (dense NAPLs). LNAPLs have a specific gravity less than water and will float on the water table. DNAPLs, on the other hand, have specific gravities greater than water and can sink deep into the saturated zone. This simple classification system, although based only on density, is a very useful framework for evaluating the migration pathway, related chemicals and industries, and the ultimate fate of NAPLs in the subsurface.

LNAPLs (Light Nonaqueous-phase Liquids)

LNAPLs are primarily associated with the production, refining, and distribution of petroleum products. Spills and accidental releases of gasoline, kerosene, diesel, and associated condensates are common sources of LNAPLs to ground water. When LNAPL is released at the surface, it first migrates downward through the unsaturated zone under the force of gravity. Upon encountering a water-bearing unit, the LNAPL forms a pancakelike layer in the capillary fringe and top portion of the saturated zone (see Figure 11.4). Ground water flowing past the floating LNAPL dissolves soluble components of the LNAPL, forming a dissolved plume downgradient of the LNAPL zone. Typical chemicals of interest resulting from the dissolution of petroleum products include benzene, toluene, ethyl benzene, and xylene (BTEX), as well as other aromatic hydrocarbons (see Table 11.1).

Because LNAPLs do not penetrate very deeply into the water table and are relatively biodegradable under natural conditions, they are generally thought to be a more manageable environmental problem than DNAPL releases. With small-scale releases (such as those found at many service stations), natural processes will usually attenuate the environmental impacts within a span of several years.

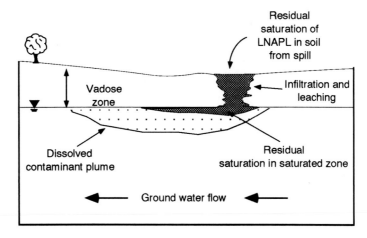

Residual saturation of LNAPL in soil from spill

Vadose zone

Infiltration and leaching

Dissolved contaminant plume

Residual saturation in saturated zone

Ground water flow

Figure 11.4 Typical LNAPL release.

DNAPLs (Dense Nonaqueous-phase Liquids)

DNAPLs are related to a wide variety of industrial activities, including almost all facilities where degreasing, metal stripping, chemical manufacturing, or other activities involving chlorinated solvents were performed. In addition, pesticide manufacturing, wood-treating operations, and spills of transformer oils are common sources of DNAPLs. As with LNAPLs, the main chemicals of concern are DNAPL dissolution products. Chlorinated aliphatic hydrocarbons comprise important dissolution products of solvent-type DNAPLs. Polyaromatic hydrocarbons (PAHs) and pentachlorobiphenyls (PCBs) are the primary dissolution products of other types of DNAPL (see Table 11.2).

After being released at the surface, DNAPLs move vertically downward through the vadose zone. If large quantities of DNAPL have been released, the DNAPL will continue to move downward until all is trapped as residual hydrocarbon or until low-permeability stratigraphic units are encountered that create DNAPL pools in the soil–aquifer matrix (see Figure 11.5). In this figure, a perched DNAPL pool is shown to fill up and then spill over the lip of a low-permeability stratigraphic unit. The spill-over point (or points) can be some distance away from the original source, greatly complicating the process of tracking the DNAPL migration.

DNAPLs, particularly if comprised of chlorinated solvents, usually present

TABLE 11.1 Chemicals Typically Associated with an LNAPL Release

Aromatics
Benzene
Ethyl benzene
Toluene
Xylenes
Naphthalene

TABLE 11.2   Chemicals Typically Associated with a DNAPL Release

Halogenated Volatiles	Nonhalogenated Semivolatiles
Chlorobenzene	
1,2-Dichloropropane	2-Methyl naphthalene
1,1-Dichloroethane	*o*-Cresol
1,1-Dichloroethylene	*p*-Cresol
1,2-Dichloroethane	2,4-Dimethylphenol
trans-1,2-Dichloroethylene	*m*-Cresol
cis-1,2-Dichloroethylene	Phenol
1,1,1-Trichloroethane	Naphthalene
Methylene chloride	Benzo[*a*]anthracene
1,1,2-Trichloroethane	Fluorene
Trichloroethylene	Acenaphthene
Chloroform	Anthracene
Carbon tetrachloride	Dibenzo[*a,h*]anthracene
1,1,2,2-Tetrachloroethane	Fluoranthene
Tetrachloroethylene	Pyrene
Ethylene dibromide	Chrysene
	2,4-Dinitrophenol
Halogenated Semivolatiles	**Miscellaneous**
1,4-Dichlorobenzene	Coal tar
1,2-Dichlorobenzene	Creosote
Aroclor 1242, 1254, 1260	
Chlordane	
Dieldrin	
2,3,4,6-Tetrachlorophenol	
Pentachlorophenol	

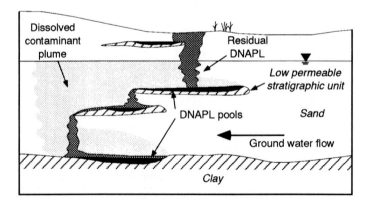

Figure 11.5 Typical DNAPL release. Source: Waterloo Centre for Ground Water Research, 1991.

a much more formidable remediation challenge than LNAPLs for three reasons: (1) chlorinated solvents do not biodegrade very rapidly and persist for long periods of time in the subsurface; (2) the greater density of DNAPLs causes the contaminated zone to spread much deeper than LNAPL releases; and (3) chlorinated solvents have physical properties that allow movement through very small fractures in the soil and downward penetration to great distances. Aquifers contaminated with large quantities of DNAPLs are almost impossible to restore to original conditions using any current proven ground water cleanup technologies (U.S. EPA, 1992b).

11.3 NAPL TRANSPORT: GENERAL PROCESSES

NAPL Transport at the Pore Level

NAPL movement occurs at the pore level when enough pressure is available to force free-phase NAPL through a small pore throat, thereby displacing air and/ or water that previously occupied the pore. The amount of pressure required depends on the **capillary forces** acting on the fluids on either side of the pore throat. The way that capillary forces act on these two fluids is partially explained by wettability; the fluid drawn into the pore is referred to as the **wetting fluid,** while the fluid repelled by capillary forces is the **nonwetting fluid** (Bear, 1972).

Wettability is defined as the tendency of one fluid to preferentially spread over a solid surface in favor of the second fluid. This property is measured by observing the contact angle of a test fluid on a surface when surrounded by a larger volume of a background fluid (see Figure 11.6). A **contact angle** of less than 70° indicates that the test fluid is wetting and the background fluid is nonwetting. A contact angle of greater than 110° indicates that the test fluid is nonwetting, and contact angles between 70° and 110° are characteristic of neutral-wetting systems (Mercer and Cohen, 1990).

Wettability is closely related to another physical property, **interfacial tension,** which is defined as the free surface energy at the interface between two immiscible fluids (Villaume, 1985). The interaction between similar molecules on one side of the interface and dissimilar molecules on either side of the interface produces capillary forces when the two fluids come in contact with a solid phase, such as a pore throat. Interfacial tension values have been reported for many organic hydrocarbons in contact with water (Arthur D. Little, 1981), and curves relating capillary pressure, interfacial tension, and wettability have been developed by Mercer and Cohen (1990).

To apply the concept of wettability to NAPL transport in specific field situations requires detailed information regarding the physical properties of each fluid in the system as well as the soil grains or aquifer media. However, the following generalizations can be made:

- Water is almost always the wetting fluid when mixed with air or NAPLs in the subsurface.

Test fluid: water
Background fluid: NAPL
_____

$\phi < 70°$

Conclusion: Water is
wetting fluid

Test fluid: NAPL
Background fluid: water
_____

$\phi > 110°$

Conclusion: NAPL is
non-wetting fluid

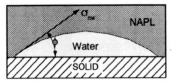 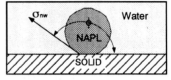

ϕ = contact angle

Fluid relationships:

System	Wetting fluid	Non wetting fluid
air:water	water	air
air: NAPL	NAPL	air
water:NAPL	water	NAPL
air:NAPL:water	water>organic>air	air

Figure 11.6 Wettability configurations.

- NAPLs serve as the wetting fluid when mixed with air, but act as the nonwetting fluid when combined with water in the subsurface (Domenico and Schwartz, 1990).

These general rules can be illustrated using three examples of NAPL movement: downward migration of DNAPLs in the saturated zone, NAPL movement through the vadose zone, and LNAPL behavior at the water table.

Downward Migration of DNAPLs in the Saturated Zone

When DNAPLs migrate through the saturated zone, a three-phase system consisting of solid, water, and DNAPL is formed. The DNAPL acts as a nonwetting fluid that must overcome capillary forces in order to squeeze through the throat of a pore filled with a wetting fluid, in this case water. The capillary pressure that must be exceeded is called the **entry pressure.** To enter the pore, sufficient driving force must be provided by the combination of DNAPL density forces and pressure forces present in the continuous DNAPL mass. If the driving force is greater than the entry pressure, then the DNAPL will move forward and enter the pore. When the driving force is removed, the continuous DNAPL mass will split apart into countless individual blobs of residual NAPLs located mostly in the larger pores of the aquifer matrix (see Figure 11.1).

Unlike the lighter-than-water chemicals, DNAPLs will continue to move vertically downward until (1) a stratigraphic barrier is reached or (2) the original supply of DNAPL is exhausted and the DNAPL mass breaks up into residual DNAPL blobs (see Figure 11.5). In some cases, DNAPL can migrate downward hundreds of feet below the surface.

NAPL Movement through the Vadose Zone

NAPL infiltration into the vadose zone forms a four-phase system consisting of the air, water, solid, and NAPL phases (U.S. EPA, 1990). Water on the soil grains serves as one wetting fluid, while NAPL acts as a wetting fluid with respect to air on the water film and a nonwetting fluid with respect to the water (Domenico and Schwartz, 1990). Thus, downward migrating NAPL infiltrates relatively easily into the media whose pores are largely filled with air, because as a partial wetting fluid in this four-phase system, capillary forces do not repel the movement of NAPL. If sufficient NAPL is available, it continues to move downward from the source area by forming (1) films between the gas and water phases, and/or (2) blobs of NAPL that replace gas in the pores or water in the pore throats (U.S. EPA, 1990). The tendency of a certain chemical to form films or blobs depends on the physical properties of the chemical. Chlorinated solvents, for example, will generally not form large, continuous films when released to the vadose zone and so respond differently to remediation techniques such as soil venting and in situ biodegradation than do chemicals that will form films (U.S. EPA, 1990).

After the original supply of NAPL is exhausted, 10% to 20% of the pore space in the vadose zone will be occupied by the remaining residual NAPL in films and/or blobs, and the rest of the pore space is filled by an air vapor mixture located primarily in the larger pores, with water spread out on the surface of the soil grains and filling the smaller pores.

The residual saturation of either LNAPLs or DNAPLs in the saturated zone is usually higher than in the unsaturated zone by a factor ranging from 2 to 5 (Domenico and Schwartz, 1990). The reason for the higher concentration of residual NAPLs in saturated media is due to the following factors (Mercer and Cohen, 1990; U.S. EPA, 1990):

- The NAPL–air density ratio is greater than the NAPL–water density ratio, favoring drainage of NAPLs from the vadose zone.
- NAPLs are the nonwetting fluid in the saturated zone and so are trapped in the larger pores.
- NAPLs tend to spread out farther in the vadose zone as a result of favorable capillary conditions.

LNAPLs Behavior at the Water Table

The interaction between water, air, and NAPLs becomes more complex once the NAPLs approach the water table and associated capillary fringe. In a completely saturated system, water is the wetting fluid and NAPL is the nonwetting fluid whose movement is repelled by capillary forces. The capillary fringe represents the transition zone between a system where NAPLs are a partial wetting fluid (the vadose zone) and where they are a nonwetting fluid (the saturated zone). If the density of a NAPL is less than water, it will accumulate on top of the capillary fringe and eventually flow in thin sheets along the water table after a certain minimum thickness is achieved (Domenico and Schwartz, 1990). If the LNAPL flow rate through the vadose zone is fast enough, the capillary fringe will collapse and the LNAPL will float on the water table. Once the LNAPL supply is exhausted, however, the continuous mass of LNAPL in the capillary fringe will begin to break up into individual blobs and form residual LNAPL.

NAPL Transport through Fractures and Heterogeneities

The previous section discussed traditional models for the movement of immiscible fluids through a porous media by moving through the primary porosity of the system, the open pore spaces. Recently, more attention has focused on another transport pathway: NAPL migration through small-scale fractures and heterogeneities in the porous media. At most sites, NAPL will migrate preferentially through secondary porosity features in the soil–aquifer matrix (such as larger fractures, partings, bioturbations, root holes, slickensides, coarse-grained layers and other microstratigraphic features), rather than saturating the open pore volume directly below the source area. In general, heterogeneities will have the following effects on NAPL movement in the subsurface:

- *Increase the size of the NAPL zone:* Even small amounts of clay or silt in a soil (as little as 2% by weight) can have a dramatic effect on NAPL migration by creating much more complex, branching migration pathways and by increasing the volume of the NAPL-contaminated zone (U.S. EPA, 1990).

- *Permit deeper penetration into the subsurface:* The presence of heterogeneities permit the NAPL to penetrate much deeper into the subsurface than would be predicted using typical residual saturation values and assuming uniform saturation of the aquifer media (see Figure 11.7). One research field study indicated that even small releases on the order of one or two drums of some DNAPLs can penetrate tens to hundreds of feet through the vadose zone before reaching the water table (Poulson and Kueper, 1991; Cherry and Feenstra, 1991).

- *Make fine-grained confining units ineffective barriers to NAPL flow:* Clay aquitards that serve as effective confining units in the context of ground water flow may not serve as effective barriers to migration of some NAPLs because of microscale heterogeneities (Waterloo Centre for Ground Water

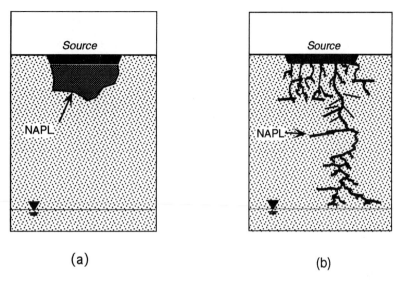

(a)   (b)

Figure 11.7   NAPL migration in primary versus secondary porosity. (a) NAPL migration through primary porosity. (b) NAPL migration through secondary porosity features (fractures, root holes, etc.).

Research, 1991; U.S. EPA, 1992b). Modeling studies have indicated that some low-viscosity DNAPLs can penetrate and migrate through hairline fractures as small as 20 μm wide. At one Superfund site, for example, large volumes of DNAPL were observed to have moved significant distances (~50 ft vertically and ~400 ft laterally) in the secondary porosity features of various clay and silt units underlying the site (Connor, Newell, and Wilson, 1989). Kueper and McWorter (1991) provide calculation techniques that relate DNAPL movement down fractures to fracture size and DNAPL properties.

NAPL Transport at the Site Level

Beyond the scale of the individual pore, NAPL migration becomes much more difficult to predict. At the scale of an entire hazardous-waste site, for example, the path taken by a NAPL as it migrates through the subsurface is dominated by factors related to the NAPL release, physical properties of the NAPL, and most importantly geological factors (Feenstra and Cherry, 1988; U.S. EPA, 1992a):

- Volume of NAPL released
- Properties of the soil–aquifer media, such as pore size and permeability
- Microstratigraphic features, such as root holes, small fractures, and slickensides found in silt–clay layers
- General stratigraphy, such as the location and topography of low-permeability units

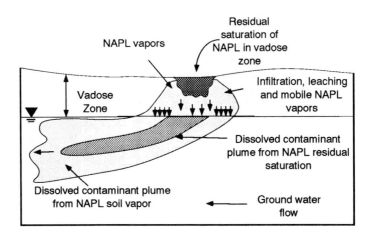

Figure 11.8 LNAPL release to vadose zone only. Source: Waterloo Centre for Ground Water Research, 1991.

- Properties of the NAPL, such as wettability, density, viscosity, and interfacial tension
- Duration of release, such as a one-time slug event or a long-term continual discharge
- Area of infiltration at the NAPL entry point to the subsurface

To illustrate the movement of NAPLs at hazardous-waste sites, several conceptual models are presented in Figures 11.8 to 11.12.

LNAPL Conceptual Models

After a spill or leak near the surface, both LNAPLs and DNAPLs move vertically downward under the force of gravity and soil capillarity. When only a small amount of NAPL is released, however, all the free-phase NAPL is eventually trapped in pores and fractures in the unsaturated zone (see Figure 11.8). Infiltration through the NAPL zone dissolves some of the soluble organic constituents in the NAPL, carrying organics to the water table and forming a dissolved organic plume in the aquifer. Migration of vapors can also act as a source of dissolved organics to ground water (Mendoza and McAlary, 1990).

A release of LNAPL sufficient to reach the water table will spread along the capillary fringe and the top of the water table because of buoyancy forces. Ground water moving past this thin, pancakelike LNAPL layer will dissolve soluble contaminants to form a dissolved hydrocarbon plume (see Figure 11.4).

If free-phase LNAPL is present, a falling water table will lower the floating LNAPL layer and leave an emplaced source of residual NAPL behind in the saturated zone when the water table recovers (see Figure 11.9). Similarly, a rising water table will lift the free-phase layer into the vadose zone and can increase the residual saturation of the affected soils after the water table drops. This movement is a cause for concern because it will increase the zone containing residual hydrocarbons, thereby increasing the concentration of dissolved organics.

As shown in Figure 11.10, LNAPL that is forced into confined aquifers by high entry pressures can migrate under buoyancy forces in directions other than the direction of ground water flow. This phenomenon is very similar to the trapping of oil in petroleum reservoirs. To increase the recovery of LNAPL, recovery wells

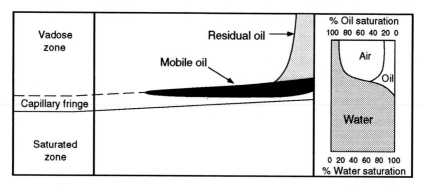

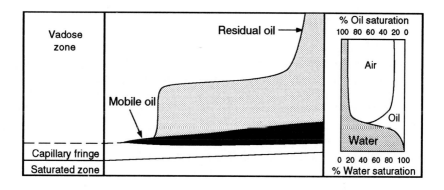

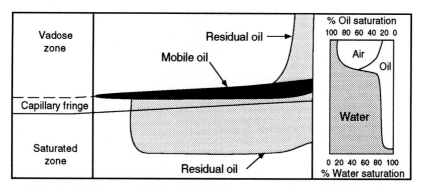

Figure 11.9 Effect of a falling and rising water table on the distribution of mobile and residual phases of an LNAPL. Source: Fetter, 1993. © 1993 by Macmillan College Publishing Company. Reprinted with permission.

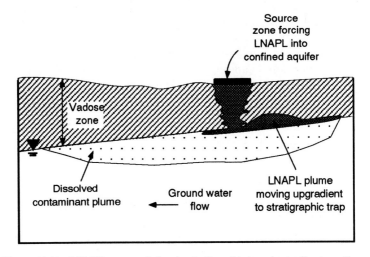

Figure 11.10 LNAPL accumulating in stratigraphic trap in confined aquifer.

should be installed near the tops of the stratigraphic traps and screened near the confining unit–aquifer boundary.

DNAPL Conceptual Models

DNAPL released at the surface can migrate all the way through the unsaturated zone and then continue downward until the mobile DNAPL is exhausted and trapped as a residual hydrocarbon in the porous media (see Figure 11.5). Ground water flowing past the trapped residual DNAPL dissolves soluble components of the DNAPL, forming a dissolved plume downgradient of the DNAPL zone.

Mobile DNAPL will continue to move downward until converted into a residual state or until low-permeability stratigraphic units are encountered that create DNAPL pools in the soil–aquifer matrix (see Figure 11.5).

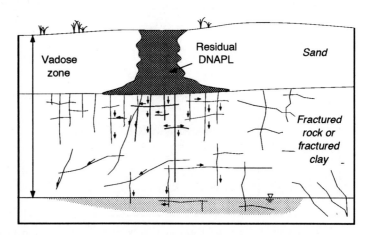

Figure 11.11 Fractured rock or fractured clay system. Source: Waterloo Centre for Ground Water Research, 1991.

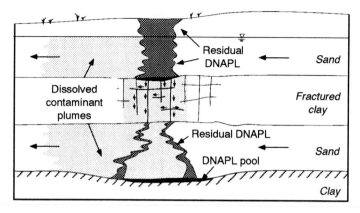

Figure 11.12 Composite DNAPL site. Source: Waterloo Centre for Ground Water Research, 1991.

DNAPL introduced into a fractured rock or fractured clay system follows a complex pathway based on the distribution of fractures in the original matrix (see Figure 11.11). The number, density, size, and direction of the fractures usually cannot be of determined due to the extreme heterogeneity of a fractured system and the lack of economical aquifer characterization technologies. Relatively small volumes of DNAPL can penetrate deeply into fractured systems due to the low retention capacity of the fractures and the ability of some DNAPLs to migrate through very small (less than 20 μm) fractures. Many clay units, once considered to be relatively impermeable to DNAPL migration, often act as fractured media with preferential pathways for vertical and horizontal DNAPL migration.

In the case of a composite DNAPL site, mobile DNAPL migrates vertically downward through the unsaturated zone and the first saturated zone, producing a dissolved constituent plume in the upper aquifer (see Figure 11.12). Although a DNAPL pool is formed on the fractured clay unit, the fractures are large enough to permit downward migration to the deeper aquifer. DNAPL then pools in a topographic low in the underlying impermeable unit, and a second dissolved constituent plume is formed.

Note that these diagrams depict general patterns of NAPL behavior only. The actual path taken by NAPLs at actual hazardous-waste sites is usually much more complex due to the large- and small-scale heterogeneities found at the site. At many sites the final distribution of NAPLs in the subsurface may never be fully characterized due to the complex distribution of countless small-scale heterogeneities in the subsurface.

11.4 NAPL TRANSPORT: COMPUTATIONAL METHODS

Although the exact path of NAPL movement through the subsurface is very difficult or impossible to predict, several general relationships are available for planning and conceptual design purposes.

Pore Level Calculations

Capillary Pressure

Several researchers have summarized the relationships between capillary forces and NAPL movement under hydrostatic and hydrodynamic conditions (Mercer and Cohen, 1990; Bear, 1979). A simple definition of capillary pressure is that it is the pressure difference between the nonwetting and wetting fluid:

$$P_c = P_{nw} - P_w \qquad (11.2)$$

where P_c = capillary pressure
P_{nw} = pressure of nonwetting fluid (such as a NAPL)
P_w = pressure of wetting fluid (such as water)

Bear (1979) provides a more detailed definition of capillary pressure for the case of a nonwetting NAPL sphere in a water-filled porous media:

$$P_c = \frac{2\sigma \cos \phi}{r} \qquad (11.3)$$

where P_c = capillary pressure
σ = interfacial tension between NAPL and water
ϕ = contact angle
r = radius of the water-filled pore that the NAPL must move through to exit or enter pore

We see from this relationship that as the radii of the soil pores get smaller the capillary pressure required to force NAPL into or out of the pore increases. In other words, NAPLs have more difficulty moving through fine-grained soils such as silts or clays than through coarse-grained sands or gravels, corresponding to observations from numerous field sites.

The capillary pressure relationship shown in Eq. (11.3) can be expanded to provide estimates of the hydrostatic and hydrodynamic conditions that are required for NAPL migration. Some of the more important NAPL migration relationships, listed next, are presented in Table 11.3 (Cohen and Mercer, 1993):

- Critical NAPL height required for NAPL penetration into the vadose zone; see Eq. (11.4)

- Critical NAPL height required for DNAPL penetration into a fine-grained unit (capillary barrier) in the saturated zone when residual DNAPL is above DNAPL body; see Eq. (11.5), Example 11.2

- Critical NAPL height required for DNAPL penetration from a coarse-grained material into a fine-grained unit (capillary barrier) in the saturated zone when no DNAPL is above DNAPL body; see Eq. (11.6)

TABLE 11.3 Additional Capillary Pressure Relationships

$$Z_n \text{ (est)} = \frac{2\sigma \cos \phi}{r_t g \rho_n} \qquad \text{Eq. (11.4)}$$

where Z_n (est) = critical NAPL height required for NAPL penetration into the vadose zone

$$Z_n \text{ (est)} = \frac{2\sigma \cos \phi}{r g (\rho_n - \rho_w)} \qquad \text{Eq. (11.5)}$$

where Z_n (est) = critical NAPL height required for DNAPL penetration into a saturated aquifer

$$Z_n \text{ (est)} = \frac{2\sigma \cos \phi (1/r_t - 1/r)}{g \, \text{abs}(\rho_n - \rho_w)} \qquad \text{Eq. (11.6)}$$

$$= \frac{P_{c\text{(fine)}} - P_{c\text{(coarse)}}}{g \, \text{abs}(\rho_n - \rho_w)}$$

where Z_n (est) = critical NAPL height required for downward DNAPL penetration or upward LNAPL migration from a coarse-grained material into a fine-grained material

$$\Delta h / \Delta z_n = \frac{\rho_n - \rho_w}{\rho_w} \qquad \text{Eq. (11.7)}$$

where $\Delta h / \Delta z_n$ = minimum hydraulic gradient required to prevent downward DNAPL migration or upward LNAPL migration

σ = interfacial tension between NAPL and water

ϕ = contact angle

r = radius of the water-filled pore that the NAPL must move through to exit or enter pore

r_t = radius of the water-filled pore throat that the NAPL must move through to exit or enter pore

ρ_n = density of NAPL (g/cm^3)

ρ_w = density of water (1 g/cm^3)

g = force of gravity (980 cm/sec^2)

μ = dynamic viscosity (centipoise)

- Minimum hydraulic gradient required to prevent downward DNAPL migration or upward LNAPL migration; see Eq. (11.7)

Example 11.2 CRITICAL DNAPL HEIGHT BELOW WATER TABLE REQUIRED TO PENETRATE FINE-GRAINED UNIT (AFTER COHEN AND MERCER, 1993)

Assume a mobile DNAPL pool is perched above a fine-grained silt layer with a pore radius of 0.005 mm. The DNAPL has a specific gravity of 1.3, an interfacial tension with water of 0.040 N/m, and a contact angle to a mineral solid phase of 35°. How thick does the DNAPL pool have to be before it penetrates the fine-grained silt layer? Assume that residual DNAPL is located above the DNAPL pool.

Solution

$$Z_n \text{ (est)} = \frac{2\sigma \cos \phi}{r_{\text{fine}} g (\rho_n - \rho_w)}$$

$$= \frac{2(0.040 \text{ N/m})(\cos 35°)}{(0.000005 \text{ m})(9.8 \text{ m/sec}^2)(1300 - 1000 \text{ kg/m}^3)}$$

$$= 4.5 \text{ m}$$

Analysis. Substantial thicknesses may be required to penetrate fine-grained units such as clays, silts, and fine sands. Because the capillary pressure relationships are so dependent on pore radius, microscale heterogeneities (such as root holes and slickensides) will often control the overall migration pathway taken by a mobile DNAPL body

Darcy's Law

Movement of free-phase NAPLs through a porous medium can be evaluated by using Darcy's equation. If all or almost all of the open pore space is filled with a continuous, free-phase NAPL mass, Darcy's equation can be applied to NAPL movement in a way similar to ground water flow. The most important exception, however, is that hydraulic conductivity (K) must be replaced by intrinsic permeability (k) to account for the different hydraulic characteristics related to the NAPL fluid. For the simple one-dimensional case,

$$v = -\frac{k\rho g}{\mu}\frac{dh}{dl} \tag{11.8}$$

where v = Darcy velocity (cm/sec)
k = intrinsic permeability (darcies; 1 darcy = 10^{-8} cm^2)
ρ = density of NAPL (g/cm^3)
g = force of gravity (980 cm/sec^2)
μ = dynamic viscosity (centipoise)
dh/dl = hydraulic gradient of NAPL mass

$$h = z + \frac{P}{\rho g} \tag{11.9}$$

z = reference elevation
P = pressure (atmospheres)

For more sophisticated applications, conventional ground water flow models can be adapted to simulate continuous NAPL flow in aquifers by adjusting the hydraulic conductivity in the model to reflect the different viscosity and density of a continuous NAPL phase.

Relative Permeability

When both water and NAPL are present in the aquifer, a multiphase flow regime is established. Multiphase flow occurs when two different fluids flow through a porous medium, compete for available pore space, and thereby reduce the mobility of both fluids (Mercer and Cohen, 1990). This reduction in mobility is defined by the **relative permeability,** the ratio of the effective permeability of a fluid at a given saturation to the intrinsic permeability of the medium. In a specific porous medium, relative permeability of a fluid ranges from 0 to 1.0 based on a complex function of the saturation of the fluid, whether the fluid is wetting or nonwetting, and whether a wetting fluid is displacing a nonwetting fluid (called **imbibition**) or a nonwetting fluid is displacing a wetting fluid (called **drainage**). NAPL moving into the saturated zone is an example of drainage, as the wetting fluid is draining away from the pore. Forcing NAPL out of the pore with water pressure is an example of imbibition.

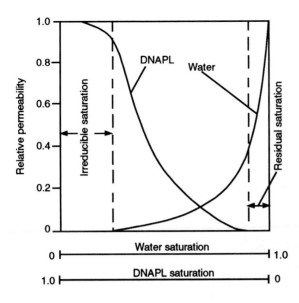

Figure 11.13 Residual saturation curve. Source: Schwille, 1988. © 1988 by Lewis Publishers, a subsidiary of CRC Press, Boca Raton, FL. Reprinted with permission.

Figure 11.13 shows a typical relative-permeability diagram for a two-fluid system comprised of NAPL and water and illustrates how the two fluids interfere with each other to reduce mobility. At most parts of the relative-saturation curve, the sums of the relative permeabilities of the NAPL (the nonwetting fluid in a NAPL–water mixture) and the water (the wetting fluid) do not equal 1 because interference reduces the overall mobility of both fluids in the porous medium (see Example 11.3). One additional point not shown on this relative-permeability curve is the difference in relative permeability for NAPLs undergoing drainage versus imbibition. As we might expect, it is easier for NAPL to enter a pore than it is to leave a pore, meaning that a relative-permeability diagram for drainage will show higher relative permeability than the relative-permeability diagram for imbibition.

Example 11.3 RELATIVE PERMEABILITY

A NAPL has entered an aquifer, and a laboratory study of cores taken from the site yielded the relative-permeability curves shown in Figure 11.13. If the average NAPL saturation (S_{ir}) is 0.20, what is the relative permeability of the NAPL and the relative permeability of water? What are the implications of these relative-permeability numbers?

Solution. Using Figure 11.13, it can be seen that, for a NAPL saturation (S_{ir}) of 0.20, the relative permeability for the NAPL is 0.02. The fraction of water in the open pore space is 0.80, so that the figure indicates the relative permeability of the water is 0.20.

For the case of water, a relative permeability of 0.20 means that the water will flow through the current mixture of sand and trapped NAPL at only 20% of the rate that water would flow through the sand alone (no NAPL). Similarly, the NAPL will flow through the sand–water mixture at a rate that is only 20% of the flow rate through a sand that is completely saturated with NAPL.

Analysis. The movement of NAPL and water through a porous media slows down significantly when oil and water are mixed together. To maximize the removal of a large free-phase NAPL pool, the continuous NAPL pool should be maintained for as long as possible to avoid the interference caused by pumping a combined NAPL and water mixture.

Using the same relative permeability curve, the type of flow regime can be described as a function of saturation, as shown in Figure 11.14 (Williams and Wilder, 1971). When the NAPL saturation is high (zone I), the NAPL mass is continuous and NAPL transport dominates. Because the water is trapped in small, isolated pores, the water phase is noncontinuous and the relative permeability of water is very low or zero. In zone II, both NAPL and water are continuous, but do not share the same pore spaces. Due to interference, however, the relative permeabilities of both fluids are greatly reduced. Zone III represents the case where the NAPL is discontinuous and is trapped as a residual hydrocarbon in isolated pores. Flow is dominated by water movement in this region, and there is little or no NAPL flow under most conditions. The saturation at which the relative permeability of the NAPL phase becomes zero is defined as the residual saturation for NAPL; all the NAPL is discontinuous, and no NAPL flow is observed.

Typical relative-permeability data are available in the technical literature (Saraf and McCafferty, 1982; Honarpour and Mahmood 1987; Lin et al., 1982) and are useful for illustrating the general relationship between permeability and saturation. The shape of the relative-permeability curve is related to the (1) intrinsic permeability, (2) pore-size distribution, (3) ratio of fluid viscosities, (4) interfacial tension, and (5) wettability (Domenico and Schwartz, 1990; Demond and Roberts, 1987). In actual practice, however, relative-permeability relationships are rarely developed from individual field studies (Mercer and Cohen, 1990).

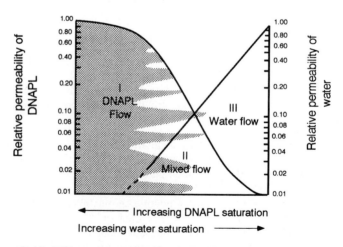

Figure 11.14 Different flow regimes in NAPL–water system. Source: Williams and Wilder, 1971.

NAPL Computations at the Site Level

Computer Models for Continuous-phase NAPL Migration

The petroleum industry pioneered the use of computer models to simulate the migration of NAPLs in the subsurface in the 1960s, and during the 1980s environmental scientists and engineers began to adapt these types of codes to NAPL contamination problems. While the current family of environmental multi-phase models can simulate a variety of ground water problems, the data requirements are significant and difficult to estimate accurately (Mercer and Cohen, 1990). Few if any field sites can be characterized accurately enough to define the medium- and small-scale heterogeneities in pore size and permeability that can dominate DNAPL migration. Therefore, most models are used for conceptual studies only or for prediction of gross migration patterns of DNAPL in the subsurface. LNAPL models can be applied with more confidence because the pooled LNAPL layer is easier to locate and characterize. Reviews of existing computer models for simulating NAPL flow for ground water problems are provided by Pinder and Abriola (1986) and Abriola (1988).

Residual NAPL Relationships for Design of Remediation Systems

While the capillary forces that hold residual NAPL in pores are relatively strong, they can be overcome by gravity forces associated with buoyancy–density forces or by viscous forces associated with ground water flow. The ratio of capillary forces to gravity forces is known as the **bond number** (N_b), while the ratio of capillary forces to viscous forces (water pressure) is called the **capillary number** (N_c):

$$N_b = \frac{k \Delta \rho g}{\sigma} \tag{11.10}$$

$$N_c = \frac{-k \rho g \; \Delta h}{\sigma \; \Delta l} \tag{11.11}$$

or

$$N_c = \frac{-K \; \Delta h \mu}{\sigma \; \Delta l} \tag{11.12}$$

where k = intrinsic permeability (cm^2)
ρ = density of water (g/cm^3)
σ = interfacial tension (dyne/cm)
g = gravitational acceleration (g/cm^2)
$\Delta \rho$ = fluid–fluid density difference (g/cm^3)
$\Delta h / \Delta l$ = hydraulic gradient
μ = dynamic viscosity of water (g/cm-sec)
K – hydraulic conductivity (cm/sec)

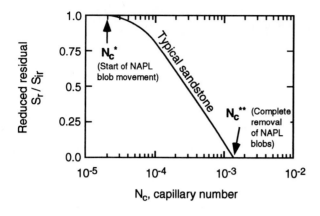

Figure 11.15 Residual hydrocarbon saturation ratio, relating final residual saturation S_r to initial residual saturation S_{ir} as a function of capillary number N_c. Source: Wilson and Conrad, 1984.

The bond and capillary number relationships can be used with empirical data to estimate the change in residual saturation due to gravity or viscous forces. Figure 11.15 shows an example of a residual saturation–capillary number curve for a sandstone material. As the capillary number increases, either from an increase in ground water velocity or a reduction in interfacial tension, the amount of residual saturation decreases. On this curve, the change in residual saturation is represented by the ratio of current saturation (S_i) to initial residual saturation (S_{ir}). The first residual hydrocarbon blobs are mobilized when $N_c = 2 \times 10^{-5}$, and all the residual hydrocarbon blobs are mobilized when $N_c = 2 \times 10^{-3}$. Note that this curve is for a particular sandstone and that other porous media will have curves with different shapes and end points.

Capillary numbers can be used to estimate the potential efficiency of a hydro-

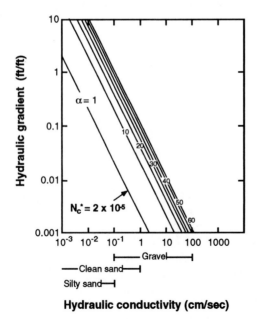

Figure 11.16 Hydraulic gradient necessary to initialize blob mobilization (at N_c^*) in soils of various permeabilities, for hydrocarbons of various interfacial tensions, α. Source: Wilson and Conrad, 1984.

carbon removal scheme employing either high hydraulic gradients or reduced interfacial tension, or both. With the empirical data from the sandstone material in Figure 11.15, Wilson and Conrad (1984) showed the magnitude of the hydraulic gradient required to mobilize NAPL blobs under various conditions (see Figure 11.16). These data show that complete mobilization of residual hydrocarbons is very difficult or impossible to achieve in most aquifers using hydraulic gradient alone (Wilson and Conrad, 1984). The required hydraulic gradients are so high for most aquifers (greater than 1 ft/ft) that no reasonable configuration of pumping and injection wells could be designed to sweep all the residual NAPL trapped in the pores of the aquifer. Only in a few cases where permeability is exceptionally high, such as gravel aquifers, is it possible to remove all or most of the residual hydrocarbons using the force of water alone.

Addition of surfactants (agents such as polymers, alcohols, or detergents) can also increase the capillary number by reducing the interfacial tension of the NAPL–water blobs. (See the discussion of Emerging Technologies in Chapter 13 for information regarding the use of surfactant agents for NAPL removal.)

11.5 FATE OF NAPLs IN THE SUBSURFACE

As indicated by the transport relationships described in the preceding section, much of the NAPL that is released in the subsurface becomes tightly trapped by capillary forces in the pores and fractures of the soil–aquifer matrix. Eventually, however, the hydrocarbons that comprise the NAPLs are either (1) transferred to the air, (2) transformed to basic chemical compounds such as carbon dioxide and water, or (3) transferred to the water. These three processes are called **volatilization, in situ biodegradation–hydrolysis,** and **dissolution.**

Volatilization

In the vadose zone, volatilization can occur when contaminated water, soil containing sorbed hydrocarbons, or NAPL comes in contact with air. As described in Chapter 7, Henry's law is used to relate the equilibrium conditions between soluble constituents dissolved in water and the vapors in air. NAPLs in the subsurface will also volatilize, but in a different way. For a single-component NAPL, the vapor pressure defines the equilibrium between NAPL and air: the vapor pressure found in chemical reference books is equal to the concentration of the vapor in air that is in equilibrium with the single-component NAPL. For NAPL mixtures, equilibrium conditions for a hydrocarbon are described by principles from Raoult's law:

$$P_a = X_a P_0^a \qquad (11.13)$$

where P_a = vapor pressure of the NAPL mixture (atm)

P_0^a = vapor pressure of the hydrocarbon of a as a pure-phase component (atm)

X_a = mole fraction of hydrocarbon a in the NAPL mixture

Note that if the NAPL is a single component the mole fraction equals 1.0 and the pure-phase vapor pressure can be used to estimate the vapor pressure in the system of interest. See Example 11.4 for more details on the application of Raoult's law.

Example 11.4 VAPOR CONCENTRATIONS AND RAOULT'S LAW

A NAPL of pure benzene is released to the unsaturated zone. (a) What is the *theoretical* concentration of benzene in an air stream that passes through the contaminated zone? (b) What is the concentration of benzene in the air if the NAPL is comprised of only 20% by mole fraction NAPL and 80% nonvolatile organics?

Solution

(a) The theoretical concentration of benzene in air for the pure benzene NAPL can be expressed in two ways, either as a volumetric concentration in units of parts per million volume (ppmv) or as a unit of mg/L (note that this is not equivalent to ppm when air is the media). To calculate a volumetrically based concentration for the pure benzene, convert from the vapor pressure expressed in atmospheres:

$$P_a = X_a P_0^o$$

$$P_a = (1)(0.10 \text{ atm})$$

$$P_0 = 0.10 \text{ atm}$$

$$C \text{ (ppmv)} = P_a \text{ (atm) } 1{,}000{,}000 = 100{,}000 \text{ ppmv}$$

where X_a = mole fraction of benzene in NAPL (1 for a pure benzene NAPL)
P_0^o = vapor pressure of pure-phase component (0.10 atm or 76 mm of mercury for benzene)
C = volumetric concentrations of benzene in air (parts per million volume)

To calculate the concentration in milligrams of benzene per liter of air, use this expression:

$$C \text{ (mg/L)} = \frac{X_a P_0^o MW}{RT}$$

$$= \frac{1.0 \text{ atm } 78.1 \text{ g/mol } 1000 \text{ mg/g}}{0.0821 \text{ 1-atm/mol}^\circ \text{ K } (293^\circ \text{ K})}$$

$$= 325 \text{ mg/L (at } 20^\circ \text{C)}$$

(b) If the NAPL contains only a 20% mole fraction of benzene, use Raoult's law to estimate the actual vapor pressure:

$$P_a = X_a P_0^o$$

$$= 0.20(0.10 \text{ atm})$$

$$= 0.02 \text{ atm}$$

To convert to a mass-based concentration, use this expression:

$$C \text{ (mg/L)} = \frac{X_a P_0^o MW}{RT}$$

$$= \frac{0.2 \ 0.10 \text{ atm } 78.1 \text{ g/mol } 1000 \text{ mg/g)}}{0.0821 \text{ 1-atm/mol}^\circ \text{ K } (293^\circ \text{ K})}$$

$$= 65 \text{ mg/L (at } 20^\circ \text{C)}$$

Analysis. Different concentrations have different practical applications: volumetric concentrations are often compared against health-based air standards, while the mass-based concentrations are important for remediation design. In the field, these theoretical concentrations are rarely observed, largely because much of the air in the subsurface never contacts NAPL due to the effect of heterogeneities.

Vapor transport through the soil under ambient conditions is usually limited by diffusion, because a large number of "dead-end" spaces in the subsurface, partially filled with NAPL, do not allow air to pass by directly. Therefore, volatilization occurs very slowly as the vapors leave the NAPL and diffuse toward zones of lower concentration, that is, the pathways carrying air. Several empirical equations are available to describe diffusion-limited volatilization from contaminated soils that are exposed to the atmosphere (Lyman et al., 1982). For example, under conditions where the contaminated soil layers can be assumed to be semi-infinite (the total depth of the zone undergoing removal of organics by volatilization is small compared to the total contaminated depth), the Hamaker method (Lyman et al., 1982) can be used to model volatilization from the soil:

$$Q_t = 2C_0 \left(\frac{D_t}{\pi}\right)^{-0.5}$$ (11.14)

where Q_t = mass loss over time per unit area of exposed soil for some time t (M/L^2)
 C_0 = initial concentration of soil (M/L^3)
 D_t = diffusion coefficient of vapor through soil (L^2/T)

Chemical and Biological Degradation

Hydrolysis

Hydrolysis is probably the most important nonbiological reaction for many NAPL-related chemicals, particularly the chlorinated solvents that comprise many DNAPLs. Hydrolysis can be defined as a nonbiological transformation in which an organic chemical reacts with water (or a component ion of water) to form a derivative organic chemical (Domenico and Schwartz, 1990):

$$R–X \Rightarrow R–OH + X^- + H^+$$ (11.15)

In this reaction R–X is an organic compound and X represents an attached halogen, carbon, phosphorus, or nitrogen group. Lyman et al. (1982) list functional groups that are susceptible to hydrolysis reactions and groups that are resistant to hydrolysis. When considering the effect of hydrolysis on NAPLs, note that only the outside of the NAPL mass actually contacts water, thereby preventing hydrolysis of the organic molecules on the inside of the blob. Therefore, hydrolysis of NAPLs is an indirect process controlled by the rate of dissolution of organics from NAPL to the dissolved phase (see Chapter 7 and the discussion of dissolution in the next section).

In Situ Biodegradation

Aerobic biological transformation is one of the key processes that controls the fate of many LNAPL releases such as petroleum hydrocarbon spills from leaking underground storage tanks. Indigenous aerobic bacteria can react with dissolved oxygen to consume some of the immiscible-phase LNAPL directly and form a biosurfactant that increases the dissolution rate of hydrocarbon into ground water. The supply of dissolved oxygen to the bacteria is almost always the limiting process for in situ biodegradation of NAPLs, as shown by Example 11.5.

Example 11.5   AEROBIC IN SITU BIODEGRADATION OF LNAPLs

How long does it take to naturally biodegrade a 100-gal (approximately 250-kg) gasoline spill in the saturated zone? Use the following assumptions:

Depth of LNAPL penetration into saturated zone: 2 m

Width of LNAPL release in saturated zone: 10 m

Ground water Darcy velocity: 1 m/day

Background oxygen concentration: 5 mg/L

Oxygen–hydrocarbon consumption ratio: 2 mg oxygen/1 mg hydrocarbon

Solution

$$\text{Time to degrade} = [250 \text{ kg LNAPL}] \left[\frac{10^6 \text{ mg}}{\text{kg}} \right] \left[\frac{2 \text{ mg O}_2}{1 \text{ mg LNAPL}} \right]$$

$$\left[\frac{L}{5 \text{ mg O}_2} \right] \left[\frac{1}{(2 \text{ m})(10 \text{ m})} \frac{\text{Day}}{1 \text{ m}} \right] \left[\frac{M^3}{10^3 \text{ L}} \right]$$

Time to degrade LNAPL = 5,000 days or about 14 yr

Analysis. This time estimate assumes that all the LNAPL will be converted through biodegradation only. At actual field sites, the soluble components in the gasoline (such as benzene, toluene, ethyl benzene, and xylene) are removed relatively rapidly by a combination of dissolution, volatilization, and in situ biodegradation.

Although a dominant factor in the consideration of the fate of most LNAPL releases, in situ biodegradation is a minor process at most DNAPL sites. DNAPLs that are comprised of chlorinated solvents are very resistant to degradation by aerobic bacteria and only degrade slowly under most anaerobic conditions. At creosote–wood preservation DNAPL sites, however, in situ biodegradation can degrade significant quantities of PAHs (polynuclear aromatic hydrocarbons) (see Chapter 8).

Dissolution

The transfer of soluble organics from an immiscible liquid (such as a NAPL) to the water is called **dissolution** (see Figure 11.17). In the vadose zone, infiltration water moving past NAPL will dissolve soluble hydrocarbons and transport them

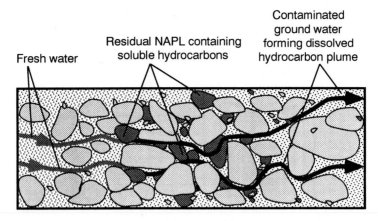

Figure 11.17   NAPL dissolution of residual NAPL, forming dissolved hydrocarbon plume.

to the saturated zone. In aquifers, dissolution occurs when ground water dissolves residual components of NAPL blobs or a large continuous-phase NAPL pool. The current state of knowledge about ground water remediation indicates that most sites are probably affected by NAPLs and that the dissolved plumes are a symptom of NAPL contamination in either the vadose or saturated zone.

The most important factors controlling the dissolution rate are the saturation of the NAPL in the subsurface and the **effective solubility** of the dissolving hydrocarbons. Simply put, higher dissolution rates are associated with more soluble hydrocarbons in a NAPL and/or large amounts of NAPL in the subsurface. Secondary parameters that influence the dissolution rate include ground water seepage velocity and porosity. Particle size, on the other hand, has been shown to have no significant relationship to dissolution (Miller et al., 1990). With several of these key values estimated from engineering judgment or obtained from field studies, an estimate of dissolution rate can be obtained and used for the purpose of predicting how long the NAPL will serve as an active source of dissolved hydrocarbons to ground water.

Effective Solubility Relationship

For single-component NAPLs, the pure-phase solubility of the organic constituent can be used to estimate the theoretical upper-level concentration of organics in aquifers or for performing dissolution calculations. For NAPLs comprised of a mixture of chemicals, however, the **effective solubility** concept should be employed. Effective solubility is defined as the theoretical aqueous solubility of an organic constituent in ground water that is in chemical equilibrium with a mixed NAPL (a NAPL containing several organic chemicals). The effective solubility of a particular organic chemical can be estimated by multiplying its mole fraction in the NAPL mixture by its pure-phase solubility, as shown in the following effective solubility relationship based on principles from Raoult's law (see Borden and Kao, 1992; Feenstra et al., 1991):

$$S_i^e = X_i S_i \gamma_i \tag{11.16}$$

where S_i^e = effective solubility of indicator constituent i in the water phase (mg/L)

X_i = mole fraction of indicator constituent i in the DNAPL phase (unitless)

S_i = solubility of indicator constituent i in water (mg/L)

γ_i = correction factor to adjust for constituent concentrations observed in the field (unitless; thought to be in the 0.10 to 0.50 range)

Effective solubilities can be calculated for all components in a DNAPL mixture (see Example 11.6). The initial mole fraction of the soluble organic in the NAPL phase is determined through laboratory analysis. Insoluble organics in the mixture (such as long-chained alkanes) will reduce the mole fraction and effective solubility of more soluble organics, but will not contribute dissolved-phase organics to ground water. Note that this relationship is approximate and does not account for nonideal behavior of mixtures, such as cosolvency and other factors.

The effective solubility relationship is capable of explaining only some of the gap between theoretical contaminant solubilities in ground water and observed values. It is extremely rare to observe dissolved contaminant concentrations approaching the effective solubility at actual hazardous-waste sites. Typically, concentrations less than 10% of effective solubility are observed, even in NAPL-rich zones. To account for the remaining difference, the effective solubility relationship presented here includes a correction factor to account for additional and poorly understood factors that also control the observed concentrations of contaminants in ground water. Anderson et al. (1992) evaluated several possible mechanisms to explain the observed discrepancy and hypothesized that many ground water streamlines that flow through the NAPL zone do not come in contact with NAPLs. Therefore, contaminated water was being diluted by clean water, reducing the observed concentration of the sample sent to the laboratory. In practice, the value of the correction factor is estimated by comparing actual concentrations in ground water in DNAPL zones to the concentrations predicted by the effective solubility relationship.

Example 11.6 EFFECTIVE SOLUBILITY

Calculate the effective solubility of a DNAPL with a mole fraction of trichloroethylene (TCE) equal to 0.10 and compare to observed maximum concentrations in ground water of 25 mg/L. Assume the correction factor is equal to 1.

Solution

$$S_i^e = X_i S_i \gamma_i$$

$$= 0.10(1100 \text{ mg/L})(1 \text{ assumed})$$

$$= 110 \text{ mg/L}$$

Analysis. Soluble organics found in mixed DNAPLs have a lower theoretical solubility in water than if the organics were found as a pure-component DNAPL. This is one reason why near-solubility concentrations are never found in the field, because many of the DNAPLs causing dissolved plumes are mixed DNAPLs. Another reason for not observing near-saturation concentrations in contaminated aquifers is dilution of contaminated zones with clean water during the sampling process. Even in an area of high DNAPL saturation, some streamlines never come into contact with the DNAPL, particularly for monitoring wells with long screen lengths (see Anderson et al., 1992).

Predicting Dissolution Rates Over Time

As with the volatilization of NAPLs, the dissolution rate of a mixed NAPL will decline over time due to the reduction in mole fraction of soluble components in the NAPL. Finite-difference models (see Borden and Kao, 1992) or computer spreadsheets can be used to simulate the slow decline in dissolved hydrocarbon concentration over time that is originating from a NAPL zone.

11.6 CHARACTERIZING NAPLs AT REMEDIATION SITES

Although the presence of NAPLs can dominate the remediation process at a site, it is often difficult to locate NAPL zones or even to confirm that NAPL is present at a site. The following sections summarize the most important techniques, both direct and indirect, that can be employed to characterize the distribution of NAPLs at a hazardous-waste site. Cohen and Mercer (1993) provide a rich collection of procedures for evaluating DNAPL sites, many of which can also be applied to investigation of LNAPL problems. The U.S. EPA (1992a) provides a worksheet and flow charts to help determine the potential for occurrence of DNAPL at hazardous-waste sites.

Direct Measure

Apparent versus Actual LNAPL Thickness

Analysis of free-phase hydrocarbons in a well is the best evidence that NAPLs are present in the subsurface. As would be expected, a floating layer in a monitoring well indicates the presence of a LNAPL layer floating on top of the water table, while a dense layer at the bottom of a well indicates a DNAPL in the formation. The thickness of the LNAPL layer in the well can provide an order-of-magnitude indication of the thickness of the actual LNAPL layer in the formation. The apparent thickness in the well is usually 2 to 10 times thicker than the actual product thickness in the aquifer (Mercer and Cohen, 1990). The primary reason for this difference is illustrated in Figure 11.18, in which the LNAPL in the formation is shown floating on top of the capillary fringe, while the LNAPL in the well floats on top of, and partially depresses, the actual water table.

The actual relationship between the apparent and true LNAPL thickness is very difficult to predict. Based on a review of many of these methods, Hampton and Miller (1988) proposed the following simplified relationship for obtaining crude approximations of the true thickness of LNAPL-saturated formation. LNAPL thickness in a formation (h_f) may be approximated by

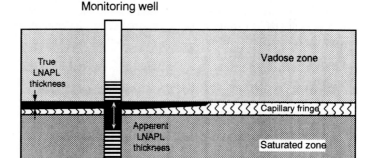

Figure 11.18 LNAPL floating on top of capillary fringe in formation and accumulating on top of water table in well.

$$h_f \approx \frac{h_w(\rho_{\text{water}} - \rho_{\text{LNAPL}})}{\rho_{\text{LNAPL}}} \tag{11.17}$$

where h_f = thickness of LNAPL in formation
h_w = thickness of LNAPL in well
ρ_{water} = density of water
ρ_{LNAPL} = density of LNAPL

Note that this relationship is only an approximation of actual conditions in the field. Some specialized field procedures, such as dielectric logging, may be more useful for estimating the true product thickness in the formation (Kemblowski and Chiang, 1989; Keech, 1988).

DNAPL Thickness in Wells

While apparent product thickness can provide a general indication of the amount of LNAPL in the field, DNAPL thickness data are much more difficult to interpret and use. Unless the DNAPL observation well is screened directly within a DNAPL pool or lens and is constructed correctly, the apparent DNAPL thickness can give erroneous indications of the quantity of DNAPL in the formation (see Figure 11.19). In general, DNAPL accumulation in wells only proves that mobile DNAPL is present in the subsurface and does not provide any useful information about the amount of DNAPL present (Waterloo Centre for Ground Water Research, 1991).

One important point to remember is that the absence of NAPL in a well does not mean that there is no NAPL in the aquifer. If a screen is set below the surface of the water table, for example, the floating LNAPL layer will ride above the screen, and buoyancy forces will prevent the LNAPL from entering the well. Similarly, a well that does not intersect any DNAPL-flowing zones or is screened above a DNAPL pool or lens will not indicate the presence of DNAPL. Considerable care must be taken to make sure that the well screen will intersect the zone that may have NAPLs. Most importantly, a well will not exhibit any NAPL if the

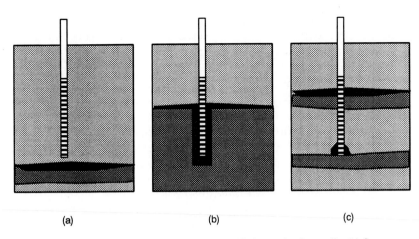

Figure 11.19 Problems with measuring DNAPL in monitoring wells. (a) Screen too high above confining unit. No accumulation. (b) Screen too deep below confining unit. Too much accumulation. (c) Screen penetrates confining unit. Accumulation in wrong place.

screened interval contains only residual NAPL. The residual NAPL is trapped in the porous media and is prevented by capillary forces from migrating into the well.

Visual Examination of Soil Samples

In cases of gross NAPL contamination, a visual inspection of a soil core can prove that NAPLs are present in the subsurface. NAPL in a saturated core, for example, is usually fairly easy to discern using the naked eye. NAPL that is present in small secondary porosity features or tied up as small residual blobs may be very difficult to observe, however, particularly if the NAPL has a color similar to the soil matrix. Some field investigators recommend shaking the soil core in a jar with water to separate the DNAPL from the soil or performing a paint filter test to increase the chance for visual detection. In addition, a black light may help identify many NAPLs containing aromatic and other fluorescent hydrocarbons, and the use of hydrophobic dyes such as Sudan IV can help make NAPLs more visible to the naked eye (Cohen and Mercer, 1993).

Indirect Measures

Ground Water Concentration and Distribution

Although we might expect that concentrations of dissolved chemicals in NAPL zones would be near the solubility of these chemicals, in actual practice near-solubility concentrations are rarely observed in the field. As described previously, two main explanations are often given for this phenomenon: (1) the effective solubility relationship limits the actual concentration of dissolved constituents in ground water in contact with mixed NAPLs, and (2) the typical monitoring well, constructed with a 10-20-ft-long screen, intersects many "clean" flow lines that do not contact NAPL, resulting in a dilution of the "dirty" NAPL-related flow lines.

Because of these factors, a general rule of thumb for NAPL states that, if concentrations of NAPL-related chemicals in ground water are greater than 1% of the pure-phase or effective solubility, then NAPLs are probably present at the site (Waterloo Centre for Ground Water Research, 1991; U.S. EPA, 1992a). In general, the more wells that exceed this trigger level and the higher the concentrations observed, the greater the likelihood that NAPLs are present at a site.

The distribution of dissolved constituents at a site can also be used to indicate the presence of NAPLs. If dissolved NAPL-related chemicals appear in anomalous upgradient–across-gradient locations, a free-phase floating LNAPL layer or DNAPL mass may have moved across the site and now serves as a source of dissolved constituents to ground water. Finally, if DNAPL-related chemical concentrations appear to increase with depth, the presence of DNAPLs is indicated (Cherry and Feenstra, 1991).

Chemical Analysis of Soil–Aquifer Matrix Samples

A general rule of thumb for soil samples is that, if greater than 10,000 mg/kg of hydrocarbon contamination (1% of soil mass) is observed in the soil or aquifer matrix sample, then that sample probably contains some NAPL (Waterloo Center for Ground Water Research, 1991). A more exact method employs a partitioning calculation based on the chemical analyses of soil samples from the saturated zone and the effective solubility concept. This method tests the initial hypothesis that all the organics in the subsurface are either dissolved in ground water or adsorbed to soil (assuming dissolved-phase sorption, not the presence of NAPL). By using the concentration of organics on the soil and the partitioning calculation, a theoretical pore-water concentration of organics in ground water is determined. If the theoretical pore-water concentration is greater than the estimated solubility of the organic constituent of interest, NAPL may be present at the site (Feenstra et al., 1991). This procedure requires the following steps:

Step 1: Calculate S_i^e, the effective solubility of organic constituent of interest.

Step 2: Determine K_{oc}, the organic carbon–water partition coefficient from literature sources or from empirical relationships based on K_{ow}, the octanol–water partition coefficient, which is also found in the literature. For example, K_{oc} can be estimated from K_{ow} using the following expression developed for polyaromatic hydrocarbons:

$$\log K_{oc} = (1.0) (\log K_{ow}) - 0.21 \qquad (11.18)$$

This and other empirical relationships between K_{oc} and K_{ow} are presented in Chapter 7.

Step 3: Determine f_{oc}, the fraction of organic carbon on the soil, from a laboratory analysis of clean soils from the site. Values for f_{oc} typically range from 0.03 to 0.00017 mg/mg (Domenico and Schwartz, 1990). Convert values reported in percent to mg/mg.

Step 4: Determine or estimate ρ_b, the dry bulk density of the soil, from a soil's analysis. Typical values for ρ_b range from 1.8 to 2.1 g/mL (kg/L). Determine or estimate φ_w, the water-filled porosity.

Step 5: Determine K_d, the partition (or distribution) coefficient between the pore water (ground water) and the soil solids:

$$K_d = K_{oc} f_{oc} \qquad (11.19)$$

Step 6: Using C_t, the measured concentration of the organic compound in saturated soil in mg/kg, calculate the theoretical pore-water concentration assuming no DNAPL (that is, C_w in mg/L):

$$C_w = \frac{C_t \rho_b}{K_d \rho_b + \varphi_w} \qquad (11.20)$$

Step 7: Compare C_w and S_i^e (from step 1):

$$C_w < S_i^e \text{ suggests possible absence of DNAPL} \qquad (11.21)$$

$$C_w > S_i^e \text{ suggests possible presence of DNAPL} \qquad (11.22)$$

Geophysics

Because of the difficulty in locating NAPLs in the subsurface, borehole and surface geophysics are becoming more popular for site characterizations. Ground-penetrating radar, complex resistivity, and electromagnetic induction have been applied successfully to the detection of aqueous and nonaqueous hydrocarbons at some sites (U.S. EPA, 1992b). The value of these methods increases with the number of applications at a particular site; these geophysical techniques are better suited for detecting subtle changes in subsurface composition, such as a moving free-phase NAPL mass, than in assessing static conditions. Currently, the application of geophysics at hazardous-waste sites is limited by the paucity of results from research sites and by the small number of personnel that are trained to use geophysics for ground water remediation problems.

Integrated Approach for Determining Potential Occurrence of DNAPLs

In practice, data from several sources are used together to determine if NAPLs are present at a hazardous-waste site. One example of an integrated approach for determining if DNAPL is present at Superfund sites uses both direct and indirect evidence (U.S. EPA, 1992a). This preliminary screening approach can be employed to determine the need for implementation of a full-scale DNAPL detection and delineation program. Evaluation of low, moderate, or high potential for DNAPL presence involves the following methodology:

Step 1: Historical Site Use Information

Certain industries, industrial processes, and chemicals correlate strongly with the presence of DNAPL in the saturated zone and/or the vadose zone at a hazardous waste site, such as the following:

- Wood preservation (creosote)
- Old coal gas plants (mid-1800s to mid-1900s)
- Electronics manufacturing
- Solvent production
- Pesticide manufacturing
- Herbicide manufacturing
- Airplane maintenance
- Commercial dry cleaning
- Instrument manufacturing
- Transformer oil production
- Transformer reprocessing
- Steel industry coking operations (coal tar)
- Pipeline compressor stations
- Metal cleaning and degreasing
- Metal machining
- Tool-and-die operations
- Paint removing and stripping
- Storage of solvents in underground storage tanks
- Storage of drummed solvents in uncontained storage areas
- Solvent loading and unloading
- Disposal of mixed chemical wastes in landfills
- Treatment of mixed chemical wastes in lagoons or ponds

Chemicals used at the site that are indicative of the presence of DNAPL include halogenated volatiles (such as dichloroethane, methylene chloride, and trichloroethylene), halogenated semivolatiles (such as 1,2-dichlorobenzene or aroclor), polyaromatic hydrocarbons (PAHs), and chemical mixtures such as creosote or coal tar.

Step 2: Site Characterization Data

Both direct measures of DNAPL presence and indirect indicators are included to help determine if DNAPL is present. Example data that are used include evidence of DNAPL accumulation in wells, concentrations of ground water samples, and concentrations of soil samples (see the methods described previously).

Step 3: DNAPL Detection Decision Matrix

Following completion of the historical site use and site characterization evaluation, a DNAPL decision matrix is used to define the potential for DNAPL presence in the site soil or the underlying ground water system. If a confirmed high potential or moderate potential for DNAPL occurrence is identified, a DNAPL detection and delineation field program should be implemented at the site.

Special Considerations for Designing DNAPL Delineation Programs

Special precautions are required to avoid inadvertent creation or enhancement of DNAPL migration pathways during the course of hydrogeologic investigations conducted at DNAPL-contaminated sites. Given the ability of DNAPL materials to move downward through very small fractures (20 μm) within the soil mass, conventional grouting procedures may not prove effective for soil borings or wells drilled directly through a DNAPL perching stratum and into an underlying, clean unit. In developing a DNAPL detection and delineation work plan, the following concepts should be applied:

1. *Outside-in approach:* Prior to penetrating a suspected DNAPL zone, critical stratigraphic features should first be identified by investigations conducted outside the area of concern. These preliminary data should be analyzed to identify potential perching strata or "safety nets" beneath the DNAPL zone. All drilling subsequently conducted within the area of DNAPL occurrence must be terminated at the depth of the uppermost, continuous safety-net stratum. In thick, fractured rock settings, such perching strata may not be present, and the risks associated with drilling inside the DNAPL zone must be carefully weighed against the need for vertical delineation data.

2. *Soil borings and wells:* To avoid inadvertent penetration of perching layers within DNAPL zones, soil borings should be sampled continuously with depth and terminated at or near the surface of such confining strata. For detection of free-phase DNAPL, observation wells must be screened across the upper surface of a perching layer (see Figure 11.5). However, under no conditions should the well boring penetrate the full thickness of the base stratum. Wells should be constructed with short screen sections and double-cased through the depth of any overlying DNAPL-contaminated sections.

3. *Noninvasive site investigation techniques:* Where applicable, noninvasive techniques should be employed to indicate DNAPL presence or to characterize site stratigraphy. Geophysical methods can be applied to define the presence and topography of perching strata. In addition, shallow soil gas sampling above the zone of DNAPL occurrence has proved to be a useful indicator of DNAPL presence in some cases.

In general, the risks involved with LNAPL delineation techniques are much smaller than for DNAPL, largely because the downward migration of LNAPL is restricted by the water table and associated capillary fringe.

SUMMARY

Nonaqueous phase liquids (NAPLs) are contaminants that exist in a separate immiscible (nondissolved) phase in the subsurface. Because they are a separate phase, NAPLs have a much different fate and transport properties than dissolved contaminant plumes. NAPLs migrate through the subsurface under gravity and buoyancy forces and do not move along the general flow of ground water. While NAPLs can migrate through the subsurface as a continuous organic mass, it is almost impossible to pump all the NAPL back out. The NAPL breaks up into countless blobs in individual pores that are trapped so tightly by capillary forces that typical pumping measures cannot dislodge them. Because of this phenomenon, it is common to never observe NAPL during site investigations. Therefore, indirect indicators must be used to determine if NAPL is present or not.

NAPLs can be divided into two major types based on density: LNAPLs (lighter-than-water NAPLs) will move downward and then float on top of the water table, whereas DNAPLs (denser-than-water NAPLs) can sink far below the water table. Although this classification system is based on density, it is also a good predictor of where NAPLs come from and what happens to them in the subsurface. In general, LNAPLs are associated with petroleum product spills and have chemicals that are biodegraded in aquifers by naturally occurring microbes. Most DNAPL releases are related to the manufacture and use of chlorinated solvents; these chemicals do not readily biodegrade and are very persistent in ground water.

Sites containing NAPLs usually have many times the contaminant mass as a site with only a dissolved plume. Therefore, it is much more difficult to remediate NAPL sites, particularly if a DNAPL comprised of chlorinated solvents is present. These solvent sites cannot be restored to drinking-water standards using any proven technologies, so the trapped DNAPL can act as a continuing source of ground water contamination for tens or hundreds of years. To effectively manage ground water problems at hazardous-waste sites, the effects of NAPLs should be considered very carefully during site characterization and remediation system design.

REFERENCES

ABRIOLA, L. M., 1988, *Multiphase Flow and Transport Models for Organic Chemicals: A Review and Assessment,* Electric Power Research Institute, Palo Alto, CA, EA-5976.

ANDERSON, M. R., JOHNSON, R. L., and PANKOW, J. F., 1992, "The Dissolution of Dense Chlorinated Solvents into Ground Water: (1) Dissolution from a Well Defined Source," *Ground Water* 30(2):250–255.

ARTHUR D. LITTLE, INC., "The Role of Capillary Pressure in the S Area Landfill," Arthur D. Little, Inc., Boston, MA, Report to Wald, Harkrader & Ross, Washington, D.C., prepared for U.S. EPA.

BEAR, J., 1972, *Dynamics of Fluids in Porous Media,* American Elsevier, New York.

BEAR, J., 1979, *Hydraulics of Groundwater,* McGraw-Hill, New York.

BORDEN, R. C., and KAO, C., 1992, "Evaluation of Groundwater Extraction for Remediation of Petroleum Contaminated Groundwater," *Water Environ. Res.* 64(1):28–36.

CHATZIS, I., and MORROW, N. R., 1984, "Correlation of Capillary Number Relationships for Sandstone," *Soc. Petr. Eng. J.* 24(5):355–562.

CHERRY, J. A., and FEENSTRA, S., 1991, "Identification of DNAPL Sites: An Eleven Point Approach," draft document in *Dense Immiscible Phase Liquid Contaminants in Porous and Fractured Media,* University of Waterloo Short Course, Kitchener, Ont.

COHEN, R. M., MERCER, J. W., 1993 *DNAPL Site Evaluation*, C. K. Smoley, Boca Raton, FL.

CONNOR, J. A., NEWELL, C. J., and WILSON, D. K., 1989, "Assessment, Field Testing, and Conceptual Design for Managing DNAPL at a Superfund Site," *Proc. Petroleum Hydrocarbon and Organic Chemicals in Ground Water,* National Well Water Association, Houston, TX.

DEMOND, A. H., and ROBERTS, P. V., 1987, "An Examination of Relative Permeability Relations for Two-phase Flow in Porous Media," *Water Resources Bull.* 23(4):616–628.

DOMENICO, P. A., and SCHWARTZ, F. W., 1990, *Physical and Chemical Hydrogeology,* Wiley, New York.

FEENSTRA, S., and CHERRY, J. A., 1988, "Subsurface Contamination by Dense Nonaqueous Phase Liquids (DNAPL) Chemicals," *International Groundwater Symposium, International Assoc. of Hydrogeologists,* Halifax, Nova Scotia.

FEENSTRA, S., MACKAY, D. M., and CHERRY, J. A., 1991, "A Method for Assessing Residual NAPL Based on Organic Chemical Concentrations in Soil Samples," *Groundwater Monitoring Rev.* 11(2):128–136.

FETTER, C. W., 1993, *Contaminant Hydrogeology,* Macmillan, New York.

HAMPTON, D. R., and MILLER, P. D. G., 1988, "Laboratory Investigation of the Relationship between Actual and Apparent Product Thickness in Sands," *Petroleum Hydrocarbons and Organic Chemicals in Groundwater: A Conference and Exposition,* National Water Well Association, Houston, TX.

HONARPOUR, M., and MAHMOOD, S. M., 1987, "Relative-permeability Measurements: An Overview," *J. Pet. Technology* 40(8):963–966.

HUNT, J. R., SITAR, N., and UDELL, K. D., 1991, "Nonaqueous Phase Liquid Transport and Cleanup, Analysis of Mechanisms," *Water Resources Res.* 24 No. (8):1247–1258.

KEECH, D. A., 1988, "Hydrocarbon Thickness on Groundwater by Dielectric Well Logging," *Proc. Petroleum Hydrocarbons and Organic Chemicals in Groundwater: A Conference and Exposition,* National Water Well Association, Houston, TX.

KEMBLOWSKI, M. W., and CHIANG, C. Y., 1989, "Hydrocarbon Thickness Fluctuations in Monitoring Wells," *Ground Water* 28(2):244–252.

KUEPER, B. H., and FRIND, E. O., 1988, "An Overview of Immiscible Fingering in Porous Media," *J. Cont. Hydrology* 2:95–110.

KUEPER, B. H., and McWORTER, D. B., 1991, "The Behavior of Dense Nonaqueous Phase Liquids in Fractured Clay and Rock," *Ground Water* 29(5):716–728.

LIN, C., PINDER, G. F., and WOOD, E. F., 1982, Water Resources Program Report 83-WR-2, *Water Resource Prog.,* Princeton University, Princeton, NJ, p. 33.

LYMAN, W. J., REEHL, W. F., and ROSENBLATT, D. H., 1982, *Handbook of Chemical Property Estimation Methods: Environmental Behavior of Organic Compounds,* McGraw-Hill, New York.

MACKAY, D. M., and CHERRY, J. A., 1989, "Ground-water Contamination: Pump and Treat Remediation," *Environ. Sci. Technology* 23(6):630–636.

MACKAY, D. M., ROBERTS, P. V., and CHERRY, J. A., 1985, "Transport of Organic Contaminants in Ground Water," *Environ. Sci. Technology* 19(5):384–392.

MENDOZA, C. A., and McALARY, T. A., 1990, "Modeling of Ground Water Contamination Caused by Organic Solvent Vapors," *Ground Water* 28(2):199–206.

MERCER, J. W., and COHEN, R. M., 1990, "A Review of Immiscible Fluids in the Subsurface: Properties, Models, Characterization and Remediation," *J. Cont. Hydrology* 6: 107–163.

MILLER, C. T., POIRIER-MCNEILL, M. M., and MAYER, A. S., 1990, "Dissolution of Trapped Nonaqueous Phase Liquids: Mass Transfer Characteristics," *Water Resources Res.* 26(11):2783–2796.

NEWELL, C. J., and CONNOR, J. A., 1992, "Detection and Delineation of Subsurface DNAPLs," Detection and Restoration of DNAPLs at Hazardous Waste Sites Preconference Seminar, Water Environment Federation Annual Conference, New Orleans, LA, pp. 51–67.

NEWELL, C. J., CONNOR, J. A., WILSON, D. K., and McHUGH, T. E., 1991, "Impact of Dissolution of Dense Non-Aqueous Phase Liquids (DNAPLs) on Groundwater Remediation," *Petroleum Hydrocarbons and Organic Chemicals in Groundwater: A Conference and Exposition,* National Water Well Association, Houston, TX.

PINDER, G. F., and ABRIOLA, L. M., 1986, "On the Simulation of Nonaqueous Phase Organic Compounds in the Subsurface," *Water Resource Res.* 22(9):109S–119S.

POULSON, M. M., and KUEPER, B. H., 1991, "A Field Experiment to Study the Behavior of Tetrachloroethylene in Unsaturated Porous Media," *Environ. Sci. Technology* 26(5): 889–895.

SARAF, D. N., and McCAFFERTY, F. G., 1982, "Two- and Three-phase Relative Permeabilities: A Review," Petroleum Recovery Institute, Calgary, Alberta, Rep. No. 81-8.

SCHWILLE, F., 1988, *Dense Chlorinated Solvents in Porous and Fractured Media: Model Experiments* (English translation), Lewis Publishers, Ann Arbor, MI.

U.S. EPA, 1990, *Laboratory Investigation of Residual Liquid Organics from Spills, Leaks, and the Disposal of Hazardous Wastes in Groundwater,* EPA/600/6-90/004, U.S. Environmental Protection Agency, Washington, DC.

U.S. EPA, 1991, *Dense Nonaqueous Phase Liquids,* EPA Ground Water Issue Paper, EPA/540/4-91-002, U.S. Environmental Protection Agency, Washington, DC.

U.S. EPA, 1992a, *Estimating Potential for Occurrence of DNAPL at Superfund Sites,* EPA Quick Reference Fact Sheet, EPA Publication 9355.4-07FS, U.S. Environmental Protection Agency, Washington, DC.

U.S. EPA, 1992b, *Dense Nonaqueous Phase Liquids: A Workshop Summary,* EPA Ground Water Issue Paper, EPA/600-R+92/030, U.S. Environmental Protection Agency, Washington, DC.

VILLAUME, J. F., 1985, "Investigations at Sites Contaminated with Dense Non-Aqueous Phase Liquids (NAPLs)," *Ground Water Monitoring Rev.* 5(2):65–72.

Waterloo Centre for Ground Water Research, 1991, "Dense Immiscible Phase Liquid Contaminants in Porous and Fractured Media," University of Waterloo Short Course, Kitchener, Ont.

WILLIAMS, D. E., and WILDER, D. G., 1971, "Gasoline Pollution of a Ground-water Reservoir—A Case History," *Ground Water* 9(6):50–54.

WILSON, J. L., and CONRAD, S. H., 1984, "Is Physical Displacement of Residual Hydrocarbons a Realistic Possibility in Aquifer Restoration?" *Proc. Petroleum Hydrocarbon and Organic Chemicals in Ground Water,* National Water Well Association, Houston, TX.

Chapter 12 | HYDROGEOLOGIC SITE INVESTIGATIONS

By John A. Connor, P.E.

12.1 INTRODUCTION TO HYDROGEOLOGIC INVESTIGATIONS

The purpose of a hydrogeologic site investigation is to characterize soil and ground water pollution problems in sufficient detail to facilitate the design of a cost-effective corrective action program. For this purpose, the site investigation entails actual measurement of the physical processes that control subsurface contaminant transport at a given site. Geologic, hydrologic, and chemical data must be acquired and integrated to define the nature and extent of soil and ground water contamination, as well as the potential migration of these contaminants within the natural ground water flow system.

The preceding chapters of this book have reviewed the general principles of ground water occurrence and flow within geologic formations and have provided an overview of the conventional tools employed for the measurement of these site conditions. In this chapter, the engineering procedures involved in the acquisition and interpretation of such flow information for any given site will be addressed. The following sections outline a systematic approach to planning soil and ground water contamination studies and summarize engineering standards for data evaluation and presentation. Case studies are provided that review the procedures and results of hydrogeologic investigations conducted at various soil and ground water contamination sites. Once a site has been properly characterized, only then can ground water remediation schemes be successful (Chapter 13).

12.2 DEVELOPMENT OF CONCEPTUAL SITE MODEL

Hydrogeologic processes are by nature very complex due to the heterogeneities of geologic formations and the transient effects of aquifer recharge and discharge phenomena. Consequently, detailed characterization of contaminant transport patterns throughout every inch of an aquifer system is impractical. From an engineering perspective, our objective is therefore to define subsurface contaminant transport processes to the degree necessary to allow us to design effective measures for the control or reversal of these processes, as needed to protect public health and the environment.

Ultimately, the protection of drinking-water resources may require us to extract or "mine" the contaminated ground water mass from the affected aquifer. Therefore, it is helpful to approach a ground water contaminant delineation study in much the same manner as prospecting for a precious metal. We do not need to know each twist and turn of every minor "ore" seam, but we do want to know how wide and how deep the play runs and, because our "ore" is a fluid, which way it is moving and how fast.

The hydrogeologic site investigation is the procedure by which we develop our understanding or our "working model" of contaminant plume migration within the ground water flow regime. In all cases, this model of the subsurface environment is constructed of three principal components of information:

1. *Geology:* the physical framework within which subsurface fluids collect and flow
2. *Hydrology:* the movement of fluids through this physical framework
3. *Chemistry:* the nature of the chemical constituents that are entrained in this flow system

We build our model of the site by systematically addressing each of these principal components in turn. First, we must characterize the stratigraphic profile beneath the site and identify the strata serving as potential conduits for fluid flow. Second, we must measure the fluid hydraulic head distribution within the zone of saturation to determine the actual rate and direction of ground water movement through these conduits. Third, water samples are collected and analyzed to map the lateral and vertical extent of contaminant migration within the ground water flow regime.

In actuality, geologic, hydrologic, and chemical information is collected simultaneously during most soil and ground water investigations. It is then the responsibility of the project engineer or scientist to sort this information into a meaningful and accurate picture of subsurface ground water flow and contaminant transport processes.

12.3 GENERAL STRATEGY FOR SOIL AND GROUND WATER CONTAMINATION STUDIES

Basis for Strategic Approach

General Considerations

For the purpose of economy and efficiency, every field and laboratory measurement conducted during a hydrogeologic site investigation must contribute to our conceptual model of the ground water flow regime. Ten monitoring wells clustered in the center of a ground water contaminant plume cost as much to install and sample as 10 wells spatially distributed throughout and beyond the full plume area, but the clustered wells provide far less information regarding plume dimensions and potential migration patterns. To achieve our project objectives in a cost-effective manner, a clear strategy for mapping of the contaminant zones must be established prior to commencement of field or laboratory work.

As a basis for a site investigation strategy, all subsurface contaminant problems should be viewed as two distinct zones of contamination: (1) contaminant source materials and contaminated soils in the unsaturated soil (or rock) zone and (2) nonaqueous phase liquids (NAPLs) and/or ground water containing dissolved contaminants within the zone of saturation (see Figure 12.1). For practical purposes, we can define the vertical boundary between these two zones as the surface of the uppermost water-bearing unit beneath the site (for example, a water-saturated stratum with hydraulic conductivity $K \geq 1 \times 10^{-4}$ cm/sec). These two zones differ significantly in terms of their operant mechanisms of contaminant transport and the requisite corrective actions and therefore should be addressed individually in the course of the hydrogeologic site investigation.

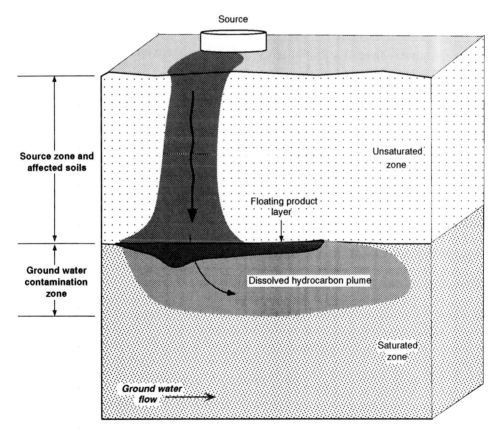

Figure 12.1 Zones of contamination for two-stage site investigation approach.

12.3 General Strategy for Soil and Ground Water Contamination Studies   **391**

Unsaturated Source Zone Characteristics

Most incidents of hazardous chemical release to the subsurface environment occur as surface spills of products or wastes or leachate percolation from the base of waste landfills, surface impoundments, or material stockpiles. As the wetting front percolates downward through the unsaturated soil (or rock) zone underlying the source area, a significant portion of the contaminant mass may be retained in the unsaturated soil matrix due to the effects of filtration, sorption, or capillary retention. For many years thereafter, this contaminated soil can serve as a source of continuing contaminant release to stormwater flowing across the site surface or percolating downward through the unsaturated soil zone to the depth of underlying ground water.

Depending on the size of this residual source zone and the nature of the contaminants, protection of surface water and ground water resources could involve either complete excavation and removal of the contaminated soils, capping of the site to minimize rainfall contact and precipitation, or contaminant extraction by means of in situ soil venting or rinsing. To support the design of appropriate corrective measures, the hydrogeologic site investigation must therefore address the full lateral and vertical extent of residual contaminants within the unsaturated soil zone and the potential for future release of contaminants to local water resources.

Ground Water Plume Characteristics

Dissolved contaminants contained in waste leachate fluids penetrating to the depth of ground water occurrence will become entrained in the natural ground water flow system and spread laterally and vertically in accordance with local ground water flow gradients (see Figure 12.1). Free-phase liquid contaminants may be subject to an additional "density gradient," with light nonaqueous-phase liquids (LNAPLs, such as gasoline) floating atop the zone of saturation and collecting in the structural highs of confined water-bearing units. Alternatively, dense nonaqueous-phase liquids (DNAPLs) can percolate downward through the water-bearing stratum to perch and spread atop underlying confining units (Chapter 11).

In all cases, ground water contamination problems are fluid problems. The contaminant entered the ground water system as a fluid and can therefore be removed or controlled as a fluid. Unlike contamination within the unsaturated soil zone, excavation and removal of the soil or rock mass from the zone of ground water contamination is neither practical nor necessary. Rather, for the purpose of corrective action, hydraulic measures can be implemented to control or reverse the migration of fluid contaminants within the ground water flow system. The hydrogeologic site investigation must therefore provide definitive information on the current lateral and vertical extent of dissolved and free-phase contaminants within the ground water, as well as on the hydraulic processes controlling contaminant migration.

Two-stage Site Investigation Approach

In practice, then, the hydrogeologic site investigation proceeds as a two-stage process: (1) delineate the unsaturated source zone, comprised of the chemical waste or product mass and the associated contaminated soils within the unsaturated soil column, and (2) investigate the presence and extent of contaminant migration within the underlying ground water system. Step-by-step strategies for implementation of these source zone and ground water contamination delineation studies are outlined next and illustrated in Figures 12.2 through 12.4.

Procedures for Unsaturated Source Zone Characterization

Objectives. The objectives of the source zone characterization study are to locate the site of the release, identify the contaminants of concern, and delineate the source material or unsaturated soil mass that may act as a continuing source of contaminant release to surface water or underlying ground water. The principal steps required for delineation of the source zone are illustrated in Figure 12.2 and described next.

Chemical Characterization and Indicator Parameter Selection. As shown on the task flow chart in Figure 12.2, to commence the delineation study, available chemical information regarding the suspected source of the subsurface release (for example, waste or product spill) must be compiled to provide a basis for the design of the laboratory testing program. If such information is unavailable or inadequate, representative contaminated soil samples should be collected from the release site and analyzed for a broad suite of chemical compounds, as appropriate, to identify the principal contaminants of concern. Appropriate laboratory indicator parameters and field testing procedures should then be selected on the basis of the prevalence, mobility, and toxicity of the principal constituents identified.

Lateral and Vertical Source Zone Characterization. In step 2 of the source delineation (see Figure 12.2), a field sampling and testing program is conducted to define the apparent lateral and vertical extent of contamination within the unsaturated soil zone. At each soil sampling location, sampling and field testing should be conducted continuously with depth until either clean soil or ground water infiltration is encountered. As discussed in Chapter 5, typical field test methods for hydrocarbon contamination include organic vapor headspace analyses and various colorimetric indicator tests.

To confirm the apparent lateral and vertical extent of contamination observed in the field, samples of the uppermost clean soils encountered at each sampling location should be submitted for laboratory analysis of indicator parameter content (see step 3, Figure 12.2). Representative samples from within the contamination zone should also be submitted for analysis of total and leachable contaminant

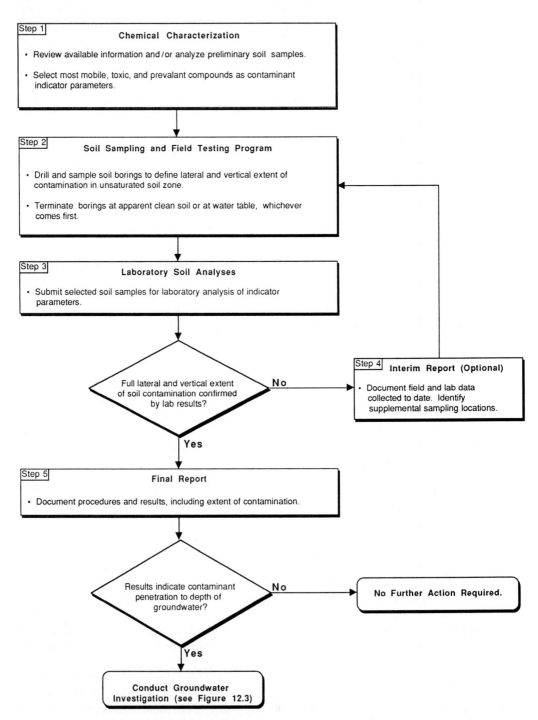

Figure 12.2 Procedures for source zone characterization.

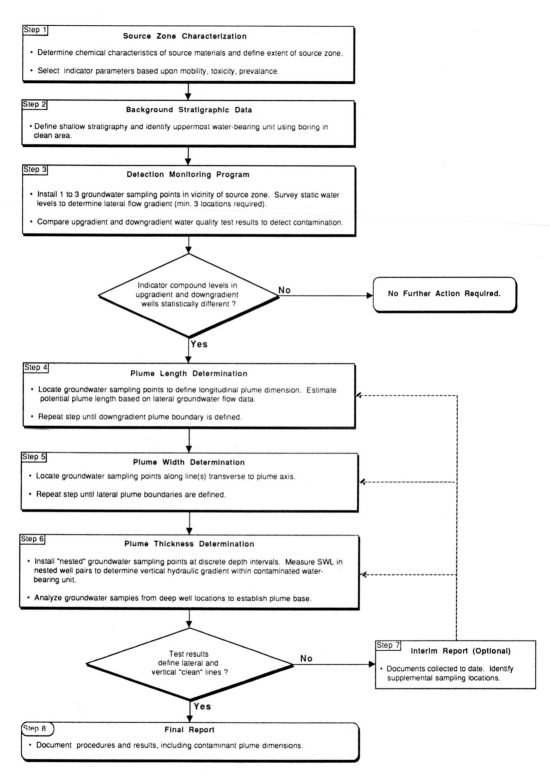

Figure 12.3 Procedures for ground water contaminant plume detection/delineation.

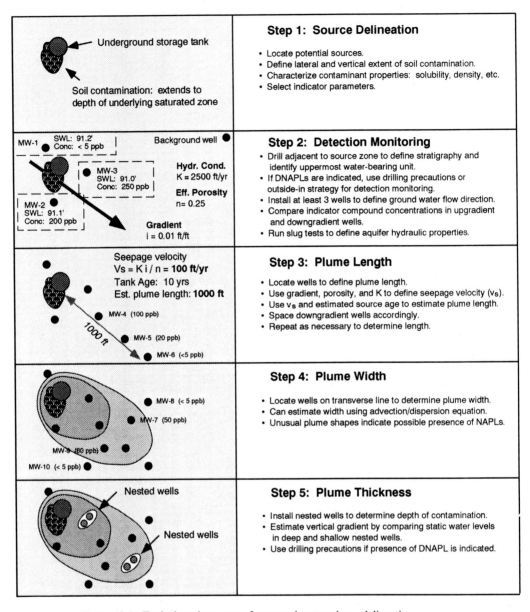

Step 1: Source Delineation

Underground storage tank

Soil contamination: extends to depth of underlying saturated zone

- Locate potential sources.
- Define lateral and vertical extent of soil contamination.
- Characterize contaminant properties: solubility, density, etc.
- Select indicator parameters.

Step 2: Detection Monitoring

MW-1 SWL: 91.2' Conc: < 5 ppb

Background well

MW-3 SWL: 91.0' Conc: 250 ppb

MW-2 SWL: 91.1' Conc: 200 ppb

Hydr. Cond. K = 2500 ft/yr

Eff. Porosity n = 0.25

Gradient i = 0.01 ft/ft

- Drill adjacent to source zone to define stratigraphy and identify uppermost water-bearing unit.
- If DNAPLs are indicated, use drilling precautions or outside-in strategy for detection monitoring.
- Install at least 3 wells to define ground water flow direction.
- Compare indicator compound concentrations in upgradient and downgradient wells.
- Run slug tests to define aquifer hydraulic properties.

Seepage velocity
$V_s = K i / n = 100$ ft/yr
Tank Age: 10 yrs
Est. plume length: **1000 ft**

MW-4 (100 ppb)
1000 ft
MW-5 (20 ppb)
MW-6 (<5 ppb)

Step 3: Plume Length

- Locate wells to define plume length.
- Use gradient, porosity, and K to define seepage velocity (v_s).
- Use v_s and estimated source age to estimate plume length.
- Space downgradient wells accordingly.
- Repeat as necessary to determine length.

MW-8 (< 5 ppb)
MW-7 (50 ppb)
MW-9 (60 ppb)
MW-10 (< 5 ppb)

Step 4: Plume Width

- Locate wells on transverse line to determine plume width.
- Can estimate width using advection/dispersion equation.
- Unusual plume shapes indicate possible presence of NAPLs.

Nested wells

Nested wells

Step 5: Plume Thickness

- Install nested wells to determine depth of contamination.
- Estimate vertical gradient by comparing static water levels in deep and shallow nested wells.
- Use drilling precautions if presence of DNAPL is indicated.

Figure 12.4 Typical work program for ground water plume delineation.

indicator concentrations in order to characterize contaminant mass and mobility.

As indicated in Figure 12.2 and discussed in further detail later in this chapter, delineation of the contaminated soil zone is an iterative process, often requiring two or more field and laboratory cycles for completion. Should the results of the source zone investigation show contaminants to have penetrated to the depth of underlying ground water at concentrations exceeding relevant cleanup standards, a ground water contamination study will also be required.

Procedures for Ground Water Contaminant Plume Characterization

Project Objective.   The objective of a ground water contaminant investigation is to determine the presence and extent of dissolved or free-phase contaminants, as well as the likely rate and direction of contaminant migration within the ground water flow system. The principal steps to be followed are shown in Figures 12.3 and 12.4.

Ground Water Plume Detection.   As indicated on the task flow chart provided in Figure 12.3, the ground water investigation must be preceded by identification and characterization of all potential source zones in the study area. A detection monitoring program, involving installation of one to three ground water sampling points at each known or suspected source location, should then be completed to identify all sites of hazardous constituent release to ground water.

Lateral and Vertical Plume Delineation.   As indicated in Figures 12.3 and 12.4, ground water plume delineation should be conducted in a step-wise procedure in order to minimize the number of ground water sampling points required. First, based on the suspected age of the release and the lateral ground water seepage velocity determined during the detection monitoring study, estimate the potential length of the contaminant plume (seepage rate × time = length) and space sampling points accordingly along the plume axis to locate the actual downgradient boundary. Second, to define the width of the contaminant plume, complete additional sampling points on one or two lines running transverse to the plume axis. Finally, to determine the vertical limit of contaminant migration, collect and analyze ground water samples from nested sampling points (that is, samples collected in close lateral proximity, for example, <10 ft distance, but from different discrete depths within the water-bearing unit).

Investigation of Deeper Water-bearing Units.   If the contaminant plume is found to extend through the full thickness of the uppermost water-bearing unit, sampling and analysis of ground water from the next underlying water-bearing stratum may be necessary to establish the vertical limit of contamination. In such cases, it is critical that any observation points penetrating the confining layer separating the upper and lower water-bearing strata be completed in a manner not providing an artificial conduit for contaminant migration. Appropriate protective measures are discussed in Chapter 5.

Final Product.   The final product of the ground water investigation should be an integrated geologic, hydrologic, and chemical model of the site, depicting the stratigraphic profile, the rate and direction of natural ground water flow, and the current lateral and vertical extent of contaminant migration.

12.4 DEVELOPMENT OF A DETAILED SITE INVESTIGATION WORKPLAN

Prior to commencing the hydrogeologic site investigation, the specific field and laboratory tasks required to implement the site investigation strategy described previously should be identified and appropriate resources allocated. Preplanning activities should include specification of the number, location, and depth of soil and ground water samples to be collected; sampling procedures and associated equipment requirements; field safety protocol; and field and laboratory test methods. The proposed sampling plan should be reviewed in advance to ensure that the information obtained will be adequate to meet the geologic hydrologic, and chemical data objectives of the source characterization or ground water plume detection and delineation study.

General guidelines for development of a site investigation work plan are outlined next. Detailed information regarding the selection of appropriate sampling and testing methods is provided in Chapter 5.

Review of Background Information

To provide a technical basis for planning of the site investigation, available site information must be compiled and reviewed to define the general location and duration of the suspected release, the probable contaminants of concern, and general stratigraphic conditions. Typical sources of information include site operating records, regulatory agency records, employee interviews, historical aerial photographs, published geologic references, and prior foundation studies or hydrogeologic site investigation reports.

On the basis of such information, we need to determine *what* we are looking for and *where*. It is generally not feasible to prove that there are no chemical contaminants of any nature at any location on a property. Rather, background information must be reviewed to define the specific focus of the study.

Project Objectives

The specific objectives of the site investigation must be clearly defined in advance. In general, the project scope should correspond to one or more of the tasks shown on the procedural flow charts presented in Figures 12.2 and 12.3. Both the source zone investigation and the ground water plume investigation should be approached in a step-by-step manner. In addition, the project engineer or scientist must clearly anticipate the end point of the site investigation, as well as the actions to be taken in the event that unexpected site conditions are encountered. For example, is the purpose of the sampling merely to confirm the presence or absence of a specific compound or to delineate its full extent? If ground water is encountered during a source characterization study, will a sample be collected for the purpose of contaminant detection?

It is not necessary or even advisable to complete a full hydrogeologic site investigation in one step. Rather, it will generally prove more economical to conduct the project in a phased manner, with each work stage having a predefined objective and end point.

Field Program Specifications

Based on available information, a preliminary plan should be developed regarding the number, location, and depth of samples required to meet the project objectives. All proposed drilling locations should be staked and cleared in advance for the presence of buried utilities. Appropriate sample collection and handling procedures must be specified and relevant equipment provided in working order. Sample kits containing the sample containers and preservatives required for the specified analytical methods should be ordered from the laboratory.

The field supervisor should be provided with a written copy of the sampling plan and project safety plan, specifying project objectives, proposed sampling locations, field test procedures, laboratory analyses, and the basis for modification of the proposed work plan during implementation. General guidelines for the design of the field program are provided next.

Design of Unsaturated Source Zone Characterization Study

General guidelines regarding the number, location, and depth of sampling points required for a source zone characterization study are as follows:

Number and Location of Samples. For initial chemical characterization of the residual waste materials or affected soil zone, analysis of one to four samples collected from known contaminated areas will generally suffice. To define the lateral limits of the source zone, samples can either be located on a rough grid pattern across the suspected contaminant area or completed on a "step-out" pattern, whereby samples are collected at even distances along lines extending radially from the known source area until clean soil conditions are encountered.

To minimize the number of samples required for the source zone delineation, the field program should be focused on establishing the *"clean line"* (the perimeter of the contaminant area), rather than defining variations in contaminant concentrations within the source zone. In general, "clean" soil conditions will correspond to either (1) the natural background concentrations of the contaminant compounds occurring in site soils or (2) other cleanup standards established by the state or federal regulatory authority.

Sampling Depth. At each sampling location, soil samples should be collected continuously to the depth of clean soil conditions or to the depth of ground water occurrence, whichever comes first. However, care must be taken not to puncture confining layers (clay, shale, or other low-permeability strata) that might be serving as a "safety net" against downward migration of contaminants beneath the source zone. For this purpose, at sites where soil contamination may extend beyond the depth of the surface soil stratum, it is advisable to drill at least one background soil boring at a known clean location to define site stratigraphy prior to drilling through the contaminant zone.

Sampling and Field Testing Methods.   Initial chemical characterization of the source zone materials will typically involve collection of wastes, spilled products, or affected soils for laboratory analyses of a broad range of hazardous chemical constituents potentially associated with the site. Thereafter, sample analyses may be limited to key indicator parameters, including the use of various field tests (such as organic vapor analyses or colorimetric methods). Hand augers may be used to collect soil samples at depths less than 5 ft below grade. For delineation of large contaminant areas or buried waste sites, backhoes are effective to depths of 10 to 15 ft. In general, conventional drilling rigs represent the most cost-effective means of soil sample collection at depths greater than 10 ft.

Ground Water Plume Detection and Delineation Studies

General guidelines regarding the number, location, and depth of sampling points required for a ground water plume study are as follows:

Number and Location of Samples.   For the purpose of a plume detection study, a minimum of three monitoring wells or piezometers is required to establish the lateral hydraulic flow gradient in the uppermost water-bearing stratum underlying the source area. Water quality measurements from upgradient sampling locations should be compared to data from downgradient locations to confirm the presence or absence of ground water contamination. The strategy for lateral and vertical delineation of the contaminant plume is outlined in Figure 12.4.

Sampling Depth.   Wells should be screened within the uppermost water-bearing stratum underlying the suspected source area. Good practice calls for limiting the length of the wellscreen to no more than 15 ft to minimize potential dilution of dissolved or free-phase contaminants. Consequently, in thick aquifers, installation of nested wells (adjacent wells screened at different depth intervals) may be required at selected locations to define the vertical limit of contaminant migration. To detect free-phase contaminants, wellscreens should be positioned to intersect either the top (for floating fluids) or the base (for sinking fluids) of the water-bearing stratum. If, upon completion of the plume delineation in the uppermost aquifer unit, sampling of deeper, underlying water-bearing units is required, special care must be taken to avoid inadvertent interconnection of contaminated and uncontaminated layers.

Ground Water Investigation Methods.   Drilling and sampling of monitoring wells provides information on site stratigraphy, static water-level elevations, and water quality. Consequently, wells are a common component of most ground water detection and delineation studies. However, as discussed in Chapter 5, many other site investigation technologies can be employed to obtain discrete-depth ground water samples or supplementary information regarding site stratigraphy or aquifer hydraulics. The utility of these alternative methods depends on the specific type of information required to complete the picture of the ground water contamination problem.

TABLE 12.1 Summary of Laboratory Test Methods for Analysis of Soil and Ground Water Samples

Soil and solid-waste analytical methods

Reference method	Parameter	Technique	Container type, no., volume	Preservation/storage requirements	Maximum holding time from collection	Maximum holding time (analysis)
EPA 8240	Volatile organics	GC/MS	(2) 40-mL VOA vials	Refrigerated at 4°C	NA	14 days[a]
EPA 8270	Semivolatile extractable organics	GC/MS	(1) 250-mL glass jar	Refrigerated at 4°C	14 days	40 days
EPA 8010	Halogenated volatile organics	GC/HSD	(2) 40-mL VOA vials	Refrigerated at 4°C	NA	14 days[a]
EPA 8020	Aromatic volatile organics	GC/PID	(1) 500-mL glass jar	Refrigerated at 4°C	NA	14 days[a]
EPA 8040	Phenols	GC/FID	(1) 500-mL glass jar	Refrigerated at 4°C	7 days	40 days
EPA 8080	Organochlorine pesticides and PCBs	GC/ECD	(1) 500-mL glass jar	Refrigerated at 4°C	14 days	40 days
EPA 8100 or 8310	Polynuclear aromatic hydrocarbons	GC/FID HPLC	(1) 500-mL glass jar	Refrigerated at 4°C	14 days	40 days
EPA 8120	Chlorinated hydrocarbons	GC/ECD	(1) 500-mL glass jar	Refrigerated at 4°C	14 days	40 days
EPA 6010	Metals: Cd, Cr, Pb, Mn, Ba, Si, Fe, Al, Sb, Be, Co, Cu, Mo, Ni, Ag, Tl, V, Zn	ICP, emission spectroscopy	(1) 250-mL glass jar	Refrigerated at 4°C	NA	6 months
EPA 7060	Arsenic	Furnace AA	(1) 250-mL glass jar	Refrigerated at 4°C	NA	6 months
EPA 7740	Selenium	Furnace AA			NA	6 months
EPA 7471	Mercury	Cold Vapor AA			NA	28 days
EPA 7421	Lead	Furnace AA			NA	6 months
ASTM 3237	Organic lead	AA after extraction	(1) 500-mL glass jar	Refrigerated at 4°C	14 days	14 days[a]
EPA 9070/9071	Oil and grease	Gravimetric	(1) 500-mL glass jar	Refrigerated at 4°C	NA	28 days
SM 209F	% Solids	Gravimetric	(1) 250-mL glass jar	Refrigerated at 4°C	NA	Analyze immediately
EPA 9045	pH	Electrometric	(1) 500-mL glass jar	Refrigerated at 4°C	NA	Analyze immediately
EPA 1010	Flashpoint	Pensky–Marten closed cup	(1) 250-mL glass jar	Refrigerated at 4°C	NA	As soon as possible[a]
EPA Reactivity	Cyanide/sulfide	Acidification, colorimetry/titration	(1) 250-mL glass jar	Refrigerated at 4°C	NA	As soon as possible[a]
EPA 340.1/340.2	Fluoride	Bellack distillation/SIE	(1) 500-mL glass jar	Refrigerated at 4°C	NA	28 days

(continued)

TABLE 12.1 (continued)

Ground water analytical methods

Reference method	Parameter	Technique	Container type, no., volume	Preservation/storage requirements	Maximum holding time from collection	Maximum holding time (analysis)
EPA 8240 624	Volatile organics	GC/MS	(2) 40-mL VOA vials	Refrigerated at 4°C; HCl pH <2	NA	14 days[a,b]
EPA 8270 625	Semivolatile extractable organics	GC/MS	(1) 1-L glass bottle; TFE-lined cap	Refrigerated at 4°C	7 days[a]	40 days
EPA 8280	Polychlorinated dibenzodioxins (PCDD) and polychlorinated dibenzofurans (PCDF)	GC/MS	(1) 1-L amber glass bottle; TFE-lined cap	Refrigerated at 4°C	30 days	40 days
or EPA 613			(1) 1-L amber glass bottle; TFE-lined cap	Refrigerated at 4°C	7 days[a]	40 days
EPA 8010 601	Halogenated volatile organics	GC/HSD	(2) 40-mL VOA vials	Refrigerated at 4°C	NA	14 days[a]
EPA 8020 602	Aromatic volatile organics	GC/PID	(2) 40-mL VOA vials	Refrigerated at 4°C; HCl pH <2	NA	14 days[a,b]
EPA 8040 604	Phenols	GC/FID	(1) 1-L glass bottle; TFE-lined cap	Refrigerated at 4°C	7 days[a]	40 days
EPA 8080 608	Organochlorine pesticides and PCBs	GC/ECD	(1) 1-L amber glass bottle; TFE-lined cap	Refrigerated at 4°C	7 days[a]	40 days
EPA 680	PCBs	GC/MS	(1) 1-L amber glass bottle; TFE-lined cap	Refrigerated at 4°C	7 days[a]	40 days
EPA 8120 (modified)	Chlorinated hydrocarbons	GC/ECD	(1) 1-L amber glass bottle; TFE-lined cap	Refrigerated at 4°C	7 days[a]	40 days
EPA 8140	Organophosphate pesticides	GC/FPD	(1) 1-L amber glass bottle; TFE-lined cap	Refrigerated at 4°C	7 days[a]	40 days
EPA 8150	Chlorinated herbicides	GC/ECD	(1) 1-L amber glass bottle; TFE-lined cap	Refrigerated at 4°C	7 days[a]	40 days
EPA 8310	Polynuclear aromatic hydrocarbons	HPLC	(1) 1-L amber glass bottle; TFE-lined cap	Refrigerated at 4°C	7 days[a]	40 days
EPA 6010 200.7	Metals	ICPES	(1) 1-L glass bottle	HNO_3; pH <2	NA	6 months
EPA 7060 206.2	Arsenic	Furnace AA	(1) 500-mL plastic bottle	HNO_3; pH <2	NA	6 months
EPA 7740	Selenium	Furnace AA	(1) 500-mL plastic bottle	HNO_3; pH <2	NA	6 months

EPA Method	Parameter	Method	Container	Preservation		Holding time
EPA 270.2 EPA 7470 EPA 245.1	Mercury	Cold Vapor AA	(1) 500-mL plastic bottle	HNO_3 pH <2	NA	28 days
EPA 7421 EPA 239.2	Lead	Furnace AA	(1) 500-mL plastic bottle	HNO_3 pH <2	NA	6 months
EPA 7196	Chromium (VI)	Colorimetric	(1) 500-mL plastic bottle	Refrigerated at 4°C	NA	24 hours[a]
ASTM 3237	Organic lead	AA after extraction	(1) 500-mL plastic bottle	Refrigerated at 4°C	14 days	30 days
EPA 9030 EPA 376.1	Sulfide	Titrimetric	(1) 250-mL glass or plastic bottle	Refrigerated at 4°C; add 2 mL of zinc acetate and NaOH to pH >9	NA	7 days[a]
EPA 9012 EPA 353.3	Cyanide (CN)	Colorimetric	(1) 500-mL plastic bottle	Refrigerated at 4°C; NaOH to pH >12	NA	14 days[a]
EPA 340.2	Fluoride	Selective ion electrode	(1) 500-mL plastic bottle	NA	NA	28 days
EPA 300.0	Chloride, nitrate, sulfate	IC	(1) 500-mL plastic bottle	NA	NA	28 days
EPA 353.1	Nitrate/nitrite	Colorimetric	(1) 500-mL plastic bottle	Refrigerated at 4°C; H_2SO_4 pH <2	NA	14 days[a]
EPA 9066	Total phenolics	Colorimetric	(1) 500-mL plastic bottle	Refrigerated at 4°C; H_2SO_4 pH <2	NA	28 days
EPA 420.2 EPA 415.1	Total organic carbon	Oxidation/NDIR	(1) 500-mL plastic bottle	Refrigerated at 4°C; H_2SO_4 pH <2	NA	28 days
EPA 418.1	Petroleum hydrocarbons	IR	(1) 1-L glass bottle	Refrigerated at 4°C; HCl pH <2	NA	28 days
EPA 450.1	Total organic halogens	Dohrmann DX-20	(1) 1-mL amber glass bottle	Refrigerated at 4°C; H_2SO_4 pH <2	NA	14 days[a]
EPA 9040 EPA 150.1	pH	Electrometric	(1) 500-mL plastic bottle	Refrigerated at 4°C	NA	72 hours[a]
EPA 9050 EPA 120.1	Conductance	Wheatsone bridge	(1) 500-mL plastic bottle	Refrigerated at 4°C	NA	28 days
EPA 410.1	Chemical oxygen demand	Titrimetric	(1) 500-mL plastic bottle	Refrigerated at 4°C	NA	28 days
EPA 413.2	Oil and grease	IR	(1) 1-L glass bottle	Refrigerated at 4°C; HCl pH <2	NA	28 days
EPA 160.1	Total dissolved solids	Gravimetric	(1) 500-mL plastic bottle	Refrigerated at 4°C	NA	7 days[a]
EPA 160.2	Total suspended solids	Gravimetric	(1) 500-mL plastic bottle	Refrigerated at 4°C	NA	7 days[a]
EPA 310.1	Alkalinity	Titrimetric	(1) 500-mL plastic bottle	Refrigerated at 4°C	NA	14 days[a]

Source: Laboratory Tech Notes, Radian Analytical Services, Vol. 1, No. 2, August 1988.

[a] Samples must be shipped within 24 hours of collection.

[b] + = Preserve for aromatics only; 7 days if no HCl.

HSD = halogen specific detector
SM = standard methods
NA = not applicable

Laboratory Specifications

The appropriate analytical method to be employed for the measurement of a specific contaminant or group of contaminants in a soil or ground water sample depends both on the level of contamination anticipated and the detection limit required. To demonstrate "clean" conditions, an analytical method having a detection limit that is less than the anticipated cleanup standard must be employed. For example, to show that benzene concentrations in a ground water sample are less than the U.S. Primary Drinking Water Standard for this contaminant (5 μg/L), a gas-chromatography (GC) or gas-chromatography mass spectroscopy (GCMS) method, providing a part-per-billion (ppb) sensitivity, must be employed. Within samples containing elevated contaminant concentrations, the use of gross indicator parameters or less sensitive analytical methods (for example, a part-per-million sensitivity level) may be required to characterize the total contaminant mass within the affected soil or ground water zone.

Procedures for selection of laboratory samples and the specific analytical methods to be employed must be defined in advance of sample collection. For each test method specified, appropriate sample containers and preservatives must be obtained and arrangements made for completion of the laboratory analysis within the holding time specified under EPA protocol. Commonly used analytical methods and associated sample containers and preservatives are listed on Table 12.1. Detailed information regarding test methods applicable to soil and ground water samples is provided in EPA Publication SW-846 (U.S. EPA, 1986).

Data Evaluation and Report Specifications

The project workplan should define what the final product of the study will be, including the specific determinations to be made, the procedures to be employed to make such determinations (for example, statistical analyses and calculations), and the manner in which such findings will be presented (cross sections, data plots, and the like). The proposed field and laboratory plan should be reviewed to ensure that the data required for completion of the final report will be obtained and properly recorded during the work program.

12.5 DATA EVALUATION PROCEDURES

Overview

Upon completion of the field and laboratory programs, site data must be integrated to construct a working model of contaminant plume migration within the ground water flow regime. The key components of this hydrogeologic site characterization are the geologic, hydrologic, and chemical data defining the occurrence and movement of fluids in the subsurface and the entrainment of contaminants in this natural flow system. Procedures for data organization and presentation are outlined next.

Geologic Data Evaluation

Available geologic data must be compiled to define the lateral and vertical configuration of permeable and impermeable strata comprising the framework within which subsurface fluids collect and flow. Conventional data presentation techniques include the following:

Log and As-built Diagrams

Scaled geologic logs should be prepared for each soil boring, monitoring well, or geophysical profile completed on the site. As shown in Figure 12.5, in addition to a description of the type and thickness of each stratum encountered, the log should indicate the drilling location, ground-surface elevation, drilling method, depth of ground water occurrence, driller identification, and date of completion (Hunt, 1984). For monitoring wells, as-built diagrams should be prepared showing well construction and static water elevation with respect to stratigraphy. Log preparation is discussed in further detail in Chapter 5.

Geologic Cross Sections

Cross-section diagrams should be prepared to illustrate the bedding and lateral continuity of the principal stratigraphic units underlying the site. General guidelines for cross-section preparation are provided in Hunt (1984) and Tearpock and Bischke (1991). Given the extreme variability of shallow strata, care should be taken to avoid extrapolation of stratigraphic interpretations beyond the area of available geologic logs.

Structure and Isopach Maps

Structure maps and isopach drawings can be employed to characterize the physical constraints on ground water or contaminant flow through a water-bearing unit. Within a confined aquifer unit, a structure map (a topographic map of the upper surface of the water-bearing unit) can be used to identify topographic highs or traps wherein floating hydrocarbons might collect. Isopach drawings (contour maps of the thickness of the water-bearing stratum encountered at each drilling location) clearly illustrate lateral discontinuities or pinch-outs of the aquifer unit, as well as preferential flow paths.

Hydrologic Data Evaluation

Hydrologic data collected during the site investigation should be evaluated to define the direction and rate of natural ground water flow through the principal water-bearing strata underlying the study area. Data evaluation requirements include the following:

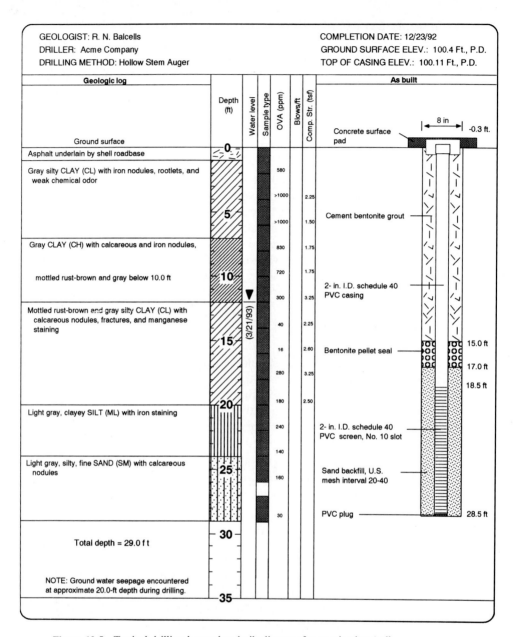

Figure 12.5 Typical drilling log and as-built diagram for monitoring well.

Aquifer Characterization

Geologic logs and cross sections should be reviewed to identify the upper-most water-bearing stratum underlying the site. Typically, this will be the first water-saturated stratum encountered beneath the site with sufficient hydraulic conductivity to yield water to wells (for example, $K > 10^{-4}$ cm/sec). Static water levels should be superimposed on geologic log and cross-section diagrams to classify the aquifer unit as either confined (static water level above top of water-bearing unit), unconfined (static water level below top of water-bearing unit), or partially confined (confined at some locations and unconfined at other locations).

Potentiometric Surface Maps

Within permeable strata, ground water movement is primarily horizontal, with a slight dip in the direction of flow. To define lateral ground water flow patterns, static water-level elevations measured within monitoring wells or piezometers screened within the same water-bearing stratum should be plotted on a scaled site plan, and the values contoured to develop a potentiometric surface map of the water-bearing unit (see Figure 12.7 later). Ground water flow is perpendicular to such equipotential lines, in the direction of lower hydraulic head. Using a potentiometric map, the lateral hydraulic flow gradient ($\Delta h/L$) is calculated as the change in head (Δh) divided by the lateral distance (L) between equipotential lines and is commonly expressed as a dimensionless value (ft/ft). Alternatively, hydraulic gradient can be determined by means of triangulation between any three well locations as described in Heath (1983).

Erroneous results regarding ground water flow patterns will be obtained if static water-level data from different or discontinuous strata are plotted on the same potentiometric surface map. Similarly, within thick units exhibiting a significant vertical hydraulic gradient (see the next section), care must be taken to correlate only those static water-level measurements from similar depths in order to accurately characterize lateral flow gradients. Distortions in the potentiometric surface contours should be carefully evaluated in light of all available site data to determine whether such apparent hydraulic head variations are indicative of actual aquifer discharge or recharge phenomena or represent a possible measurement error or misinterpretation.

Vertical Gradient Calculation

The vertical hydraulic flow gradient within a single water-bearing stratum can be determined by comparison of static water-level elevations from nested well pairs screened at different depths within the same unit. In such cases, the vertical gradient is estimated as the difference in static water-level elevations (Δh) divided by the vertical distance between the midpoints of the wellscreens. Vertical gradients between separate water-bearing strata can be determined either from nested well measurements or comparison of potentiometric surface maps at any one point. Most of the hydraulic head difference observed between two water-bearing strata represents head loss occurring within the intervening low-permeability confining unit. Consequently, vertical gradients between water-bearing strata should be estimated as the hydraulic head difference (Δh) at any location divided by the thickness (L) of the confining unit at that location.

Aquifer Hydraulic Properties

The hydraulic conductivity of a water-bearing stratum can be measured by means of single-well permeability tests or constant-rate aquifer pumping tests conducted on wells screened within that unit. Aquifer pumping tests can also provide direct measurement of the aquifer transmissivity (T) and storage (S) coefficients. (However, note that values of S calculated from pumping tests in unconfined aquifers are not valid.) Test procedures and calculation methods are discussed in Chapters 3 and 5. In the absence of direct field measurements, rough approximations of hydraulic conductivity can be made on the basis of aquifer soil or rock characteristics (see Chapters 2 and 3).

Ground Water Flow-rate Calculation

The lateral ground water seepage velocity within a water-bearing stratum, representing the actual rate of ground water movement, can be estimated using Darcy's law. Note that hydraulic conductivity measurements are usually determined from field tests, while porosity is typically estimated on the basis of soil or rock type.

Chemical Data Evaluation

Results of chemical analyses or indicator tests conducted on soil and ground water samples in either the field or laboratory must be plotted on scaled site plans and correlated with available geologic logs and cross sections to define the lateral and vertical limits of contaminant migration. General procedures for laboratory data review and interpretation are as follows:

Data Validation Procedures

Upon receipt from the laboratory, all test results should be carefully reviewed to confirm laboratory accuracy and precision and compliance with relevant quality-control standards. Formal data validation procedures are outlined by the EPA (U.S. EPA 1986, 1988, 1989a). Following confirmation of data validity, contaminant concentrations reported for soil and ground water samples should be corrected for those compounds detected in field, trip, or laboratory blank samples. Special attention should be paid to low concentrations of laboratory solvents (for example, acetone and methylene chloride) or plasticizers (for example, phthalate esters), which may represent inadvertent laboratory contamination of the soil or ground water samples.

Data Interpretation

Soil and ground water test results should be used to establish vertical and lateral "clean lines," the boundary beyond which contaminant concentrators are either less than natural background levels or less than the applicable cleanup standards specified by the regulatory authority. Various statistical methods can be employed to characterize background conditions and make comparisons to concentrations detected at individual sampling points (EPA, 1989b). Due to the variability of individual compound concentrations, it is usually instructive to define source zone or plume dimensions on the basis of the total organic or total inorganic contaminant levels detected at each sampling location. Data should be plotted on scaled site plans and cross sections and clean lines defined on the basis of clean sampling points. The delineation study is complete when a clean line can be drawn around all sides of the source zone or ground plume area, and the depth to clean soil or ground water has been established. These data should be sufficient to define the full volume of contaminated soil and ground water at the site.

Following completion of the delineation study, contaminant concentrations within the source zone or ground water plume should be inspected (and, if feasible, contoured according to lines of equal concentration) to confirm the location of the suspected contaminant source. In general, total contaminant concentrations should decrease with increasing distance from the point of release. Irregular concentration patterns may suggest either (1) variable rates of contaminant release over time or (2) the presence of multiple sources. In such case, available site data should be carefully reviewed to confirm that all potential source areas have been identified and adequately outlined.

Final Data Correlation

The geologic, hydrologic, and chemical data collected during the site investigation should fit together as an overall picture that makes sense. Geologic cross sections and isopach maps should define stratigraphic conditions that are consistent with general depositional patterns observed in the site region. The apparent variability of ground water flow patterns indicated on potentiometric surface maps should correlate with actual stratigraphic variations (that is, pinch-outs or K variations) or with the locations of actual ground water recharge or discharge features (ponds, streams, and the like). Hydraulic conductivity measurements should be generally consistent with the soil or rock types observed during the drilling program. Dissolved contaminants should move in the direction of ground water flow, diminishing in concentration with distance from the source area. Geologic, hydrologic, and chemical data should be correlated in this manner, and inconsistencies resolved as needed to provide an accurate technical basis for corrective action design.

12.6 CASE STUDIES

Case 1: Source Zone and Dissolved Plume Delineation Study

Project Background

For a 20-yr period ending in 1984, a petrochemical manufacturing plant operated a wet well facility to collect and route process wastewater to a wastewater treatment plant. The wet well consisted of a 40-ft by 40-ft reinforced concrete

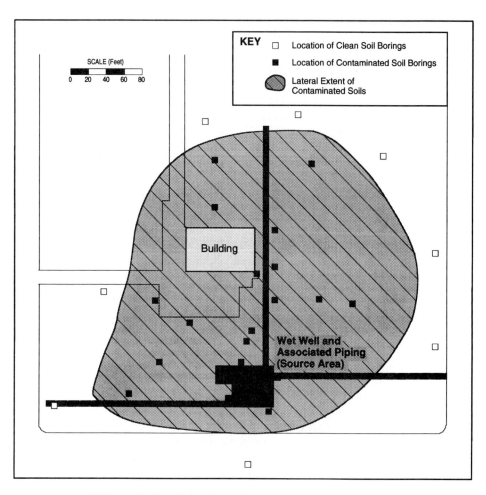

KEY
- □ Location of Clean Soil Borings
- ■ Location of Contaminated Soil Borings
- ▨ Lateral Extent of Contaminated Soils

SCALE (Feet)
0 20 40 60 80

Building

Wet Well and Associated Piping (Source Area)

Figure 12.6 Soil sampling locations, case study 1.

basin extending to a depth of 15 ft below site grade, with associated subgrade inlet and outlet piping (see Figure 12.6). During demolition operations commenced in 1984, chemical odors and discoloration were noted within a shallow water-bearing sand stratum occurring at a depth of 12 ft below grade at this site. A hydrogeologic site investigation was therefore implemented to delineate the full extent of affected soil and ground water.

Unsaturated Source Zone Delineation Program

Affected Soil Detection.   To confirm the presence of affected soils, three soil borings were drilled adjacent to the wet well facility using a truck-mounted drilling rig equipped with a hollow-stem auger assembly (see Figure 12.6). Soil samples were collected continuously through the surface clay stratum to the depth of the underlying sand unit using Shelby tube core samplers. The samples were logged in accordance with the Unified Soil Classification System (USCS), inspected for hydrocarbon staining or odor, and analyzed for organic vapor content using a portable organic vapor analyzer (OVA). Selected soil samples were submitted for laboratory analysis of volatile organic compounds associated with the waste materials previously managed in the wet well facility.

Lateral Delineation of Affected Soils.   Drilling and sampling of the three initial soil borings indicated the presence of nonaqueous-phase liquids (NAPL) within the surface clay soils adjacent to the former wet well facility. To determine the lateral extent of this affected soil zone, 25 hand-auger soil borings were drilled to depths of 3 to 5 ft below grade at the locations shown in Figure 12.6. The results of this affected soil delineation study are discussed below.

Ground Water Plume Delineation Program

Lateral Plume Delineation Procedures.   To define the lateral extent of affected ground water within the shallow water-bearing sand unit underlying the wet well, 14 observation wells were installed to an average depth of 30 ft below grade at the locations shown in Figure 12.7, using a hollow-stem auger drilling rig. At each well location, soil samples were collected and logged at maximum 5-ft-depth intervals using either Shelby tube or split-spoon core samplers. The wells were constructed of 2-in. diameter, schedule 40 PVC casing fitted with 10- to 15-ft lengths of slotted PVC wellscreen positioned to intersect the top of the sand unit.

The lateral plume delineation program was completed in two stages. Ten wells were first installed over the estimated lateral extent of the plume, and ground water samples were collected and analyzed in the field using a field test kit for aromatic hydrocarbon content. On the basis of the preliminary test results, four additional wells were installed to establish a downgradient clean line. Groundwater samples were then collected from all 14 observation wells and submitted for laboratory analysis of volatile organic compound content to confirm lateral plume dimensions.

Vertical Plume Delineation Procedures.   To define the vertical limit of affected ground water, three vertical water-quality profiles were conducted along the downgradient axis of the plume at the locations indicated in Figure 12.7. At each location, temporary ground water sampling points were installed, and ground water samples were collected at 10-ft-depth intervals through the sand unit. The temporary sampling points consisted of 1.5-in.-diameter steel pipes fitted with 1-ft lengths of stainless steel screen encased in a watertight sheath. The sampling points were driven to the desired sampling depth using a 21-ton truck-mounted cone penetrometer rig, and the sheaths were retracted to facilitate ground water sample collection. To identify the vertical clean line, samples were immediately analyzed for aromatic hydrocarbon content using a field test kit. Duplicate samples were submitted for laboratory analysis. The results of this plume delineation study are discussed next.

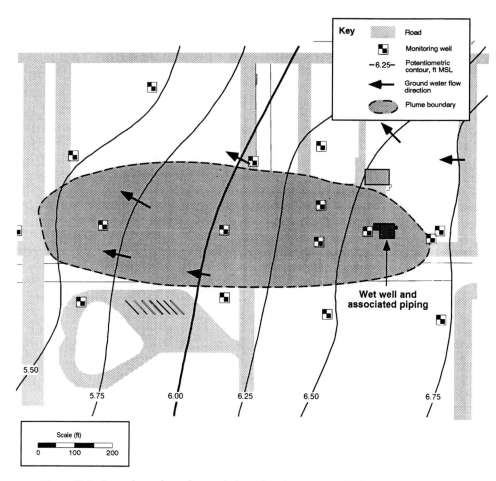

Figure 12.7 Potentiometric surface and plume location, case study 1.

Results of Hydrogeologic Investigation

Site Stratigraphy. Geologic data collected during the site investigation show shallow stratigraphy beneath the study area to consist of the following units, in order of increasing depth (see Figure 12.8):

1. *Surface clay:* Red, high-plasticity clay, ranging from 10 to 20 ft in thickness.
2. *Upper sand:* Tan, silty, fine sand occurring at an average depth of 14 ft and ranging from 55 to 65 ft in thickness.
3. *Lower clay:* Gray, silty clay occurring at an average depth of 77 ft below grade, with approximate 12-ft thickness.
4. *Lower sand:* Gray, silty, fine sand encountered at a depth of 90 ft below grade.

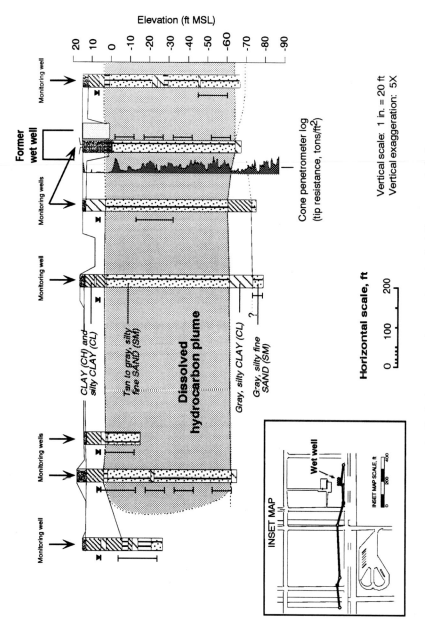

Figure 12.8 Geologic cross section, west–east orientation, case study 1.

413

Ground Water Flow Patterns.   Static water-level measurements in site wells show the upper silty sand unit to be hydraulically confined, with a potentiometric surface rising above the top of the sand to within 5 ft of ground surface. As shown in Figure 12.7, a potentiometric surface contour map shows shallow ground water to be moving in a west–northwest direction beneath the site, at an average hydraulic gradient of 0.001 ft/ft. Results of rising-head slug tests conducted in the observation wells indicate the average hydraulic conductivity of the sand unit to be 1.5×10^{-2} cm/sec. Based on an estimated effective porosity of 0.25, these measurements correspond to a lateral ground water seepage velocity of roughly 60 ft/yr.

Results of Affected Soil Zone Delineation

The presence of NAPL materials was observed within the surface clay soil stratum over a 70,000-ft^2 area surrounding the former wet well (see Figure 12.6). The depth of this hydrocarbon penetration was found to average less than 3 ft below grade. Total volatile organic compound concentrations measured in the affected soil zone ranged from 1 to 2000 mg/kg. Rainwater percolation through this zone could act as a continuing source of dissolved constituent release to the underlying ground water.

Results of Ground Water Plume Delineation

The affected ground water plume associated with the wet well was found to extend approximately 900 ft downgradient of the source location and to a depth of 75 ft below grade within the upper silty sand layer (see Figure 12.7). Total volatile organic compound concentrations within the plume area range from less than 0.1 mg/L to over 300 mg/L. Under natural ground water flow conditions, this plume could be expected to continue to move in a westward direction beneath the site at an average lateral seepage velocity of 60 ft/yr.

Case 2: DNAPL Plume Delineation Study

Project Background

The Motco Superfund Site is a former used-oil reclamation facility that operated in the vicinity of La Marque, Texas, during the period from 1958 to 1968. Reclamation efforts involved the collection and reprocessing of petrochemical residues within an 11-acre system of earthen pits (see Figure 12.9). Pit contents included styrene tars, refinery tank bottoms, and chlorinated solvents. Following abandonment of the site by the property owners, environmental investigations were commenced in 1980 by the EPA and industrial parties.

A principal concern of the Motco study is control of both dissolved and free-phase organic compounds detected within shallow water-bearing strata underlying the site. Hydrogeologic site investigations have shown this area to be underlain by a surface deposit of interbedded sands, silts, and clays designated the transmis-

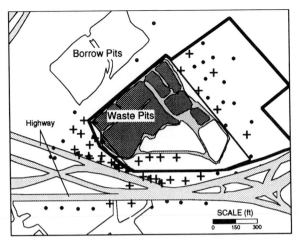

LEGEND

+   Cone penetrometer test, soil boring, or well
location where DNAPL noted

●   Cone penetrometer test, soil boring, or well
location where DNAPL not noted

━━━   Property Line

Figure 12.9 Base map of Motco site showing locations of cone penetrometer logs, soil borings, and monitoring wells.

sive zone (TZ), extending from ground surface to a depth of approximately 50 ft below grade. A stiff, high-plasticity clay layer, termed the UC-1 Clay, forms the base of the transmissive zone. Previous investigations have shown this clay unit to have an average thickness of approximately 34 ft, separating the transmissive zone from the underlying Upper Chicot aquifer. An idealized geologic cross section is presented in Figure 12.10. As shown, the on-site pits penetrate the upper 20 ft of the transmissive zone.

The TZ strata comprise the principal conduits of shallow ground water movement beneath this site. Water quality investigations have detected the presence of dissolved organic compounds in TZ unit ground water within the near vicinity of the pits. In addition, dense nonaqueous-phase liquids (DNAPL) have been detected within these shallow water-bearing strata, typically occurring near the base of the TZ unit, perched atop the underlying clay layers. Due to their relatively high density and low water solubility, DNAPL fluids accumulating at the base of the TZ unit appear to migrate laterally atop the underlying UC-1 Clay surface, in a general downslope direction. Procedures and results of the DNAPL delineation study completed at this site are described next (Connor, Newell, and Wilson, 1989). Remediation efforts are described in Section 13.9.

DNAPL Delineation Program

To define potential pathways of DNAPL migration beneath the Motco Site, an extensive cone penetrometer survey was conducted to characterize the continuity and base topography of the TZ water-bearing units in the immediate vicinity

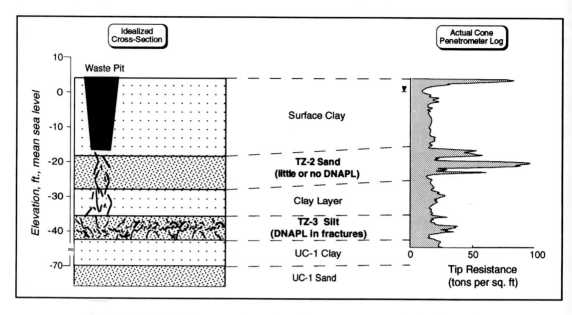

Figure 12.10 Idealized cross section and actual cone penetrometer log for Motco site.

of the waste pits. Within complex geologic environments similar to that of the Motco Site, the electric cone penetrometer has proved a cost-effective tool for such detailed stratigraphic analysis. At each sampling location, the cone yields a continuous log of minor silt, sand, and clay units without generating drill cuttings or fluids (see Figure 12.10). The presence or absence of perched hydrocarbon within this stratigraphic column can then be determined by means of discrete-depth soil cores or borehole fluid samples.

During March 1989, a total of 73 cone penetrometer tests were completed in the vicinity of the Motco Site at the locations shown in Figure 12.9. At each test location, cone tip resistance and sleeve friction were recorded continuously with depth, providing a detailed log of sand, silt, and clay units through the full thickness of the transmissive zone. As an indication of DNAPL presence, the return flow of grout tremied into each completed cone borehole was inspected for evidence of an oil sheen or visible oil content.

On the basis of preliminary geologic cross sections and structure maps developed from cone penetrometer data, 15 soil boring locations were selected to confirm the presence or absence of perched DNAPL at the base of the TZ permeable units. Based on these soil core analyses, several wells were installed in an area of known DNAPL accumulation.

DNAPL Distribution within TZ Units

The presence of DNAPL fluid within the transmissive zone extends beneath the pit area and includes a broad area immediately east of the Motco Site. Analysis of soil cores shows the principal pathway of this DNAPL migration to be the TZ-

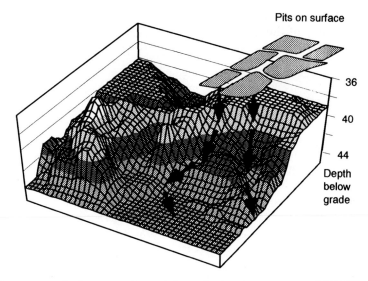

Pits on surface

36

40

44

Depth
below
grade

Figure 12.11 Surface map of top of clay unit, showing movement of DNAPLs down topographic valleys in TZ-3 unit.

3 silt layer comprising the base of the TZ unit. Vertical penetration of the underlying UC-1 Clay by DNAPL was not observed to exceed 2 ft at any drilling location.

Careful inspection of soil cores from the TZ-3 unit found DNAPL to be preferentially concentrated within the slickensides, fractures, and silt partings of the predominately silty clay soil matrix, rather than uniformly distributed throughout the soil mass, a phenomenon likely related to the interfacial tension of the DNAPL fluid and the fine pore size of the silt unit. Within several TZ-3 soil cores, such DNAPL concentrations were observed perched above minor clay seams. On a macroscopic level, lateral migration of DNAPL does appear to be strongly influenced by the topography of the UC-1 Clay layer, which forms the base of the TZ-3 unit. Specifically, the point of maximum lateral transport as seen in Figure 12.11 corresponds to a significant structural trough in the base of TZ-3.

Summary of Significant Findings

Methodology. The cone penetrometer provided a cost-effective means of defining permeable thickness and base structure of water-bearing strata within a complex alluvial environment. Furthermore, the presence or absence of an oil sheen within the return flow of grout tremied into each penetrometer hole correlated perfectly with the occurrence of DNAPL, as determined by subsequent soil core analyses.

DNAPL Migration. The distribution of DNAPL fluids within the transmissive zone strata suggest the principal factors in DNAPL migration to be the downgradient fluid density gradient and the secondary porosity features of the fine-grained sediments underlying the site. DNAPL has seeped downward from the source pits, migrating laterally via fractures, partings, and other secondary porosity features of the TZ-3 silt stratum, in general accordance with the topography of the underlying UC-1 Clay, a relatively low-permeability, massive clay deposit.

REFERENCES

CONNOR, J. A., NEWELL, C. J., and WILSON, D. K., 1989, "Assessment, Field Testing, and Conceptual Design for Managing Dense Non-Aqueous Phase Liquids (DNAPL) at a Superfund Site," *Proc Conference on Petroleum Hydrocarbons and Organic Chemicals in Groundwater,* National Water Well Association, Houston, TX.

HEATH, R. C., 1983, *Basic Groundwater Hydrology,* U.S.G.S. Water Supply Paper 2220, U.S. Government Printing Office, Washington, DC.

HUNT, R. E., 1984, *Geotechnical Engineering Investigation Manual,* McGraw-Hill, New York.

TEARPOCK, D. J. and BISCHKE, R. E., 1991, *Applied Subsurface Geologic Mapping,* Prentice Hall, Englewood Cliffs, NJ, p. 648.

U.S. EPA, 1986, *Test Methods for Evaluating Solid Waste, SW-846,* Volumes 1A, 1B, 1C: Laboratory Manual, Volume 2: Field Manual, 3rd Edition, NTIS No. PB88239223. U.S. Environmental Protection Agency, Washington, DC.

U.S. EPA, 1988, *Laboratory Data Validation, Functional Guidelines for Evaluating Organic Anayses,* Data Review Workshop, Hazardous Site Evaluation Division. U.S. Environmental Protection Agency, Washington, DC.

U.S. EPA, 1989a, *Laboratory Data Validation, Functional Guidelines for Evaluating Inorganic Analyses,* Data Review Workshop, Hazardous Site Evaluation Division, U.S. EPA, Contract No. 68-01-7443. U.S. Environmental Protection Agency, Washington, DC.

U.S. EPA, 1989b, *Statistical Analysis of Groundwater Monitoring Data at RCRA Facilities—Interim Final Guidance,* February, EPA/530/SW-89/026, NTIS No. PB89151047. U.S. Environmental Protection Agency, Washington, DC.

GROUND WATER REMEDIATION AND DESIGN

13.1 *INTRODUCTION TO REMEDIATION METHODS*

During the past decade, ground water scientists and engineers have devised a number of methods to contain and/or remediate soil and ground water contamination. This technology has been largely driven by ground water regulations (RCRA, CERCLA, and HSWA) relating to the transport and fate of contaminants at waste sites (see Chapter 14). Ground water remediation is still in its infancy due to a number of complicating issues that were discovered at waste sites only in the past few years. Many of the original pump-and-treat systems for removing soluble contaminated plumes from the subsurface simply failed to clean up the ground water to acceptable levels. These problems have been documented by the EPA (U.S. EPA, 1989), but most relate to difficulties in site characterization and the lack of recognition of the nonaqueous-phase liquid (NAPL) problem described in detail in Chapter 11.

The remediation of a site must address two major issues: the contaminant source area and the associated soluble contaminant plume. It is now recognized that these source areas and soluble plumes may have to be addressed in very different ways, with greater attention given to the remediation or containment of the source area. Examples of contaminant sources include leaking landfills, pipes, and tanks, spills that have sorbed to the subsurface near the water table, and NAPLs either floating near the capillary fringe or residing on clay lenses below the water table (Chapter 11). The control of these complex source areas is a major challenge at every hazardous-waste site. Migrating soluble plumes, on the other hand, which have been studied extensively over the past decade, can often be adequately controlled by traditional pump-and-treat systems.

The goals of a ground water remediation effort may cover a range of options, from halting the migration of a soluble plume from its source, to isolating and containing a source area, to treating a contaminated ground water aquifer to a safe drinking-water standard. Through efforts to control the water quality of surface lakes and streams in the 1970s, it was discovered that it may not be technologically or economically feasible to make all streams fishable and swimmable. Similarly, we may not be able to remediate a contaminated aquifer to remove all traces of contamination. Rather, some level of protection at the property fence line to control off-site migration of the contaminant combined with intensive source con-

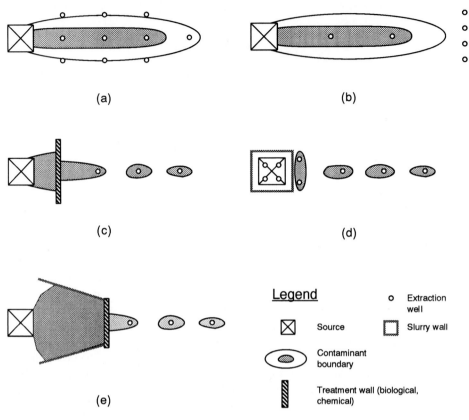

Figure 13.1 Remedial options and source control. (a) Standard pump and treat. (b) Fence line pump and treat. (c) Treatment wall system. (d) Slurry wall system. (e) Funnel and gate system.

trols may be acceptable in some cases. The EPA is beginning to define a new set of goals for many of the nation's Superfund remediation sites as ground water engineers and scientists learn more about the actual performance of standard remediation methods such as pump and treat (EPA, 1992). Figure 13.1 summarizes several remedial options currently being practiced.

13.2 REMEDIAL ALTERNATIVES

Once a site has been characterized for hydrogeology and contaminant concentrations, control and remediation options can be selected and combined to provide an overall cleanup strategy. Choosing a remedial technology is a function of contaminant type, site hydrogeology, source characteristics, and the location of the contaminant in the subsurface. The variation of hydraulic conductivity (K) or transmissivity (T) of a subsurface formation is one of the most important parameters of interest. The ultimate success or failure of any remediation system is a direct function of the ability of the aquifer to transport fluids (both water and contaminant), nutrients, NAPLs, and vapor or air. Well mechanics dictate that pumping and injection rates of liquids and vapors are directly related to aquifer properties. Many of these issues will be addressed in detail in this chapter.

Knowledge of the reactivity or biodegradability of a contaminant in the subsurface is vital for determining the viability of an in situ treatment process. Chapter 8 indicated that the availability of an electron acceptor, such as dissolved oxygen, is crucial to the design of a bioremediation system. The ability of an injection system or infiltration system to deliver nutrients or additional electron acceptors to a desired plume location for bioremediation is a function of both the hydraulic conditions and the reactions that occur in the subsurface. Chemical reactions and sorption of organics onto the soil matrix, which might occur during a remediation effort, must be addressed in detail as part of the remediation design phase (Chapter 7).

If pure product or NAPL exists at or near the water table in the form of separate-phase fluid, the problem of contaminant removal may be greatly complicated. As described in Chapter 11, floating product on the water table (LNAPL) is easier to remove than denser water contaminants (DNAPLs), which can sink into lower regions of the aquifer and cannot be easily removed. Thus, depending on the site hydrogeology, it may be necessary to combine a pumping system with other techniques (bioderemediation, soil vapor extraction, skimming of NAPL) in order to complete remediation of the saturated and vadose zones at a hazardous-waste site.

Generally accepted remediation alternatives for subsurface contamination problems include the following:

1. Excavation and disposal
2. Containment (physical barriers)
3. Pumped removal of product or contaminated water and aboveground treatment (pump and treat with activated carbon or air stripping)
4. In situ biological or chemical treatment
5. Soil vapor extraction

These methods are not the only ones in use. However, they are the most prevalent and successful techniques currently being applied at many hazardous-waste sites. The remainder of this chapter describes these methods in detail. Section 13.8 discusses some of the new and emerging technologies briefly. Thomas et al. (1987) and Mercer et al. (1990) provide very thorough reviews of remedial methods for ground water and soil contamination. The U.S. EPA (1992) describes an evaluation of effectiveness of ground water extraction for remediating contamination at 24 sites across the United States.

13.3 SOURCE CONTROL

The objective of source control is to eliminate or reduce the spread of contaminants in the subsurface by the use of physical containment methods or hydrodynamic controls. Hydraulic or hydrodynamic controls, discussed in Section 13.4, usually involve the injection or pumping of ground water via a series of wells surrounding the source or in the immediate plume area to manipulate the natural ground water gradient in such a way as to inhibit the migration of contaminants. Physical containment measures, as discussed next, are designed to isolate contaminated soil and ground water from the local environment. Isolation techniques include excavation and removal of the contaminated soil, physical barriers to ground water flow, and surface-water controls.

Excavation Methods

For complete source removal, a pit may be excavated in the vicinity of the source and the excavated soil transported to a secured site, such as a landfill or surface impoundment, for disposal (Ehrenfeld and Bass, 1984). A more recent process is to remove the contaminated soil to a hazardous-waste incinerator for complete thermal destruction of organic contaminants. Contaminated ground water may also be pumped out and transported off site for storage, treatment, or disposal. The inherent problem in excavation and removal to another location is that a new hazardous-waste site is often created. However, removal of contaminated soil and ground water to a more environmentally appropriate location may be necessary if in situ containment or treatment poses special problems or initiates litigation. An obvious difficulty associated with excavation and removal is that the complete removal of subsurface soil and ground water may be impossible when the contamination extends deep into the subsurface or the contaminants spread beneath a facility or building.

Barriers to Ground Water Flow

Physical barriers used to prevent the flow of ground water include slurry walls, grout curtains, sheet piling, and compacted liners or geomembranes (Ehrenfeld and Bass, 1984; Canter and Knox, 1986). These impermeable barriers may be used to contain contaminated ground water or leachate or to prevent the flow of clean ground water into a zone of contamination. These barriers may need to be combined with a well system or a leachate collection system for complete hydraulic control (Figure 13.1).

Aquifers with sandy surficial soil less than 60 ft in depth and underlain by an impermeable layer of fine-grain deposits or bedrock are most amendable to **slurry wall** construction (Need and Costello, 1984). Construction of a slurry wall entails excavating a narrow trench (2 to 5 ft in width) surrounding the contaminated zone (Figure 13.2). The trench is generally excavated through the aquifer and into the underlying bedrock or impermeable zone. The slurry acts to maintain the trench during excavation and is usually a mixture of soil or cement, bentonite, and water (Ehrenfeld and Bass, 1984). Several recent books are available on detailed design issues related to slurry wall construction. Johnson et al. (1985) provide a useful series of papers on slurry walls based on a symposium sponsored by the American Society for Testing and Materials (ASTM).

To date, there are two basic methods of slurry wall construction. Trenches constructed using a cement–bentonite (C–B) mixture are allowed to set, whereas

Figure 13.2 Trenching in progress. Source: Ryan, 1985. © ASTM. Reprinted with permission.

those constructed with a soil–bentonite (S–B) mixture are backfilled and solidified with appropriate materials. Solidification of the trench may be accomplished by backfilling with soil mixed with bentonite; soil mixed with cement, concrete, and asphaltic emulsion; or a combination of these with synthetic membranes (Lynch et al., 1984; Tallard, 1985). The chosen materials should be compatible with the in situ soil and contaminant regime, and laboratory tests should be conducted to make this determination. Depending on the backfill material used, the permeability of the resulting barrier may range from 10^{-6} to 10^{-8} cm/sec (Nielson, 1983). An S–B slurry cutoff wall was chosen for the first cleanup financed by the Environmental Protection Agency's Superfund program (Ayres et al., 1983).

Grout curtains are another type of physical barrier that is constructed by injecting grout (liquid, slurry, or emulsion) under pressure into the ground through well points (Canter and Knox, 1986). Ground water flow is impeded by the grout that solidifies in the interstitial pore space. The curtain is made contiguous by injecting the grout into staggered well points that form a two- or three-row grid pattern (Ehrenfeld and Bass, 1984). Spacing of the well points for grout injection and the rate of injection are critical. Premature solidification occurs when the

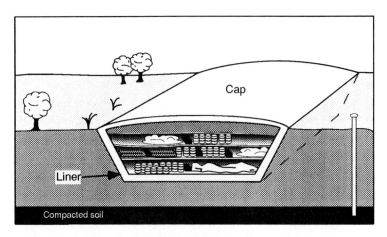

Figure 13.3 Example of a synthetic liner for a hazardous-waste landfill.

injection rate is too slow, whereas the soil formation is fractured when the rate is too fast. Soil permeability is decreased and soil-bearing capacity is increased after the grout properly solidifies. Chemical or particulate grouting is most effective in soils of sand-sized grains or larger. The expense of grouting and the potential for contamination-related problems in the grout tend to limit its application to hazardous-waste containment problems.

Sheet piling is a physical barrier constructed by driving interlocking sections of sheet steel into the ground (Canter and Knox, 1986). The sheets are assembled before use by slotted or ball-and-socket types of connections and are driven into the soil in sections using a pile hammer (Nielsen, 1983). After contacting an impermeable strata beneath the zone of isolation, the length of piling remaining aboveground is usually cut off. The connections between the steel sheets are not initially watertight; however, fine-grained soil particles eventually fill the gaps and the barrier generally becomes impermeable to ground water flow. Sheet piling may be ineffective in coarse, dense material because the interlocking web may be disrupted during construction (Nielsen, 1983).

Liners are another type of physical barrier; they are often used in conjunction with surface-water controls and caps (Canter and Knox, 1986). Liners may be used to protect ground water from leachate resulting from landfills containing hazardous materials. The type of liner used depends on the type of soil and the contaminants that are present. Types of liners include polyethylene, polyvinyl chloride (PVC), many asphalt-based materials, and soil–bentonite or cement mixtures. Polyvinyl chloride liners have permeabilities of less than 3.0×10^{-11} ft/sec; however, little is known about the service life of the PVC membranes (Threlfall and Dowiak, 1980). The membrane should be installed over fine-grained soil to prevent punctures. A typical liner for a landfill is depicted in Figure 13.3.

Surface-water Controls

The vertical migration of contaminants in the subsurface is known to be induced by the infiltration of surface water through the vadose zone. Caps, dikes, terraces,

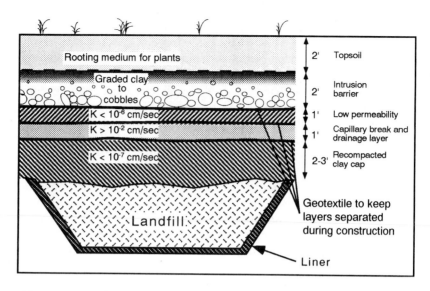

Figure 13.4   Typical multilayered cap constructed of natural soil materials.

vegetation, and grading are used to reduce the amount of infiltration into a site and to control erosion. Channels, chutes, downpipes, seepage basins, dikes, and ditches may be used to divert uncontaminated surface water away from waste sites, collect contaminated leachate, or direct contaminated water away from clean areas (Ehrenfeld and Bass, 1984; Thomas et al., 1987). Many of these techniques may be used in combination with each other.

Surface capping is a three-point process that usually involves covering the contaminated area with an impermeable material, regrading to provide positive drainage and minimize infiltration of surface water, and revegetating the site (Canter and Knox, 1986). Surface caps are usually constructed using materials from one of three groups: (1) natural soils, (2) commercially designed materials, or (3) waste materials. The material used should be compatible with the soil type and contaminant regime. Examples include clay, concrete, asphalt, lime, fly ash, and mixed layers or synthetic liners. Often, several types of materials are used in combination to create a multilayer cap (Figure 13.4).

13.4 HYDRAULIC CONTROLS AND PUMP-AND-TREAT SYSTEMS

The design objective in the hydraulic control of ground water contamination can be to generally alter the ground water flow regime to prevent off-site discharge, to reduce the rate of plume migration by removing contaminated waste, or to confine the plume to a potentiometric low created by a combination of pumping and injection wells. Maintenance of wells and pumps is particularly important for the successful implementation of this method, and as ground water levels change, the system design may have to be altered accordingly. Aboveground biological or physical treatment units may be necessary to handle any contaminated water pumped by the wells. Hydraulic control systems can include interceptor systems and/or pumping and injection well systems.

Interceptor systems are used to collect contaminated ground water close to the water table and may consist of drains, a line of buried perforated pipe, and/or an open trench, usually backfilled with gravel (Figure 13.5). These systems, operating as an infinite line of wells near the shallow water table, are efficient at removing contamination near the surface. Trenches are often used to collect nonaqueous-phase liquids (particularly LNAPLs) like crude oil or gasoline, which are light and tend to move near the capillary fringe just above the water table.

Pumping wells are used to extract water from the saturated zone by creating a capture zone for migrating contaminants (Figure 13.6). A major problem with pumped wells is the proper treatment and disposal of the contaminated water once removed from the subsurface. On-site treatment facilities are usually required before water can be reinjected to the aquifer or released to the surface. The number of wells, their locations, and the required pumping rates are the key design parameters of interest. Methods of analysis for wells are described in more detail in Chapter 3. The analysis of a well system that is pumping water containing dissolved contaminants can be addressed using standard well mechanics and capture zone theory (Mercer et al., 1990). If the hydrogeology is conducive to an injection or pumping system, several design approaches can be used to develop an efficient and reliable system for contaminant removal. The most successful designs have been developed for relatively sandy or silty sand soils that are well characterized.

Pilot-scale systems or small field demonstration projects have been used at a number of sites to evaluate the pump rates and the placement of wells in a small area of the site before expanding to the entire site. In this way, operational policies, mechanical problems, and costs can be evaluated before attempting the larger cleanup. Careful monitoring of the system is the key to understanding how the injection pumping pattern will respond over time. The EPA reviewed a number of pump-and-treat systems in detail and found generally mixed performance (U.S. EPA, 1989, 1992). However, many of these systems were originally designed without any additional remediation methods applied to the source area (such as vapor extraction or NAPL controls).

Capture Zone Techniques

The pumping of water containing dissolved contaminants can be addressed using standard well mechanics and capture zone theory. If the hydraulic conductivity is too low or the geology is very complex and heterogeneous, pumping may not be a feasible alternative for hazardous-waste cleanup. If the hydrogeology is permeable enough for an injection/pumping system, then both analytical and numerical models are available to evaluate well placement and the efficiency of remediation.

Javandel and Tsang (1986) developed a useful analytical method for the design of recovery well systems, based on the concept of a capture zone (Figure

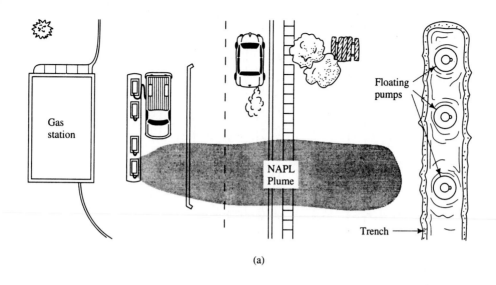

(a)

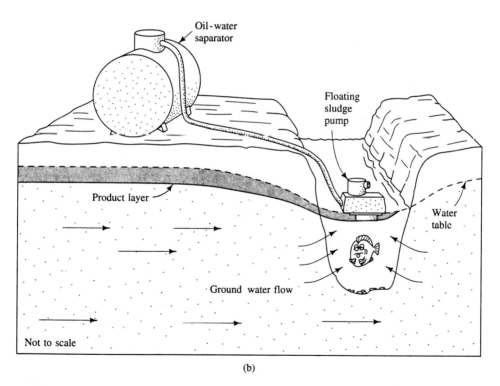

(b)

Figure 13.5 (a) Top view of LNAPL plume and interceptor trench. (b) Cross section of trench and floating pump used to capture floating product and depress the water table. Source: Fetter, 1993. © Macmillan College Publishing Co., Inc. Reprinted with permission

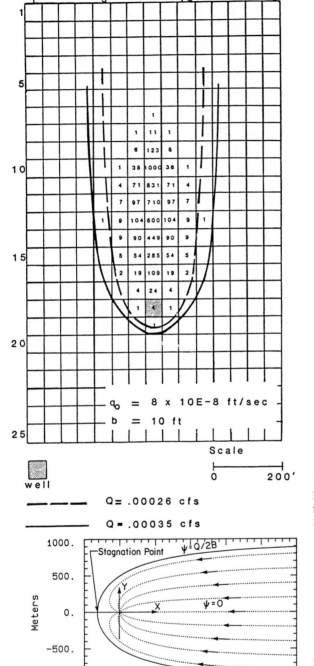

$q_o = 8 \times 10E{-}8$ ft/sec

$b = 10$ ft

Scale

0   200'

well

Q = .00026 cfs

Q = .00035 cfs

Figure 13.6 Single-well capture zones for different pumping rates. Source: Satkin and Bedient, 1988.

Stagnation Point

$\psi = Q/2B$

$\psi = 0$

$\psi = -Q/2B$

Meters

Meters

Figure 13.7 Paths of some water particles within the capture zone with $Q/BU = 2000$, leading to the pumping well located at the origin. Source: Javandel and Tsang, 1986.

13.7). The capture zone for a well depends on the pumping rate and the aquifer conditions. Ideally, the capture zone should be somewhat larger than the plume to be cleaned up so that wells can be added until sufficient pumping capacity is provided to create a useful capture zone. However, with more wells, well spacing becomes an important parameter, as well as pumping rate. The greater the pumping rate is, the larger the capture zone, so the closer the wells are placed, the better the chance of complete plume capture. The method from Javandel and Tsang works to minimize the pumping/injection rates through a proper choice of well location and well spacing.

Javandel and Tsang (1986) use complex potential theory as the basis for a simple graphical procedure to determine the pumping rate, the number of wells, and the distance between wells. The method requires type curves for one to four wells (Figure 13.8) and values for two parameters: B the aquifer thickness (assumed to be constant) and U the specific discharge or Darcy velocity (also assumed constant) for the regional flow system. The method involves the following five steps:

1. A map of the contaminant plume is constructed at the same scale as the type curves. The edge or perimeter of the plume should be clearly indicated, together with the direction of regional ground water flow.

2. Superimpose the type of curve for one well on the plume, keeping the x axis parallel to the direction of regional ground water flow and along the midline of the plume so that approximately equal proportions of the plume lie on each side of the x axis. The pumped well on the type curve will be at the downstream end of the plume. The type curve is adjusted so that the plume is enclosed by a single Q/BU curve.

3. The single-well pumping rate (Q) is calculated using the known values of aquifer thickness (B) and the Darcy velocity for regional flow (U), along with the value of Q/BU indicated on the type curve (TCV) with the equation

$$Q = B \cdot U \cdot TCV$$

4. If the pumping rate is feasible, one well with pumping rate Q is required for cleanup. If the required production is not feasible due to a lack of available drawdown, it will be necessary to continue adding wells (see step 5).

5. Repeat steps 2, 3, and 4 using the two-, three-, or four-well type curves in that order, until a single-well pumping rate is calculated that the aquifer can support. The only extra difficulty comes from having to calculate the optimum spacing between wells using the following simple rules,

$$\text{Two wells} \qquad \frac{Q}{\pi BU} = \text{center-line spacing}$$

$$\text{Three wells} \qquad \frac{1.26Q}{\pi BU}$$

$$\text{Four wells} \qquad \frac{1.2Q}{\pi BU}$$

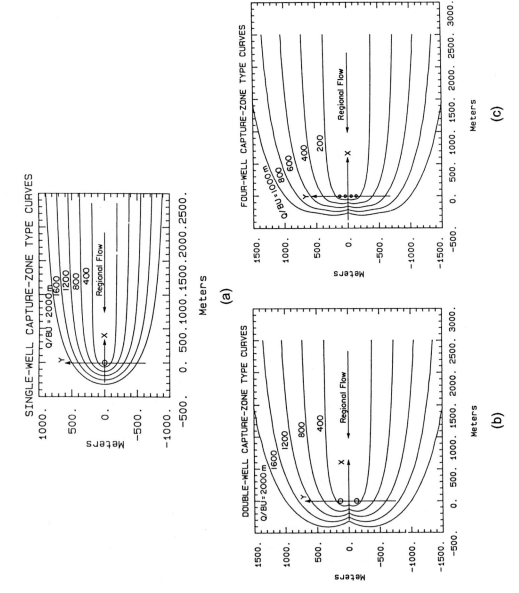

Figure 13.8 Set of type curves showing the capture zones for various values of Q/BU. (a) Single pumping well at the origin. (b) Two pumping wells located on the Y axis. (c) Four pumping wells located on the Y axis.   Source: Javandel and Tsang, 1986.

and having to account for the interfacing among the pumped wells when checking on the feasibility of the pumping rates. The wells are always located symmetrically around the x axis, as the type curves show.

Reinjecting the treated water produced by the wells will accelerate the rate of aquifer cleanup. The procedure for locating and designing injection wells is essentially the same as that just discussed, except that the type curves are reversed and the wells are injecting instead of pumping. The authors suggest that the injection wells should be moved slightly upstream of the calculated location to avoid causing parts of the plume to follow a long flow path. Their rule of thumb is to place the wells half the distance between the theoretical location and the tail of the plume. The following example taken from Javandel and Tsang (1986) illustrates how the technique is used.

Example 13.1   CAPTURE ZONE

Figure 13.9 shows a plume of trichloroethylene (TCE) in a shallow confined aquifer having a thickness of 10 m, a hydraulic conductivity of 10^{-4} m/sec, an effective porosity of 0.2, and a storativity of 3×10^{-5}. The hydraulic gradient for the regional flow system is 0.002, and the available drawdown for wells in the aquifer is 7 m. Given this information, design an optimum collection system.

Solution. Values of B and U are required for the calculation. B is given as 10 m, but U needs to be calculated from the Darcy equation

$$U = \frac{K\,dh}{dl} = 10^{-4} \times (0.002) = 2 \times 10^{-7} \text{ m/sec}$$

Now we are ready to work with the type curves following the steps just outlined. Superposition of the type curve for one well on the plume provides a Q/BU

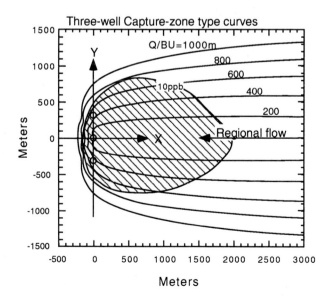

Figure 13.9 The 10-ppb contour line of TCE at the matching position with the capture zone type curve of Q/BU = 800.   Source: Javandel and Tsang, 1986.

curve of about 2500. Using this number and the values of B and U, the single-well pumping rate is

$$Q = B \cdot U \cdot TCV$$

or

$$10 \times (2 \times 10^{-7}) \times 2500 = 5 \times 10^{-3} \text{ m}^3/\text{sec}$$

A check is required to determine whether this pumping rate can be supported for the aquifer. The Cooper–Jacob (Chapter 3) equation provides the drawdown at the well, assuming $r = 0.2$ m and the pumping period is 1 yr:

$$s = \frac{2.3Q}{4\pi T} \log \frac{2.25Tt}{r^2 S}$$

where $Q = 5 \times 10^{-3}$ m³/sec, $T = KU = 10^{-3}$ m²/sec, $t = 1$ yr or 3.15×10^7 sec, $r = 0.2$ m, and $S = 3 \times 10^{-5}$.

The pumping period represents some preselected planning horizon for cleanup. Substitution of the known values into the Cooper–Jacob equation gives a drawdown of 9.85 m. Even without accounting for well loss, the calculated drawdown exceeds the 7 m available. Thus, a multiwell system is necessary.

Superimposing the plume on the two-well type curve provides a Q/BU value of 1200, which in turn gives a Q for each of the two wells of $10 \times (2 \times 10^{-7}) \times 1200$ or 2.4×10^{-3} m³/sec. The optimum distance between wells is $Q/(\pi BU)$ or $2.4 \times 10^{-3}/[\pi \times 10 \times (2 \times 10^{-7})] = 382$ m. Again we check the predicted drawdown at each well after 1 yr against the available 7 m. Because of symmetry, the drawdown in each well is the same. The total drawdown at one of the wells includes the contribution of that well pumping plus the second one 382 m away, based on the principle of superposition.

The calculated drawdown is 6.57 m, which is less than the available drawdown. However, well loss should be considered, which makes the two-well scheme unacceptable. Moving to a three-well scheme, Q/BU is 800 (Figure 13.9), which translates to a pumping rate of 1.6×10^{-3} m³/sec for each well. Carrying out the drawdown calculation for three wells located $1.26Q/(\pi BU)$ or 320 m apart provides an estimate of 5.7 m for the center well, which is comfortably less than the available drawdown. Thus, we have been able to ascertain the need for three wells located 320 m apart and each pumped at 1.6×10^{-3} m³/sec.

Assumptions are built into the formulation, such as constant aquifer transmissivity, fully penetrating wells, no recharge, and isotropic hydraulic conductivity, which have to be satisfied by the field problem or the method will not necessarily yield a correct result. Actual field sites where boundary conditions and site variabilities are important issues may require analysis using numerical models, as described in the next section.

Analysis of Pumped Systems with Numerical Models

Ground water flow and transport modeling performed during the remedial investigation can be a powerful tool to estimate plume movement and response to various remedial schemes (see Chapter 10). Flow and contaminant transport models should be calibrated to a measured plume of contamination to the greatest extent possible. However, caution should be used when applying models at hazardous-waste sites because there can be great uncertainty whenever subsurface transport is modeled, particularly when the results of the model are based on estimated parameters. The purposes of modeling ground water flow for remediation include the following:

1. Predict concentrations of contaminants at receptor points.
2. Estimate the effect of source-control actions on remediation.
3. Guide the placement of monitoring wells and hydrogeologic characterization when the remedial study is conducted in phases.
4. Evaluate expected remedial performance under a variety of alternative designs so that the efficiency and time of cleanup to some specified level can be predicted.

The determination of whether or not to use modeling and the level of effort that should be expended is made on the basis of the objectives of the modeling, the ease with which the subsurface can be conceptualized mathematically, and the availability of data. Field data are collected to characterize the variables that govern the hydrologic and contaminant response of the site in question. Estimates based on literature values or professional judgment are frequently used as well.

Models such as the three-dimensional, finite difference flow code MOD-FLOW (McDonald and Harbaugh, 1984) and the semianalytical flow code RESSQ (Javandel et al., 1984) can be used to simulate flow patterns and changes resulting from the operation of a pump-and-treat system. Other models are available to analyze contaminant transport, such as the two-dimensional MOC model from Konikow and Bredehoeft (1978) and BIOPLUME II from Rifai et al. (1988). Several detailed modeling exercises simulating a pump-and-treat system and bio-remediation of an aquifer are described in Section 13.9.

Optimizing Pumping Injection Systems

Many investigators have used numerical ground water models as a tool in the design of aquifer restoration strategies because they provide a rapid means of predicting or assessing the effects of different remedial alternatives. Andersen (1984) used a finite difference ground water model as an aid in selecting an appropriate remedial action at the Lipari Landfill in New Jersey. Althoff et al. (1981) used a ground water flow model to test a variety of well configurations, well locations, and pumping rates for hydraulic capture of a 100-ft-long 1,1,1-trichloroethane plume. Freeberg et al. (1987) delineated a trichloroethylene plume and used the USGS MOC model to evaluate different withdrawal schemes at an industrial-waste site (Section 13.9).

To select the best well configuration for a particular withdrawal scheme requires the ability to predict changes in flow and chemical concentration in the aquifer for each possible management alternative (Konikow and Thompson, 1984).

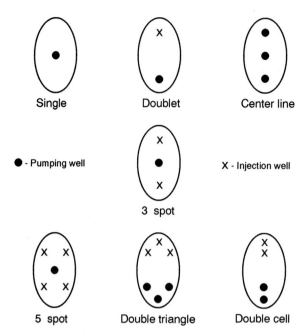

Single   Doublet   Center line

● - Pumping well   X - Injection well

3 spot

5 spot   Double triangle   Double cell

Figure 13.10 Well patterns used in this study. Source: Satkin and Bedient, 1988.

The best pumping arrangement at a field site is developed generally by a tedious trial-and-error process (Glover, 1982). The trial-and-error approach suffers because it is inefficient; however, the heuristic knowledge gained by the user is invaluable and allows the modeler to steadily improve on future trials. Alternatively, several investigators (Gorelick, 1983, Gorelick et al., 1984; Molz and Bell, 1977; and Shafer, 1984) have demonstrated how linear or nonlinear programming (optimization techniques) can be combined with a ground water transport model to efficiently arrive at an optimal design strategy. However, optimization techniques have suffered because of ill-posed ground water transport problems.

Satkin and Bedient (1988) used the USGS MOC model to investigate various pumping and injection patterns to remediate a contaminant plume. Seven different well patterns were studied for various combinations of hydraulic gradient, maximum drawdown, and aquifer dispersivity (Figure 13.10). Various cleanup levels were evaluated, along with the volume of water circulated and the volume of water requiring treatment. Eight hydrogeologic conditions (Table 13.1) were modeled for the various remediation schemes. The location of a single well or multiple pumping wells requires that the capture zone encompass the entire plume. Generally, the closer a single well can be placed to the center of the contamination mass, the faster the cleanup time. Additional wells aligned with the axis of the plume will increase the rate of cleanup by pumping a greater volume of water.

The key hydrogeologic variables that control the rate of cleanup are well locations, pumping rates, transmissivity, dispersivity, and hydraulic gradient. The three-spot, doublet, and double-cell well patterns were effective under low hydraulic gradient conditions. These well patterns require on-site treatment and reinjection. The three-spot pattern performed best under a high hydraulic gradient.

TABLE 13.1   Different Generic Hydrogeologic Conditions Modeled

| | Generic hydrogeologic conditions | | |
Condition	Maximum drawdown	Hydraulic gradient	Longitudinal dispersivity
A	High *s*	Low *i*	Low *a*
B	Low *s*	Low *i*	Low *a*
C	Low *s*	Low *i*	High *a*
D	High *s*	Low *i*	High *a*
E	Low *s*	High *i*	Low *a*
F	High *s*	High *i*	Low *a*
G	High *s*	High *i*	High *a*
H	Low *s*	High *i*	High *a*

High s = ≤ 10 ft
Low s = ≤ 5 ft
High i = 0.008
Low i = 0.00008
High a = 30 ft
Low a = 10 ft

Source: Satkin and Bedient, 1988.

The center-line well pattern performed reasonably well under both low- and high-gradient conditions, but may present a water-disposal problem in an actual case. For a given set of well locations, by varying transmissivity and maintaining drawdown, dispersivity, and hydraulic gradient constant, the cleanup time was found to be inversely related to the pumping rate. Figure 13.11 indicates typical results for hydrogeologic condition A (Table 13.1).

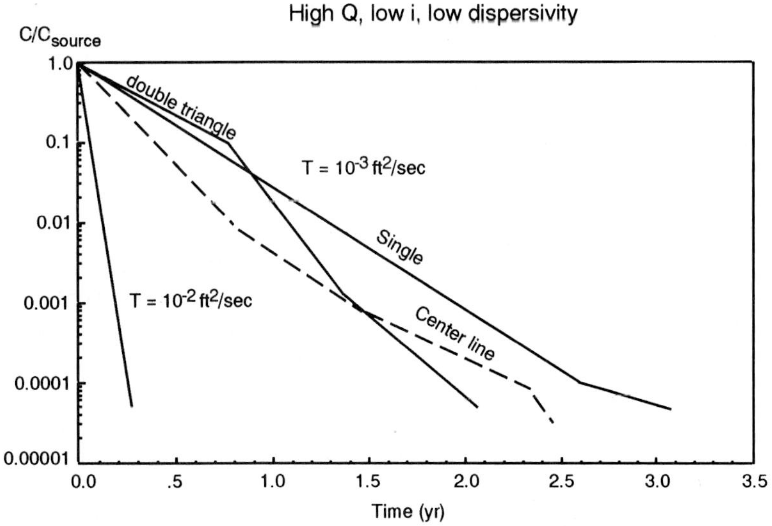

Figure 13.11 Comparison of relative concentrations versus cleanup time for various well patterns for condition A. Source: Satkin and Bedient, 1988.

13.5 BIOREMEDIATION

The practical application of biodegradation discussed in Chapter 8 for the remediation of hazardous-waste sites is termed **bioremediation.** The process, when carried out in situ, usually involves stimulating the indigenous subsurface microorganisms by the addition of nutrients and an electron acceptor to biodegrade the contaminants of concern. The process can also be utilized ex situ to biodegrade a variety of waste streams. Ex situ biological treatment, however, requires extraction of the ground water from the subsurface or excavation of the contaminated soils.

In contrast with other remedial techniques that transfer the contaminants from one phase in the environment to another, in situ biorestoration offers partial or complete destruction of the contaminants (Chapter 8). Partial destruction takes place usually because the chemicals may not all be available to the microorganisms due to transport limitations or because the chemicals may not be completely mineralized, but rather biotransformed to other organic chemicals.

The in situ bioremediation process might offer attractive economics for remediation because it precludes the need for excavation, transportation, and disposal costs associated with other remediation alternatives. In situ bioremediation also offers an advantage where physical limitations due to the presence of structures might inhibit removal of the contaminants. Additionally, in situ bioremediation can be used to treat contamination that is sorbed to the aquifer matrix and dissolved in the ground water simultaneously.

Bioremediation is not without its problems, however, the most important being the lack of well-documented field demonstrations that show the effectiveness of the technology and what, if any, are the long-term effects of this treatment on ground water systems. The lack of documented field studies may be attributed in part to the significant commercial potential of the technology, which basically means that the information is proprietary. Other problems include the possibility of generating undesirable intermediate compounds during the biodegradation process that are more persistent in the environment than the parent compound.

Two essential criteria must be satisfied before bioremediation can be considered a viable remediation alternative for a field site. First, the subsurface geology must have a relatively large hydraulic conductivity ($>10^{-4}$ cm/sec) to allow the transport of the electron acceptor and nutrients through the aquifer. Second, microorganisms must be present in sufficient numbers and types to degrade the contaminants of interest. It is important to keep in mind that any bioremediation project at a field site needs to be preceded by laboratory experiments of microbial stimulation and modeling studies of nutrient transport to ensure efficient performance of the system.

Enhanced aerobic bioremediation for a petroleum spill, for example, is essentially an engineered delivery of nutrients and oxygen to the contaminated zone in an aquifer. The main constraints on the rate of delivery of nutrients and oxygen are (1) hydrogeologic and (2) oxygen sources. Bioremediation, as discussed earlier, is difficult in aquifers with an average hydraulic conductivity (K) that is less than 10^{-4} cm/sec (Thomas and Ward, 1989). Oxygen sources include air, pure oxygen (gaseous and liquid forms), and hydrogen peroxide. Sparging the ground water with air and pure oxygen can supply only 8 to 40 mg/L of oxygen depending on the temperature of the injection fluid (Lee et al., 1988).

Hydrogen peroxide, which dissociates to form water and one-half molecule of oxygen, is infinitely soluble in water (Thomas and Ward, 1989); however, hydrogen peroxide can be toxic to microorganisms at concentrations as low as 100 ppm. A stepping-up procedure is usually utilized to allow the microorganisms to adapt to the higher concentrations of the oxidant. Other problems have to do with the stability of hydrogen peroxide. The key to success in using hydrogen peroxide as an oxygen source is to add a relatively large quantity to water and have oxygen released in a controlled manner as it advances through the aquifer. If hydrogen peroxide is destabilized, oxygen will come out of solution as a gas, and the process becomes less efficient (Hinchee et al., 1987). Proprietary techniques have been developed to stabilize hydrogen peroxide.

Design Issues

The basic steps involved in an in situ biorestoration program (Lee et al., 1988) are the following:

1. Site investigation
2. Free product recovery
3. Microbial degradation enhancement study
4. System design
5. Operation
6. Maintenance

It is important to define the hydrogeology and the extent of contamination at the site prior to the initiation of any in situ effort. The parameters of interest include the direction and rate of ground water flow, the depths to the water table and to the contaminated zone, the specific yield of the aquifer, and the heterogeneity of the soil. In addition, other parameters such as hydraulic connections between aquifers, potential recharge and discharge zones, and seasonal fluctuations of the water table should be considered. The pumping rate that can be sustained in the aquifer is an important consideration because it limits the amount of water that can be circulated in the system during the bioremediation process.

After defining the hydrogeology, recovery of free product, if any, at the site should be completed. The pure product can be removed using physical recovery techniques such as a single-pump system that produces water and hydrocarbon or a two-pump, two-well system that steepens the hydraulic gradient and recovers the accumulating hydrocarbon. Physical recovery often accounts for 30% to 60% of the hydrocarbon before yields decline (Lee et al., 1988).

Prior to the initiation of a bioremediation activity, it is important to conduct a feasibility study for the biodegradation of the contaminants present at the site. First, contaminant-degrading microorganisms must be present and, second, the response of these native microorganisms to the proposed treatment method must be evaluated. In addition, the feasibility study is conducted to determine the nutrient requirements of the microorganisms. These laboratory studies provide a reliable basis for performance at the field level only if they are performed under conditions that simulate the field.

The chemistry of a field site will obviously affect the types and amounts of required nutrients. Limestone and high-mineral-content soils, for example, will affect nutrient availability by reacting with phosphorus. Silts and clays at sites may induce nutrient sorption on the soil matrix and hence decrease the amount of nutrients available for growth. In general, a chemical analysis of the ground water provides little information about the nutrient requirements at a field site; it is mostly the soil composition that is of significance.

Nutrient requirements are usually site specific. Nitrogen and phosphorus were required at the Ambler site (Raymond et al., 1976); however, the addition of ammonium sulfate, mono- and disodium phosphate, magnesium sulfate, sodium carbonate, calcium chloride, manganese sulfate, and ferrous sulfate was required at other sites (Raymond et al., 1976; Minugh et al., 1983). The form of the nutrient may also be important; ammonium nitrate was less efficient than ammonium sulfate in one aquifer system.

Feasibility studies can be completed using several different techniques. Batch culture techniques are used to measure the disappearance of the contaminant; electrolytic respirometer studies are utilized to measure the uptake of oxygen. Tests designed to measure an increase in microbial numbers are not sufficient indicators of metabolization of the contaminant in question. Instead, studies that measure the disappearance of the contaminant or mineralization studies that confirm the breakdown of the contaminant to carbon dioxide and water need to be conducted. Controls to detect abiotic transformation of the pollutants and tests to detect the toxic effects of the contaminants on the microflora should be included (Flathman et al., 1984).

A system for injection of nutrients into the formation and circulation through the contaminated portion of the aquifer must be designed and constructed (Lee and Ward, 1985). The system usually includes injection and production wells and equipment for the addition and mixing of the nutrient solution (Raymond, 1978). A typical system is shown in Figure 13.12. Placement of injection and production wells may be restricted by the presence of physical structures. Wells should be screened to accommodate seasonal fluctuations in the water-table level. Nutrients can also be circulated using an infiltration gallery; this method provides an additional advantage of treating the residual gasoline that may be trapped in the pore spaces of the unsaturated zone. Oxygen can be supplied by a number of methods, including oxygen in water, pure oxygen, hydrogen peroxide, ozone, or by soil venting.

Well installation should be performed under the direction of a hydrogeologist to ensure adequate circulation of the ground water (Lee and Ward, 1985). Produced water can be recycled to recirculate unused nutrients, to avoid disposal of potentially contaminated ground water, and to avoid the need for makeup water. Inorganic nutrients can be added to the subsurface once the system is constructed.

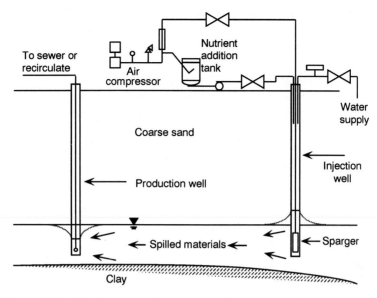

Figure 13.12 Injection system for oxygen.

Continuous injection of the nutrient solution is labor intensive but provides a more constant nutrient supply than a discontinuous process. Continuous addition of oxygen is recommended because the oxygen is likely to be a limiting factor in hydrocarbon degradation.

The performance of the system and proper distribution of the nutrients can be monitored by measuring the organic, inorganic, and bacterial levels. Carbon dioxide levels are also an indicator of microbial activity in the formation. A number of field demonstrations of the bioremediation process have been implemented over the past few years.

Documented Biorestoration Field Demonstrations

Researchers from Suntech, Inc., were among the earliest pioneers in the use of bioremediation at sites contaminated with gasoline. Important field experiments are discussed by Jamison et al. (1975), Raymond et al. (1976), and Raymond (1978). The field study at a site in Ambler, Pennsylvania, involved a leak in a gasoline pipeline that had caused the township to abandon its ground water supply wells. The free product was physically removed prior to the initiation of biodegradation studies at the site. Laboratory studies showed that the natural microbial population at the site could use the spilled high-octane gasoline as the sole carbon source if sufficient quantities of the limiting nutrients, in this case oxygen, nitrogen, and phosphate, were supplied. Pilot studies carried out in the field in several wells confirmed the laboratory findings.

Canada Forces Base Borden

In a controlled field experiment at the Canada Forces Base Borden site, two plumes of gasoline-contaminated ground water were introduced into the aquifer. Immediately upgradient of one plume, ground water spiked with nitrate was added so that a nitrate plume would overtake the organic plume (Berry-Spark et al., 1986). The success of the field experiment was limited (Berry-Spark and Barker, 1987). The dissolved organic contaminant mass (BTEX) decreased rapidly due to residual oxygen concentrations in the aquifer prior to the nitrate overlap. Insufficient organic mass left in the aquifer was not adequate to evaluate anaerobic biotransformation.

Bioremediation Demonstration Project, Traverse City, Michigan

A pilot-scale biorestoration experiment was conducted at the Traverse City field site. A nutrient mix of phosphorus, nitrogen, and oxygen source was injected into a portion of the contaminated aquifer (Wilson et al., 1988). The objective was to stimulate the indigenous microorganisms and enhance the natural biodegradation activity occurring at the site.

A schematic diagram of the well system is shown in Figure 13.13. The system design consisted of five water injection wells and five chemical feed wells, which were screened at different depths. The chemical feed wells had shallow screens, 14 to 19 ft below natural ground, and were designated for injecting the nutrient and oxygen solution (11 gpm). The injection wells had a deeper screen set at 24 to 29 ft below natural ground and were utilized for injecting additional water (31 gpm) to raise the water table to the desired elevations.

The injection of nutrients and oxygen was initiated on March 1, 1988. There were two phases to the project. Phase 1 included a controlled chloride tracer test and pretreatment of the aquifer with liquid oxygen at an average concentration of 4 ppm. A total of 480 lb of liquid oxygen was injected in phase 1 over a period of 3 months. Phase 2, which began on June 1, 1988, utilized hydrogen peroxide as the oxygen source at an initial concentration of 50 ppm, which increased stepwise to 500 ppm. The nutrient solution containing NH_4Cl (250 ppm), KH_2PO_4 (75 ppm), and NA_2HPO_4 (100 ppm) was injected in both phases.

Analysis of the field data collected from the demonstration test between March 1988 and December 1988 indicated the following:

1. The water table was raised to its design levels between March and July of 1988. The water table receded by approximately 1.0 ft in the summer months.
2. The core data prior to the test (August 1987) verified the presence of high levels of fuel hydrocarbons in the test area. Core data from March and June 1988 were not very useful for quantifying the soil contaminants. The October core data showed the persistence of the fuel hydrocarbons in the test area, but with the absence of BTEX from the soils.

3. It was not clear when the BTEX contaminants were removed from the soil. The amount of dissolved oxygen injected between March and July does not satisfy the demand of BTEX, and the water table had dropped in the summer months, so that a large percentage of the soil contaminants were not accessible to the microorganisms.

4. The dissolved oxygen data indicates that the oxygen breakthrough front was located between the 7-ft cluster and the 31-ft cluster between the months of March and June 1988. The dissolved oxygen data between July and December are not a good indicator of cleanup activity because of the drop in water levels in the test area. The amount of contaminants available for the microorganisms was significantly reduced.

5. The ground water BTEX data collected between March and December verify a distinct correlation between leaching mechanisms associated with infiltration. The BTEX concentrations in the ground water declined in the dry summer months and only increased after a rain event in November 1988.

6. The BTEX data in ground water declined to below detection limits in most of the cluster wells. The BTEX persisted in the shallow sampling wells in the BD-62, BD-83, and BD-109 clusters (see Figure 13.13).

The BIOPLUME II model that was used for the selection of parameters for the demonstration test was modified to include the changes that were implemented in the operation of the test. The modeling results closely matched the field data between March and July 1988, using a retardation factor of 10. The modeling results also indicate that if one were to treat the period for the drop in the water table a retardation factor of 100 would be more appropriate. Overall, it was concluded that modeling of the demonstration test was a useful tool for designing and predicting the behavior of the proposed test. Final results from the test indicated that soluble BTEX in ground water was removed in the pilot area.

Anaerobic Bioremediation at Moffet Naval Air Station, California

Roberts et al. (1990) and Semprini et al. (1990) have assessed the capacity of native methanotrophic bacteria to metabolize halogenated organics at the Moffett Naval Air Station in California. The growth of the bacteria was stimulated by injecting dissolved methane and oxygen into the aquifer. Under biostimulation conditions, the transformation of trichlorethylene (TEC), *cis*-1,2-dichlorethylene (*cis*-DCE), *trans*-1,2-dichlorethylene (*trans*-DCE), and vinyl chloride (VC) was assessed by means of controlled addition, frequent sampling, quantitative analysis, and mass-balance comparisons.

The basic approach of the evaluation experiments was to create a test zone in the subsurface. A series of injection, extraction, and monitoring wells was installed within the shallow confined aquifer. Injection and extraction wells were 6 m apart. The organics mentioned previously were injected into the system until a steady state was established. After biostimulation with dissolved oxygen and

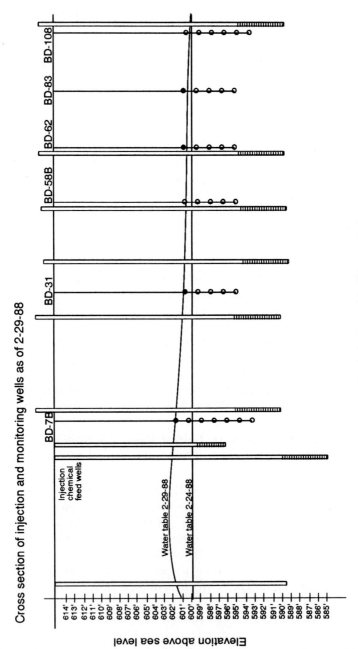

Figure 13.13   Vertical schematic of bioremediation test.

methane, the steady-state transformations observed during the experiment were TCE, 10% to 30%; *cis*-DCE, 45% to 55%; *trans*-DCE, 80% to 90%; and VC, 90% to 95%. These results are in agreement with the laboratory studies, which indicate that the rate of biotransformation is more rapid when the molecules are less halogenated. The lower end of the range represents the nearest observation point (1 m distant), whereas the upper end of the range represents more distant observation points with longer residence times (2 to 4 m distant). Residence times in the aquifer were only 1 to 2 days for the preceding biodegradation rates. There appears to be an optimal concentration of methane beyond which it inhibits the process.

13.6 SOIL VAPOR EXTRACTION SYSTEMS

Vacuum extraction is an important new technology that has emerged to treat spills of volatile organic compounds in the unsaturated zone. It is suited to removing volatile nonaqueous-phase liquids (NAPLs) in the unsaturated zone that are trapped at residual saturation and free-product layers. Experience has shown that, in comparison to remedial strategies like excavation or pump and treat, soil venting can be more efficient and cost effective (Johnson et al., 1988, 1990). Vacuum extraction involves passing large volumes of air through or close to a contaminated spill using an air circulation system. The organic compounds or various fractions of a mixture of organic compounds volatilize or evaporate into the air and are transported to the surface. Thus, just like the other contaminant transport problems considered earlier in the book, the vapor extraction method involves fluid flow (air in this case) and the transport of mass dissolved in the fluid.

Among the advantages of the soil vapor extraction process are that it creates a minimal disturbance of the contaminated soil, it can be constructed from standard equipment, there is demonstrated experience with the process at pilot and field scale, it can be used to treat larger volumes of soil than are practical for excavation, and there is a potential for product recovery. With vapor extraction, it is possible to clean up spills before the chemicals reach the ground water table. Soil vapor extraction technology is often used in conjunction with other cleanup technologies to provide complete restoration of contaminated sites (Malot and Wood, 1985; Oster and Wenck, 1988, CH$_2$M-Hill, 1987).

Unfortunately, there are few guidelines for the optimal design, installation, and operation of soil vapor extraction systems (Bennedsen, 1987). Theoretically based design equations that define the limits of this technology are especially lacking. Because of this, the design of these systems is mostly empirical. Alternative designs can only be compared by the actual construction, operation, and monitoring of each design. A large number of pilot- and full-scale soil vapor extraction systems have been constructed and studied under a wide range of conditions. The information gathered from the experience has been used to deduce the effectiveness of this technology in a major state-of-the-technology review (Hutzler et al., 1989).

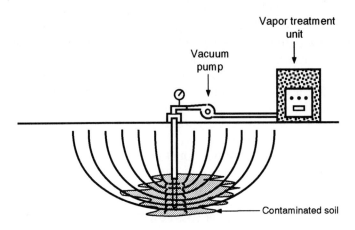

Figure 13.14 Simple vapor extraction system and the resulting pattern of airflow. Source: Johnson et al., 1990.

System Components

A typical soil vapor extraction system such as the one shown in Figure 13.14 consists of (1) one or more extraction wells, (2) one or more air inlet or injection wells (optional), (3) piping or air headers, (4) vacuum pumps or air blowers, (5) flow meters and controllers, (6) vacuum gauges, (7) sampling ports, (8) air–water separator (optional), (9) vapor treatment (optional), and (10) a cap (optional).

Extraction wells are typically designed to fully penetrate the unsaturated zone to the capillary fringe. If the ground water is at a shallow depth or if the contamination is confined to near-surface soils, the extraction wells may be placed horizontally. Extraction wells usually consist of slotted, plastic pipe placed in permeable packing. The surface of the augered column for vertical wells or the trench for horizontal wells is usually grouted to prevent the direct inflow of air from the surface along the well casing or through the trench.

It may be desirable to also install air inlet or injection wells to control airflow through zones of maximum contamination. They are constructed similarly to the extraction wells. Inlet wells or vents are passive and allow air to be drawn into the ground at specific locations. Injection wells force air into the ground and can be used in closed-loop systems (Payne et al., 1986). The function of inlet and injection wells is to enhance air movement in strategic locations and promote horizontal airflow to the extraction wells.

The pumps or blowers reduce gas pressure in the extraction wells and induce airflow to the wells. The pressure from the outlet side of the pumps or blowers can be used to push the exit gas through a treatment system and back into the ground if injection wells are used. Gas flow meters are installed to measure the volume of extracted air. Pressure losses in the overall system are measured with vacuum gauges. Sampling ports may be installed in the system at each well head, at the blower, and after vapor treatment. In addition, vapor and pressure monitoring probes may be placed to measure soil vapor concentrations and the radius of influence of the vacuum in the extraction wells.

To protect the blowers or pumps and to increase the efficiency of vapor treatment systems, an air–water separator may need to be installed. The condensate may then have to be treated as a hazardous waste depending on the types and concentrations of contaminants. The need for a separator may be eliminated by covering the treatment area with an impermeable cap or by designing the extraction wells to separate water from air within the well packing. An impermeable cap serves to cover the treatment site to minimize infiltration and controls the horizontal movement of inlet air.

Vapor treatment may not be required if the emission rates of chemicals are low or if they are easily degraded in the atmosphere. Typical treatment systems include liquid–vapor condensation, incineration, catalytic conversion, or granular activated carbon adsorption.

Patterns of air circulation to extraction wells have been studied in the field by direct measurements (Batchelder et al., 1986) and by mathematical and experimental modeling (Johnson et al., 1988, 1990; Krishnayya et al., 1988). Most of the theoretical work to date has considered that any density differences in the vapor can be neglected under the forced convective conditions created by the vacuum extraction. The following section presents the governing equations that describe the flow and transport of vapor in the subsurface.

Governing Equations of Flow

Johnson et al. (1988, 1990) have exploited the analogy to ground water flow in developing simple screening models to describe the distribution in pressure around venting wells. For conditions of radial flow, the governing equation can be written as

$$\frac{1}{r}\frac{\partial}{\partial r}\left(r\frac{\partial P'}{\partial r}\right) - \left(\frac{e\mu}{kP_{atm}}\right)\frac{\partial P'}{\partial t} \tag{13.1}$$

where P' is the deviation of pressure from the reference pressure P_{atm}, k is soil permeability, μ is the vapor viscosity, e is porosity, and t is time. When Eq. (13.1) is solved with appropriate boundary conditions, with m as the thickness of the unconfined zone and r as the radial distance from the well to the point of interest,

$$P' = \frac{Q}{4\pi m(k/\mu)}W(u) \tag{13.2}$$

where $u = r^2 e\mu/4kP_{atm}t$ and $W(u)$ is the well function of u, which is a commonly calculated function.

Calculations with Eq. (13.2) show that for sandy soils ($10 < k < 100$ darcys) the pressure distribution approximates a steady state in a day to one week. Thus, it is appropriate to model pressure distributions using a steady-state solution to the governing flow equation, for the following set of boundary conditions: $P = P_w$ at $r = R_w$ and $P = P_{atm}$ at $r = R_I$, where P_w is the pressure at the well of radius R_w and P_{atm} is the ambient pressure at the radius of influence R_I.

Johnson et al. (1988) provide the following solution to the steady-state equation for radial flow:

$$P(r) = P_w \left\{ 1 + \left[1 - \left(\frac{P_{\text{atm}}}{P_w} \right)^2 \right] \frac{\ln(r/R_w)}{\ln(R_w/R_I)} \right\}^{1/2} \qquad (13.3)$$

As Johnson et al. (1988) point out, while not explicitly represented in Eq. (13.3), the properties of the soil influence the steady-state pressure distribution because the radius of influence (R_I) varies as a function of permeability and layering. Johnson et al. (1988) developed corresponding solutions for radial darcy velocity and volumetric flow rate for this steady-state case. The latter of these solutions provides a useful way to determine what the theoretical maximum airflow is to a vapor extraction well and is written as

$$Q = H\pi \, \frac{k}{\mu} \, P_w \, \frac{1 - (P_{\text{atm}}/P_w)^2}{\ln(R_w/R_i)} \qquad (13.4)$$

where H is the total length of the screen. Just as is the case with well hydraulics, these kinds of analytical equations not only form the basis for predictive analysis, but also for well testing methods for permeability estimations.

Analytical solutions are most useful for screening purposes and for exploring the relationships among variables; however, their practical applicability is limited to simple problems. An alternative way of solving differential equations like (13.1) is with powerful numerical models like the one developed by Krishnayya et al. (1988). Numerical models are effective in modeling complicated problems of the type commonly encountered in practice.

By analogy with problems involving ground water flow to wells, the general pattern of air circulation will be influenced by features of the fluid being circulated, the geologic system, and the well system used to withdraw and inject air. The permeability is the most important of all the parameters that influence flow. Ultimately, permeability determines the efficacy of vapor extraction because flow rates at steady state for a well under a specified vacuum are a direct function of permeability. Vapor extraction, to be practically useful, requires some minimal rate of air circulation, which may not be feasible in some low-permeability units.

From an analytical model, Johnson et al. (1990) developed a series of relationships between permeability and flow rate (Figure 13.15). For a given vacuum in the extraction well, the steady-state rate of airflow is a linear function of permeability (log scales). An increase in the vacuum (smaller P_w) at a given permeability will increase the air flow. However, the maximum change that might be expected is about an order of magnitude in flow rate.

It is well known from ground water flow theory that variability in permeability plays an important role in controlling the pattern of flow. This is also the case for patterns of air circulation. Most circulation will occur through the most permeable zones for a layered system, and this can have a profound effect on the rate and efficiency of cleanup schemes.

Features of the design of the system as a whole also control air circulation. The most important factors in this respect include (1) the flow rates in the injection–extraction wells, (2) the types and locations of wells in a multiwell system, and (3) the presence of a surface seal. For a single vacuum well, Figure 13.15 illustrates how reducing P_w increases the air withdrawal rate. Similarly, adding

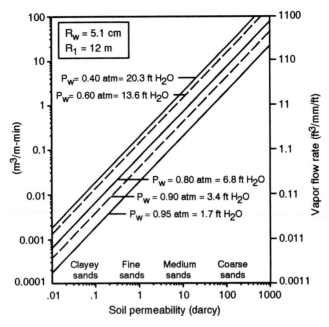

Figure 13.15 Predicted steady-state rates of airflow per unit length of well screen from a vapor extraction well for a range of permeabilities and applied vacuums (P_w). Source: Johnson et al., 1990.

more wells to the system will do the same. Furthermore, the pattern of air circulation can be controlled by the types and locations of the wells.

Organic Vapor Transport

The vapor-phase transport of mass has much in common with the transport of dissolved contaminants in ground water. The processes of advection and dispersion operate to physically transport the mass, while the chemical processes are involved with the generation of the contaminants through volatilization and subsequent interactions with the water and solid phases in the system.

Removal of mass from the spill depends on the process of volatilization, which is a phase partitioning between a liquid and a gas. For solvents, the process is described in terms of equilibrium theory by a form of Raoult's law,

$$P_i = x_i t_i P_i^0 \tag{13.5}$$

where P_i is the vapor pressure of component i (atm) in the soil gas, x_i is the mole fraction of the component in the solvent, and P_i^0 is the vapor pressure of the pure solvent at the temperature of interest; these are tabulated constants (see, for example, Verschueren, 1983). An activity coefficient t_i for the ith component in the mixture is added to account for nonidealities (Hinchee et al., 1986). By applying the ideal gas equation, the partial pressures calculated in Eq. (13.5) can be expressed in terms of concentration, or

$$C_s = \frac{PV}{RT} \sum MW_i \qquad (13.6)$$

where C_s is the saturation concentration in mg/L, V is volume, which is assumed equal to 1000 cm^3 to calculate concentration, T is temperature in Kelvins, R is the universal gas constant (82.04 atm cm^3 mol^{-1}), MW_i is the number of milligrams per mole, and P is the partial pressure of the compound.

The equilibrium model provides the basis for most analyses carried out in practice. The equilibrium approach applies in cases where the rate of volatilization is large relative to the rate of physical transport through the medium. When the flow of air contacts residual product in every pore, phase equilibrium is achieved very rapidly, at least for the more volatile compounds at the start of venting. One important reason for less-than-equilibrium concentrations of vapor is geologic heterogeneities that cause most of the airflow to avoid most of the contaminated zones.

With spills of complex solvent mixtures like gasoline, soil venting removes the components with higher vapor pressure first. The residual contamination thus becomes progressively enriched in the less volatile compounds. Because of this compositional change, the overall rate of mass removal decreases with time. This process has been described mathematically by Marley and Hoag (1984) and Johnson et al. (1988) using forms of Eq. (13.6) and information on the most volatile components of gasoline.

Vapor pressures for organic compounds increase significantly as a function of increasing temperature. For example, the vapor pressure for a volatile compound like benzene increases from 0.037 atm at 32°F to 0.137 atm at 80°F (Johnson et al., 1988). Over the same temperature range, the vapor pressure for n-dodecane increases from 2.8×10^{-5} to 2.3×10^{-4} atm. The main implication of this result is that the overall time required for cleanup will change depending on temperature.

The physical transport process, advection and dispersion, plays an important role with respect to vapor transport. The simplest and most extensively analyzed model of physical transport assumes that advection transports contaminants from the point of generation to the vapor extraction well (Figure 13.16a). The distribution of contaminants is reasonably homogeneous, and much of the airflow is assumed to move through the bulk of the spill. Ideally, the vapor concentration in that fraction of the air moving through the spill is the equilibrium concentration determined by Raoult's law. In Figure 13.16a, about 25% of the air passes through the spill, giving a vapor concentration in the well that is 25% of equilibrium value and a removal rate of $0.25QC_{\text{tot}}$.

However, in heterogeneous media or when pure product is present, the airflow may not pass directly through the spill. Specific examples cited by Johnson et al. (1988, 1990) include (1) airflow across the surface of a free-liquid floating on the water table or low-permeability layer (Figure 13.16b), or (2) product trapped in a lower-permeability lens (Figure 13.16c). In both cases, the mass loss rate of contaminants is controlled by the rate at which mass can diffuse into the moving

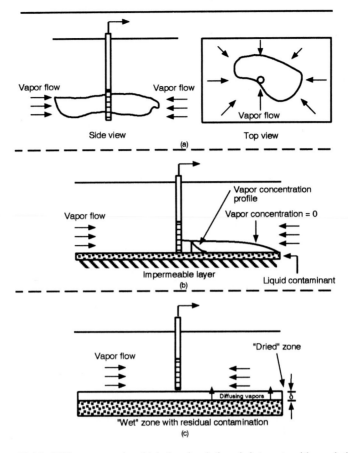

Figure 13.16 Different ways in which the circulating air interacts with a volatile contaminant source.   Source: Johnson et al., 1990.

vapor stream. Thus, when flowing air bypasses the spill, the rate of mass removal may be much lower than for the homogeneous case. The result may be a vapor extraction system that circulates considerable quantities of air without removing much of the contaminant. Mathematical approaches for estimating the contaminant removal rates under these more complex conditions are presented by Johnson et al. (1990).

Example 13.2 VAPOR EXTRACTION COMPUTATIONS

Johnson et al. (1988) examined a hypothetical case in which 400 gal (approximately 1500 L) of gasoline was spilled into 1412 ft^3 of soil at a 10% moisture content. This spill provides a residual saturation of 2% gasoline by dry weight or 20,000-ppm total petroleum content. Other conditions specified include a soil bulk density of 1.5 g/cm^3, a porosity of 0.40, an organic carbon fraction (f_{oc}) of 0.01, an airflow rate of 20 cfm, a soil temperature of 60°F, and a relative humidity for the incoming air of 100%. The problem assumes that 25% of the circulating air passes through the spill. Thus, component concentrations in the vapor extraction well will be at equilibrium

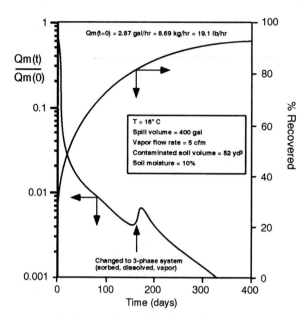

Figure 13.17 Predicted mass loss rates for a hypothetical venting operation. Source: Johnson et al., 1988.

for an airflow of 5 cfm, which is the value used in subsequent calculations. Also required for the calculation is information about the compounds found in gasoline, their molecular weights, mass fractions, and physical properties (see Johnson et al., 1988).

The total rate of mass loss from vapor extraction for the entire gasoline mixture is the sum of the mass loss rates $C_i Q$ for the compounds. The total mass loss rate as a ratio of the initial mass loss rate is plotted versus time in Figure 13.17. Also depicted in the figure is the cumulative percentage of the initial spill recovered as a function of time. The example shows clearly how the rate of mass loss decreases with time as the most volatile components are removed from the mixture. Figure 13.18 displays the time variation in the soil concentrations of a number of the most volatile compounds in gasoline. The pattern of removal rates (benzene > toluene > xylenes) is explained by the respective vapor pressures (0.10 > 0.029 > 0.0066 to 0.0088 atm). At the end of 400 days, the residual product consists mainly of the larger-molecular-weight compounds in gasoline having vapor pressures in general less than 0.005 atm. Thus, about 25 days are required to recover half of the original spill mass, but about 400 days are required to collect 90%.

System Variables

A number of variables characterize the successful design and operation of a vapor extraction system. They may be classified as site conditions, soil properties, chemical characteristics, control variables, and response variables (Anastos et al., 1985; Enviresponse, 1987; Hutzler et al., 1989).

The extent to which VOCs are dispersed in the soil, vertically and horizontally, is an important consideration in deciding if vapor extraction is preferable to other methods. The depth to ground water is also important. Where ground

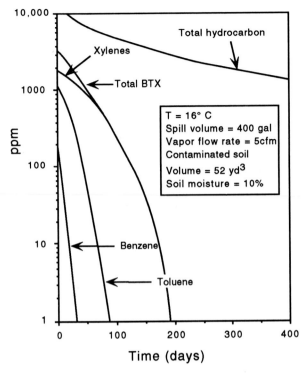

Figure 13.18 Predicted soil concentrations of hydrocarbons for a hypothetical venting operation. Source: Johnson et al., 1988.

water is at depths of more than 40 ft and the contamination extends to the ground water, the use of soil vapor extraction systems may be one of the few ways to remove VOCs from the soil (Malot and Wood, 1985).

Heterogeneities influence air movement as well as the location of contaminants, and the presence of heterogeneities makes it more difficult to position extraction and inlet wells. There generally will be significant differences in the air conductivity of the various strata of a stratified soil. A horizontally stratified soil may be favorable for vapor extraction because the relatively impervious strata will limit the rate of vertical inflow from the ground surface and will tend to extend the influence of the applied vacuum horizontally from the point of extraction.

The soil characteristics at a particular site will have a significant effect on the applicability of vapor extraction systems. Air conductivity controls the rate at which air can be drawn from soil by the applied vacuum. Grain size, moisture content, soil aggregation, and stratification probably are the most important properties (Bennedsen et al., 1985; Hutzler et al., 1989). The soil moisture content or degree of saturation is also important in that it is easier to draw air through drier soils. The success of the soil vapor extraction in silty or clayey soils may depend on the presence of more conductive strata, as would be expected in alluvial settings, or on relatively low moisture content in the finer-grained soils.

Chemical properties will dictate whether a soil vapor extraction system is feasible. A vapor-phase vacuum extraction system is most effective at removing compounds that exhibit significant volatility at the ambient temperatures in soil.

Low-molecular-weight volatile compounds are favored, and vapor extraction is likely to be most effective at new sites where the more volatile compounds are still present. Examples of compounds that have been effectively removed by vapor extraction include trichloroethene, trichloroethane, tetrachloroethene, and most gasoline constituents. Compounds that are less applicable to removal include trichlorobenzene, acetone, and heavier petroleum fuels (Payne et al., 1986; Bennedsen et al., 1985; Texas Research Institute, 1980).

Soil vapor extraction processes are flexible in that several variables can be adjusted during design or operation. These variables include the air withdrawal rate, the well spacing and configuration, the control of water infiltration by capping, and the pumping duration. Higher airflow rates tend to increase vapor removal because the zone of influence is increased and air is forced through more of the air-filled pores. More wells will allow better control of airflow but will also increase construction and operation costs. Intermittent operation of the blowers will allow time for chemicals to diffuse from immobile water and air and permit removal at higher concentrations.

Parameters responding to soil vapor extraction system performance include air-pressure gradients, VOC concentrations, moisture content, and power usage. The rate of vapor removal is expected to be primarily affected by the chemical's volatility, its sorptive capacity onto soil, the airflow rate, the distribution of airflow, the initial distribution of chemical, soil stratification or aggregation, and the soil moisture content.

Design Issues for Vapor Extraction Systems

Johnson et al. (1990) present a detailed and comprehensive approach that can be followed in practice for implementing a vapor extraction scheme. Having arrived at vapor extraction as a candidate strategy for remediating a contamination problem, the following four major activities constitute a vapor extraction project.

1. Feasibility analysis based on available data to establish whether a vapor system is an appropriate remedial strategy
2. Testing with a pilot vapor extraction system and ground water wells to determine physical parameters and confirm the accuracy of the feasibility analysis
3. Design of the complete system
4. Monitoring to confirm that the designed level of cleanup is being achieved

The feasibility analysis requires preliminary information about the geology and hydrogeology of the site, chemical and physical characteristics about the liquid contaminant, and any regulations applicable to the spill. Several areas need to be addressed for any soil vapor extraction design. Johnson et al. (1990) recommend that the following data be collected on an ongoing basis: (1) airflow rates, (2) pressure at each extraction and injection well, (3) both ambient and soil temperatures, (4) water-table elevation, (5) vapor concentrations and composition at each extraction well, and (6) soil–gas vapor concentrations at various distances from the extraction wells. The vapor extraction system will be shut down when target levels of a cleanup are reached. These targets are often site specific, depending on particular water-quality, health standards, and safety considerations. Johnson et al. (1990) discuss in detail how the monitoring data are evaluated in making this decision.

The general literature should be consulted for some of the recent advances in soil vapor extraction systems and results from specific site studies.

13.7 REMEDIATING NAPL SITES

Proven Technologies for Removing NAPLs

Chapter 11 indicated the difficulties associated with characterizing and remediating sites with NAPLs in the subsurface. It is now known that NAPLs can create a source of contamination for years or decades due to the slow dissolution process after continuous-phase fluid has been removed. For this reason, standard remediation methods used extensively during the 1980s, such as pump and treat, only address the soluble plume problem and do very little for NAPLs control. The following sections describe the proven approaches for removal of NAPLs from the subsurface. A remediation technology is defined as proven if (1) it is commonly practiced in the field and (2) sufficient design methodologies are available so that practioners can apply the technology and obtain the predicted system response. An emerging technology, on the other hand, is one for which some bench-scale or field-scale tests have been conducted, but detailed design procedures are not generally available. Section 13.8 describes some emerging techniques for remediation.

Continuous-phase and some residual-phase NAPLs can generally be removed from the saturated zone by means of ground water pumping. Under long-term operating conditions, NAPL production at a recovery well location can be expected to attenuate as the mobile fraction of the NAPL is drained from the aquifer media. After the artificial gradient is removed, the residual NAPL will be immobilized under background gradient conditions.

General Remediation Strategy for NAPL Sites

As restoration of ground water to drinking-water standards may prove infeasible at many NAPL-affected sites, alternative approaches to managing NAPL problems should be considered. The following general NAPL management strategy divides a NAPL site into three zones:

1. Dissolved-phase zone
2. Confirmed free-phase NAPL zone
3. Potential NAPL zone (either free-phase or residual NAPL)

As shown in Figure 13.19, this classification system is based on the confirmed or suspected presence of NAPL in the aquifer. Areas where site data indicate a low probability of NAPL in the aquifer would be designated as "dissolved-phase zones" where conventional cleanup standards and remediation technologies would be applied. At many sites, the applicable cleanup standards could be at or near current drinking-water standards for these zones, so pump-and-treat systems could be used to reach the required concentration levels.

In areas where the presence of NAPL is confirmed, due to the presence of free-phase NAPL in monitoring wells, a NAPL containment strategy could be applied in place of conventional ground water remediation standards. Containment of NAPL would be achieved by imposing a relatively high hydraulic gradient across the site to remove free-phase material and to ensure that residual NAPL will remain immobilized under natural flow conditions. Long-term hydraulic containment of ground water within the residual NAPL zone would then be required to control NAPL dissolution products. Operation of such a system could require a longer-term financial commitment than is currently afforded these sites.

As the presence of NAPL is difficult to confirm in many areas at a site, a third zone has been defined for areas where residual and/or free-phase NAPL may be present. For example, the presence of relatively high concentrations of dissolved organics or proximity to NAPL entry points would be sufficient to designate an area as a potential NAPL zone. The remediation approach for these areas would be identical to that for confirmed NAPL zones: long-term containment of NAPL and all dissolved constituents.

The remediation goals outlined here can be achieved with existing, proven technologies (for example, ground water recovery wells) and would provide a significant level of protection to both human and environmental receptors. Implementation of this remediation approach would require modification of current regulatory policies regarding cleanup standards and financial assurance for corrective action programs. Note that the containment period for LNAPL sites is probably much shorter than the period of time required for containment of DNAPL sites due to the beneficial effects of in situ biodegradation for many floating fuel spills.

Pumping to Remove Continuous-phase LNAPLs

The most common strategy for maximizing the recovery of free-phase LNAPL is to pump the LNAPL layer relatively slowly in order to keep the LNAPL mass as a continuous flowing mass (Charbeneau et al., 1989; and Abdul, 1992). Water pumping is avoided or carefully controlled to minimize smearing of the LNAPL layer, although some water pumping may be helpful to prevent upconing of the water table. Although high water pumping rates will initially yield a higher hydrocarbon recovery rate, the ultimate recovery will be greatly reduced because much of the LNAPL that was originally "flowable" (that is, a free phase) is smeared into uncontaminated soil as the water table is depressed, converting it to an "unflowable" residual LNAPL.

Charbeneau et al. (1989) simulated low-yield pumping strategies for an

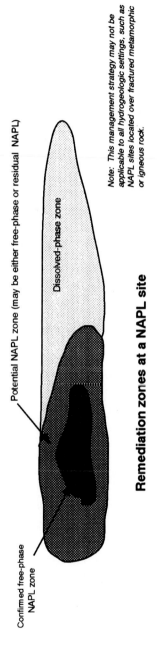

Confirmed free-phase NAPL zone

Potential NAPL zone (may be either free-phase or residual NAPL)

Dissolved-phase zone

Remediation zones at a NAPL site

Note: This management strategy may not be applicable to all hydrogeologic settings, such as NAPL sites located over fractured metamorphic or igneous rock.

Remediation zone	Characteristics	Remediation goal	Remediation technology	Potential remediation standard
Dissolved phase zone	• Low dissolved phase concentrations • No NAPL detected in monitoring wells or soil cores • Downgradient of potential NAPL zone	Ground water restoration of leading edge of plume	Ground water pump and treat	Existing ground water remediation standards. For example: CERCLA: Based on risk assessment RCRA: MCL, background, or ACL
Confirmed free-phase NAPL zone	• NAPL observed in monitoring wells or • Free-phase NAPL observed in soil cores	Prevent NAPL migration Contain dissolved plume	Use pumping/injection wells to: • Remove free-phase NAPL • Remove potentially mobile NAPL (residual NAPL that could be mobilized under natural flow conditions at site) Operate pumping wells for long-term hydraulic containment of NAPL dissolution products	A hydraulic gradient, at least 2 - 4 times greater than highest expected natural gradient at site, must be imposed on the site until: • No further NAPL accumulation in monitoring wells is observed **and** • NAPL recovery rate from extraction wells reaches asymptotic response. Operate pumping wells indefinitely (until ground water reaches current cleanup standards)
Potential NAPL zone (may be either residual or free-phase NAPL)	• Located near potential NAPL entry points **or** • High dissolved-phase concentrations **or** • NAPL observed in soil cores			

Figure 13.19 General management strategy for NAPL sites.

455

LNAPL layer approximately 2500 ft by 3500 ft in size using an LNAPL recovery model called TWOLAY. The aquifer for this case study was located in California and consisted of sand, silty sand, and sandy silt having an average hydraulic conductivity of 1.5×10^{-3} cm/sec. The most effective low-yield system consisted of 207 dual-pump wells that removed 60% of the total LNAPL and 90% of the recoverable LNAPL from the site. (Recoverable LNAPL is defined as the difference between the original saturation and the residual saturation.) The average LNAPL pumping rate was 0.25 gpm. Water production for each of the wells was determined using an optimization routine and ranged between 0.004 and 10 gpm. Thirty-five of the wells had water flow rates of less than 1 gpm.

LNAPLs can be removed from the subsurface by pumping recovery wells screened at the water table or by pumping an interceptor system such as a French drain or interceptor trench. Two alternative approaches can be employed: (1) pumping the combined ground water/LNAPL mixture with a single pump or (2) using two separate pumps working under a control system to remove the aqueous phase and nonaqueous phase separately. In the case of pumping wells, a single well creates a cone of depression that allows floating product to move toward the well, accumulate in the well bore, and then be slowly sucked into the pump intake along with ground water. This system is preferred where removal of both NAPL and control of contaminated dissolved hydrocarbon plumes is required and has the advantage of being simpler, less expensive, and easier to maintain than a two-pump system. One drawback is that the combined NAPL/ground water mixture can be emulsified by the pump, thereby making treatment of the resulting fluid mixture more expensive.

For systems with significant accumulations of LNAPL, a two-pump system will sometimes provide a more economical remediation approach. A control system comprised of two water–LNAPL interface probes is used to operate dedicated LNAPL and ground water pumps. The first interface probe is set below the intake of the LNAPL pump and keeps the LNAPL pump operating as long as there is still hydrocarbon near the pump intake. The lower interface probe turns the ground water pump off if the hydrocarbon level approaches the intake of the ground water pump and on again when there is sufficient water above the pump intake (see Figure 13.20). An alternative design uses a floating "skimmer" pump and floating interface probe that move up and down on the floating hydrocarbon layer.

Pumping to Remove Continuous-phase DNAPL

In practice, DNAPL removal is more difficult than LNAPL removal. The DNAPL pools and lenses are much harder to locate, and the installation of recovery wells increases the risk of mobilizing DNAPL accumulations. Limited field experience indicates that DNAPL can be mobilized by pumping and that pumping ground water from the aquifer above the DNAPL recovery will increase yields by creating a water drive for the DNAPL pool. A detailed case study for the overall management of DNAPLs at a waste site is presented in Section 13.9.

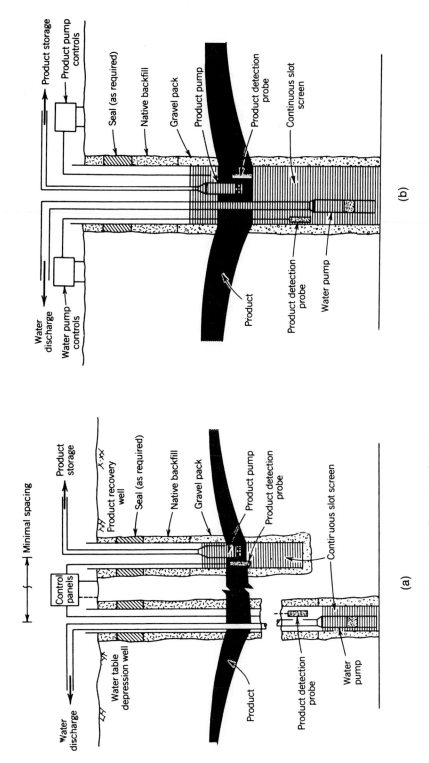

Figure 13.20 Examples of NAPL recovery systems. (a) Two-well, two-pump system. (b) One-well, two-pump system. Source: Blake and Lewis (1982).

457

Pumping to Remove Residual NAPLs

After the continuous-phase NAPL is broken up and converted to a trapped residual phase, NAPL removal becomes increasingly difficult. In general, removal of residual hydrocarbons will be hindered by high capillary forces in the aquifer, and only small reductions in residual saturation will be realized during ground water remediation efforts. As shown by the capillary number calculations presented in Chapter 11, increasing the hydraulic gradient can mobilize some of the trapped residual NAPLs in some cases. These include sites with extremely high permeabilities or sites where very high hydraulic gradients can be imposed across the affected area.

Soil Vapor Extraction

While soil vapor extraction is primarily directed at remediation of NAPLs in unsaturated soils, it can be adapted to the removal of volatile hydrocarbons in NAPLs from the saturated zone. First, there will be some transport of volatile organics from contaminated ground water into an air stream being forced across the water table by a soil vapor extraction system. However, a much more important process occurs when the water table is depressed, either intentionally or inadvertently, by a ground water pumping system, thereby exposing the former saturated zone to the effects of a soil vapor extraction system. In most cases, a soil vapor extraction system will be many times more effective at removal of volatile hydrocarbons compared to a conventional ground water pumping system removing dissolved hydrocarbons. The key design parameters are the vapor pressure of the volatile chemicals in the NAPLs and the permeability of the zone that will be used for vapor extraction (see Section 13.6).

Dissolution (Pump and Treat)

In cases where the water table cannot be depressed to allow soil venting, clean ground water can be drawn across the NAPL zone to dissolve soluble organics for subsequent removal by a ground water pump-and-treat system. In most cases this approach will remove some quantities of contaminant mass, but will be impracticable for restoring aquifers to drinking-water standards. An example of a dissolution field test is presented in Section 13.9.

Aerobic in Situ Biodegration

In situ biodegradation can be considered to be a polishing step to the NAPL dissolution process. As described in Section 13.5, injection water is amended with oxygen and nutrients to enhance the aerobic degradation of trapped hydrocarbons in the subsurface. In cases where large amounts of NAPL are trapped in the subsurface, in situ biodegradation will proceed relatively slowly because of the limited solubility of oxygen in the injection water. Application of biodegradation is most effective near the end stages of a remediation process, where large volumes of ground water are required to remove small masses of hydrocarbon. Aerobic biodegradation is applicable to sites contaminated with most petroleum-related LNAPLs and some polyaromatic hydrocarbons (PAHs). For sites contaminated with chlorinated compounds in high enough concentrations to create DNAPL pools, anaerobic processes dominate, and rates of biodegradation are much slower than under aerobic conditions. The key design principles are covered in detail in Chapter 8.

13.8 EMERGING REMEDIATION TECHNOLOGIES

Enhanced Oil Recovery Technologies

Many researchers are currently evaluating the applicability of enhanced oil recovery (EOR) technology to the remediation of LNAPL sites. EOR technologies under consideration include the use of hot water or steam flooding, cosolvents (ethanol), surfactants, and polymers. Primary LNAPL recovery systems (for example, drains and pumping wells) may result in the removal of approximately 30% to 40% of the total LNAPL volume. Enhanced oil recovery may remove up to an additional 50% of the LNAPL volume. However, significant quantities of LNAPL may remain in place following EOR applications (Mercer and Cohen, 1990).

The use of heat to increase the mobility of NAPLs in the subsurface is one EOR technique being used in the petroleum industry. Specific technologies include hot-water flooding or the addition of steam to reduce NAPL density and NAPL viscosity. Use of steam will also condense volatile NAPLs in front of the steam front, thereby increasing NAPL saturation and mobility (U.S. EPA, 1992). Although these technologies have removed significant amounts of NAPL in both laboratory and field pilot tests, complete NAPL removal has not been demonstrated in the field.

Air Sparging

Air sparging is a relatively new technology that is being implemented at numerous sites around the country (Mercer and Cohen, 1990). This remediation technique involves the forced introduction of air into a trench or well bore under sufficient pressure to form bubbles in the ground water. The bubbles sweep through the aquifer to (1) strip volatile organic hydrocarbons from the dissolved phase and from any NAPLs present along the path of the bubbles, (2) to add oxygen to the water to spur the in situ bioremediation process, and (3) for certain designs, to establish large circulation cells in the subsurface, which tend to move contaminated water to the surrounding wells for extraction. After the bubbles make their way to the unsaturated zone, a soil vapor extraction system is used to remove the vapors for treatment prior to release to the atmosphere. The overall performance of this approach has not been verified in the field.

An application to air sparging in gate wells and trenches for controlling volatile organic compounds is described in Pankow et al. (1993). They present a well–trench sparging design (WTS) and an in situ sparging design (ISS). With the WTS design, cutoff walls are needed to force contaminated water through the gate wells (Figure 13.1). Interest in sparging is growing because of its possible use in low-velocity systems and where highly volatile organic chemicals (VOCs) are involved. However, the technology is still unproven and has not yet been rigorously field tested, and its ultimate success will depend on the ability to deliver the air and bubbles to the contaminated zone. One limitation is that the VOCs contained in the rising air bubbles may have to be treated prior to release to the atmosphere.

Johnson et al. (1993) present an overview of the process and conclude that adequate monitoring methods do not exist and need to be developed. Significant new risks include the potential off-site migration of vapors and contaminant plumes. Migration of previously immobile liquid hydrocarbons can also be induced. There is a need to better understand and quantify the benefits of air sparging. Evidence exists that air flow mostly occurs in small channels rather than as bubbles, and is very sensitive to formation structure.

Funnel and Gate Barrier Systems

A funnel-and-gate system is the term applied to a hydraulic device designed to funnel contaminated ground water through some type of in situ treatment system. Such a system would consist of a series of slurry-wall ''wings'' that would capture and divert a contaminant plume through a permeable treatment zone installed in the subsurface (Figure 13.1). The chemical treatment zone could be designed with activated carbon for organic contaminants. A biological treatment zone, for example, would be designed to slowly bleed oxygen and nutrients into the moving ground water, thereby allowing in situ biodegradation to proceed at an accelerated pace. Some funnel-and-gate systems have already been tested for DNAPL applications, and several biological systems are now being proposed for testing.

Enhanced Dissolution

Enhanced dissolution refers to a remedial technique that uses surfactants for an unconventional purpose: the surfactants are not designed to mobilize residual NAPL blobs, but to increase the effective solubility of the NAPL components. Controlled field experiments at DNAPL sites have indicated that the effective solubility of NAPL chemicals can be increased by up to one or two orders of magnitude, thereby greatly increasing the performance of a pump-and-treat dissolution system. A recent assessment of surfactant washing studies was completed by Raghavan et al. (1990). They reviewed processes that involved soil excavation, mixing with cleaning solution, and solid–liquid separation. It is clear that more research and field testing are needed before surfactant-based remediation processes become accepted practice (Sabatini and Knox, 1992). Surfactant delivery systems are now being tested at larger field sites.

Bioventing

Bioventing refers to a combined soil vapor extraction (SVE) and in situ biodegradation process for removing degradable contaminants from the unsaturated zone. The SVE system delivers oxygen to the indigenous microbes, thereby promoting degradation of organics in the pore space. Research efforts are currently focusing on design issues such as the need for nutrient addition and the kinetics of the biological reactions.

Wellpoint Systems

A relatively new recovery technology for the environmental field, but a proven technology for construction dewatering, is the wellpoint system. A large number of driven or drilled wells, usually 2 in. in diameter, spaced closely together are connected to a manifold that is in turn connected to a wellpoint pump. The wellpoint pump consists of an air–water separator, a vacuum pump, and a water pump. The vacuum is used to lift the fluids from the subsurface, therefore eliminating the need for an individual pump in each well.

   Wellpoint systems are good for low-permeability units where the water table is within 18 ft of the surface (the practical lift of a field vacuum system). For systems containing LNAPLs, particular care must be taken in the selection of the water pump, as centrifugal pumps may emulsify the water–LNAPL mixture, while a diaphragm pump will reduce the potential for emulsification. The wellpoint system has three main advantages over conventional pumping systems: (1) only one pumping system is used for many wells, greatly reducing maintenance costs, (2) a large number of small-diameter wells can be installed much more cheaply than the same number of downhole pumping wells, and (3) by using a high-volume vacuum pump, the system can double as a fluid recovery–soil vapor extraction system. The main disadvantage is that flow rates are more difficult to control. In addition, some amount of the vapor produced by the vacuum pump may need treatment.

Natural Attenuation–Assimilation

Because residual NAPLs are difficult to remove from the subsurface, many researchers have begun to assess the long-term assimilative capacity of contaminated aquifers. LNAPLs are particularly well suited for this approach, as most are amenable to relatively rapid biodegradation by naturally occurring aerobic microorganisms. Several site assessment studies have been or are now being performed to evaluate this process. A site at Traverse City, Michigan, where natural attenuation was quantified is described in Section 13.9. A number of sites have been studied and modeled in detail and are presented in Section 8.5.

Practicality of Restoring Aquifers Containing NAPLs

Recently, ground water scientists and engineers have been investigating the ability of current ground water remediation technology to meet specified cleanup standards, typically specified to be drinking-water quality. One such study, completed by the EPA in 1989 and updated in 1992, evaluated the performance of 24 ground water remediation systems that had been in operation at least 5 years (U.S. EPA, 1989, 1992). Although containment and removal of significant hydrocarbon mass was realized, the decrease in contaminant concentration over time was much slower than originally anticipated. Factors that were identified as major impediments to ground water restoration include the following:

- Contaminant factors such as NAPLs, high sorption potential, and continued leaching from source areas
- Hydrogeologic factors such as heterogeneities, low permeability units, and fractures
- Design factors such as pumping rates, recovery well locations, and screened intervals

Additional EPA research and directives have focused on the problems associated with NAPLs in the subsurface and have concluded that ''NAPLs will have significant influence on the time frame required and/or the likelihood of achieving cleanup standards'' (U.S. EPA, 1992).

The results of the phase II study reinforced the main conclusions of phase I and led to some additional conclusions concerning impacts of NAPLs on ground water remediation (U.S. EPA; 1989, 1992). Data collected, both site characterization data prior to system design and subsequent operational data, were not sufficient to fully assess contaminant movement or ground water system response to extraction. In the majority of cases studied (15 of the 24 sites), the ground water extraction systems were able to achieve hydraulic containment of the dissolved-phase contaminant plume. Extraction systems were often able to remove a substantial mass of contamination from the aquifer. When extraction systems were started up, contaminant concentrations usually showed a rapid initial decrease, but then tended to level off or decrease at a greatly reduced rate.

Based on the available information, potential NAPL presence was not addressed during site investigations at 14 of the 24 sites. At 5 sites they were ''addressed'' because they were encountered unexpectedly during the investigation. As a result, it is difficult to determine NAPL presence conclusively from available site data. At 20 of the 24 sites, chemical data collected during remedial operation exhibited trends consistent with the presence of dense nonaqueous phase liquids (DNAPLs). However, even where substantial soil and water quality data were available, a separate immiscible phase was rarely sampled or observed. This is consistent with DNAPL behavior; i.e., they can move preferentially through very discrete pathways that may easily be missed even in thorough sampling schemes. DNAPL was observed at sites where contaminant concentrations in ground water were less than 15% of the respective solubilities. The importance of treating ground water remediation as an iterative process, requiring ongoing evaluation of system design, remediation time frames, and data collection needs, was recognized at all of the sites where remedial action was continuing.

These studies have indicated that it may not be practicable to restore many difficult hazardous-waste sites using current proven and emerging technology. For sites where conditions preclude restoration to mandated cleanup standards, the general remediation strategy described earlier might be applied. Many current sites are being managed in a way to control and contain any off-site contamination, while the emerging technologies discussed here are being developed and field tested to deal with complex NAPL transport issues.

Containment alternatives for NAPL sites must address several potential contaminant migration pathways. First, the migration of mobile NAPL must be stopped by (1) pumping continuous-phase NAPL until it is all trapped in the aquifer as a residual phase; (2) operating some type of NAPL interception system, such as a NAPL interceptor trench, and/or (3) constructing some type of physical barrier, such as a slurry wall. Second, the vertical and lateral migration of the dissolution products in the aquifer must be curtailed using hydraulic controls or physical barriers. In some cases, natural attenuation (from dispersion, dilution, chemical degradation, and biological degradation) will stop migration of dissolved contaminants and may be sufficient to ensure that potential receptors are protected. Some design considerations for these approaches are included in the following case studies.

13.9 CASE STUDIES OF REMEDIATION

Case 1: Modeling Natural in Situ Biodegradation

Description of Traverse City Field Site

The Traverse City field site is a U.S. Coast Guard Air Station located in Grand Traverse County in the northwestern portion of the lower peninsula of Michigan. The ground water at the site is contaminated with organic chemicals from a source near the Hangar/Administration building (Figure 13.21). In the upper reach of the plume, hydrocarbons occur at the surface of the water table and move downward in the aquifer as the plume migrates toward east Grand Traverse Bay (Twenter et al., 1985). The soils at the site are generally free of contamination except near the source area. The major contaminants at the site include benzene, toluene, and xylene (BTX).

The contaminant plume ranges from 50- to 120-m wide and is about 1300-m long. An interception (interdiction) well field (pump and treat) was installed in April 1985 to prevent the migration of the plume off the site. The captured contaminated water is renovated by carbon adsorption. The U.S. Geological Survey performed an extensive study of the geology and hydraulic characteristics of the site (Twenter et al., 1985). Glacial deposits, as much as 100 m thick, underlie the site. The upper 10 to 40 m is a sand and gravel unit that is underlain by impermeable clay that is about 30 m thick. The water table varies from 3 to 6.5 m below land surface, with ground water velocities ranging from 1 to 2 m per day. Hydraulic conductivities in the aquifer range from 28 to 50 m per day. Biodegradation activity at the site has been identified and discussed by several researchers (Wilson et al., 1986; Rifai et al., 1988).

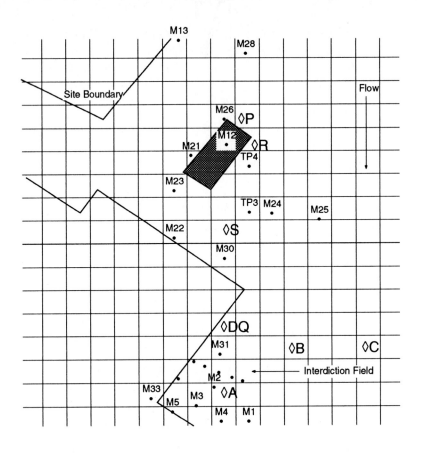

M33
• Well location and number

◊P Sample locations from July 1986 trip

Hangar/Administration building

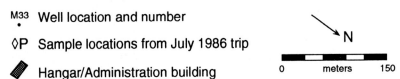

0   meters   150

N

Figure 13.21 Traverse City field site map. Source: Rifai et al., 1988. Reproduced with permission of ASCE.

Site Characterization Studies

The Traverse City field site is well monitored and has an extensive data base that was utilized in a modeling study. Two sampling trips to the field site were undertaken by the U.S. EPA Robert S. Kerr Environmental Research Laboratory (RSKERL) and Rice University for data collection. Data collected during the first trip (July 1985) indicated that aerobic biodegradation was occurring at the site.

The data from wells indicated an absence of oxygen in monitoring wells, where high concentrations of total alkylbenzenes were observed. Dissolved oxygen concentrations were elevated in monitoring wells with negligible levels of hydrocarbons. This phenomenon, which is characteristic of biodegradation sites, seems to indicate that it is appropriate to assume that hydrocarbons react with dissolved oxygen for modeling purposes. It was noted that pristine zones in the aquifer had background levels of 8.0 mg/L of dissolved oxygen.

The second field trip (June 1986) involved more extensive sampling longitudinally and along a transect across the lower portion of the plume (Figure 13.21, locations A, B, C, P, R, S, D, and Q). The data collected indicated that anaerobic activity was taking place in the center of the plume near locations S, D, Q, and A. Methane concentrations at locations D and Q were higher than at S and A. No methane was detected near the source (locations P and R) or at the outer edge of the plume (locations B and C).

In addition to the above-mentioned field trips, the U.S. Coast Guard collected data on a regular basis from a large number of wells. The data collected from April 1985 through 1986 from approximately 25 selected wells were analyzed to (1) define the areal extent of the contaminant plume and (2) estimate the amount of mass biodegraded over the period of sampling.

The total BTX from monitoring well data were averaged over 3-month intervals beginning with the second quarter of 1985 through the fourth quarter of 1986. The data were also averaged vertically. This was a necessary approach due to the large number of data points at each well. A statistical analysis was performed with the averaged data to determine the mean and standard deviation for each well over each quarter. Data from wells with relatively high concentrations of BTX show a small amount of spreading, whereas some of the wells with contaminant levels of less than 100 μg/L exhibit more variation. It is noted that some of the variation correlates well with rainfall and snowmelt events and water-table fluctuations in the aquifer. Overall, it was concluded that the average values over each 3-month period were representative values for the period in question. Concentration contours were developed from the averaged data and are shown in Figure 13.22 for 1986.

In Situ Biodegradation and Change in Organic Mass Over Time

Based on analysis of plume characteristics, it can be seen that the BTX plume is undergoing significant changes with time: (1) the plume exhibits a significant loss of mass and (2) the interception field has succeeded in halting the migration of the contaminant plume across the site boundaries. In general, three basic mechanisms could have contributed to the observed mass loss: biodegradation, volatilization, and dispersion. The flow system at the site is highly advective, so it is unlikely that dispersion could account for the observed loss of mass. A sensitivity analysis on the dispersion coefficients at the site supported this conclusion.

Estimating the rate of mass loss due to volatilization is more difficult. Chiang et al. (1987) calculate mass loss due to volatilization using Henry's law for benzene to be 5% of the total mass loss at a site similar to the U.S. Coast Guard facility. Based on extensive lab analyses of cores and modeling of transport, it was reasonable to conclude that biodegradation is the most important mechanism that accounts for the observed mass loss at the site.

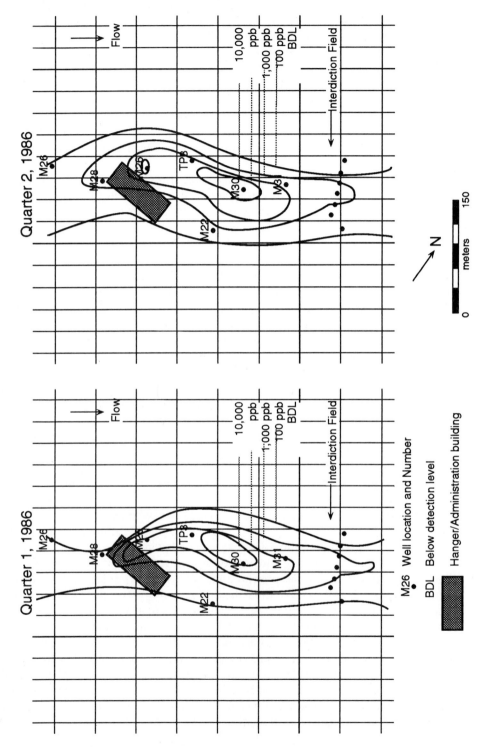

Figure 13.22 Traverse City field site: BTX plume. Source: Rifai et al., 1988. Reproduced by permission of ASCE.

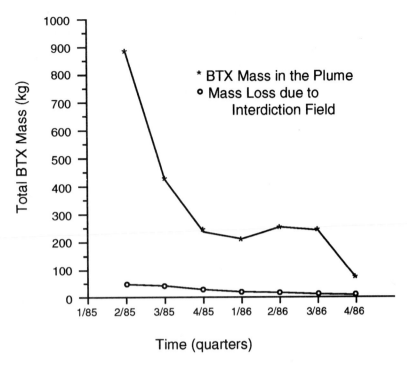

Figure 13.23 Variation in total BTX with time at Traverse City field site. Source: Rifai et al., 1988. Reproduced by permission of ASCE.

The contour plots such as shown in Figure 13.22 were utilized to estimate the mass of contaminants in the system for each time period. These computations are very rough estimates since the data have been averaged over time in the vertical direction. The contour level that represents concentrations less than 10 μg/L was not included in these computations because there was less certainty and more interpolation in its development. The mass loss due to biodegradation for each quarter was calculated by comparing the BTX mass in the system for successive quarters and subtracting the BTX mass captured by the interception field during the quarter.

The data in Figure 13.23 show the change in total BTX mass with time in the system. The data also show the mass removed by the pump-and-treat system. It is obvious that biodegradation had a much more significant effect than the pumping well field on removing contaminant mass. The data indicated a slight increase in mass for the second quarter of 1986. This could be due to the leaching of contaminants from the soil due to seasonal recharge.

Ground Water Modeling Studies

The modeling effort at the site was performed in order to calibrate the BIO-PLUME II model (Chapters 8 and 10) to the April 1985 conditions, the time the interception field was turned on (Rifai et al., 1988). The simulation grid used in this effort extends well beyond the site to cover the area downstream of the interception field and into the Grand Traverse Bay. The aquifer thickness data at the finite difference grid cells were obtained from well logs and were used to calculate the transmissivity assuming a constant hydraulic conductivity.

The source was represented with injection wells near the hangar building (Figure 13.22) at a combined flow rate of 6.85 m^3/day, which is equivalent to half the annual rate of precipitation. This source definition is rather arbitrary and can account for some of the calibration error that is presented later. The source concentrations varied for the seven wells with an average concentration of about 1380 mg/L. The interception field was simulated with pumping wells at the specified flow rates located near the base boundary.

Three sources of oxygen into the contaminated aquifer were simulated: (1) mixing with uncontaminated water containing dissolved oxygen of 8.0 mg/L, (2) natural recharge of dissolved oxygen from the upgradient constant-head boundary at the site (also at 8.0 mg/L), and (3) reaeration from the unsaturated zone. The reaeration coefficient was estimated from computations performed at another site (Borden et al., 1986) and in this case would be another source of error in calibration results.

The boundary conditions were developed using constant-head cells and diffuse recharge. The constant-head data were obtained from water-table measurements and from water levels in the Grand Traverse Bay. Diffuse recharge was assumed to be 50% of the average annual precipitation except at cells along the ditch on the northern side of the site, where the recharge rate was increased to match the hydraulics.

The calibrated water-table elevations matched the observed data reasonably well for most wells except M28, M30, and TP3. The differences between calibrated and observed water-table elevations are attributed to variations in seasonal recharge. In the calibration process, recharge was assumed to be constant for the simulation period.

Results of Simulation

The model predictions matched the observed concentrations at the monitoring wells reasonably well. BIOPLUME II predictions overestimate the observed plume in the vicinity of well M31. This is due to the fact that the simulations to this point have not accounted for anaerobic biodegradation processes. Data from the June 1986 sampling trip show elevated levels of methane near well M31, but the coefficient of anaerobic decay at the site has not yet been estimated. It should be noted that anaerobic biodegradation activity at the site was verified in laboratory microcosms. The reader is referred to Wilson et al. (1986) for details of these studies.

The mass-balance computations for the calibrated plume indicate that the total BTX dissolved mass in the system for the April 1985 conditions is 1470 kg. This is higher than the BTX mass computed from the contour plots (878 kg for the second quarter of 1985). The difference is partly due to the anaerobic biodegradation component, which was not simulated at this point, and partly due to the approximation in the mass computations presented earlier.

Significant Findings

- Site characterization data indicated that, although high concentrations of dissolved oxygen (8 mg/L) were present in clean areas around the site, the dissolved oxygen concentrations in the plume area were negligible. Laboratory data support the premise that naturally occurring microbial activity was responsible for consuming the oxygen when a sufficient carbon source (in this case BTX) was present in the aquifer.
- The most important removal process at the site was natural aerobic in situ biodegradation. The total mass removal by the pump-and-treat remediation system was a small fraction of the total mass removed by natural biodegradation processes.

Case 2: Modeling of TCE Recovery in a Shallow Aquifer Using Pump and Treat

Freeberg et al. (1987) applied the USGS MOC solute transport model (Konikow and Bredehoeft, 1978) to an industrial site where a mixture of solvents (primarily trichloroethylene, TCE) leaked from an underground storage tank to the underlying sand aquifer. The model in its two-dimensional form for steady-state flow is capable of simulating the horizontal movement of a soluble contaminant plume under both ambient flow conditions and conditions produced by pumping activity. For this reason, the model is well suited to application at industrial sites where soluble contaminant movement and recovery systems are monitored.

Hydrogeologic Setting

Investigation of near-surface hydrogeologic conditions showed that the site is situated on sand deposits extending 10 to 35 ft (3 to 10.7 m) below ground surface. Geologic logs of 15 monitoring wells at the site revealed continuous clay layers ranging in thickness from 20 to 100 ft (6 to 30 m) separating the surface sands from the deeper aquifer, which is pumped for municipal water supply. Because contaminant migration occurred primarily in a lateral direction within a shallow, confined, relatively homogeneous sand aquifer, the site provided an ideal opportunity for application of the transport model. Boreholes and PVC well casings were screened over the entire saturated sand thickness, which was both underlain and overlain by confining clay layers.

Monitoring of the system, using the configuration of wells shown in Figure 13.24, began after the leaking tank had been removed and continued through the installation and operation of a four-well recovery system. Water-level measurements and analyses of samples taken from the wells over a 2-yr period provided an extensive data base for calibrating the computer model. Because the wells were

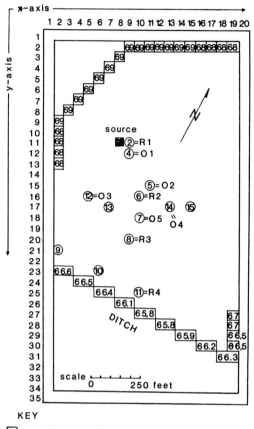

KEY

68 | constant head node with potentiometric value indicated in box

R1-4 = recovery well

② | monitoring well

O1-5 = designated observation well in simulations

Figure 13.24 Grid of study area.
Source: Freeberg et al., 1987.

fully screened through the entire thickness of the saturated sand, TCE concentrations in ground water samples taken from the wells represented vertically averaged concentrations.

Water-level measurements, as well as data from a constant-discharge pumping test, provided information on the hydraulic regime. Ground water movement in the system occurs as an essentially linear flow field in a southeastern direction. Some bending of flow lines occurs in the vicinity of a ditch that bounded the site along its southern edge.

The ground water velocity in the sand aquifer was relatively high and, consequently, the movement of contaminants was readily detectable during the monitoring period. Results from the pumping test indicated that the transmissivity of the sand unit was approximately 206 m²/day. With an average saturated thickness of 25 ft (7.6 m), the hydraulic conductivity of the aquifer was calculated to be 3.14 × 10⁻² cm/sec. A storage coefficient of 0.0075 was determined from pumping-test data by Theis analysis, indicating that the aquifer was confined. Assuming a typical value for effective porosity of 0.3 and an average hydraulic gradient of 0.002, seepage velocity at the site was found to be approximately 60 m/yr. Negligible sorption was assumed for this sandy site.

The 19-m³ (5000-gal) capacity concrete tank contained primarily a mixture of varying proportions of TCE and toluene. The tank was repeatedly filled and drained during the service period. Consequently, the release of contaminants, through periodic accidental spillage and from continuous leakage through the vinyl seams of the tank, which had gradually dissolved away, occurred at an unknown rate. Chemical data from the midway point in the monitoring period allowed delineation of the areal extent of contamination (Figure 13.24). Calculation of the total volume of contaminated water (46.0×10^3 m³) and determination of the average concentration of contaminant exhibited in the plume at that time (176 μg/1 TCE) allowed the total mass of contaminant to be estimated. The leaking storage tank had released an estimated total load of 8 kg of TCE.

Remedial Activity

Remedial activity at the industrial site began approximately 1 year after the solvent tank was excavated and monitoring had begun. The recovery system was designed (1) to produce an area of drawdown sufficient to encompass the contaminant plume, (2) to minimize the total flow rate and thereby reduce the volume of contaminated ground water to be treated, and (3) to minimize installation and operating costs. The recovery system consisted of four monitoring wells, which were to be pumped continuously at a design flow rate of 30 gpm (160 m³/day) each.

Application of the Model

The study area was simulated using a 20 × 35 cell grid with each cell representing a 15- × 15-m area. Five of the 15 wells monitored at the site were designated as observation wells in the model. Actual water-quality data from wells MW-4, 5, 12, 14, and 7 provided the most information about movement of the plume and distribution of contaminants. Predicted data from observation wells were useful in calibrating the model predictions to observed data.

Other parameters describing the hydrogeologic regime at the site were adjusted during calibration of the model. Ranges of values were used in calibration for parameters such as effective porosity, longitudinal dispersivity, transverse dispersivity, saturated thickness, and background contaminant concentration. The length of the injection period, which simulated the period of leakage prior to removal of the tank, and the concentration of TCE in the injection fluid were chosen so that a total mass of 8 kg of TCE was released into the aquifer.

Modeling Results

Figure 13.25 shows the predicted plume compared to measured values at the end of 1 year of sampling. Simulation of the recovery system followed simulation of the monitoring period described. Four recovery wells corresponding to R1, R2,

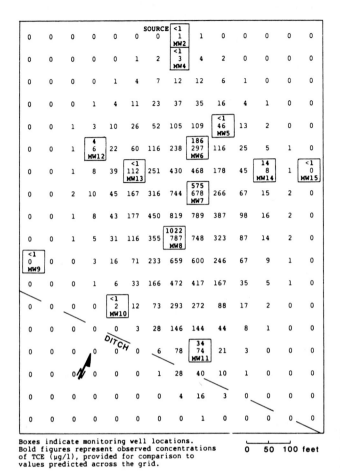

Boxes indicate monitoring well locations.
Bold figures represent observed concentrations
of TCE (µg/l), provided for comparison to
values predicted across the grid.

0   50   100 feet

Figure 13.25 TCE plume predicted at end of monitoring period. Source: Freeberg et al., 1987.

R3, and R4 were specified in the model (Figure 13.24). Pumping rates corresponding to the average pumping rates actually applied at the site were also specified. The recovery wells were turned on at the start of the simulation and allowed to operate continuously until the end of the simulation period. Calibration of the model, which required more than 100 computer runs, continued until the error between the predicted and measured contaminant plumes was minimized.

Simulation of the recovery system showed that operation of the four wells produced a marked decrease in the extent of the plume and in the predicted concentrations of TCE. Figure 13.26 shows contaminant distributions predicted by the model after 0.5, 1.0, 1.5, and 2.0 yr of pumping. At 2.0 yr in the simulation, a mass-balance error of 0.46% was computed, again indicating good numerical accuracy in the solution.

In September 1984, 0.5 yr after pumping began, nine of the monitoring wells were sampled, providing a check on the accuracy of the recovery system simulation. When the concentrations of TCE measured at these nine monitoring wells

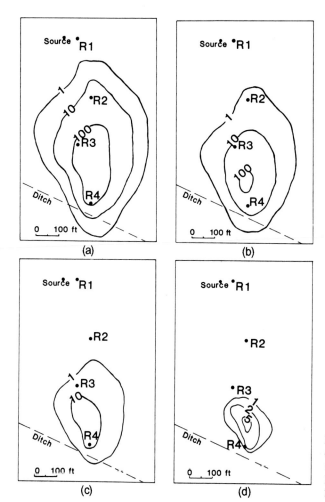

Figure 13.26 Predicted concentrations of TCE (μg/L) at points in time after start-up of pumping at wells Rl through R4. (a) 0.5 yr. (b) 1.0 yr. (c) 1.5 yr. (d) 2.0 yr. Source: Freeberg et al., 1987.

are totaled, the chemical data show a 90% reduction from the total concentration of TCE measured at the same nine wells prior to pumping. The simulation predicts an 89% reduction in TCE concentrations at the nine wells during the 0.5 yr of pumping, suggesting that the simulation of the recovery system is fairly accurate.

The model shows that a recovery period on the order of 2.0 yr is necessary to reduce concentrations of TCE to 6 μg/L or less. It should be noted that NAPLs were not initially observed at this site in the sand aquifer, which contributed to the relatively good match with the model predictions. Predictions made by the model at the four wells matched actual observations at three points in time, before and after startup of the recovery system. However, the site has since had problems due to the discovery of NAPLs in deeper zones, and additional remediation will be required.

Case 3: Conceptual Design for Managing a DNAPL Superfund Site

The presence of dense nonaqueous-phase liquids (DNAPLs) can greatly complicate ground water remediation efforts due to the downward migration of these heavy oils within an aquifer system and the difficulty of extracting DNAPL using conventional pumping methods. These problems are now being observed at the Motco Superfund Site near Houston, Texas, introduced as case 2 in Chapter 12, where DNAPL is present in a shallow surficial aquifer (Connor et al., 1989; Newell et al., 1991). To develop an effective scheme for DNAPL management, the aquifer remediation program for this site includes these elements:

- Detailed stratigraphic interpretation of the aquifer to delineate the zone of DNAPL accumulation.
- Pilot recovery test to determine the effectiveness of enhanced oil recovery technologies (EOR) for mobilizing DNAPL.
- Combined modeling and field study to evaluate the effect of DNAPL dissolution on future remediation activities at the site.
- Conceptual remedial design for extracting mobile DNAPL and for managing residual DNAPL and DNAPL dissolution products.

Given the difficulty of achieving conventional ground water cleanup standards in DNAPL-affected portions of an aquifer, an alternative approach for the management of DNAPL sites is proposed in the following. This approach is based on restoration of ground water quality in areas that are considered to be free of DNAPL and long-term containment of DNAPL and DNAPL dissolution products in zones where DNAPL presence is either confirmed or suspected based on evaluation of site data.

Delineation of DNAPL Zones

The Motco Site is a former reclamation facility that operated in the vicinity of La Marque, Texas, during the period from 1958 to 1968 (Figure 12.9). Reclamation efforts involved collection and reprocessing of petrochemical residues within an 11-acre system of unlined pits. Hydrogeologic investigations have shown this area to be underlain by a surface deposit of interbedded sands, silts, and clays designated the transmissive zone. Extending from the ground surface to a depth of approximately 50 ft below grade, the transmissive zone is in turn underlain by a stiff, high-plasticity clay layer, termed the UC-1 Clay.

DNAPL accumulations have been observed in wells screened in low spots in the shallow transmissive zone immediately atop the underlying clay stratum (Figure 12.10). To map the DNAPL zone, 60 boring and well logs at the site were supplemented with an additional 73 cone penetrometer logs (see Chapter 5) to generate a continuous and laterally extensive stratigraphic record of the shallow aquifer system. The data were used to develop detailed topographic maps of the base of the transmissive zone to find DNAPL accumulation, to select locations for soil borings to confirm the presence of pools atop this clay surface, and to design a pilot test for recovery of DNAPL and affected ground water. Results of this investigation showed DNAPL to be moving through fractures and other secondary porosity features of the silt stratum in general accordance with the base topography of the unit (Figure 12.11).

DNAPL Recovery Pilot Test

To evaluate the feasibility of pumped withdrawal of DNAPL fluids, a pilot test was conducted at the Motco Site in 1989 to compare three recovery technologies: pumping, water-flooding, and vacuum-enhanced recovery. To test the performance of pumping alone, a recovery well was installed in a zone of DNAPL accumulation and equipped with a submersible, positive-displacement pump. For the vacuum-enhanced pumping scheme, the downhole pump was augmented by a wellbore vacuum to increase the available drawdown and maximum yield of the recovery well. For the water-flooding scheme, a freshwater injection well was operated at a distance of 100 ft from a pumping well to increase the hydraulic gradient. A 3-week testing program demonstrated that some DNAPL could be removed by pumping alone, but that water-flooding and vacuum-enhanced recovery greatly increased recovery rates.

Results of the pilot test demonstrated that mobile DNAPL could be recovered from the transmissive strata by means of ground water pumping. As evidenced by DNAPL accumulation within nearby observation wells during recovery well operation, hydraulic gradients 40 to 60 times greater than normal are capable of overwhelming density forces and inducing DNAPL flow within the aquifer.

Test results indicate that any of the three pumping schemes should prove effective for ground water/DNAPL recovery from the transmissive zone. However, due to improved well yield and higher induced flow gradients, vacuum-enhanced pumping and water flooding offer significant advantages in terms of the rate and potential degree of DNAPL removal, with over a 100% increase in DNAPL production rate when both enhancements were used (50 to 100 gal DNAPL/day versus 25 to 50 gal/day with pumping alone).

DNAPL Dissolution Study

The pilot recovery test demonstrated that the mobile fraction of the DNAPL fluids could be removed from the water-bearing stratum using enhanced oil recovery technologies. However, a significant percentage of the DNAPL mass would remain immobilized within the aquifer matrix, possibly acting as a continued source of organic constituent release to ground water due to dissolution of soluble DNAPL components.

To quantify the effect of DNAPL dissolution on the aquifer remediation program now being planned for the Motco Site, a modeling study and in situ field test were designed to simulate the performance of the full-scale ground water

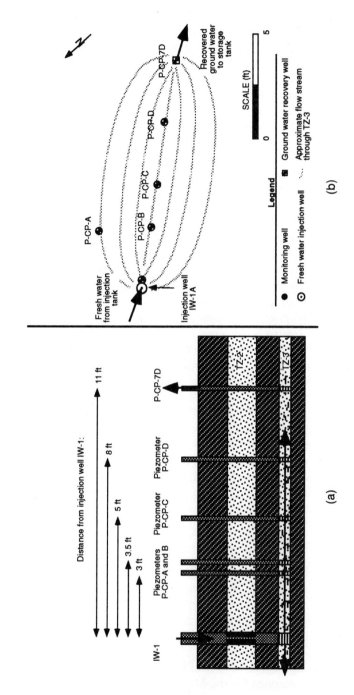

Figure 13.27 DNAPL dissolution pilot study project design. (a) System layout. (b) Ground water flow diagram.

recovery system. The modeling study employed an existing dissolution relationship (Borden and Kao, 1989) that assumed equilibrium partitioning of soluble organics between water and DNAPL. On the basis of this analysis, a field test was designed to simulate the actual hydraulic conditions that might be expected during a full-scale pumping and water-flooding program for DNAPL-affected areas (Figure 13.27). One injection well and one production well, spaced 11 ft apart, were used to flush freshwater through the test area during the 30-day test period. As the test progressed, over 11,000 gal of freshwater was drawn through the test area, resulting in significant reductions in the concentrations of key organic constituents (> 85%; see Figure 13.28).

At this site, the field dissolution test showed that approximately 50 to 100 pore volumes of freshwater may be required to achieve the current ground water recovery standards (ranging from 2 to 50 mg/L for key indicator constituents in the class III saline aquifer) in DNAPL-affected areas. These empirical results are in agreement with modeling predictions, which indicated that roughly a 50-pore volume flush would be required to adequately reduce dissolution products from the residual DNAPL. Because this level of flushing is probably impracticable to achieve at the site, a long-term containment system was specified to protect human health and the environment.

Long-term Containment Modeling Study

To assist in the design of a long-term containment system for dissolved constituents at the Motco Site, a modeling study was conducted using the MODFLOW ground water model (Macdonald and Harbaugh, 1984). To simulate the various transmissive zones and clays at the site, a three-dimensional model consisting of four different layers was created. Transmissivity and vertical conductance values

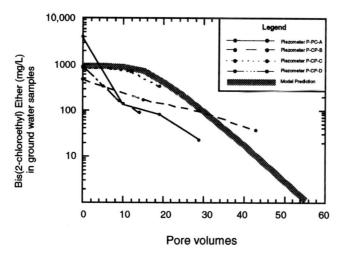

Figure 13.28 Concentration of indicator parameter versus pore volumes flushed during DNAPL dissolution pilot test.

were assigned to the 9200 finite difference cells using data from pump tests, slug tests, soil borings, and cone penetrometer surveys. To ensure accurate simulations, the model was calibrated using data from natural flow conditions and from the DNAPL recovery test (see Chapter 10).

Ground water pumping rates required to achieve hydraulic control of the affected areas at the site were determined for three containment system designs: (1) existing conditions, (2) partial enclosure of affected areas, and (3) complete enclosure of affected areas. The existing conditions include a 900-ft-long slurry wall currently in place along one side of the site. A number of computer runs were made to evaluate these options. Complete enclosure of affected areas with a slurry wall reduces pumping rates required to achieve hydraulic control by approximately ten-fold, from 2.0 to 3.0 GPM to 0.3 to 0.6 GPM. Lower pumping rates will reduce the cost of treating recovered ground water and may lead to significant cost savings during operation of the long-term containment system.

SUMMARY

Chapter 13 reviews the important concepts of aquifer remediation as applied to contaminated hazardous-waste sites. Many of the flow and transport issues discussed in previous chapter are used here. Various methods of containment, hydraulic control, pump and treat, bioremediation, soil vapor extraction, and NAPL control are presented in detail, along with three complete case studies from the authors' experiences. Modern approaches to site remediation include consideration of two related but distinct problems: (1) source control near the area of greatest contamination and (2) plume control either at some downgradient boundary or for complete removal of soluble components. Figure 13.1 depicts several of the conceptual models available for site cleanup; earlier pump-and-treat systems for soluble plume removal are combined with more intensive methods near the source area. Finally, hazardous-waste site remediation is a rapidly evolving field with new technologies emerging rapidly. New remediation approaches include funnel and gate systems, treatment walls, air sparging, surfactant flushing, and cosolvent flushing for NAPLs removal. However, due to the performance problems of original pump-and-treat systems, new and combined remediation methods are currently being tested at a number of field sites. It is clear that no single method will emerge as superior due to the heterogeneous nature of field sites, the complexity of the contaminants involved, the NAPL issue, and the expense involved in cleaning up a site.

REFERENCES

ABDUL, A. S., 1992, "A New Pumping Strategy for Petroleum Product Recovery from Contaminated Hydrogeologic Systems: Laboratory and Field Evaluations," *Ground Water Monitoring Rev.* XIII, 105–114.

ALTHOFF, W. F., CLEARY, R. W., and ROUX, P. H., 1981, "Aquifer Decontamination for Volatile Organics: A Case History," *Ground Water* 19:495–504.

ANASTOS, G. J. MARKS. P. J., CORBIN, M. H., and COIA, M. F., 1985, "Task 11. In Situ Air Stripping of Soils, Pilot Study, Final Report." Report No. AMXTH-TE-TR-85026. U.S. Army Toxic and Hazardous Material Agency. Aberdeen Proving Grounds, Edgewood, MD.

ANDERSEN, M. P., 1984, "Movement of Contaminants in Groundwater: Groundwater Transport—Advection and Dispersion," *Studies in Geophysics, Groundwater Contamination,* National Academy Press, Washington, DC, pp. 37–45.

AYRES, J. E., LAGER, D. C., and BARVENIK, M. J., 1983, "The First EPA Superfund Cutoff Wall: Design and Specifications," *Proc. 3rd Natl. Symp. on Aquifer Restoration and Ground Water Monitoring,* NWWA, Columbus, OH, p. 13.

BATCHELDER, G. V., PANZERI, W. A., and PHILLIPS, H. T., 1986, "Soil Ventilation for the Removal of Adsorbed Liquid Hydrocarbons in the Subsurface," *Proc. Conf. on Petroleum Hydrocarbons and Organic Chemicals in Ground Water,* NWWA, Houston, TX, pp. 672–688.

BENNEDSEN, M. B., 1987, "Vacuum VOC's from Soil," *Pollution Engineering* 19(2):66–68.

BENNEDSEN, M. B., SCOTT, J. P., and HARTLEY, J. D., 1985, "Use of Vapor Extraction Systems for in Situ Removal of Volatile Organic Compounds from Soil," *Proc. Natl. Conf. on Hazardous Wastes and Hazardous Materials,* HMCRI, pp. 92–95.

BERRY-SPARK, K., and BARKER, J. F., 1987, "Nitrate Remediation of Gasoline Contaminated Ground Waters: Results of a Controlled Field Experiment," *Proc. Conf. on Petroleum Hydrocarbons and Organic Chemicals in Ground Water,* NWWA, Houston, TX, pp. 3–10.

BERRY-SPARK, K., BARKER, J. F., MAJOR, D., and MAYFIELD, C. I., 1986, "Remediation of Gasoline-contaminated Ground Waters: A Controlled Field Experiment," *Proc. Conf. on Petroleum Hydrocarbons and Organic Chemicals in Ground Water,* NWWA, Houston, TX, pp. 613–623.

BLAKE, S. B., and LEWIS, R. W., 1982, "Underground Oil Recovery." *Proc. of Second National Symposium on Aquifer Restoration and Ground Water Monitoring,* D. M. Nielsen, ed., National Water Well Association, Dublin, OH, pp. 69–75.

BORDEN, R. C., BEDIENT, P. B., LEE, M. D., WARD, C. H., and WILSON, J. T., 1986, "Transport of Dissolved Hydrocarbons Influenced by Oxygen Limited Biodegradation: 2. Field Application," *Water Resources Res.* 22:1983–1990.

BORDEN, R. C., and KAO, C., 1989, "Water Flushing or Trapped Residual Hydrocarbon: Mathematical Model Development and Laboratory Validation," *Proc. Conf. on Petroleum Hydrocarbons and Organic Chemicals in Ground Water,* NWWA, Houston, TX, pp. 175–189.

CANTER, L. W., and KNOX, R. C., 1986, *Ground Water Pollution Control,* Lewis Publishers, Chelsea, MI.

CHARBENEAU, R. J., WANUKULE, N., CHIANG, C. Y., NEVIN, J. P., and KLEIN, C. L., 1989, "A Two-layer Model to Simulate Floating Free Product Recovery: Formulation and Applications," *Conf. on Petroleum Hydrocarbons and Organic Chemicals in Ground Water,* NWWA, Houston, TX, pp. 333–345.

CH$_2$M-HILL, INC., 1987, "Operable Unit Remedial Action, Soil Vapor Extraction at Thomas Solvents Raymond Road Facility, Battle Creek, MI, Quality Assurance Project Plan," U.S. Environmental Protection Agency, Chicago, IL.

CHIANG, C. Y., 1987, "Effects of Dissolved Oxygen on the Biodegradation of BTX in a Sandy Aquifer," *Proc. Conf. on Petroleum Hydrocarbons and Organic Chemicals in Ground Water,* NWWA, Houston, TX, pp. 451–469.

COHEN, R. M., and MERCER, J. W., 1993, *DNAPL Site Evaluation,* C. K. Smoley, Boca Raton, FL.

CONNOR, J. A., NEWELL, C. J., and WILSON, D. K., 1989, "Assessment, Field Testing, and Conceptual Design for Managing Dense Nonaqueous Phase Liquids (DNAPL) at a Superfund Site," *Proc. Conf. on Petroleum Hydrocarbons and Organic Chemicals in Ground Water,* NWWA, Houston, TX.

COOPER, R. M., and ISTOK, J. D., 1988a, "Geostatistics Applied to Groundwater Pollution—I: Methodology," *J. Environ. Eng.* 114(2):270–286.

COOPER, R. M., and ISTOK, J. D., 1988b, "Geostatistics Applied to Groundwater Contamination—II: Methodology," *J. Environ. Eng.* 114(2):287–299.

EHRENFELD, J., and BASS, J., 1984, *Evaluation of Remedial Action Unit Operations at Hazardous Waste Disposal Sites,* Noyes Publications, Park Ridge, NJ.

ENVIRESPONSE, INC., 1987, "Demonstration Test Plan, in Situ Vacuum Extraction Technology, Terra Vac Inc., SITE Program, Groveland Wells Superfund Site, Groveland, MA." EERU Contract No. 68-03-3255, Work Assignment 1-R18, Enviresponse No. 3-70-06340098, Edison, NJ.

FETTER, C. W., 1988, *Applied Hydrogeology,* Merrill Publishing, Co., Columbus, OH.

FETTER, C. W., 1993, *Contaminant Hydrogeology,* Macmillan College Publishing Company, Inc.

FLATHMAN, P. E., QUINCE, J. R., and BOTTOMLEY, L. S., 1984, "Biological Treatment of Ethylene Glycol-contaminated Ground Water at Naval Engineering Center in Lakehurst, New Jersey," *Proc. 4th Natl. Symp. on Aquifer Restoration and Ground Water Monitoring,* Nielsen, D. M., and Curl, M., eds., NWWA, Houston, TX.

FREEBERG, K. M., BEDIENT, P. B., and CONNER, J. A., 1987, "Modeling of TCE Contamination and Recovery in a Shallow Sand Aquifer," *Ground Water* 25:70–80.

GLOVER, E. W., 1982, "Containment of Contaminated Groundwater—An Overview," *Proc. Second National Symposium on Aquifer Restoration and Groundwater Monitoring,* NWWA, Houston, TX, pp. 6–13.

GORELICK, S. M., 1983, "A Review of Distributed Parameter Ground Water Management Modeling Methods," *Water Resources Res.* 19:305–319.

GORELICK, S. M., VOSS, C. I., GILL, P. E., MURRAY, W., SAUNDERS, M. A., and WRIGHT, M. H., 1984, "Aquifer Reclamation Design: The Use of Contaminant Transport Simulation Combined with Nonlinear Programming," *Water Resources Res.* 20:415–427.

HINCHEE, R. E., DOWNEY, D. C., and COLEMAN, E. J., 1987, "Enhanced Bioremediation, Soil Venting, and Ground-water Extraction; A Cost-Effectiveness and Feasibility Comparison," *Proc. Conf. on Petroleum Hydrocarbons and Organic Chemicals in Ground Water,* NWWA, Houston, TX, pp. 147–164.

HINCHEE, R. E., REISINGER, H. J., and BURRIS, D., 1986, "Underground Fuel Contamination, Investigation, and Remediation—A Risk Assessment Approach to How Clean Is Clean," *Proc. Conf. on Petroleum Hydrocarbons and Organic Chemicals in Ground Water,* NWWA, Houston, TX, pp. 539–563.

HUTZLER, N. J., MURPHY, B. E., and GIERKE, J. S., 1989, "State of Technology Review Soil Vapor Extraction Systems," Cooperative Agreement No. CR-814319-01-1. Risk Reduction Engineering Lab, U.S. Environmental Protection Agency, Cincinnati, OH.

JAMISON, V. W., RAYMOND, R. L., and HUDSON, J. O., JR., 1975, "Biodegradation of High-octane Gasoline in Groundwater," *Dev. Ind. Microbiol.,* pp. 16, 305.

JAVANDEL, I., DOUGHTY, C., and TSANG, C. F., 1984, *Groundwater Transport: Handbook of Mathematical Models,* American Geophysical Union, Water Resources Monograph 10, Washington, DC.

JAVANDEL, I., and TSANG, C. F., 1986, "Capture-zone Type Curves: A Tool for Aquifer Cleanup," *Ground Water* 24(5):616–625.

JEPSEN, C. P., and PLACE, M., 1985, "Evaluation of Two Methods for Constructing Vertical Cutoff Walls at Waste Containment Sites," *Hydraulic Barriers in Soil and Rock,* ASTM STP 874, A. I. Johnson, R. K. Frobel, N. J. Cavalli, and C. B. Pettersson, eds., American Society for Testing and Materials, Philadelphia, pp. 45–63.

JOHNSON, A. I., FROBEL, R. K., CAVALLI, N. J., and PETTERSSON, C. B., eds., 1985, *Hydraulic Barriers in Soil and Rock,* American Society for Testing and Materials, Philadelphia.

JOHNSON, P. C., KEMBLOWSKI, M. W., and COLTHART, J. D., 1988, "Practical Screening Models for Soil Venting Applications," *Proc. Conf. on Petroleum Hydrocarbons and Organic Chemicals in Ground Water,* NWWA, Houston, TX, pp. 521–546.

JOHNSON, P. C., STANLEY, C. C., KEMBLOWSKI, M. W., BYERS, D. L., and COLTHART, J. D., 1990, "A Practical Approach to the Design, Operation, and Monitoring of In-situ Soil-venting Systems," *Ground Water Monitoring Rev.* 10(2):159–178.

JOHNSON, R. L., JOHNSON, P. C., MCWHORTER, D. B., HINCHEE, R. E., GOODMAN, I. Fall 1993. "An Overview of Air Sparging," *Ground Water Monitoring and Remediation,* Vol XIII, No. 4, 127–135.

KONIKOW, L. F., and BREDEHOEFT, J. D., 1978, "Computer Model of Two-dimensional Solute Transport and Dispersion in Ground Water, Automated Data Processing and Computations," *Techniques of Water Resources Investigations of the USGS,* U.S. Geological Survey, Washington, DC.

KONIKOW, L. F., and THOMPSON, D. W., 1984, "Groundwater Contamination and Aquifer Reclamation at the Rocky Mountain Arsenal, Colorado," *Groundwater Contamination, Studies in Geophysics,* National Academy Press, Washington, DC, pp. 93–103.

KRISHNAYYA, A. V., O'CONNOR, M. J., AGAR, J. G., and KING, R. D., 1988, "Vapor Extraction Systems Factors Affecting Their Design and Performance," *Proc. Conf. on Petroleum Hydrocarbons and Organic Chemicals in Ground Water,* NWWA, Houston, TX, pp. 547–569.

LEE, M. D., THOMAS, J. M., BORDEN, R. C., BEDIENT, P. B., WARD, C. H. and WILSON, J. T., 1988, "Biorestoration of Aquifers Contaminated with Organic Compounds," *CRC Critical Rev. Environ. Control,* 18(1):29–89.

LEE, M. D., and WARD, C. H., 1985, "Biological Methods for the Restoration of Contaminated Aquifers," *Environ. Toxicol. Chem.* 4:743.

LYNCH, E. R., ANAGNOST, S. W., SWENSON, G. A., and GOLDMAN, R. K., 1984, "Design and Evaluation of In-place Containment Structures Utilizing Ground Water Cutoff Walls," *Proc. Fourth National Symposium and Exposition on Aquifer Restoration and Ground Water Monitoring,* NWWA, pp. 1–7.

MACKAY, D. M., and CHERRY, J. A., 1989, "Groundwater Contamination: Pump-and-Treat Remediation," *Environ. Sci. Technol.* 23(6):630–636.

MALOT, J. J., and WOOD, P. R., 1985, "Low Cost, Site Specific, Total Approach to Decontamination," *Conf. on Environmental and Public Health Effects of Soils Contaminated with Petroleum Products,* University of Massachusetts, Amherst, MA.

MARLEY, M. C., and HOAG, G. E., 1984, "Induced Soil Venting for the Recovery/Restoration of Gasoline Hydrocarbons in the Vadose Zone," *Proc. Conf. on Petroleum Hydrocarbons and Organic Chemicals in Ground Water,* NWWA, Houston, TX, pp. 473–503.

McDONALD, M. G., and HARBAUGH, A. W., 1984, "A Modular Three-dimensional Finite-difference Groundwater Flow Model," Open File Report 83-875, U.S. Geological Survey, Washington, DC.

MERCER, J. W., and COHEN, R. M., 1990, "A Review of Immiscible Fluids in the Subsurface," *J. Contaminant Hydrology* 6:107–163.

MERCER, J. W., SKIPP, D. C., and GIFFIN, D., 1990, *Basics of Pump-and-Treat Groundwater Remediation Technology,* EPA-600/8-90/003, R. S. Kerr Environmental Research Laboratory, U.S. Environmental Protection Agency, Ada, OK.

MINUGH, E. M., PATRY, J. J., KEECH, D. A., and LEEK, W. R., 1983, "A Case History: Cleanup of a Subsurface Leak of Refined Product," *Proc. 1983 Oil Spill Conf.—Prevention, Behavior, Control, and Cleanup,* San Antonio, TX, p. 397.

MOLZ, F. J., and BELL, L. C., 1977, "Head Gradient Control in Aquifers Used for Fluid Storage," *Water Resources Res.* 13:795–798.

NEED, E. A., and COSTELLO, M. J., 1984. "Hydrogeologic Aspects of Slurry Wall Isolation Systems in Areas of High Downward Gradients," *Proc. 4th Natl. Symp. on Aquifer Restoration and Ground Water Monitoring,* Columbus, OH, D. M. Nielsen and M. Curl, eds., NWWA, Houston, TX, p. 18.

NEWELL, C. J., CONNOR, J. A., WILSON, D. K., and McHUGH, T. E., 1991, "Impact of Dissolution of Dense Non-Aqueous Phase Liquids (DNAPLs) on Groundwater Remediation," *Proc. Conf. on Petroleum Hydrocarbons and Organic Chemicals in Ground Water,* NWWA, Houston, TX, pp. 301–315.

NIELSON, D. M., 1983, "Remedial Methods Available in Areas of Groundwater Contamination," *Proc. 6th Natl. Ground Water Quality Symp.,* Atlanta, GA, September 1982, NWWA, Houston, TX, p. 219.

OSTER, C. C., and WENCK, N. C., 1988, "Vacuum Extraction of Volatile Organics from Soils," *Proc. 1988 Joint CSCE–ASCE Natl. Conf. on Environmental Engineering,* Vancouver, B.C., Canada, ASCE, pp. 809–817.

PANKOW, J. F., JOHNSON, R. L., and CHERRY, J. A., 1993, "Air Sparging in Gate Wells in Cutoff Walls and Trenches for Control of Plumes of Volatile Organic Compounds (VOCs)," *Ground Water* 31(4):654–663.

PAYNE, F. C., CUBBAGE, C. P., KILMER, G. L., and FISH, L. H., 1986, "In Situ Removal of Purgeable Organic Compounds from Vadose Zone Soils," Purdue Industrial Waste Conference.

RAGHAVAN, R., COLES, E., and DIETZ, D., 1990, "Cleaning Excavated Soil Using Extraction Agents: A State-of-the-Art Review," EPA/600/52-89/034, U.S. Environmental Protection Agency, Cincinnati, OH.

RAYMOND, R. I., 1978, "Environmental Bioreclamation," presented at Mid-Continent Conference and Exhibition on Control of Chemicals and Oil Spills, Detroit, MI.

RAYMOND R. L., JAMISON, V. W., and HUDSON, J. O., 1976, "Beneficial Stimulation of Bacterial Activity in Groundwaters Containing Petroleum Products," *AIChE Symp. Ser.* *73,* p. 390.

RIFAI, H. S., BEDIENT, P. B., WILSON, J. T., MILLER, K. M., and ARMSTRONG, J. M., 1988, "Biodegradation Modeling at a Jet Fuel Spill Site," *ASCE J. Environ. Engr. Div.* 114:1007–1019.

ROBERTS, P. V., HOPKINS, G. D., MACKAY, D. M., and SEMPRINI, L., 1990, "A Field Evaluation of in-Situ Biodegradation of Chlorinated Ethenes: Part I, Methodology and Field Site Characterization," *Ground Water* 28(4):591–604.

RYAN, C. R., 1985, "Slurry Cutoff Walls: Applications in the Control of Hazardous Waste," *Hydraulic Barriers in Soil and Rock*, A. I. Johnson et al., eds., American Society for Testing and Materials, STP 874, Philadelphia, pp. 9–23.

SABATINI, D. A., and KNOX, R. C., EDS., 1992, *Transport and Remediation of Subsurface Contaminants: Colloidal, Interfacial, and Surfactant Phenomena*, ACS Symposium Series 491, American Chemical Society, Washington, DC.

SATKIN, R. L., and BEDIENT, P. B., 1988, "Effectiveness of Various Aquifer Restoration Schemes under Variable Hydrogeologic Conditions," *Ground Water* 26(4):488–98.

SCHMEDNECHT, F. C., and GOLDBACH, K. P., 1984, "Slurry Wall Installation by the Vibrated Beam Technique," *Proc. 4th Natl. Symp. on Aquifer Restoration and Ground Water Monitoring*, Columbus, OH, D. M. Nielsen and M. Curl, eds. NWWA, Houston, TX, p. 27.

SCHULZE, D. M., BARVENIK, M. J., and AYRES, J. E., 1984, "Design of Soil Bentonite Backfill Mix for the First Environmental Protection Agency Superfund Cutoff Wall," *Proc. 4th Natl. Symp. on Aquifer Restoration and Ground Water Monitoring*, Columbus, OH, D. M. Nielsen and M. Curl, eds., NWWA, Houston, TX, p. 8.

SEMPRINI, L., ROBERTS, P. V., HOPKINS, G. D., and McCARTY, P. L., 1990, "A Field Evaluation of in Situ Biodegradation of Chlorinated Ethenes: Part 2, Results of Biostimulation and Biotransformation Experiments," *Ground Water* 28(5):715–727.

SHAFER, J. M., 1984, "Determining Optimum Pumping Rates for Creation of Hydraulic Barriers to Ground Water Pollutant Migration," *Proc. 4th Natl. Symp. on Aquifer Restoration and Ground Water Monitoring*, Columbus, OH, D. M. Nielsen and M. Curl, eds., NWWA, Houston, TX, p. 50.

TALLARD, G., 1985, "Slurry Trenches for Containing Hazardous Wastes, *Civil Engineering ASCE* 41.

TEXAS RESEARCH INSTITUTE, 1980 (reprinted 1986), "Examination of Venting for Removal of Gasoline Vapors from Contaminated Soil," American Petroleum Institute, Washington, DC.

THOMAS, J. M., LEE, M. D., BEDIENT, P. B., BORDEN, R. C., CANTER, L. W., and WARD, C. H., 1987, "Leaking Underground Storage Tanks: Remediation with Emphasis on in Situ Biorestoration," EPA/600/2-87/008, NCGWR, R. S. Kerr Environmental Research Laboratory, U.S. Environmental Protection Agency, Ada, OK.

THOMAS, J. M., and WARD, C. H., 1989, "In Situ Biorestoration of Organic Contaminants in the Subsurface," *Environ. Sci. Technology* 23(7):760–766.

THRELFALL, D., and DOWIAK, M. J., 1980, "Remedial Options for Ground Water Protection at Abandoned Solid Waste Disposal Facilities," U.S. EPA National Conference on Management of Uncontrolled Hazardous Waste Sites, Washington, DC.

TWENTER, F. R., CUMMINGS, T. R., and GRANNEMANN, N. G., 1985, "Groundwater Contamination in East Bay Township, Michigan," U.S. Geological Survey Water Resources Investigation Report 85-4064, Washington, DC.

U.S. ENVIRONMENTAL PROTECTION AGENCY, SEPTEMBER 1989, "Evaluation of Ground Water Extraction Remedies," Vol. 1, Summary Report, EPA/540/2-89/054, Washington, DC.

U.S. ENVIRONMENTAL PROTECTION AGENCY, 1992, *Evaluation of Ground Water Extraction*

Remedies: Phase II, Vol., Summary Report, Office of Emergency and Remedial Response, Publication 9355.4-04, Washington, DC.

VERSCHUEREN, K., 1983, *Handbook of Environmental Data on Organic Chemicals,* Van Nostrand Reinhold, New York.

WILSON, D. J., CLARKE, A. N., and CLARKE, J. H., 1988, "Soil Clean-up by in Situ Aeration, I. Mathematical Modelling," *Separation Sci. Tech.* 23:991–1037.

WILSON, J. T., LEACH, L. E., HENSON, M. J., and JONES, J. N., 1986, "In Situ Biorestoration as a Ground Water Remediation Technique," *Ground Water Monitoring Rev.,* 6(4).

<table>
<tr><td>Chapter
14</td><td>LEGAL PROTECTION
OF
GROUND WATER</td></tr>
</table>

Chapter 14

LEGAL PROTECTION OF GROUND WATER

By James B. Blackburn

The protection of ground water in the United States is accomplished through a set of statutes passed at different times. No statute is comprehensive; instead, each covers specific types of ground water contamination problems. These statutes reflect political issues at the time of their passage and incorporate various relationships between the executive and legislative branches.

In this chapter, the governmental institutions that address ground water contamination will be presented, followed by the major requirements of the various acts. Then associated case studies will be presented. The goal of this chapter is to describe both the substance and process of national ground water protection.

14.1 PROCESS OF GROUND WATER PROTECTION

The federal statutes that protect the nation's ground water are the following:

- Safe Drinking Water Act (SDWA) of 1974
- Resource Conservation and Recovery Act (RCRA) of 1976
- Comprehensive Environmental Response, Compensation and Liability Act (CERCLA) of 1980
- Hazardous and Solid Waste Amendments (HSWA) of 1984
- Superfund Amendments and Reauthorization Act (SARA) of 1986.

These federal statutes are implemented under the auspices of the U.S. Environmental Protection Agency (EPA). The EPA regulates by publishing regulations in the *Federal Register* through a process known as informal rule making. Regulations are published as draft regulations and are commented on by the regulated public, environmental groups, and other interested persons. After consideration of these comments, the EPA will promulgate final regulations in the *Federal Register*. Each year, the rules of the agency are codified in a single document titled *Code of Federal Regulations* (CFR). The final regulations of the EPA are just as binding as are the terms of the statutes.

The EPA is part of the executive branch of government, which is headed by the president of the United States. The legislative branch is comprised of the

Congress, which includes the U.S. House of Representatives and the U.S. Senate. A dynamic exists between the two branches. In essence, Congress writes the policy of the United States in the form of statutes and the executive branch (for example, the EPA) implements that policy through rule making.

A key issue is the extent of the discretion that is delegated to the EPA by Congress. In the early days of ground water regulation, the EPA was granted a large degree of discretion by Congress. However, conflicts emerged between the EPA and Congress in the early days of the Reagan administration over the implementation of the newly passed CERCLA/Superfund program, as well as RCRA. These conflicts culminated in the criminal indictment of Rita LaVelle for perjury and the exit of Ann Burford as the administrator of the EPA.

Since that time, Congress has taken substantial discretion away from the EPA with regard to the implementation of ground water protection. Stated otherwise, Congress has been much more explicit in its policy statement, leaving less policy discretion in EPA headquarters. This dispute between the EPA and Congress is important because it was the driving force behind the 1984 HSWA amendments, which in turn substantially altered U.S. ground water policy. This dispute will also have a bearing on future initiatives by Congress in the ground water arena.

A second institutional issue of importance is the role between the states and the EPA. Just as there is a dynamic between Congress and the EPA, there is also a dynamic between the states and the EPA. Under the U.S. Constitution, states are the repository of governmental power, whereas the federal government has limited power. Federal environmental control is undertaken pursuant to the commerce clause of the Constitution.

The states have a very strong role in ground water protection. First, all laws about ground water supply and allocation arise under state rather than federal law. Second, each state has property rights and tort concepts that apply to ground water. For example, if your neighbor contaminated your ground water, there may be rights that you as a landowner could assert in a state court. These rights are in addition to and distinct from federal environmental law. These will not be covered in this chapter.

Every state has one or more administrative agencies that are the state counterpart to the EPA. Ground water protection programs of the federal government are implemented in whole or in part by these state agencies. As a general rule, the state program must be as strong as the federal program. It can, however, be more stringent. The state agency may be designated through a process called delegation to act on behalf of the EPA. If delegation occurs, then a separate regulatory program will not exist at the EPA for that state for those matters that have been delegated.

Confusion sometimes exists when some portions of a program have been delegated and some have not. For example, when HSWA was passed in 1984, it substantially changed the RCRA program. Most states had already been delegated the RCRA program. For that reason, the HSWA program was implemented by the EPA until the state could pass regulations sufficiently strong to allow delegation of the new HSWA programs. By 1991, most of the HSWA requirements had been delegated to the states.

In this chapter, ground water law will be presented from the standpoint of the requirements of federal law and EPA regulations. This should define the bottom line of ground water protection throughout the United States. It is important to remember that each state may have variations from this bottom line, and these state requirements should always be consulted to be confident about the status of regulation in any particular state.

Finally, it is important to note that this ground water system is constantly being reviewed, criticized, interpreted, and reinterpreted by the EPA, Congress, and the judicial system. Given that Superfund site cleanups may cost hundreds of millions of dollars and ground water contamination can paralyze a community with fear, it is inevitable that changes and fine-tuning will occur in this system.

14.2 SAFE DRINKING WATER ACT OF 1974

The Safe Drinking Water Act (SDWA) of 1974 was passed because of concerns regarding the safety of public water supplies. This act set forth two large initiatives. First, the EPA was empowered to develop drinking-water standards throughout the United States for public water-supply systems. Second, the EPA was given responsibility for implementing a broad-scale ground water protection program called the Underground Injection Control (UIC) program.

Of these two major programs, only the UIC program protects ground water. The drinking-water standards govern the quality of water delivered to the consumer, but do not regulate sources of ground water contamination. That responsibility is vested in the UIC program.

The SDWA was passed in 1974 before Congress determined that the federal government would directly regulate ground water protection activities. For this reason, the UIC program gives the states a major role in the implementation of this act. Under the SDWA, the EPA was empowered to develop general regulations for underground injection control. Then, each state was required to adopt rules and regulations implementing the UIC program, including the development of a state permit program. The EPA is authorized to implement a program in any state that does not have its own program.

The UIC program regulates underground injection of fluids into wells. A well is defined as a hole in the ground that is deeper than it is wide or long. A fluid is defined to include liquids, semisolid material, and other nonliquid substances. According to the SDWA, a state must have a regulatory program to prevent underground injection that endangers drinking-water sources. A drinking-water source is considered to be endangered if underground injection results in the placement of any contaminant into an underground source of drinking water and such contaminant results in a violation of the national primary drinking-water standards.

Under the regulations promulgated by the EPA to implement the SDWA, five classes of underground injection wells have been identified.

Class I wells are those that inject hazardous waste or industrial or municipal waste below the lowermost underground source of drinking water (USDW).

Class II wells are those that inject fluids brought to the surface in association with oil and gas production or injected as part of a secondary and tertiary recovery process.

Class III wells are those that inject fluids for the extraction of minerals, including the Frasch method of mining sulfur.

Class IV wells are those that inject hazardous waste into or above an underground source of drinking water.

Class V wells are those that are not included in classes I to IV. Class V wells include air conditioning return flow, community cesspools (not single family), cooling water, drainage wells, dry wells, recharge wells, saltwater barrier wells, backfill wells, community septic system wells (not single family), subsidence control wells, radioactive waste wells, geothermal injection wells, conventional mine solution wells, brine extraction wells, injection wells for experimental technologies, and injection wells used for the in situ extraction of lignite, coal, tar sands, and oil shale.

As a general premise, all the preceding underground injection wells may be permitted by the state except for class IV wells, which are prohibited (40 CFR 144.13). The state is not required to allow underground injection; however, if it chooses to allow underground injection, this injection must be accomplished in accordance with EPA regulations (40 CFR Part 146).

Underground injection of waste is only one aspect of the activities regulated by the UIC program. Although the SDWA was passed in 1974, the regulations implementing this act were not promulgated by the EPA until 1980, the same time the RCRA regulations were promulgated.

Underground injection is accomplished by injecting fluids at high pressure into a well. The regulations established by the EPA require that these wells be cased and that the casing be cemented into the geologic formation. The space between the well and the casing, called the annulus, must be filled with a fluid, and a positive annulus pressure must be maintained such that if a leak occurs the leak will be from the annulus into the well rather than from the well outward.

Enough controversy was generated by the oil and gas lobby over the potential effect of the UIC regulations on the extraction of oil and gas that a separate class II well for oil and gas exploration and production was established. Brine waste is generated in the production of oil and gas when water is injected into the ground to enhance production in depleted reservoirs. Permits may not have to be obtained for individual wells, but may be obtained for entire fields. The oil and gas lobby was so strong that Section 1425 was added to the SDWA in 1980 to allow an optional approach to regulating underground injection associated with oil and gas production.

A substantial amount of site-specific geologic and construction data are required to obtain a permit for underground injection of hazardous waste. First, the injection of hazardous waste must be below the lowermost source of drinking water. Second, the injection must be into a formation that is suitably permeable and confined by impermeable layers. Third, the confining zone must be free from faults, fractures, and punctures (wells). Corrective action may be required to plug abandoned wells that could allow the upward migration of hazardous waste (40 CFR 146.7). Fourth, the well must exhibit mechanical integrity, which means no significant leak in the casing, tubing, or packer and no significant fluid movement into an underground source of drinking water through vertical movement adjacent to the well (40 CFR 146.8).

One major difference between the UIC program and the RCRA program is that ground water monitoring is not required for the UIC program. Instead, the UIC program depends on remote sensing of well operation and integrity analyses of the well to determine whether the fluid is actually entering the injection zone. In this manner, the UIC program differs substantially in philosophy and in specific regulatory provisions from the RCRA program.

Substantial disagreement exists among experts as to the desirability of underground injection of wastes. Underground injection generally does not require pretreatment and results in the long-term presence of hazardous waste and other wastes in the receiving formation. As such, underground injection is not considered destruction, but simply land disposal and storage. The continuation of the practice of underground injection was in question after the passage of the Hazardous and Solid Waste Amendments of 1984, which banned land disposal of hazardous waste. Underground injection is subject to the land ban, but EPA, particularly Region 6, has reviewed many injection wells and found that there is "no migration" of the waste for 10,000 yr, thereby triggering an exemption to the land ban. For more on the land ban, see the section on HSWA.

An interesting subpart of the underground disposal issue is the disposal of hazardous waste into caverns mined in salt domes. This salt dome storage and disposal involves discharge into a hole in the ground that is deeper than it is wide or long, thereby making the injection of waste into this hole an underground injection of fluids. The EPA has passed no separate regulation governing salt dome disposal and subsequent litigation has occurred over this issue.

The UIC program covers more than waste disposal. Many economic actions may threaten ground water through mining activities or nonwaste-related activities. To the extent that these activities involve injection of fluids through a well, they are regulated under the SDWA if underground sources of drinking water are potentially affected.

14.3 RESOURCE CONSERVATION AND RECOVERY ACT OF 1976

The Resource Conservation and Recovery Act (RCRA) was passed in 1976. RCRA is the centerpiece of U.S. efforts to protect ground water by regulating solid and hazardous waste. Although RCRA was passed in 1976, regulations implementing the hazardous-waste requirements of RCRA were not promulgated by the EPA until November 1980.

The practice of midnight dumping was prevalent prior to late 1980. This practice involved the disposal of hazardous waste in unpermitted places such as roadside ditches, vacant lots, and abandoned industrial sites. Prior to regulation, transporters often handled hazardous waste on a "turn-key" basis. Transporters would come to companies generating hazardous waste and provide a waste-disposal service that included transporting and disposing of the waste. Many companies never knew where the waste was taken, choosing to let the transporter control the process. Of course, many transporters chose the cheapest process possible.

A recurring theme of U.S. hazardous-waste laws is to make the generator of hazardous waste responsible for the ultimate fate of that waste. As will be seen, RCRA set in motion a comprehensive system for tracking hazardous waste. The point here, however, is that a uniform system of regulation did not exist prior to late 1980, and many ground water problems were in existence at the time RCRA became effective. In all these ground water protection statutes and their amendments, Congress is continually trying to bring all ground water contamination sources under either the RCRA or CERCLA programs.

The Resource Conservation and Recovery Act forever changed the solid-waste disposal practices of the United States. RCRA is an extremely powerful piece of legislation. However, it is important to note that it is part of a comprehensive approach to solid-waste management with references to solid waste in the definitions and provisions of RCRA. The regulation of municipal solid-waste disposal is undertaken pursuant to subtitle D of RCRA, whereas the disposal of hazardous waste is regulated under subtitle C of RCRA. For purposes of this subpart, the manifest system of RCRA will be discussed first, followed by a discussion of hazardous-waste permitting under RCRA and then permitting of sanitary landfills under subtitle C. Subtitle D will not be discussed.

RCRA has been amended several times since its initial passage. The most far-reaching of these amendments have come from the Hazardous and Solid Waste Amendments (HSWA) of 1984. To the extent possible, the HSWA changes will be treated in a separate section because these amendments significantly altered the requirements for hazardous-waste disposal in the United States. In some instances, however, the HSWA changes will be mentioned in the discussion of RCRA. All these changes have a significant impact on what is legal at what point in time. Stated otherwise, action that was legal in 1990 may not be legal in 1994 as new provisions become applicable.

Goals and Objectives of RCRA

The RCRA is a far-reaching act that does more than control the handling and disposal of hazardous waste. Section 1003(b) states:

> (b) NATIONAL POLICY.—The Congress hereby declares it to be the national policy of the United States that, wherever feasible, the generation of hazardous waste is to be reduced or eliminated as expeditiously as possible. Waste that is nevertheless generated should be treated, stored, or disposed of so as to minimize the present and future threat to human health and the environment.

The objectives of RCRA are set forth in Section 1003(a) and include the following:

(a) OBJECTIVES. The objectives of this Act are to promote the protection of health and the environment and to conserve valuable material and energy resources by—

(3) prohibiting future open dumping on the land and requiring the conversion of open dumps to facilities which do not pose a danger to the environment or to health;

(4) assuring that hazardous waste management practices are conducted in a manner which protects human health and the environment;

(5) requiring that hazardous waste be properly managed in the first instance thereby reducing the need for corrective action at a future date;

(6) minimizing the generation of hazardous waste and the land disposal of hazardous waste by encouraging process substitution, materials recovery, properly conducted recycling and reuse, and treatment;

Additionally, Congress made specific findings with regard to solid waste, environment and health, materials, and energy in Section 1002 of the act. The bottom line is that the generation of hazardous waste is to be reduced over time, proper recycling is encouraged, and destruction and detoxification of hazardous waste are encouraged. The details of the realization of these lofty goals are contained in the specific requirements of the various programs.

RCRA Manifest Program

The structure of RCRA is that standards are established for hazardous-waste generators, hazardous-waste transporters, and facilities that treat, store, or dispose of hazardous waste (TSD facilities). Section 3002(a)(5) of "RCRA requires that a manifest system be established" to assure that all such hazardous waste generated is designated for treatment, storage, or disposal in, and arrives at treatment, storage, and disposal facilities . . . for which a permit has been issued. . . ." In other words, the manifest program is to track the hazardous waste from the generator through the transporter to the treatment, storage, and disposal facility.

Essentially, the manifest is a set of papers. The intent of the manifest program is to create a "paper trail" to follow the waste from "cradle to grave." The generator initiates the manifest and gives it to a transporter who must follow the manifest's instructions on the delivery of the hazardous waste to a TSD facility. The generator retains a copy of the manifest when the waste is picked up and gives the original to the transporter. The transporter delivers the waste to the site identified on the manifest and passes on the manifest, retaining a copy for the records. The TSD facility must be permitted to receive the waste. Upon receipt of the waste, the TSD facility retains a copy and returns the manifest to the generator, thereby completing the cycle. If the manifest is not returned within 35 days, the generator is responsible for finding the missing hazardous waste and must submit an exception report to EPA if the waste is not found within 45 days.

RCRA and the manifest program have divided the world of hazardous waste into parts. There are generators, transporters, and TSD facilities. Each generator and transporter has an identification number and each TSD facility must have a permit. RCRA is clear that the generator is responsible for determining where the waste is sent, thereby removing transporters and disposers from their pre-1980 turn-key role. Today, a prudent generator will perform an extensive investigation of the disposal company taking the waste to ensure that cleanup liability and environmental damage liability will not be realized under other statutes such as CERCLA.

Generator Responsibility

It is the responsibility of the generator to determine if its waste is hazardous. The definition of hazardous waste under RCRA is very complex and full of loopholes. A hazardous waste is a solid waste that may cause or increase mortality or serious irreversible or incapacitating reversible illness or pose a substantial present or potential hazard to humans or the environment when improperly treated, stored, transported, or disposed of or otherwise managed [RCRA section 1004(5)].

Some wastes may be excluded from being hazardous by definition and many are, including waste from oil and gas exploration and production. If a waste is not excluded, it may be listed as a hazardous waste by source or by name. If a waste is not excluded but not listed, it must be tested to determine if it is (1) ignitable, (2) reactive, (3) corrosive, or (4) toxic. If the waste meets any one of these four tests, it is a RCRA hazardous waste.

A manifest is not required for all hazardous-waste disposal. If a generator is disposing of hazardous waste on site, no manifest is required. If less than 100 kg of hazardous waste is generated per month and that waste is not acutely hazardous, no manifest is required and that waste may be disposed of in facilities not permitted for hazardous-waste disposal. Similarly, in some situations, hazardous wastes that are being recycled may be excluded from the manifest requirements. Otherwise, the generator is required to initiate a manifest.

In addition to manifesting, the generator must prepare the hazardous waste for shipment. The waste must be packaged, labeled, marked, and/or placarded prior to shipment so that the hazardous waste will be properly identified. A generator must not accumulate hazardous waste on site in too great a quantity or awaiting shipment for too long a time. Otherwise, the generator may be deemed to be a storage facility subject to permitting requirements. A generator normally is not required to obtain a hazardous-waste permit. Finally, extensive record keeping is required of the generator, and a biennial report must be submitted to the EPA.

Throughout the requirements applicable to generators, provisions of the regulations inquire as to the waste reduction accomplished by the generator. For example, the biennial report requests a comparison of the volume of hazardous waste generated in the prior reporting period to the volume generated in the current reporting period. Such provisions press the generator to reduce the volume of hazardous waste.

On the other hand, regulatory changes may substantially increase the amount of hazardous waste. For example, the EPA adopted a new procedure for testing to comply with the toxicity requirement in 1990. The toxicity characteristic leaching procedure (TCLP) increased substantially the volume of hazardous waste generated in the United States and made many facilities RCRA hazardous-waste generators overnight.

Transporter Requirements

The requirements applicable to transporters are very straightforward. Transporters must have an EPA identification number and must follow the Department of Transportation regulations for transporting hazardous materials as set forth in 49 CFR Subchapter C. Transporters must sign for the hazardous waste when they pick it up and must follow the instructions of the manifest. It is illegal for a transporter to pick up hazardous waste without a manifest unless the generator generates less than 100 kg/month. Under the RCRA scheme, the transporter simply provides a service of transportation and is removed from major decision making with regard to the place and type of disposal activity.

TSD Facility Requirements

All TSD facilities must have a permit in order to receive manifested hazardous wastes. The receiving facility must verify that the amount and type of waste received match the amount and type of waste manifested. The receiving facility must sign the manifest, leaving a signed copy with the transporter, retaining a signed copy for its own records, and sending the original back to the generator within 30 days. The owner or operator of a TSD facility must retain manifests for 3 years and must keep an operating record identifying the disposition of each waste shipment. A biennial report is required from each TSD facility.

A waste analysis plan must be developed by TSD facilities to ensure that the waste delivered indeed matches that which was manifested [40 CFR 265.13(b)]. An attempt must be made to reconcile manifest discrepancies. If the discrepancy is not resolved within 15 days, the TSD facility must report the discrepancy to the regional administrator of the EPA and identify the manifest at issue and attempts to resolve the discrepancy.

The manifest program was intended to make illegal disposal of hazardous waste very difficult. As will be seen in the next section, RCRA imposed very strict standards on facilities that treat, store, and dispose of hazardous waste. However, if hazardous waste is simply thrown in an abandoned pit or upon the side of the road, the most stringent permitting program will fail to protect ground water. Therefore, the manifest was viewed as step 1 in the national strategy under RCRA to protect ground water.

RCRA Hazardous Waste Permitting Program

Subtitle C of RCRA created a program for the permitting of facilities that treat, store, or dispose of hazardous waste in Section 3005(a). Under this section, facilities in existence on November 19, 1980, were to be treated differently from facilities constructed after that date. Facilities in existence as of November 19, 1980, were eligible for "interim status," which allowed these facilities to continue operation until a permit application could be filed and a final permit issued. Therefore, a distinction exists between interim-status facilities and final permitted facilities or new facility permits.

The rationale for this distinction is valid and created one of the most interesting of all permitting systems in federal environmental law. Unlike air and water pollution, hazardous waste deposited into the ground is not going to disappear if the dumping activity ceases. Indeed, the United States is full of sites that continue to contaminate ground water decades after the disposal activity ceased. On the other hand, the EPA had no records of the location of hazardous-waste disposal activities around the country in the late 1970s. For these two reasons, existing facilities that were storing, treating, or disposing of hazardous waste were given interim status if they would identify themselves and adhere to minimal regulations.

Essentially, interim status allowed existing sites to continue operation until the magnitude of the contamination problem was assessed and the safety of the operations evaluated. Interim status also gave the United States time to develop other technologies for hazardous-waste disposal. In the sections that follow, interim status is described first, followed by final permitting.

Interim Status

As one seasoned professional in the petrochemical industry once said, "Interim status is the easiest permit any industry ever got." And it is true. To claim interim status for a TSD facility, the hazardous-waste activities occurring at the time of the filing were identified. The applicant had to file a Part A form and was treated as if a permit had been issued. Certain rules and regulations had to be followed and data collected. But, certainly, compared to wastewater discharge and air emission permitting, interim status for hazardous-waste disposal was easily obtained.

However, the RCRA permitting concept essentially turned the traditional permitting concepts around. Instead of environmental controls being required when the permit is issued, interim status simply brings the industry into the permitting program. The important environmental controls occur when the interim-status facility moves to final permitted status after several years of ground water data collection. Stated otherwise, the controls are on the back end of the RCRA interim status program, not the front end.

All interim status facilities had to file a Part B application to matriculate from interim status to final permitted status. Most of these Part B applications were filed between 1985 and 1987. To receive a final permit, very stringent environmental performance standards have to be met. A large number of interim-status facilities, perhaps as much as 75%, are not safe enough to be granted final permits. These facilities must undertake closure activities. A significant amount of environmental cleanup occurs at the closure stage.

Interim-status Requirements

Once a TSD facility qualifies for interim status, the EPA regulations (40 CFR Part 265) become applicable. These regulations apply to treatment, storage, and disposal facilities and contain specific requirements with regard to activities such as tanks, surface impoundments, waste piles, land treatment, landfills, incinerators, thermal treatment, and chemical, physical, and biological treatment.

Interim status was granted to all TSD facilities that filed a Part A application without regard to how well they were designed or whether they were contaminating the ground water. The regulatory presumption was, however, that many of these facilities would have to be shut down over time. Therefore, the regulatory system attempted to anticipate the cost of cleaning up these facilities and removing the contamination. This led to the development of ground water monitoring and detection requirements, as well as closure and postclosure care requirements.

As a general proposition, all TSD facilities must comply with general facility standards that require trained personnel, facility security, waste analysis, and separation of various incompatible wastes. These facility standards are an attempt to prevent leakage and accidental releases that will contaminate the soil and ground water and to protect human health and the environment generally. All TSD facilities are required to meet certain preparedness and prevention requirements relating to proper maintenance and access to alarms and other communications equipment. Each facility must have a contingency plan for responding to emergency conditions such as fires or explosions, and a copy of the contingency plan must be maintained at the facility. Emergency procedures are therefore identified in advance and ready to be activated.

Each facility claiming interim status must develop a plan for shutting down the facility (a closure plan) and a plan for monitoring the closed facility for 30 years after closure (a postclosure care plan). A written closure plan had to be developed by May 1981 for facilities granted interim status in November 1980, and that closure plan must be amended if operating conditions change. Closure is essentially the steps necessary to isolate or decontaminate the hazardous portions of the TSD site (40 CFR 265.111). As part of the closure plan, the expected cost of closure must be estimated (40 CFR 265.142). Similarly, a postclosure care plan must be developed that identifies the monitoring and maintenance activities that will be undertaken to ensure the integrity of the closed facility (40 CFR 265.118). Then, financial assurances must be executed by the TSD facility to ensure that the estimated cost of closure and postclosure care will be available when the facility must be closed.

In summary, interim status allowed existing hazardous-waste treatment, storage, and disposal operations to proceed while putting in place safety and emergency and handling procedures. Interim status was granted with the full intention of many facilities being shut down, and plans and money were required to be put in place at the beginning of interim status. However, the most important process of interim status was the ground water monitoring and protection requirements.

Ground Water Requirements

If land disposal alternatives such as surface impoundments, landfills, or land treatment facilities were used to manage hazardous waste and were granted interim status, then ground water monitoring was required to be implemented within 1 year. This ground water monitoring program (40 CFR 265.90-94) must be carried

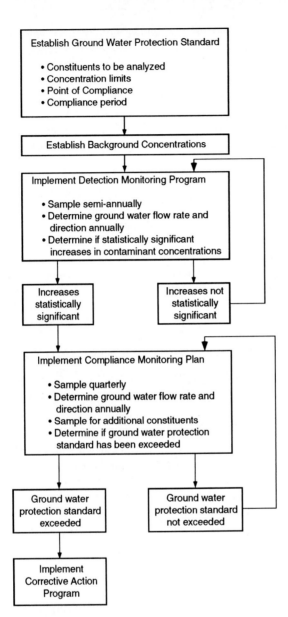

Figure 14.1 Requirements for a RCRA ground water monitoring program.

out during the life of the facility and during the postclosure care period for these disposal facilities.

The requirements for a RCRA ground water monitoring program are summarized in Figure 14.1. An upgradient well is required to test the uppermost aquifer for background levels that will be used for comparison purposes with downgradient wells. Downgradient, at least three wells are required, although more may be necessary to immediately detect any statistically significant concentrations of hazardous waste or hazardous-waste constituents in the uppermost aquifer (40

CFR 265.91). The number, location, and depths of wells will vary to reflect the geometric and geologic complexity of the site.

If there are multiple land disposal units on site, then the waste-management area must be adequately monitored. The boundaries of the waste-management area are determined on a case-by-case basis but, once determined, at least one upgradient and at least three downgradient wells must be present at the boundaries of the waste-management area to immediately detect leakage of contaminants. All ground water monitoring wells must be cased to maintain the integrity of the monitoring-well bore hole.

Once these ground water monitoring wells are completed, the owner or operator must develop and follow a ground water sampling plan. This plan will identify how samples are collected, preserved and shipped, how the samples are analyzed, and the chain of custody of the samples (40 CFR 265.92). These wells must be sampled for EPA interim drinking-water standards in Appendix III, as well as for parameters of ground water quality such as chlorides, iron, manganese, phenols, sodium, and sulfate. Furthermore, monitoring is required of ground water contamination indicators such as pH, specific conductance, total organic carbon (TOC), and total organic halogen (TOH). Background levels must be established for each of these parameters. The sampling frequency is more intense in the first year, with quality parameters being measured annually thereafter and contamination indicators being measured at least semiannually thereafter. The ground water elevation must be determined each time the well is sampled.

The owner or operator must prepare a ground water-quality assessment program that includes a comprehensive ground water monitoring program. This ground water-quality assessment program must be capable of determining whether hazardous constituents have entered the ground water, the rate and extent of contaminant migration, and the concentrations of hazardous waste or hazardous-waste constituents in the ground water. Ground water contaminant indicators of pH, specific conductance, TOC, and TOH must be measured by the Students' t-test to determine whether statistically significant increases have occurred over the monitored background (upgradient) concentrations.

Under some circumstances, upgradient well concentrations may significantly increase. In this case, the information must be reported to the regional administrator as part of the annual report required to be filed. Consequently, it is often determined that the upgradient well is indeed not upgradient, and a new upgradient well may need to be identified.

Usually, however, the downgradient concentrations are those that show statistically significant increases. When a statistically significant increase is detected, the owner or operator must immediately resample to make sure that laboratory error was not involved. Unless laboratory error is found, the statistically significant increase must be immediately reported to the regional administrator of the EPA.

Within 15 days of notifying the EPA, the facility must develop a plan certified by a qualified geologist or geotechnical engineer for a ground water assessment program. This assessment plan must include number, location, and depth of wells, including the development of new wells. Sampling must be increased to include all hazardous waste and hazardous-waste constituents at the facility. Furthermore, evaluation procedures must be specified and a schedule set forth identifying the implementation of the program over time.

Upon development, the new ground water assessment plan must be implemented. The goal of this assessment plan is to determine the concentration and rate and extent of migration of hazardous waste and hazardous-waste constituents in the ground water. These determinations must be made as soon as technically feasible and must be reported to the regional administrator within 15 days.

If the ground water assessment indicates that there really is no contamination, then the owner or operator may continue indicator monitoring and disband the more detailed ground water assessment program. However, if the presence of contamination is confirmed, the owner or operator must continue to undertake quarterly monitoring until the facility is closed. The results of the various monitoring steps must be reported to the regional administrator on an annual basis.

The importance of ground water monitoring to the RCRA scheme cannot be overemphasized. Interim-status facilities were allowed to continue in operation but only long enough to determine the ground water problems and issues associated with the facility. If the facility has no major ground water problems, the facility may be granted final permitted status. If, however, the ground water contamination is severe enough, the owner or operator may have to close the facility and undertake corrective action to remediate the ground water contamination.

Note, however, that EPA regulations do not require immediate cleanup or corrective action in the case where contamination of the ground water is found. As long as the facility is active, ground water contamination can be identified, studied, and evaluated for a substantial amount of time. Indeed, a shortcoming of the RCRA structure is that the ground water analysis may take an extremely long time to be completed, with substantial discretionary authority being given to the agency to allow continued study prior to action.

Closure and Postclosure Care Requirements

Perhaps no provisions distinguish RCRA from other statutes so much as the requirements for closure, postclosure care, and corrective action (Figure 14.2). These requirements reflect the fact that contamination of soil and ground water is not dissipated as is air and surface-water pollution. Even though the hazardous waste storage, treatment, and disposal may be completed, the impacts of that activity remain after cessation of the TSD activity.

Closure plans are required of all facilities having interim status. A three-part performance standard is established for closure. First, all facilities must be closed in a manner that minimizes the need for further maintenance. Second, facilities must be closed in a manner that controls, minimizes, or eliminates—to the extent of protecting human health and the environment—postclosure escape of hazardous waste, hazardous constituents, leachate, contaminated runoff, or hazardous-waste decomposition products into the ground or surface waters or into the air. Third, the closure must meet requirements specific to various types of facilities.

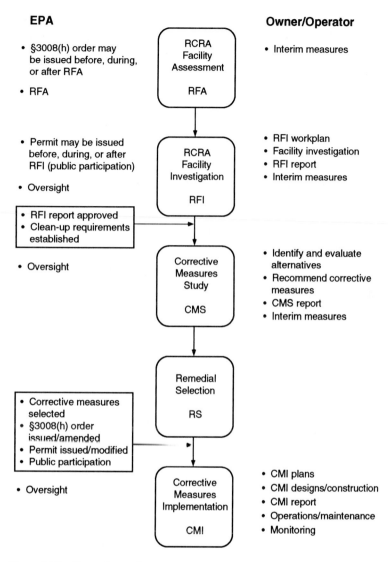

EPA

- §3008(h) order may be issued before, during, or after RFA
- RFA

- Permit may be issued before, during, or after RFI (public participation)
- Oversight

- RFI report approved
- Clean-up requirements established

- Oversight

- Corrective measures selected
- §3008(h) order issued/amended
- Permit issued/modified
- Public participation

- Oversight

Owner/Operator

- Interim measures

- RFI workplan
- Facility investigation
- RFI report
- Interim measures

- Identify and evaluate alternatives
- Recommend corrective measures
- CMS report
- Interim measures

- CMI plans
- CMI designs/construction
- CMI report
- Operations/maintenance
- Monitoring

RCRA Facility Assessment — RFA

RCRA Facility Investigation — RFI

Corrective Measures Study — CMS

Remedial Selection — RS

Corrective Measures Implementation — CMI

Figure 14.2 Corrective action process.

As can be seen, the key terms are controls, minimizes, or eliminates. The closure requirement, therefore, will vary depending on the facts of a particular situation.

A closure plan must be prepared that identifies how long the facility will continue to operate and how the waste on site will be handled after closure. Obviously, waste will have to be removed from treatment and storage facilities, and the closure plan requires identification of how such removal is to be accomplished. In the case of land disposal facilities, waste may be left on site and contained or removed from the site. These plans are to be maintained during the life of the interim-status facility and shown to the regional administrator on request. If changes occur in the operation of the facility, the closure plan must be amended.

One hundred-eighty days prior to closure of a land disposal facility and 45 days before the closure of other facilities, the owner or operator must submit the closure plan to the regional administrator, who will provide the opportunity for comments on the closure plan and may conduct a public hearing on the closure. The regional administrator is required to approve, modify, or disapprove the plan within 90 days.

No regulations exist that specify the criteria for approval or disapproval of a closure plan other than that the closure performance standard that requires protection of public health and the environment must be met.

Permitted Facility Requirements

Interim-status facilities matriculate to final permitted status over time. Initially, the time for submitting a Part B application for final permitted status was unclear. However, the Hazardous and Solid Waste Amendments of 1984 specified that interim status would expire for land disposal facilities that failed to submit Part B applications within 12 months of the passage of HSWA. In addition to submitting a Part B application, the land disposal facilities had to certify that they were in compliance with the ground water monitoring and financial responsibility requirements of interim-status regulations.

Once the Part B application has been submitted, the agency conducts a review to determine if the final RCRA permit should be issued or denied. If issued, the detailed permit will identify the terms and conditions under which that facility must operate.

An important part of the permit process is the RCRA facility assessment (RFA). The RFA is a study of existing (for example, interim status) facilities to determine the status of their RCRA compliance, including the results of their ground water monitoring and evaluation. If substantial ground water contamination exists, a major hurdle will exist in moving to final permitted status.

The draft permit is developed during the permit review process. This permit contains substantial detail and includes specific elements of ground water monitoring and analysis, as well as closure and postclosure care. Once this permit is issued, it is legally binding on the facility and the permittee. If the terms and conditions of this permit are violated, the permittee may be civilly or criminally prosecuted.

General standards exist that are applicable to facilities that have been permitted to store, treat, and dispose of hazardous waste (40 CFR Part 264). The facility must comply with these standards as well as the requirements of the permit. These standards include requirements for good housekeeping to minimize the potential for ground water contamination, as well as specific aspects associated with closure and postclosure care. Specific standards are identified for various types of disposal activities, including incinerators (40 CFR Part 264, Subpart O) and landfills (40 CFR Part 264, Subpart N), as well as several other storage or disposal alternatives.

These standards include design as well as operating requirements. For example, in the landfill section, elements associated with landfill design are specified to protect the ground water. Here, a liner is required that has been designed to prevent migration of the waste from the landfill to the surrounding soil, ground water, and surface water. This liner system is required to contain a top liner, a composite bottom liner, a leachate collection and removal system, and a leak-detection system (40 CFR 264.301). The permit will contain provisions to implement these and other requirements.

Detailed in the section discussing the Hazardous and Solid Waste Amendments of 1984, many disposal options have been eliminated or substantially restricted by the ban on land disposal of hazardous wastes. Therefore, a significant portion of the interim-status facilities never received final permitted status and instead went straight to closure and postclosure care. Also, many interim-status disposal facilities identified ground water contamination and went to closure, postclosure care, and remediation.

Facility standards have been adopted by most states. In such a situation, RCRA provides that the federal program may be delegated to the state for implementation if the state program is "consistent" with the federal program (40 CFR 271.4). Therefore, as a practical matter, much of the actual business of protecting ground water under RCRA is undertaken by state agencies implementing a state program that is consistent with the RCRA requirements and EPA guidelines, rather than by the EPA itself.

14.4 HAZARDOUS AND SOLID WASTE AMENDMENTS OF 1984 (HSWA)

The EPA developed serious personnel problems between 1981 and 1984. Not only were budget and staff drastically reduced amid political upheaval, but questionable environmental decisions were also being made. When the EPA decided to allow the disposal of free liquid hazardous waste in land disposal facilities, Congress stepped in. The nation's legislature no longer trusted the EPA to implement the general policy directives of RCRA. Therefore, Congress signaled a major change of national environmental policy with the passage of the Hazardous and Solid Waste Amendments of 1984, which took policy control over hazardous waste back from the executive branch.

The changes of HSWA were swift and far-reaching. First, a ban on the land disposal of hazardous waste was implemented. Second, the small-generator exemption was substantially reduced. Third, underground storage tanks became regulated. And, fourth, an overlooked area of hazardous-waste disposal, solid-waste management units or SWMUs ("smoos"), became regulated.

The Land Ban

Section 101(a)(7) of HSWA created a new section 1002(b)(7) of RCRA, which states:

(b) ENVIRONMENT AND HEALTH.—The Congress finds with respect to the environment and health, that:

(7) certain classes of land disposal facilities are not capable of assuring long-term containment of certain hazardous wastes, and to avoid substantial risk to human health and the environment, reliance on land disposal should be minimized or eliminated, and land disposal, particularly landfill and surface impoundment, should be the least favored method for managing hazardous wastes.

To implement this congressional finding, certain prohibitions on land disposal were enacted by Congress. Over time, successive prohibitions would apply. Land disposal was defined for purposes of the land ban in a new subsection 3004(k) to include "any placement of such hazardous waste in a landfill, surface impoundment, waste pile, injection well, land treatment facility, salt dome formation, salt bed formation or underground mine or cave."

Initially, the disposal of free liquids into salt domes was prohibited. Six months after the date of enactment of HSWA, a prohibition against the disposal of bulk or noncontainerized liquid hazardous waste into landfills took effect. More generally, the placement of any noncontainerized liquids into a hazardous-waste landfill was prohibited in section 3004(c)(3), as amended.

The prohibition of land disposal of hazardous liquids was a priority of Congress due to the high potential for ground water contamination associated with free liquid hazardous waste. The HSWA land ban also included a prohibition of various types of land disposal activities at varying time increments.

For example, the land disposal of certain types of solvents and dioxin-containing material was banned 24 months after the passage of HSWA (except for deep well injection). The land disposal of certain heavy metals, liquid hazardous waste with a very low pH, liquid hazardous waste containing polychlorinated biphenyls greater than 50 ppm, and organic halogenated compounds greater than 1000 mg/kg was banned 32 months after passage of HSWA.

All other types of hazardous waste proposed for land disposal were to be analyzed and evaluated by the EPA administrator, with certain of the wastes to be analyzed within 48 months of the passage of HSWA and all wastes to be analyzed within 66 months of the passage of HSWA. This requirement came to be divided into thirds, with the first third of EPA hazardous wastes evaluated in 48 months, the second third evaluated within 55 months, and the third evaluated within 66 months.

The goal of this evaluation was to determine if one or more types of land disposal should be banned. If the EPA failed to act within a specified time frame, the ban was automatically imposed. This is the "hammer provision." If EPA failed to act (which was common during the Reagan administration), the prohibition would automatically take place.

This land ban was not absolute. In many cases, treatment could be undertaken that would alter the hazardous waste to such an extent that land disposal was allowed. The goal of this new section 3004(m) was to either substantially diminish the toxicity of the waste or substantially reduce the likelihood of migration to drinking-water sources. Subsequent EPA rules have identified some treatment standards that allow land disposal, but these requirements can be difficult and expensive in certain situations and unavailable in others. Rules concerning each waste type must be consulted to determine the exact situation with regard to treatment prior to land disposal.

By adopting the land ban, Congress rejected land disposal and endorsed virtually all other concepts of waste disposal. Only if a "no-migration" petition was granted could land disposal without pretreatment occur. This petition must show that the disposal concept will be "protective of human health and the environment for as long as the waste remains hazardous." The petitioner therefore must demonstrate to the administrator with a reasonable degree of certainty that there will be no migration of hazardous constituents from the disposal unit or injection zone for as long as the waste remains hazardous (for example, 10,000 years). Although such a finding may seem to be impossible to uphold, the EPA has granted several no-migration petitions for underground injection.

The net result of the imposition of the land ban was to shift national disposal preferences to technologies other than land disposal. Almost immediately upon passage, many generators and disposers prepared applications for incinerators. Indeed, incineration appeared to be the technology favored by Congress in passing HSWA. Unfortunately, cost considerations kept more innovative treatment and disposal concepts from coming to the forefront. In many respects, the implementation of the land ban transferred our hazardous-waste disposal problems from ground water to the air.

Small-quantity Generators

Since the inception of the RCRA program, a distinction was made between generators of hazardous waste on the basis of volume. Utilizing discretionary authority, the EPA's initial RCRA regulations separated generators of over 1000 kg of hazardous waste a month from those generating less than 1000 kg/month. This division regulated 60,000 generators accounting for a large majority of the volume of hazardous waste in the country.

Small-quantity generators are not required to manifest their waste and are not required to dispose of their hazardous waste in permitted hazardous-waste disposal facilities. Instead, this waste may be deposited in municipal landfills that are not designed to contain hazardous waste. This small-quantity-generator provision has been controversial because it allows hazardous waste to be disposed in facilities where ground water contamination may very likely result. In the 1984 HSWA, Congress changed the small-quantity-generator exemption to lower the volume requirements. By March 1986, the EPA was mandated to promulgate regulations for facilities generating more than 100 kg/month but less than 1000 kg/month. These regulations could vary from those required for generators of 1000 kg/month, but at the least had to include manifesting and disposal in permitted hazardous-waste facilities. With a few minor differences, the regulations for the approximately 130,000 generators of from 100 to 1000 kg/month of hazardous waste are the same as those for the greater than 1000 kg/month generators.

Solid-waste Management Units

A major loophole was closed by HSWA through the incorporation of provisions addressing "solid-waste management units" or SWMUs. SWMUs are defined as currently inactive but formerly used hazardous-waste disposal sites that are within the boundaries of a RCRA-permitted facility.

For example, assume that a site within a refinery had been used from 1975 to 1980 for hazardous-waste disposal, but usage ceased when a new land disposal facility was created for submission for interim status in 1980. According to EPA regulations, the owner or operator of a RCRA permitted (interim status) facility only had to comply with ground water monitoring and corrective action requirements for units that received hazardous waste after January 26, 1983. No other regulatory program of either RCRA or CERCLA addressed this loophole.

HSWA changed this loophole by adding a section to RCRA which requires that any permit issued after the date of HSWA enactment establish a requirement for undertaking corrective action for all release of hazardous waste or constituents from solid-waste management units at a storage treatment or disposal facility regardless of when that release occurred. An additional section extends the requirement for corrective action outside the boundaries of the storage, treatment, or disposal facility unless the owner/operator can demonstrate that permission to take corrective action on adjacent property was sought and denied.

The inclusion of SWMUs within the RCRA regulatory program substantially expands the scope of the RCRA program. The program's focus changed dramatically from manifesting and permitting to include remediation of old disposal sites. Furthermore, the SWMU concept goes far beyond disposal to include sites where hazardous waste or hazardous constituents have been released to the land surface. In a major industrial facility, it would not be uncommon to have one RCRA land disposal site, a RCRA incinerator, several RCRA storage facilities, and 40 to 80 solid-waste management units.

From a regulatory standpoint, the major requirement for SWMUs is that they be identified and evaluated to determine the type of waste contained therein. Furthermore, a determination must be made as to whether the waste has contaminated the ground water. These efforts require testing and ground water monitoring. Once the leakage has been detected, the extent of the leakage must be characterized and a corrective action plan developed. Then corrective action must be undertaken, ultimately resulting in a cleanup of the SWMU.

The remediation of SWMUs represented a large volume of the hazardous-waste consulting business in the 1990s. Extensive ground water monitoring and analysis are required in the determination of the appropriate corrective action. And, generally, the remediation alternative will be to some extent dictated by the type of hazardous-waste options available within the facility.

Leaking Underground Storage Tanks

The last major program initiated by HSWA involved the regulation of leaking underground storage tanks. It is important to note that this program regulates hazardous product, not waste, thereby differentiating it from other RCRA programs.

Underground storage tanks containing hazardous waste were subject to regulation under RCRA. However, the estimated 2.8 to 5 million underground storage tanks containing hazardous products such as gasoline were not. Of these, as many as 450,000 were estimated to be leaking by 1989.

This underground storage tank regulatory program unfolded between 1984 and 1986 when the EPA promulgated implementing regulations. First, each state had to designate an agency to implement the program. Second, each tank owner had to identify the existence of new or old tanks removed from operation after 1973 to the state agency by May 8, 1986. Then EPA promulgated regulations that set forth several steps that had to be followed for regulated storage tanks.

These EPA regulations required that regulated tank owners test the integrity of their tanks. The existence of any releases had to be reported and corrective action had to be undertaken in response to releases. In some cases, the sites had to be closed in accordance with general closure requirements.

In essence, the site of an underground storage tank became regulated in a manner similar to a SWMU or even an existing disposal site. The leakage of contaminants had to be identified, the extent of the damage to soil and ground water had to be assessed, and the problem had to be remediated.

Additionally, the EPA was required to promulgate standards for new underground storage tank construction under section 9003(e) of RCRA. These "New Tank Performance Standards" shall include requirements for "design, construction installation, release detection and compatibility standards." In the resulting regulations, EPA opted for noncorroding tank shells and/or leachate collection systems, making the design of underground storage tanks similar in many ways to landfill design requirements.

Conclusion

HSWA forever changed the relationship between Congress and the EPA. Not only did Congress reassert control over the nation's hazardous waste program, it did so decisively. The EPA was forced into responding to a series of HSWA deadlines. When the EPA failed to meet statutory deadlines, the regulatory hammer fell, eliminating the activity. If a mistake was to be made, Congress had decided the mistake would be in regulating too much rather than too little.

HSWA also marked a new, get-tough attitude on underground contamination. Leak-prone land disposal methods were banned and old hazardous-waste disposal sites were brought into the corrective action and remediation program. Underground storage tanks were regulated, analyzed, and remediated. Congress had declared its clear desire to regulate and clean up ground water contamination with the passage of HSWA.

14.5 Comprehensive Environmental Response, Compensation, and Liability Act of 1980, As Amended by the Superfund Amendment and Reauthorization Act of 1986

The Comprehensive Environmental Response, Compensation, and Liability Act of 1980, known as CERCLA, was passed to provide the legal and regulatory basis to clean up releases of hazardous substances, as well as to introduce a concept of hazardous substance reporting. Section 101(14) of CERCLA defines hazardous substance as:

> (A) any substance designated pursuant to section 311(b)(2)(A) of the Federal Water Pollution Control Act, (B) any element, compound, mixture, solution or substance designated pursuant to section 102 of this Act, (C) any hazardous waste having the characteristics identified under or listed pursuant to Section 3001 of the Solid Waste Disposal Act. . . , (D) any toxic pollutant listed under section 307(a) of the Federal Water Pollution Control Act, (E) any hazardous air pollutant listed under section 112 of the Clean Air Act, and (F) any imminently hazardous chemical substance or mixture with respect to which the Administrator has taken action pursuant to Section 7 of the Toxic Substances Control Act. The term does not include petroleum, including crude oil or any fractions thereof which is not otherwise specifically listed or designated under subparagraphs (A) through (F) of this paragraph, and the term does not include natural gas, natural gas liquids, liquefied natural gas or synthetic gas usable for fuel (or mixtures of natural gas and such synthetic gas).

During the decades preceding the passage of RCRA and CERCLA, hazardous substances had been dumped around the United States in thousands of places. No permits were required in most cases. When permits were required, the state of the art was simply not sufficient to contain the waste and protect ground water. By the time CERCLA was passed, Congress knew that a major problem existed in the United States with regard to past disposal practices and ongoing releases.

The concept of cleanup that Congress adopted had several parts. First, CERCLA had a companion piece of legislation called the Superfund Tax Act. This act created an excise tax on oil and certain hazardous substances. This tax was paid into a fund that was initially bankrolled by Congress. This fund was called the Superfund and was to be used to study and clean up non-RCRA sites where hazardous substance releases were occurring. As the companion to the Superfund Tax Act, CERCLA provided controls on the use of Superfund monies. The EPA was required to establish rules for the expenditure of Superfund money and establish liability provisions to ensure that the parties responsible for the release ultimately paid for the cleanup. A reporting requirement was also created for past disposal operations and current releases.

From a conceptual standpoint, CERCLA is very different from RCRA and HSWA. Figure 14.3 provides a comparison. CERCLA is not a true regulatory act. Instead, it created a process for identifying releases and cleaning up sites that pose a hazard to health and the environment. No permits are required. No application is made. Instead, the EPA identifies the site and prepares a cleanup plan. When the cleanup has been completed, the potentially responsible parties (PRPs) are sued by the federal government to reimburse the Superfund for the money spent to clean up the problem that they created. In this respect, CERCLA is unique among environmental laws.

CERCLA is a harsh statute. It imposes statutory strict liability for cleanup costs upon generators, transporters, and owners responsible for releases. Conceptually, the statute can be viewed in three parts. The first part concerns the identification, analysis, and remediation of releases. The second part concerns the rules

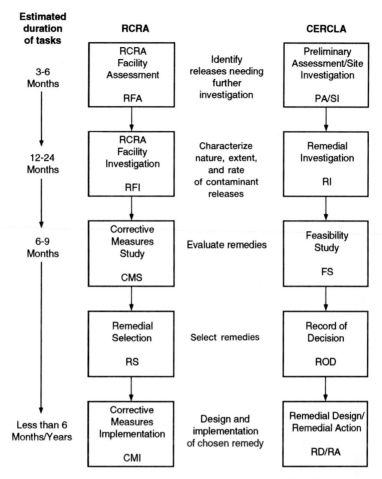

Figure 14.3 Comparison of RCRA corrective action and CERCLA remedial process.

of liability associated with the remediation of these releases. The third part addresses the more general reporting requirements.

The common thread through all parts of CERCLA is the focus on releases. Releases are to be reported and releases are to be remediated. As will be seen, the concept of release is very broad, covering an extremely wide range of actions. In virtually all cases, these releases either have affected or have the potential to affect ground water. For the most part, the release of greatest concern will involve the leaching of chemicals from an old disposal site into the ground water. For the student of ground water, knowledge of the CERCLA process is essential. This process is summarized in Figure 14.4.

It is also important to note that the Superfund process has a tremendous potential to anger and frighten the general public. Most Superfund sites were forgotten over the years, existing in various states of decay, disrepair, and anonymity. In many Superfund cases, people live in the vicinity of the sites. Children

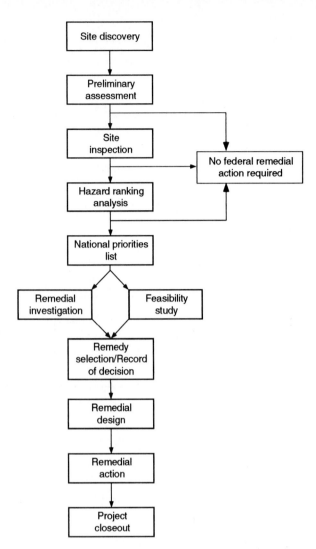

Figure 14.4 CERCLA remedial response process.

play on and around them. Nearby homes often rely upon ground water wells for drinking water. People in the area may have some vague knowledge about the hazard of a particular site but generally lack detailed information.

The Superfund process changes residents' perception of a release site. First the analysts in "moon suits" appear. Then the chain link fence is put up with the warning signs. The site becomes "listed" and appears on the front page of the newspaper. The EPA may conduct public hearings. A health risk study is conducted. And people are told how dangerous the site is.

In virtually all site remediations, liaison and communication with the public becomes more and more difficult. The neighbors change from being concerned, to being scared, to being angry and scared. They must decide whether to move out or not. Many discover that the process of Superfund listing has caused their property to become classified as undesirable. And, in fact, property values are negatively affected by adjacency to Superfund sites. Public involvement and participation are extremely important, yet extremely difficult, aspects of Superfund cleanup processes.

The Superfund process and its harsh liability provisions have forever changed the real estate industry in the United States. Because a property owner may be liable for the cleanup of a release under CERCLA, prospective property owners have become concerned about purchasing cleanup liability. Banks, savings and loans, and large institutional real estate investors are now on notice that the purchase of real estate may be accompanied by liability. The owners of shopping centers and commercial buildings are now on notice that they may be liable for the release of hazardous substances by their tenants.

This liability was initiated by the CERCLA provisions and enhanced and amplified by the Superfund Amendment and Reauthorization Act of 1986 (SARA), which reauthorized CERCLA. SARA created a defense for real estate purchasers if they have "no reason to know" about hazardous substances on the property. This "no reason to know" defense can be perfected by conducting an environmental audit of property prior to purchase. Such audits are now routinely required throughout the Untied States for property transactions and represent a major consulting practice for engineers and scientists around the country.

Finally, the 1986 SARA amendments created a new reporting requirement under the Title III provisions. The SARA Title III program required that certain industries report all their releases of hazardous substances, both permitted and unpermitted, to EPA. This report includes an inventory of hazardous substances in the wastewater, air emissions, underground injection, and land disposal. Never before had such a compilation been required. No legal provision stands for the power of information more than does SARA Title III.

Because of the concise reporting format, the SARA Title III information can be accessed and compared across industrial sectors and across the United States. Lists of the "Toxic 500" companies or of the most toxic counties in the United States were compiled from the data sets and became public. No other act created information more readily accessible for media use than did SARA Title III. And the results have been phenomenal. Substantial competition exists today among industries *not* to be number one, or even in the top fifty, of the "Toxic 500" companies. As a result, substantial waste reduction has occurred in many industries originally identified as the most toxic in the United States.

CERCLA and SARA have left a legacy that will affect the country for decades to come. Extensive liability has been brought to generators and owners. Remediation of releases will continue well into the 21st century. Reporting of hazardous-substance releases are pervasive. And the public is more concerned than ever about ground water contamination, Superfund site cleanup, and personal safety. Bankers are worried about their real estate loans and companies are scrambling to escape the "Toxic 500" list.

National Contingency Plan

The overall process for identifying and cleaning up Superfund sites is contained in a set of regulations (40 CFR Part 300) titled the National Contingency Plan. Essentially, these are criteria for placing sites on the Superfund list and studying, evaluating, and remediating the sites. No action may qualify for the use of Superfund monies unless the procedures outlined in the National Contingency Plan are followed.

National Priority List

40 CRF 300.425 sets out the process for establishing remedial priorities. The structure of CERCLA is such that the focus of the statute is on the reporting and remediation of releases of hazardous substances. Section 101(22) of CERCLA defines release as

> any spilling, leaking, pumping, pouring, emitting, emptying, discharging, injecting, escaping, leaching, dumping or disposing into the environment (including the abandonment or discarding of barrels, containers, and other closed receptacles containing any hazardous substance or pollutant or contaminant) . . . [with certain exceptions].

As a practical matter, hundreds of thousands of releases of hazardous substances throughout the United States are theoretically in competition for the Superfund remediation money and agency priority action. Congress needed a manner to discriminate among these candidate sites to ensure that the Superfund money was spent properly. Therefore, Congress directed the EPA to create a National Priorities List to guide the expenditure of Superfund monies.

Under Section 105(a)(8)(B) of CERCLA, Congress stated that the president of the United States "shall list as part of the [national contingency] plan national priorities among the known releases or threatened releases throughout the United States and shall revise the list no less often than annually." In Section 105(a)(8)(A) of CERCLA, Congress states:

> Criteria and priorities under this paragraph shall be based upon relative risk or danger to the public health or welfare or the environment . . . taking into account . . . the population at risk, the hazard potential of the hazardous substances at such facilities, the potential for contamination of drinking water supplies, the potential for direct human contact, the potential for destruction of sensitive ecosystems, the damage to natural resources which may affect the human food chain and which is associated with any release or threatened release, the contamination or potential contamination of the ambient air which is associated with the release or threatened release, State preparedness to assume State costs and responsibilities, and other appropriate factors.

Under the directive of this section, EPA has established a National Priorities List (NPL). Only the releases included on the NPL are eligible for Superfund-financed remedial activities. On the other hand, removal actions that are not financed by the fund are not limited to NPL sites.

The EPA has developed a methodology for determining eligibility for the NPL. This methodology is extremely complex, taking up more than 100 pages in the *Code of Federal Regulations* (1992). This Hazard Ranking System (HRS) is found in Appendix A to 40 CFR Part 300 and has resulted in the listing of 1072 NPL sites plus 116 NPL federal facility sites. These are the Superfund sites.

Releases are added to the NPL by action of the lead agency, which is usually the EPA or the state agency working with the EPA. It is the responsibility of the lead agency to apply the HRS methodology to a particular release and to submit the results of this analysis to EPA. If EPA concurs in the HRS scoring and if the HRS score is sufficiently high (for example, greater than 28.5), a site will be added to the NPL. The NPL will be updated annually and new sites must be published in the *Federal Register* for public comment prior to being added to the NPL. Releases may be deleted from the NPL where no further response is appropriate. A release may be deleted if (1) responsible parties have undertaken appropriate response action, (2) fund-financed response under CERCLA has been implemented and no additional action by responsible parties is appropriate, or (3) investigations indicate the release poses no significant threat to public health or the environment and remedial responses are not appropriate. Notices of intent to delete must also be published in the *Federal Register*.

Remedial Investigation and Feasibility Study Process

A tremendous amount of time and effort go into the analysis of a Superfund site and the selection of a cleanup remedy. EPA regulations set out procedures that must be followed to move forward with the cleanup. There are two distinct steps in the process.

First are the remedial investigations (RI). The purpose of the RI is to collect data to adequately characterize the site for the purpose of developing and evaluating effective remedial alternatives. Information regarding the risk to the public and the environment posed by a particular site are generated by the RI. The nature and character of the threat posed by the hazardous substances and the particular conditions of the release at the site will be determined in the RI.

The second step, called the feasibility study (FS), evaluates various cleanup alternatives to determine if they meet the goals of the EPA and the needs of the public. The purpose of the FS is to develop a series of alternative remedies so that the decision maker may choose the appropriate remedy for a particular site. The development of cleanup options is intended to be integrated with site-characterization work. A large number of options may be considered and screened to produce a manageable number of possibilities for detailed analysis. Within this process, interaction with the affected community must occur and, ultimately, a remedy must be selected. All this must be undertaken under the umbrella of the ultimate goal of adequately protecting the public and the environment.

The importance of this RI/FS process cannot be overstated. Remedies for various Superfund sites can be extremely costly, ranging well into the tens if not hundreds of millions of dollars. The future of the citizens living next to a facility may hinge on remedy selection as well. To what level will cleanup be effected? Will there be air pollution residuals? Will ground water be cleaned up or contained or left contaminated in place? All these issues and many more are decided in the RI/FS process leading to remedy selection.

EPA Program Goals

The purpose of the remedy selection process is to implement remedies that eliminate, reduce, or control risks to human health and the environment [40 CFR 300.430(a)(1)]. Risk analysis is a major aspect of the remedy selection process. The purpose of this process is not only to eliminate risks, but also to reduce and control risks. Therefore, the role of risk assessment and risk management is extremely important.

The EPA has a program goal that shapes the overall direction of the remedy selection process:

> The national goal of the remedy selection process is to select remedies that are protective of human health and the environment, that maintain protection over time and that minimize untreated waste. 40 CFR 300.430(a)(i).

To implement this program goal, the EPA has identified a number of expectations to be considered in developing the appropriate response alternatives:

1. Treatment is expected to be used to address the principal threats of a site where practicable, including particularly liquids, high concentration toxic areas, and highly mobile compounds.
2. Engineering controls such as containment are expected to be used for waste that poses a relatively low long-term threat or where treatment is impracticable.
3. Combinations of cleanup controls are expected to be used where principal threats may be addressed by treatment and lesser threats addressed by containment or institutional controls.
4. Institutional controls such as water use limitations and deed restrictions limiting property use are expected to be used in association with engineering controls and not as the sole remedy unless active measures are not determined not to be feasible in the remedy selection process.
5. Innovative technology (such as bioremediation) is expected to be used when it can be shown that it will result in equal or superior treatment, performs better from an environmental impact standpoint, or costs less for the same level of performance as other alternatives.
6. Ground water is expected to be returned to beneficial uses wherever practicable within site-specific reasonable time frames. When beneficial-use restoration is not practicable, EPA expects to prevent plume migration and exposure to the contaminated ground water and to evaluate further risk reduction.

From the foregoing goals and expectations, it is clear that absolute cleanup and zero risk are not requirements of the national contingency plan. Instead, the analysis and management of the risk are the critical elements. A clear bias exists for treatment rather than containment, although containment is acceptable in certain situations, as are land- and water-use controls. Therefore, the choice of remedies is highly dependent on the specifics of the site and the analysis of the risk.

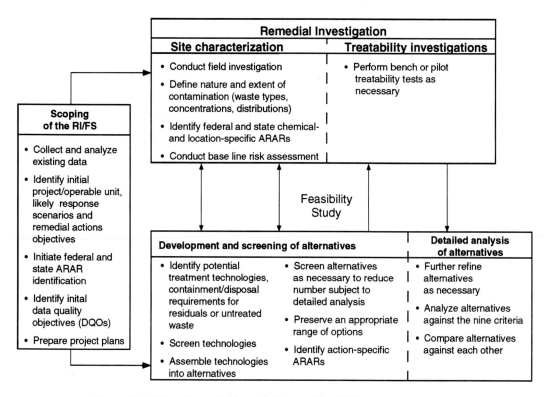

Figure 14.5 Phased remedial investigation and feasibility process.

Scoping

The first step in the RI/FS process (Figure 14.5) is to determine how to proceed with the study. This step is called scoping and sets the protocol for site investigation. Existing information is evaluated in the scoping step to determine the extent to which additional data must be collected. Of primary importance is the determination of future data-collection efforts to be undertaken in the RI step. The type of data collection and the quality and the quantity of data must all be determined in advance of the RI. This process is expected to result in a sampling and analysis plan with both field protocols and quality assurance/quality control (QA/QC) components. Also, the type of protective equipment necessary for workers needs to be determined at this juncture.

During scoping, a preliminary assessment is made of the range of potential cleanup alternatives to guide future deliberations. If natural resource damage has been identified or is anticipated, preliminary contact must be made with the state and federal resource agencies that are authorized to act as natural resource trustees to ensure protection of the natural environment in the remedy selection process. Finally, applicable or relevant and appropriate requirements (ARARs) must be determined. ARARs are regulatory requirements from other federal or state laws that apply to a particular site cleanup. ARARs have a major role in determining the acceptability of a particular remedial alternative. If no ARAR exists (or if the ARAR allows relative risk assessment), risk analysis is used to determine appropriate cleanup levels.

Community Relations

Interface with the affected community is initiated prior to any fieldwork being undertaken. Interviews should be conducted with public officials, community residents, and public-interest groups to discover their concerns and to set up a system for informing them of the progress of the RI/FS process. A community relations plan (CRP) is to be developed that identifies opportunities for citizens to participate in the decision-making process, and a local information repository must be set up so that the local affected public may review the documents utilized by the EPA.

Affected citizens are eligible for technical assistance grants (TAG) to assist them in understanding the documentation that will be developed during the Superfund process. As a general rule, these lay citizens will lack detailed technical knowledge. The TAG will provide monies so that they may hire consultants (but not lawyers) to help them understand the specifics of the Superfund process.

If the government proposes to settle its grievances with potentially responsible parties (PRPs) who may be liable for the cleanup, there is a requirement that the public be involved in that settlement. The National Contingency Plan gives citizens a role in the Superfund process. Citizens living near these sites will usually take this role seriously, and certain citizens will become very familiar with the documentation. In many cases, the EPA and the PRPs may agree with each other but disagree with the position of residents living near the Superfund site. This public participation aspect provides a challenge for the agency and PRPs along with the affected citizenry.

Source Control Actions

Source control actions are to be evaluated by the lead agency and include first and foremost a range of alternatives that utilize treatment to reduce the toxicity, mobility, or volume of hazardous substances. Here the goal is removal and/or destruction of the hazardous substance, thereby eliminating or minimizing the long-term need for management. The lead agency is authorized to consider a range from total treatment to partial treatment.

Where ground water response actions are required, a number of alternatives that produce site-specific remediation levels over different periods of time will be developed. Furthermore, the agency is required to develop one or more innovative options if such options can generate comparable or better results than existing alternatives.

The detailed analysis of alternatives is accomplished in a comparative manner by utilizing nine evaluation criteria:

1. Overall protection of human health and the environment
2. Compliance with ARARs (or meet criteria for waiver)

3. Long-term effectiveness and permanence, including consideration of residual risk resulting from remaining, untreated waste and adequacy and reliability of controls

4. Reduction of toxicity, mobility, or volume through treatment of recycling, including the type of waste remaining after the cleanup

5. Short-term effectiveness, focusing on risks to the community, workers, and the environment during the cleanup, including the length of such exposures

6. Implementability, meaning the technical and administrative feasibility of implementing the alternative, as well as the availability of off-site treatment, storage, and disposal sites

7. Cost, including capital costs, annual operation and maintenance, and net present value of capital

8. State acceptance, including state's preferred alternative and state ARARs

9. Community acceptance, including a determination of community concerns with alternatives and preferences.

Remedy Selection

The decision maker must select a remedy based on the preceding nine factors. However, each factor is not equally weighted. There are two threshold requirements: (1) to be eligible for selection, each alternative must achieve overall protection of human health and the environment (2) and must comply with applicable ARARs, unless they are waived.

All options meeting the threshold requirements are then reviewed against the five primary balancing requirements. These are (1) long-term effectiveness and permanence, (2) reduction of toxicity, mobility, and volume through treatment, (3) short-term effectiveness, (4) implementability, and (5) cost. The final two criteria, state and community acceptance, are modifying criteria.

Remedy selection is a two-step process. First, a proposed plan is put forth. Here, the agency identifies the option that best meets the evaluation criteria and proposes it to the public for review and comment. The plan must be published in a newspaper of general circulation. An administrative record of all pertinent documents, studies, and analyses, including those developed during the RI/FS process, must be made available to the public for review and inspection.

Then the public must be given time to submit comments in writing. The agency must provide the opportunity for a public meeting during the public comment period, and a transcript must be made of the meeting. A written responsiveness summary must be prepared by the agency and carried forward to the final decision. If significant new information is identified during this public comment process, the lead agency must evaluate it and incorporate it into the analysis.

The second step in remedy selection is to reconsider the proposed plan by factoring in additional information provided by the state and the public. These comments may prompt the lead agency to modify its proposed plan. When the final remedy selection is made, it is documented in a record of decision (ROD). This ROD documents all facts, analyses, and site-specific policy determinations considered in the course of determining the selection.

Remedial Design and Remedial Action (RD/RA) Stage

The selected option must be designed in sufficient detail to be implemented, and then the actual cleanup must take place. The RD/RA process must follow the ROD. Specifically, the attainment of specific cleanup levels and/or ARARs must be monitored to determine that the alternative has performed to the extent specified in the ROD. Specific attention must be focused on the QA/QC process, as was the case in the RI/FS stage. If the need arises to alter the cleanup process adopted in the ROD, the ROD must be amended through formal procedures that involve public notice and formal amendment. At this point there is a strong preference to maintain the integrity of the ROD.

When the cleanup is completed, a determination is made that the remedy is operational and functional. At that time, the site enters the operation and maintenance (O&M) phase. A site is operational and functional either 1 year after the completion of the construction or when the remedy is determined to be properly functioning. When ground water remediation is involved, the operation of the treatment for a period of up to 10 years is considered to be part of the remedial action.

Liability Under CERCLA

Liability under CERCLA is among the harshest ever adopted by the U.S. Congress. Strict liability exists under CERCLA for cleanup costs incurred by the federal government acting pursuant to the National Contingency Plan. Strict liability also exists for damages to the environment occurring as a result of the hazardous substance release.

Who Is Liable?

Section 107 of CERCLA contains the liability provisions. Section 107(a) states as follows:

(a) Notwithstanding any other provision or rule of law, and subject only to the defenses set forth in subsection (b) of this section:

(1) the owner and operator of a vessel or a facility,

(2) any person who at the time of disposal of any hazardous substance owned or operated any facility at which such hazardous substances were disposed of,

(3) any person who by contract, agreement or otherwise arranged for disposal or treatment or arranged with a transporter for transport for disposal or treatment, of hazardous substances owned or possessed by such person, by any other party or entity, at any facility or incineration vessel owned or operated by another party or entity and containing such substances, and

(4) any person who accepts or accepted any hazardous substances for transport to disposal or treatment facilities, incineration vessels or site selected by such person, from which there is a release or threatened release which causes the incurrence of response costs, of a hazardous substance, shall be liable for:

(A) all costs of removal or remedial action incurred by the United States Government or a state or an Indian tribe not inconsistent with the national contingency plan,

(B) any other necessary costs of response incurred by any other person consistent with the national contingency plan,

(C) damages for injury to, destruction of, or loss of natural resources, including the reasonable costs of assessing such injury, destruction, or loss resulting from such release; and

(D) the costs of any health assessment or health effects study carried out under section 104(i).

This Section 107(a) creates liability in the current owner and operator of the facility that is the site of the release, the past owner and operator of the facility that was associated with the release, the generator of the hazardous substance that is being released, and the transporter delivering the hazardous substance to the release site. Any of these parties are liable for *all* costs identified in (A) to (D). This liability is known as joint and several. In this manner, a single generator can be liable for an entire cleanup even though that generator may not have generated all the hazardous substances being released from a site.

Defenses to Liability

If a party did not generate or transport hazardous substances to a particular release site, a defense exists. If the hazardous substance that is present at the release site could not be the defendant's, then a defense exists. If a person is sued for owning a site and never owned it, a defense exists. However, if you are correctly identified as a generator, transporter, or owner/operator contributing to a release at a facility, the defenses are limited.

The defenses to liability under subsection (a) are very few and narrow. If the release of hazardous substance and the damages relating therefrom were caused by either (1) an act of God or (2) an act of war, then no liability will exist. An act of God is defined to include severe natural disasters, the effects of which could not have been prevented or avoided by the exercise of due care or foresight.

Because the defenses are so limited, the liability concept is considered to be statutory strict liability. Reasonable care is not a defense except in very specific situations such as third-party liability or in the utilization of the act of God defense. By and large, if a defendant generated the waste, transported the waste, currently owns and/or operates the facility, or formerly owned and/or operated the facility, liability exists.

Punitive Damages

A third aspect exists to the liability under CERCLA. Under Section 107(c)(3), a potentially responsible party (PRP) may be liable for punitive damages in the amount of three times the costs actually incurred by the Superfund. Punitive damages become applicable if that PRP is requested by the president (for example, through the EPA) to provide removal or remedial actions under Sections 104 and 106 of CERCLA and fails to do so. In this manner, a PRP may be liable for the costs of cleanup and for three times that amount. The clear intention of this punitive damages section is to force PRPs to cooperate with the government at an early stage in CERCLA process.

Process of Liability

As a practical matter, PRPs are contacted in writing by the EPA at a relatively early stage in the process and asked to cooperate with the various investigations, including the provision of documents and access. The president, acting through the regional administrator of EPA, has the authority to request individual PRPs to undertake certain removal and remedial actions under either Section 104 or 106 of CERCLA. If the PRP lacks just cause for refusing such a request, punitive damages apply. As a practical matter, most PRPs are very hesitant to refuse voluntary cooperation and risk the imposition of punitive damages.

PRPs usually take a very active role in site cleanup selection by undertaking their own RI/FS process under EPA supervision. In this manner, the PRPs have control of the National Contingency Plan process, at the same time avoiding the potential punitive damage assessment. All the PRP studies are conducted under EPA overview and control.

14.6 SUPERFUND AMENDMENT AND REAUTHORIZATION ACT OF 1986 (SARA)

The Superfund Amendment and Reauthorization Act of 1986 (SARA) is notable for several provisions. First, Congress left the major liability requirements of CERCLA unchanged, meaning that the relatively harsh cleanup liability was acceptable from a congressional point of view. Second, additional monies were placed into the Superfund to aid in the cleanup of the ever-increasing number of abandoned hazardous-waste disposal sites. However, the two most important requirements are related to (1) the liability for innocent purchasers and (2) the disclosure of annual releases of hazardous substances.

Limited CERCLA Liability through Environmental Audits

Under CERCLA, the current landowner may be liable for the cleanup of hazardous substances even if that landowner did not cause the contamination. This liability provision is particularly important to banks and lending institutions because of money lent with property as collateral for the loans. If the borrower does not pay back the money and the bank or savings and loan takes over the property used as collateral, the financial institution becomes the current property owner. In this manner, both new owners and foreclosed owners may be liable under CERCLA.

Although SARA generally left the CERCLA liability provisions unchanged, a provision was added that allows "innocent" purchasers to limit their liability under CERCLA. Under the SARA amendment, a due diligence defense was added to allow new purchasers to limit their liability. If, when the property was purchased, the buyer had "no reason to know" about the contamination of the property, then a defense to liability exists.

SARA sets out the considerations that a judge should take into account in determining whether or not the buyer had "no reason to know" about the contamination [CERCLA §9601 (35) (A)]. According to §9601 (35)(B), to establish "no reason to know," the defendant must establish that an inquiry was undertaken into previous ownership and uses of the property in an attempt to determine if there might be a reason to anticipate contamination. This section also states that the court shall consider any specialized knowledge of the purchaser, the purchase price, commonly known or easily obtained information about the contamination, the obviousness of the problem, and the ability to detect the problem by inspection.

This provision has totally changed the commercial real estate market. After the passage of SARA, virtually all lenders and smart buyers began requiring studies of proposed land purchases to determine if there was "reason to know" about contamination. These studies have evolved into a nationwide consulting practice called phase I, II, and III environmental audits.

Phase I audits typically entail a visual inspection of the property, a review of the prior ownership and use, a review of agency records and known contamination sites, a review of aerial photographs, and perhaps interviews with neighbors and/or past employees. The purpose of this initial review is to determine whether or not the purchaser has reason to know that contamination exists on the property. If no problems are detected, the audit is concluded and the land purchase goes forward with the written documentation of the phase I audit ready to be used as a defense.

If problems are detected, a phase II study is initiated to determine the extent of the problem. Phase II typically involves soil testing and ground water monitoring to determine the extent of soil and ground water contamination. Phase II includes a detailed analysis of the extent of contamination, including plumes of contamination and delineation of affected soils.

Phase III is the remediation effort. Depending on how the audit is conducted, the design of the remediation effort may be included in either the phase II or phase III analysis. Generally, a buyer requires information on the cost of remediation in order to make an informed decision about the extent of the risk that will be incurred if the property is purchased.

Often, the detection of soil or ground water contamination will stop the transaction from occurring. Many buyers and lenders require full cleanup and certification of closure prior to even considering lending money on contaminated property. More sophisticated lenders and buyers have developed criteria to assist them in understanding these risks and working with them. However, the important point is that ground water monitoring and soil contamination and remediation are now common issues in the real estate community. Such was not the case prior to the passage of CERCLA and SARA.

SARA Title III

The other major change introduced by SARA was a reporting requirement under Title III. Here, facilities that exceeded certain size thresholds in the handling and release of hazardous substances had to submit an annual report to EPA identifying the total poundage of releases of hazardous substances, both permitted and unpermitted, into the environment.

The hazardous-substance releases required to be submitted include permitted and accidental wastewater discharges, permitted RCRA land disposal, underground injection permitted under the Safe Drinking Water Act, and permitted stack and fugitive air emissions. These submissions are on a facility by facility basis and cover virtually all major industrial facilities in the Untied States.

The SARA Title III requirement is informational. On its face, it may not appear to be overly powerful, yet it is arguable that a substantial amount of environmental protection has been achieved by this provision. More than any other provision in environmental law, SARA Title III achieves environmental protection, including protection of ground water, by making information about the hazardous-substances practices of various facilities easily available.

Prior to the passage of SARA Title III, no single source existed that provided an easily accessible reference on the total emissions of hazardous substances in the United States. The SARA Title III reports provided that reference. Shortly after the submission of the initial reports, the National Wildlife Federation obtained the data and published a report titled "The Toxic 500," a summary of the 500 largest releasers of toxic chemicals in the United States in 1987. Additionally, newspapers, television, and various periodicals have taken advantage of this readily available information.

The point here is that no company wants to be number one on the list of the Toxic 500. Substantial waste-minimization efforts have been undertaken since the first year of the SARA Title III reports.

SUMMARY

The changes that occurred in ground water protection in the 1980s were staggering. RCRA and the Safe Drinking Water Act became effective in November of 1980, CERCLA was passed in 1980, HSWA was passed in 1984 and SARA was passed in 1986. The United States went from a country that had no uniform hazardous waste requirements to a comprehensive system of ground water protection.

As a result of 1980s legislation, the 1990s will focus on reduction in the generation of hazardous wastes and in the release of hazardous substances. Simply stated, the costs of waste generation and disposal are extremely high, and the negative publicity associated with hazardous wastes and substances is great. For many good reasons, the trend of the 1990s will be to minimize the production of hazardous materials and substantially reduce the release of hazardous substances into the environment.

In thinking about the future of ground water protection, it is important to keep in mind that Congress has continually revised and expanded its view of the scope of environmental law and ground water protection. Federal law has expanded from RCRA manifesting and permitting and CERCLA cleanup litigation to land disposal bans under HSWA and environmental audits and Title III reporting under SARA. Full disclosure of contamination may ultimately prove to be one of the most important ground water protection and waste-minimization tools. Therefore, while it is important to understand the site-specific requirements, it is also important not to lose sight of the larger societal goals of ultimately eliminating sources of contamination.

On the other hand, the costs of contamination remediation are extremely high. It is reasonable to question the overall cost to society regarding the gains to public health associated with many of these cleanup activities and to assess the risks associated with leaving contaminants in the ground. In many respects, the major ground water protection debate of the 1990s may be between proponents of extensive remediation and proponents of encapsulation and monitoring. The use of risk-assessment concepts and techniques will likely become more prevalent as cleanup negotiations concentrate on the relative benefits associated with extensive cleanup costs.

The use of risk assessment will raise ethical and professional questions. Who is competent to assess the risk to the public? What type of credentials should be required? And who is going to assess the risk of risk assessors?

These laws have created a framework within which professionals must work into the 21st century. These laws were designed to protect health and the environment. However, it is difficult to clearly define the balance between protecting the public and fairly regulating sources of contamination. In many respects, correctly balancing these concerns will be the ultimate task in the regulation and protection of ground water.

REFERENCES

Code of Federal Regulations (CFR) 1992 (revised July 1, 1992), Washington DC.

HOMEWORK PROBLEMS

Chapter 2

2.1 A sample of silty sand in its natural condition had a volume of 220 cm³ and weighed 481.2 g. After saturating the sample with water it weighed 546.9 g. The sample was then drained by gravity until it reached a constant weight of 445.7 g. Finally the sample was oven dried at 105°C after which it weighed 432.8 g. Note that the volume of the voids in the sample under natural conditions is the volume of water in the sample at saturation. Assuming the density of water is 1 g/cm³, compute the following:

   (a) water content (mass basis) under natural conditions, defined as the ratio of the mass of water to the mass of the sample under natural conditions.

   (b) volumetric water content under natural conditions, defined as the ratio of the volume of water to the volume of the sample under natural conditions.

   (c) saturation ratio under natural conditions, defined as the fraction of voids filled with water under natural conditions.

   (d) porosity, defined as the ratio of the volume of voids to the volume of the sample under natural conditions.

   (e) specific yield, defined as the ratio of the volume of water drained from the saturated sample due to gravity to the volume of the saturated sample.

   (f) specific retention, defined as the ratio of the volume of water a sample can retain against gravity to the volume of the saturated sample, also equal to the difference between porosity and specific yield.

   (g) dry bulk density, defined as the mass of the soil particles divided by the volume of the sample under natural conditions.

2.2 Three piezometers, A, B, and C are located 1000 m apart (in a line) in an unconfined aquifer. Fill in the blank spaces in the table below. Calculate the pressure head and elevation head at the base of the piezometer.

Piezometer	A	B	C
Ground surface (msl)(m)	450	435	430
Depth of piezometer (m)	—	—	70
Depth to water (m)	27	—	50
Pressure head (m)	53	—	—
Elevation head (msl)(m)	—	335	—
Total head (msl)(m)	—	400	—
Hydraulic gradient (m/m)	—	—	

2.3 Compute the discharge velocity, seepage velocity and flow rate for water flowing through a sand column with the following characteristics:

$$K = 10^{-4} \text{ cm/s,}$$

$$\frac{dh}{dl} = 0.01,$$

$$\text{Area} = 75 \text{ cm}^2,$$

$$n = 0.20.$$

2.4 The average water table elevation has dropped 5 ft due to the removal of 100,000 ac-ft from an unconfined aquifer over an area of 75 mi². Determine the storage coefficient for the aquifer.

2.5 A confined aquifer is 50 m thick and 0.5 km wide. Two observation wells are located 1.4 km apart in the direction of flow. The head in well 1 is 50.0 m and in well 2 the head is 42 m. The hydraulic conductivity K is 0.7 m/day.
(a) What is the total daily flow of water through the aquifer?
(b) What is the height of the piezometric surface 0.5 km from well 1 and 0.9 km from well 2?

2.6 Three piezometers located in the same aquifer are 1000 m apart from each other. Well A is located to the east of Well B and Well C which are aligned in a north-south direction. The water surface elevations are measured for the three wells and are as follows: A = 150 m, B = 100 m and C = 200 m. Find the direction and two dimensional rate of flow within the triangle using graphical methods. The hydraulic conductivity of the aquifer is 5×10^{-7} cm/s.

2.7 Answer the following questions with reference to Figure 2.10(b) for a flow net under a dam with a sheet pile. Note that the numbers beside the equipotentials are labels only and not representative of head values.
(a) What are the values of equipotentials 17, 9, 2, and 0?
(b) What is the rate of seepage under the dam if the dam is 120 ft long and the hydraulic conductivity is 5 m/d?
(c) What is the rate of seepage under the dam if the upstream water level drops to 12 ft?

2.8 Refer to Figure 12.7 which shows the potentiometric map for a shallow sand aquifer. The hydraulic conductivity of the unit is estimated to be 1.5×10^{-2} cm²/s.
(a) Which well would you expect to be most contaminated?
(b) What is the groundwater velocity and seepage velocity across the plume?
(c) Estimate how long the source has been contaminating the aquifer. You may ignore the effects of dispersion, diffusion and adsorption.
(d) Estimate the rate of flow across the plume.
(e) How would you explain contamination upgradient of the source?

2.9 Two observation wells 1700 m apart have been constructed in a confined aquifer with variable hydraulic conductivity in the x direction (see Figure P2.9). The flow rate is 0.008 m³/h per unit width of the confined aquifer. Well 1 is drilled in Soil A with hydraulic conductivity 13 m/d, and Well 2 is drilled in Soil C with hydraulic conductivity 9 m/d. The soil zone B is between the wells, 600 m from Well 1 and 300 m from Well 2. The potentiometric surface is 6 m above the upper confining unit in Well 1 and 2.5 m above the upper confining unit in Well 2. Evaluate the hydraulic conductivity of Soil B.

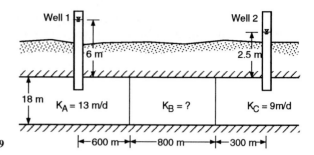

Figure P2.9

2.10 Three observation wells separated by 1000 m are drilled in a line in a confined aquifer with variable hydraulic conductivity in the x direction. The water surface elevation in Well 1 is 60 m and 52 m in Well 3. The hydraulic conductivity of the aquifer between Wells 1 and 2 is K_A, and between Wells 2 and 3 the hydraulic conductivity is K_B. Determine the water surface elevation in Well 2 for the following conditions:

(a) $K_A = 10$ m/d, $K_B = 20$ m/d
(b) $K_A = 10$ m/d, $K_B = 50$ m/d
(c) $K_A = 50$ m/d, $K_B = 10$ m/d

2.11 Two piezometers are located 1000 ft apart with the bottom located at depths of 50 ft and 350 ft, respectively, in a 400 ft thick unconfined aquifer. The depth to the water table is 50 ft in the deeper piezometer and 40 ft in the shallow one. Assume that hydraulic conductivity is 0.0002 ft/s.

(a) Use the Dupuit equation to calculate the height of the water table midway between the piezometers.
(b) Find the quantity of seepage through a section 25 ft wide.

2.12 Three geologic formations overlie one another with the characteristics listed below. A constant velocity vertical flow field exists across the three formations. The hydraulic head is 33 ft at the top of the formations (top of formation no. 1) and 21 ft at the bottom (bottom of formation no. 3). Calculate the hydraulic head at the two internal boundaries, that is, at the top and bottom of formation no. 2.

$$b_1 = 50 \text{ ft} \qquad K_1 = 0.0002 \text{ ft/s}$$

$$b_2 = 20 \text{ ft} \qquad K_2 = 0.000005 \text{ ft/s}$$

$$b_3 = 210 \text{ ft} \qquad K_3 = 0.001 \text{ ft/s}$$

2.13 A formation is composed of four horizontal, homogeneous, and isotropic strata overlying one another. The hydraulic conductivity in the top layer is 1×10^{-3} cm/s, 1×10^{-6} cm/s in the second layer, 1×10^{-2} cm/s in the third layer, and 1×10^{-4} cm/s in the fourth layer. Calculate the effective horizontal and vertical hydraulic conductivities for the entire formation if each layer is 3 m thick. Give an example of soil types that might make up this formation.

2.14 A soil sample 6 in. in diameter and 1 ft long is placed in a falling head permeameter. The falling head tube diameter is 1 in. and the initial head is 6 in. The head falls 1 in. over a two hour period. Calculate the hydraulic conductivity.

2.15 A constant head permeameter containing fine grained sand has a length of 12 cm, and an area of 30 cm². For a head of 10 cm, a total of 100 ml of water is collected in 25 min. Find the hydraulic conductivity.

2.16 Two observation wells, 450 m apart, have been constructed in a confined aquifer with a hydraulic conductivity of 5×10^{-4} cm/s and a porosity of 0.35. The average thickness of the aquifer between the wells is 45 m. Well 1 (water surface elevation of 95 m) lies to the east of Well 2 (water surface elevation of 94.6 m). A stream lies 900 m to the west of Well 2.

(a) Determine the flow rate per unit width of the aquifer.

(b) Calculate how long it would take for groundwater contaminants to travel from Well 1 to the stream.

(c) Calculate the level of the potentiometric surface at a distance 600 m to the east of Well 1.

2.17 A conservative compound is discharged into a pond with water surface elevation of 100 m. The compound moves from the pond to a river (water surface elevation of 99.3 m) 1.2 km away through a confined aquifer 12 m thick. The aquifer has a transmissivity of 0.05 m^2/s and a porosity of 0.30. Ignoring the effects of dispersion (see Chapter 6) and adsorption to the soil, and assuming immediate and complete mixing, how long would it take for the contaminant to reach the river? How would your answer differ if the aquifer was unconfined?

2.18 The base of a proposed landfill for paper-mill sludge is in a glacial till, 30 ft above an aquifer. The vertical hydraulic conductivity of the till is 1×10^{-7} cm/s, the vertical hydraulic gradient is 0.075, and the effective porosity is 0.30. If the leachate (contaminated fluid) drains from the landfill into the till, how many years would pass before the leachate reached the aquifer below? If the vertical hydraulic conductivity was 3×10^{-6} cm/s, what would the travel time be?

2.19 A 3 m thick aquitard ($K_2 = 0.5$ m/d) separates an unconfined aquifer ($K_1 = 15$ m/d) from a leaky confined aquifer ($K_3 = 10$ m/d). The water table is above the potentiometric surface for the leaky confined aquifer. Three observation wells are drilled—one in the unconfined aquifer, one in the aquitard and one in the leaky confined aquifer. The head in the leaky confined aquifer is 23 m and in the unconfined aquifer the head is 25 m. Using Darcy's Law calculate the head in the well at the top of the leaky confined aquifer (the base of the aquitard). Assume steady state conditions. The datum is at the base of the aquitard layer.

2.20 Two parallel rivers are separated by 1.5 km. The water surface elevations of the two rivers are 60 m and 45 m and the datum is defined as the base of the river beds. A confined aquifer extends between the two rivers from elevation 0 m to elevation 30 m, and has a transmissivity of 210 m^2/d. The hydraulic conductivity of an unconfined aquifer on top of this formation is 10 m/d.

(a) Determine the combined flow rate between the two rivers.

(b) Determine the shortest time of travel for a contaminant to move from one river to the other. Neglect dispersion effects and assume both aquifers have a porosity of 0.25.

2.21 A confined aquifer (hydraulic conductivity 1 m/d) flows into an unconfined aquifer (hydraulic conductivity 2 m/d) and then enters a shallow lake (as shown in Figure P2.21). The elevation of the lower confining unit for both aquifers is 130 m. The elevation of the base of the upper confining unit is 138 m. The upper confining unit is 2 m thick and extends across both aquifers. The thickness of the confined aquifer is 8 m. If the potentiometric surface at A is at an elevation of 140 m (h_A), find the elevation of the lake water surface (h_C).

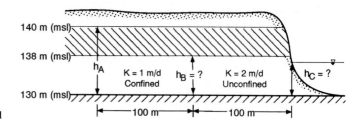

Figure P2.21

2.22 Two rivers are 1000 m apart. The water surface elevation of the first river is 30 m, and 25 m for the second river. The area between the rivers receives a recharge of 10 cm/y (after runoff). If the hydraulic conductivity is 0.8 m/d, find the location and elevation of the water divide and the daily seepage into each river. Sketch a flow net for the aquifer in the problem above.

2.23 A landfill liner is laid at elevation 50 ft msl (mean sea level) on top of a good clay unit. A clean sand unit extends from elevation 50 ft to elevation 75 ft, and another clay unit extends up to the surface located at elevation 100 ft. The landfill can be represented by a square with length of 500 ft on a side and a vertical depth of 50 ft from the surface. The landfill has a 3 ft thick clay liner with $K = 10^{-7}$ cm/s around the sides and bottom, as shown in Figure P2.23. The regional ground water level for the confined sand ($K = 10^{-2}$ cm/s) is located at depth of 10 ft below the surface. How much water will have to be continuously pumped from the landfill to keep the potentiometric surface at 60 ft elevation (msl) within the landfill? Assume mostly horizontal flow.

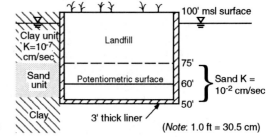

Figure P2.23

2.24 Prove the refraction law for streamlines at a geologic boundary using Darcy's law and the fact that $\theta_1 = \theta_2$ for Figure 2.11. Assume that $K_2 > K_1$ and note the following identities. Note that dh_1 is the head drop over dl_1 and dh_2 is the head drop over dl_2. Prove that $K_1/K_2 = \tan \theta_1/\tan \theta_2$.

$$a = b \cos \theta_1$$

$$c = b \cos \theta_2$$

$$dl_1/b = \sin \theta_1$$

$$dl_2/b = \sin \theta_2$$

$$\sin \theta_1/\cos \theta_1 = \tan \theta_1$$

2.25 Repeat Example 2.3 for the case where net recharge $W = 10$ cm/yr. Repeat the example for the case where $W = 0$.

2.26 Repeat Problem 2.23 for the case where the clean sand unit extends below the landfill to an elevation of 25 ft msl. Assume water can flow through the sides and bottom of the clay liner.

2.27 A confined aquifer slopes gradually from 20 m to 10 m in thickness over a length of 1 km. The slope of the potentiometric surface is 0.5 m/km and the hydraulic conductivity of the aquifer is 50 m/d. For a sloping aquifer, the continuity equation yields the following relation:

$$Q = \frac{\phi_1 - \phi_2}{L} \int \frac{dx}{T(x)}$$

What is the flow rate through a 100 m wide section of the aquifer?

2.28 Sketch a quantitative flow net for the following confined aquifer and determine the total flow per unit width if the hydraulic conductivity is 20 m/d.

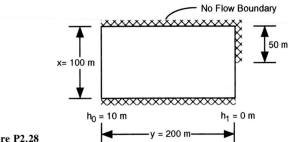

Figure P2.28

Chapter 3

3.1 A well with a diameter of 18 in. penetrates an unconfined aquifer that is 100 ft thick. Two observation wells are located at 100 ft and 235 ft from the well, and the measured drawdowns are 22.2 ft and 21 ft, respectively. Flow is steady and the hydraulic conductivity is 1320 gpd/ft². What is the rate of discharge from the well?

3.2 A well of 0.5 m diameter is pumped from confined aquifer at 0.08 m³/s. The drawdown recorded at Observation Well 1 (15 m from the pumping well) is 3.8 m; at Observation Well 2 (50 m from the pumping well) the drawdown is 2.8 m. What is the drawdown at the pumping well?

3.3 A single well pumps 1000 m³/d in a confined aquifer of thickness 10 m and porosity 0.3. At the radius of influence of the well (1000 m) the head is 20 m. Assuming steady state conditions calculate the seepage velocity towards the well at a radius of 100 m, and a radius of 500 m.

3.4 (a) A well in a confined aquifer of thickness B pumps continuously at a rate Q. Using Darcy's Law in radial coordinates and the fact that the seepage velocity, $v_s = dr/dt$, show that the travel time for contaminants to move to the well of radius r_w, from a point on the radius of influence, R, is

$$t = (\pi Bn/Q)(R^2 - r_w^2)$$

(b) A fully penetrating well of 0.8 m diameter and with a radius of influence of 100 m, is drilled in a confined aquifer 15 m thick. The aquifer has a transmissivity of 50 m²/d and a porosity of 0.3. At what rate should the well be pumped if the time of travel for a contaminant to reach the well from the radius of influence is to be not less than 200 days. If the potentiometric surface was 15 m above the upper confining layer before pumping began, what is the steady state drawdown in the well?

3.5 A confined aquifer extends 3.5 km between the West River and the East River. The water surface elevations of the rivers are 230 m and 228.5 m respectively. The radius of influence of a fully penetrating pumping well located halfway between the rivers is 90 m. The well has a diameter of 0.5 m and pumps at a rate of 750 L/min. The elevation of the upper confining unit is 227 m and the aquifer is 25 m thick. The transmissivity of the aquifer is 8 m^2/d. Determine the steady state head 20 m to the east of the well.

3.6 In a fully penetrating well, the equilibrium drawdown is 30 ft with a constant pumping rate of 20 gpm. The aquifer is unconfined and the saturated thickness is 100 ft. What is the steady-state drawdown measured in the well when the pumping rate is 10 gpm? Assume radius of influence remains constant.

3.7 A fully penetrating well in an unconfined aquifer is pumped continuously so that the steady state drawdown is 15 m. The water table is at a height of 55 m from the bottom of the aquifer which has a hydraulic conductivity of 20 m/d. The diameter of the well is 0.25 m.
  (a) If the pumping has no effect on the water table at a distance of 1.2 km, determine the pumping rate.
  (b) Determine the drawdown in an observation well located 600 m from the pumping well.
  (c) Determine the travel time from the observation well to the pumping well if the aquifer has a porosity of 0.4.

3.8 A fully penetrating well of radius r_w is located in the center of a circular confined aquifer of radius r_a and thickness B. The hydraulic conductivity of the aquifer varies with radius such that $K = K_1$ for $r_w \leq r < r_1$, and $K = K_2$ for $r_1 \leq r \leq r_a$. If the piezometric head is maintained at h_w at $r = v_w$ and h_a at $r = v_a$ at what rate would the well be pumping?

3.9 Two wells are located 100 m apart in a confined aquifer with transmissivity $T = 2 \times 10^{-4}$ m^2/s and storativity $S = 7 \times 10^{-5}$. One well is pumped at a rate of 6.6 m^3/hr and the other at a rate of 10.0 m^3/hr. Plot drawdown as a function of distance along the line joining the wells at 1 hr after the pumping starts.

3.10 At a burn pit, a Hvorslev slug test was performed in a confined aquifer with a piezometer screened over of 20 ft and a radius of 1 in. The radius of the rod was 0.68 in. The following recovery data for the well were observed. Given that the static water level is 7.58 ft and $H_0 = 6.88$ ft, calculate the hydraulic conductivity.

TIME(s)	20	45	75	101	138	164	199
h (ft)	6.94	7.00	7.21	7.27	7.34	7.38	7.40

3.11 A well casing with a radius of 6 in. is installed through a confining layer into a formation with a thickness of 10 ft. A screen with a radius of 3 in. is installed in the casing. A slug of water is injected, raising the water level by 0.5 ft. Given the following recorded data for head decline, find the values of T, K, and S for this aquifer.

TIME(s)	5	10	30	50	80	120	200
h (ft)	0.47	0.42	0.32	0.26	0.10	0.05	0.01

3.12 A small municipal well was pumped for 2 hr at a rate of 15.75 liters/s. An observation well was located 50 ft from the pumping well and the following data were recorded. Using the Theis method outlined in Example 3.3, compute T and S.

TIME (min)	DRAWDOWN (ft)	TIME (min)	DRAWDOWN (ft)
1	1.5	15	14.9
2	4.0	20	17.0
3	6.2	40	21.4
5	8.5	60	23.1
7	10.0	90	26.0
9	12.0	120	28.0
12	13.7		

3.13 A well in a confined aquifer is pumped at a rate of 833 liters/min for a period of over 8 hr. Time-drawdown data for an observation well located 250 m away are given below. The aquifer is 5 m thick. Use the Cooper-Jacob method to find values of T, K, and S for this aquifer.

TIME (hr)	DRAWDOWN (m)	TIME (hr)	DRAWDOWN (m)
0.050	0.091	1.170	1.860
0.083	0.214	1.333	1.920
0.133	0.397	1.500	2.040
0.200	0.640	1.670	2.130
0.333	0.976	2.170	2.290
0.400	1.100	2.670	2.530
0.500	1.250	3.333	2.590
0.630	1.430	4.333	2.810
0.780	1.560	5.333	2.960
0.830	1.620	6.333	3.110
1.000	1.740	8.333	3.320

3.14 Drawdown, s, was observed in a well located 100 ft from a pumping well that was pumped at a rate of 1.11 cfs for a 30 hr period. Use the Cooper-Jacob method to compute T and S for this aquifer.

TIME(hr)	1	2	3	4	5	6	8	10	12	18	24	30
s (ft)	0.6	1.4	2.4	2.9	3.3	4.0	5.2	6.5	7.5	9.1	10.5	11.5

3.15 Repeat Problem 3.14 using the Theis method.

3.16 The drawdown in three observation wells was recorded after 24 hours of pumping from a well at 2300 m³/d and the results are shown below. Calculate the transmissivity and storage coefficient of the aquifer using the Theis Method. You will need to plot drawdown versus r^2/t at a point in time for the three different radii and match the curves as before. Based on your estimate of the storativity, do you think the aquifer is confined or unconfined?

Well Number	Distance from Well (m)	Drawdown (m)
1	100	1.5
2	200	1.25
3	300	1.10

3.17 A well located at $x = 0$, $y = 0$ injects water into an aquifer at $Q = 1.0$ cfs and observation wells are located along the x-axis at $x = 10, 50, 150, 300$, and 450 ft away from the injection well. The confined aquifer with thickness of 10 ft has $T = 3200$ sq ft/day and $S = 0.005$. The injection is affected by the presence of a linear river located at $x = 300$ ft west of the injection well.

(a) Compute the head buildup along the x-axis at the observation well locations after 6 hr of injection.

(b) Compute the Darcy velocity along the x-axis using simple graphical methods at the various well locations.

3.18 Use the Theis Method equations for the functions u and $W(u)$ to characterize the expected behavior of an areally infinite, homogeneous, horizontal, isotropic, confined aquifer. The aquifer has a transmissivity of 500 m^2/d and a storativity of 1×10^{-5}. The well is pumped continuously at 2500 m^3/d.

(a) What will be the drawdown 75 m away from the pumping well at 1, 10, 100, 1000, and 10,000 minutes after pumping begins?

(b) Estimate when the system will reach steady state.

(c) Estimate the steady state drawdown in the well.

(d) What will be the drawdown after one day at distances of 2, 5, 10, 50, 200, and 1000 m from the pumping well? Estimate the distance at which the pumping will have no effect on the potentiometric surface after one day of pumping.

3.19 A well a distance d from an impermeable boundary pumps at a flow rate Q. The head at any point (x, y) is given by the following expression:

$$h(x, y) = \frac{Q}{2 \pi kb} \ln(r_1 r_2) + C$$

where C is a constant, r_1 is the straight line distance from the well to point (x, y), and r_2 is the straight line distance from the image well to the point (x, y). The y-axis lies along the impermeable boundary. Use Darcy's Law to show that the discharge across the y-axis is zero.

3.20 A well is to be placed in an unconfined aquifer at a distance x from a stream. The well will be required to pump at a constant rate of 1.3×10^{-3} m^3/s. The aquifer has an average thickness of 10 m and hydraulic conductivity of 1×10^{-2} cm/s. Regulations stipulate that the drawdown at a radius of 15 m from the well may not exceed 0.6 m. Calculate x such that it is the smallest distance that satisfies the requirements.

3.21 A fully penetrating well pumps from a confined aquifer of thickness 20 m and hydraulic conductivity 10 m/d. There is no water table above the confined aquifer. The radius of the well is 0.25 m, and the recorded drawdown in the well is 0.5 m. Assume that the radius of influence of the well is 1250 m.

(a) Calculate the pumping rate if the well is 100 m from a stream.

(b) Calculate the pumping rate if the well is 100 m from an impermeable boundary.

3.22 An excavation site is to be dewatered by using four pumping wells located at the corners of a square with sides of 500 m. The wells are fully penetrating and have diameters of 0.5 m. What is the required flow rate in each well if the objective is to lower the original water table (h_0 = 100 m) by 5 m everywhere in the square site? Note that the maximum drawdown may not occur in the center of the square. The hydraulic conductivity of the aquifer is 1.5 m/d and the radius of influence for each well is 1000 m.

3.23 (a) Three fully penetrating wells in an unconfined aquifer form an equilateral triangle of side a. Two of the wells are a distance d from a long river. Each well has a radius r_w and pumps at a rate Q. At steady state the wells have a radius of influence of R and the water table is at an elevation of h_0. If the aquifer has a hydraulic conductivity of K, derive an expression for the drawdown in the center of the triangle.

 (b) What is the drawdown in the center of the triangle if the parameters are as follows: a = 70 m, d = 100 m, K = 1.5 m/d, Q = 700 m³/d, R = 600 m, h_0 = 50 m.

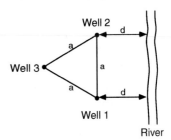

Figure P3.23

3.24 Walton (1960) gave data for an aquifer test in a well confined by a leaky aquitard 14 ft thick. Drawdown was measured in an observation well 96 ft away from a pumping well which was pumping at 25 gpm. For the following data, use Figure 3.13 to determine T, S and K'.

TIME (min)	DRAWDOWN (ft)	TIME (min)	DRAWDOWN (ft)
5	0.76	493	5.96
28	3.30	669	6.11
41	3.59	958	6.27
60	4.08	1129	6.40
75	4.39	1185	6.42
244	5.47		

3.25 An interceptor well fully penetrates a confined aquifer with a hydraulic conductivity of 0.5 cm/s. Before pumping began the aquifer had a hydraulic gradient of 0.001 and a saturated thickness of 27 m. The interceptor well pumps at a rate of 60 L/s. Define the capture zone of the interceptor well by calculating the maximum width and the distance to the stagnation point. Sketch the shape of the groundwater divide using at least ten points.

Chapter 6

6.1 Chloride was injected as a continuous source into a 1-D column 50 cm long at a seepage velocity of 10^{-3} cm/sec. The concentration measured after 1800 seconds from the beginning of the test was 0.3 of the initial concentration, and after 2700 seconds it reached 0.4 of the initial concentration. Find the coefficient of dispersion and longitudinal dispersivity.

6.2 Chloride has been injected into a 1-D column continuously. If the system has a Darcy velocity of 5.18×10^{-3} m/day, the porosity of the medium is 0.3, and the longitudinal dispersivity is 5 m.
  (a) What is C/C_0 at 0.3 m from the point of injection after 5 days.
  (b) Repeat (a) with a longitudinal dispersivity of 20 m.
  (c) Comment on the difference in results.

6.3 The estimated mass from an instantaneous spill of benzene was 107 kg per m² of the 1-D aquifer. The aquifer has a seepage velocity of 0.03 m/day and a longitudinal dispersion coefficient of 9×10^{-4} m²/day.
  (a) Calculate the maximum concentration at $t = 1$ year.
  (b) Calculate the benzene concentrations at $t = 1$ year, at one standard deviation on either side of the center mass of the plume.
  (c) Plot the breakthrough curve vs. time at $x = (v')$ $(t = 1$ year).
  (d) Plot the breakthrough curve vs. distance at $t = 1$ year.
  (e) Discuss the difference.

6.4 Match the breakthrough curves given in the diagram with the proper description given below. Curves represent responses to the injection of a tracer in 1-D soil columns. Choose the best match for each case or indicate none of the above (NA).

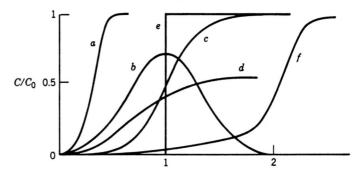

Figure P6.4

Number of pore volumes[1]

[1] A pore volume in a column is the volume of water that will completely fill all of the voids in the column. (Domenico and Schwartz, 1990)

  (a) Curve for a 1-D continuous injection where the constituents get sorbed to the medium.
  (b) Curve for a 1-D continuous injection where plug flow is observed.
  (c) Curve for an instantaneous source with longitudinal dispersion.
  (d) Curve for an instantaneous source with no dispersion.
  (e) Curve for a 1-D continuous injection with longitudinal dispersion
  (f) Curve for a 1-D continuous injection with longitudinal dispersion and biodegradation.
  (g) Curve for a 1-D continuous injection in which a fracture existed along the soil column.

6.5 Sketch (on the same graph) C/C_0 as a function of time for transport in a sand column from an instantaneous source for each of the following cases:
  (a) Advection only
  (b) Advection with low dispersion.
  (c) Advection with high dispersion.
  (d) Advection with high dispersion with biodegradation.
  (e) Advection with high dispersion with retardation.

6.6 A continuous source has been leaking contaminant for 2 years into a 1-D aquifer. The steady state dilution attenuation factor DAF for the aquifer is 100. DAF = Concentration at source/concentration at receptor well. Source concentration = 1200 mg/L.
   (a) What is the concentration of benzene at a receptor well after 2 years.
   (b) What is the concentration of benzene at a receptor well in 2 years, if there is first order decay at a rate of 0.01 (1/day).

6.7 An instantaneous release of biodegradable organics occurs in a 1-D aquifer. Assume that the mass spilled is 1.0 kg over a 10 m^2 area normal to the flow direction, α_x = 1.0 m, the seepage velocity is 1.0 m/day and the half life of the decaying contaminant is 33 years. Compute the maximum concentration at 100 m from the source.

6.8 For a 2-D aquifer, calculate the maximum concentration of a spike source of decaying Tritium, H-3 (half life = 12.26 years) at 100 m from a well injecting 1.0 kg of mass over 10 m of well screen. The plume velocity is 0.5 m/day, α_x = 1.0 m, α_y = 0.1 m.

6.9 An accidental spill from a point source introduced 10 kg of contaminant mass to an aquifer. The seepage velocity in the aquifer is 0.1 ft/day in the x direction. The longitudinal dispersion coefficient D_L = 0.01 ft^2/day, the lateral and vertical dispersion coefficient, D_y = D_z = 0.001 ft^2/day.
   (a) Calculate the maximum concentration at x = 100 ft and t = 5 years.
   (b) Calculate the concentration at point x = 200 ft, y = 5 ft, z = 2 ft, 5 years after of the spill.

6.10 Domenico & Schwartz, 1990, developed a model for a planar source that accounts for the source geometry with longitudinal, lateral and vertical spreading. The steady state model was applied at the plane of symmetry where y = z = 0, as shown in Figure P6.10.

$$C(x, y = 0, z = 0, t)$$

$$= (C_0/2)\mathrm{erfc}[(x - vt)/2(\alpha_x vt)^{0.5}]\mathrm{erf}[\,Y/4(\alpha_y x)^{0.5}]\mathrm{erf}[Z/2(\alpha_z x)^{0.5}]$$

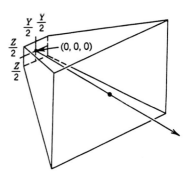

Figure P6.10

where x = distance from source in x direction
    Y = width of source
    Z = depth of source

The above model is to be applied to the case of a continuous source that has been leaking contaminant into an aquifer for 15 years. The source had a width Y of 20 m and a depth Z of 6 m, the initial concentration of the source was 10 mg/L, the seepage velocity is 0.057 m/day, and the longitudinal, transverse and vertical dispersivities were estimated at 1 m, 0.1 m, 0.01 m respectively. Calculate the present contaminant concentration at $x = 200$ m from source.

6.11 A slug of contaminant was injected into a well for a tracer test (2-D). If the initial contaminant concentration is 1000 mg/L. The background seepage velocity in the aquifer is 0.022 m/day, the well radius is 0.05 m, the longitudinal dispersion coefficient is 0.034 m²/day, and the transverse dispersion coefficient is 0.01 of the longitudinal dispersion coefficient.
(a) What is the maximum concentration reached after 24 hours.
(b) Plot the concentration vs. distance at $t = 24$ hours.

6.12 An underground storage (volume $= 1.0$ m³) spills its contents of 1.0 kg of TCE in a 10 m thick sand aquifer. The seepage velocity is 1.0 m/day in the x direction. The longitudinal dispersivity is 1.0 m, the transverse and vertical dispersivities are 0.1 m.
(a) Calculate the maximum concentration at a distance of $x = 500$ m. Note: $y = z = 0$ for maximum concentration.
(b) If the contaminant was uniformly spilled over the 10 m thickness of the aquifer, what is the maximum concentration at $x = 500$ m, $y = z = 0$, with vertical dispersion $= 0$.
(c) Comment on the results.

6.13 A finite source of chloride is released at $x = 0$ into a 1-D aquifer over a 1 year period. The plume moves at a seepage velocity of 0.2 m/day, $\alpha_x = 2$ m. Assume that the solution can be found by using an imaginary negative concentration source beginning at the end of the release period. Develop the equation for concentration $C(x, t)$ if $C_0 = 1000$ mg/l. Find the maximum concentration at $x = 100$ m.

6.14 The concentration of organic waste in a landfill is 100 mg/L. The pit is placed in a clay layer, $k = 10^{-6}$ cm/sec underlain by a sand layer, $k = 2 \times 10^{-2}$ cm/sec. The two units are separated by a thin impermeable silty layer. The top surface of the clay layer is at elevation 9.15 m above MSL for which the water table is at 8.5 m above MSL. The top surface of the sand layer is at elevation 6.1 m, for which the piezometric level is 7.6m above MSL. The pit bottom lies at elevation 8.4 m above MSL. The slope of the piezometric level is 0.01. The porosity of the 2 layers is 0.25, and the longitudinal dispersivity is estimated at 0.3 m. An observation well is located 300 m from the pit. (See Figure P6.14.)
(a) Calculate the seepage velocity for the clay layer. How long doe it take the contaminant to reach the sand layer.
(b) Calculate the seepage velocity for the sand layer. What is the contamination level in the well after 15 years from the beginning of the release, assuming a 1-D model.
(c) What would happen to the concentration in the well if lateral dispersion exists in the sand aquifer in addition to longitudinal dispersion.

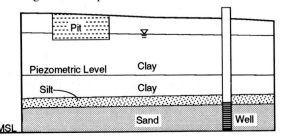

Figure P6.14

6.15 From Example 6.4, multiply the vertical scale of Figure 6.10 times 2 and repeat the problem.

6.16 Prove that the plume described in Eq. (6.29) takes the form of a two dimensional gaussian (normal) distribution as described below, and evaluate σ_x, and σ_y.

$$P = [1/(2\pi\sigma_x\sigma_y)] \exp(-[(x - x_0)^2/(2\sigma_x^2)] - [(y - y_0)^2/(2\sigma_y^2)])$$

6.17 What are D_x and D_y for Figures 6.9c and 6.9d in Example 6.3.

Chapter 7

7.1 A landfill located next to a river has been leaking chromium continuously for 20 years, creating a plume of contamination. The gradient of ground water flow is towards the river, as shown in the figure below. Well concentration data is provided along with water level readings. Assume that chromium acts as a conservative tracer in this aquifer where longitudinal dispersivity = 1.0 m. Landfill, ground water, and river data are provided in the following table and Figure P7.1.

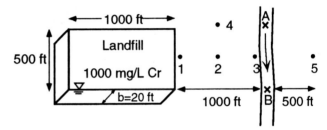

Figure P7.1

Well	Water elevation (ft)	Cr Concentration (mg/L)
1	100	1000
2	90	no data
3	80	no data
4	90	0
5	100	0

$K = 10^{-3}$ cm/sec; $n = 0.4$; $b = 20$ ft; $Q(\text{river}) = 200$ cfs; Concentration of Cr at point A (upstream) = 0 mg/L.
(a) Find the predicted concentration of Cr in well 3 (at t = 20 years).
(b) Find the concentration of Cr in the river at point B (immediately downstream). Consider the flow from both sides of the river.

7.2 A well fully penetrates an aquifer which has a uniform depth of 10 m. 100 grams of benzene is spilled into the well and is immediately dissolved and mixed into the water in the well. The seepage velocity = 30 m/yr in the x direction, the longitudinal dispersivity = 1.0 m, and the transverse dispersivity is 0.1 m.
(a) Using the following aquifer characteristics, calculate the retardation factor R for benzene in this aquifer. Bulk density = 1.8 g/cm³, porosity = 0.3, organic carbon fraction = 1%, K_{ow} for benzene = 135 L/kg.
(b) Which equation should you use for calculating benzene concentrations downgradient of the well? (1-D, 2-D, or 3-D; spike or continuous)
(c) What is the maximum benzene concentration at $t = 1$ year? Where is this maximum located?

(d) What is the benzene concentration at $t = 1$ year, 3 meters away from the center of mass in the x direction?

(e) What is the benzene concentration at $t = 1$ year, 3 meters away from the center of mass in the y direction?

(f) Comment on how the dispersivity affects the concentration of benzene 3 m away in the x direction as opposed to 3 m away in the y direction.

7.3 Repeat Example 6.1a from Chapter 6, using a retardation factor ($R = 2.5$) for benzene.

7.4 Repeat Example 6.2 from Chapter 6 for benzene (with a retardation factor, $R = 2$) for a 10,000 ug/L spill.

7.5 Several 1 L solutions of organic material were placed into beakers, each with 10 g of soil. The concentration was measured before and after adding soil, to determine adsorption properties. Which adsorption isotherm (Langmuir or Freundlich) fits the experimental test results best? Also, determine the appropriate coefficients for the selected isotherm.

CONCENTRATION, g/m³

Initial	Final
0.710	0.671
0.453	0.418
0.189	0.163
0.131	0.110
0.103	0.085

Chapter 10

10.1 Repeat Example 10.1 with the following change in the top boundary conditions: $u(x, 1) = 1.0$ for all x.

(a) Solve by elimination.

(b) Solve by iterative method.

10.2 Solve Eq. (10.20) and (10.22) for $a = 1$. A 1-D aquifer of unit length with $u = 0$ at $x = 0$ and $x = 1$ for all t, (boundary condition)

$$\left.\begin{array}{ll} u = 2x & 0 \le x \le 1/2 \\ u = 2(1 - x) & 1/2 \le x \le 1 \end{array}\right\} t = 0 \text{ (initial condition)}$$

Thus the center of this aquifer is initially contaminated by a triangular distribution with $u = 1$ at $x = 1/2$. Let $\Delta x = h = 1/10$, $\Delta t = k = 1/1000$, $a = 1$, $m = 1/10$. Solve for $u(x, t)$ by hand or write a computer program for time up to 0.02 units and $x = 0, 0.1, 0.2, 0.3, 0.4, 0.5$ since the solution is symmetric.

INDEX

FLOW CONVERSION

UNIT	m³/s	m³/day	liter/s	ft³/s	ft³/day	ac-ft/day	gal/min	gal/day	mgd
1 m³/s	1	8.64×10^4	10^3	35.31	3.051×10^6	70.05	1.58×10^4	2.282×10^7	22.824
1 m³/day	1.157×10^{-5}	1	0.0116	4.09×10^{-4}	35.31	8.1×10^{-4}	0.1835	264.17	2.64×10^{-4}
1 liter/s	0.001	86.4	1	0.0353	3051.2	0.070	15.85	2.28×10^4	2.28×10^{-2}
1 ft³/s	0.0283	2446.6	28.32	1	8.64×10^4	1.984	448.8	6.46×10^5	0.646
1 ft³/day	3.28×10^{-7}	0.02832	3.28×10^{-4}	1.16×10^{-5}	1	2.3×10^{-5}	5.19×10^{-3}	7.48	7.48×10^{-6}
1 ac-ft/day	0.0143	1233.5	14.276	0.5042	43,560	1	226.28	3.259×10^5	0.3258
1 gal/min	6.3×10^{-5}	5.451	0.0631	2.23×10^{-3}	192.5	4.42×10^{-3}	1	1440	1.44×10^{-3}
1 gal/day	4.3×10^{-8}	3.79×10^{-3}	4.382×10^{-5}	1.55×10^{-6}	11,337	3.07×10^{-6}	6.94×10^{-4}	1	10^{-6}
1 million gal/day (mgd)	4.38×10^{-2}	3785	43.82	1.55	1.337×10^5	3.07	694	10^6	1